SELECT EDITION

Computers
Tools for an Information Age

Fourth Edition

H. L. Capron

Computers
Tools for an Information Age

Fourth Edition

The Benjamin/Cummings Publishing Company, Inc.
Menlo Park, California
Reading, Massachusetts
New York
Don Mills, Ontario
Wokingham, U. K.
Amsterdam
Bonn
Singapore
Tokyo
Madrid
San Juan

Sponsoring Editor	Maureen A. Allaire
Developmental Editor	Sue Ewing
Production Management	Jean Lake
Project Editor for SELECT	Nancy E. Davis
Assistant Editor	MaryLynne Wrye
Marketing Manager	Melissa Baumwald
Marketing Communications Specialist	Liane Shayer
Custom Publishing Operations Specialist	Michael Smith
Art Supervisor	Becky Hainz-Baxter
Text Design/Composition	Mark Ong, Susan Riley
Cover and Part Opener Illustrations	Joseph Maas
Cover Designer	Yvo Riezebos
Illustrations	Illustrious, Inc.
Photo Editor & Researcher	Kelli d'Angona
Copy Editor	Barbara Conway
Manufacturing Coordinator	Janet Weaver
Film	York Graphic Services
Printing and Binding	R. R. Donnelley & Sons Company

Copyright © 1996 by The Benjamin/Cummings Publishing Company, Inc.

All rights reserved. No part of this publication may be reproduced, or stored in a database or retrieval system, distributed, or transmitted in any form or by any means, electronic, mechanical, photocopying, recording, or otherwise without the prior written permission of the publisher. Printed in the United States of America. Published simultaneously in Canada.

IBM and PS/2 are registered trademarks of International Business Machines Corporation.

Apple, Macintosh, Mac, HyperCard, Hypertalk, and the Apple logo are registered trademarks of Apple Computer, Incorporated.

MS-DOS , Windows, Windows 95, Microsoft Word, Excel, QuickBASIC, and Microsoft are registered trademarks, and Microsoft NT is a trademark of Microsoft Corporation.

WordPerfect is a registered trademark of WordPerfect Corporation.

Lotus, 1-2-3, and Lotus Notes are registered trademarks of Lotus Development Corporation.

Paradox is a registered trademark of Borland International, Incorporated.

CompuServe is a registered service mark of CompuServe, Incorporated.

America Online is a registered service mark of America Online, Incorporated.

Prodigy is a registered service mark of Prodigy, Incorporated.

Intel is a registered trademark and Pentium is a trademark of Intel Corporation.

SAT is a registered trademark of The College Board.

Many of the designations used by manufacturers and sellers to distinguish their products are claimed as trademarks. Where those designations appear in this book, and Benjamin/Cummings was aware of a trademark claim, the designations have been printed in initial caps or all caps.

Library of Congress Cataloging-in-Publication Data

Capron, H. L.
    Computers: tools for an information age/ H. L. Capron, —4th ed.
      p.   cm.
    Rev. ed. of: Computers & information systems. 3rd. ed. c1993.
    Includes index.
    ISBN 0-8053-0662-5
      1. Computers.   I. Capron, H. L.   Computers & information systems.
II. Title.
QA76.C358   1995
004–dc20   95-38716
                                                                                                     CIP

SE   ISBN 0-8053-0662-5
AIE   ISBN 0-8053-0676-5
MC   ISBN 0-8053-0686-2

Dedicated to

Angela Gill Barnes

The Capron Collection

A Complete Supplements Package

Supplements to the Text

■ **Instructor's Edition with Annotations,** by Jim Johnson with H. L. Capron. This special edition contains annotations for lecture preparation and includes supplementary material not found in the Instructor's Guide. The annotations include chapter outlines for lectures, lecture objectives, Group Work notes, Critical Thinking Questions, Infobits, Global Perspective notes, Learn by Doing notes, test bank references, transparency references, and key terms. An icon for using the CD-ROM *How Multimedia Computers Work* (HMMCW) is also included.

■ **Instructor's CD-ROM package.** This is an interactive instructional support package that includes book art and animations to illustrate complex concepts. The package allows instructors to customize lecture notes from chapter outlines. All material from the annotated Instructor's Edition and the Test Bank and Instructor's Guide is included. It can be used to create quizzes and study tutorials. This is free to instructors upon adoption of the text.

■ **Additional CD-ROM packages.** Benjamin/Cummings is pleased to offer three CD-ROM packages to adopters of the text: *How Computers Work,* by Warner New Media; *How Multimedia Computers Work,* by Mindscape; and *Computer Works,* by Software Marketing Corporation. Please contact your Benjamin/Cummings sales representative for details about these offers.

■ **Test Bank and Instructor's Guide,** by H. L. Capron (288 pages). There are four types of questions: multiple choice, true/false, matching, and completion. Each question is referenced to the text by page number, and the answers are provided. The test bank is available both as hard copy and in a computerized format for the IBM PC (and compatibles) and Macintosh computers.

The Instructor's Guide contains learning objectives, a chapter overview, a detailed lecture outline, and a list of key words for each chapter. It also includes reference guides to the Interactive Multimedia Packages and the videotapes.

- **Color Transparency Acetates.** The 50 full-color transparency acetates include photos and artwork taken directly from the text.

- **Instructor Newsletter:** *BC Link*. Benjamin/Cummings is now offering a free instructor-oriented newsletter for teaching introductory computing. This useful resource includes articles on the use of computer technology in education, teaching strategies, and a selection designed for use with students in the classroom (and is available online at the BC web site).

- **Videotapes.** Benjamin/Cummings makes available to qualified adopters free videotapes from our library of commercially produced tapes. Use this valuable resource to enhance your lectures on concepts presented in the text. Your Benjamin/Cummings sales representative has details about this offer.

Brief Table of Contents

Photo Essay: The Age of Information 1

Part 1
Hardware Tools 13

Chapter 1
Overview of a Computer System: Hardware, Software, and People 15

Chapter 2
The Central Processing Unit: What Goes on Inside the Computer 41

Chapter 3
Input and Output: The User Connection 65

Chapter 4
Storage Devices and Stored Data: Just the Facts 95

Chapter 5
Communications: Computer Connections 125

Part 2
Software Tools 161

Chapter 6
Programming and Languages: Telling the Computer What to Do 163

Chapter 7
Operating Systems: Software in the Background 201

Chapter 8
Systems Analysis and Design: The Big Picture 229

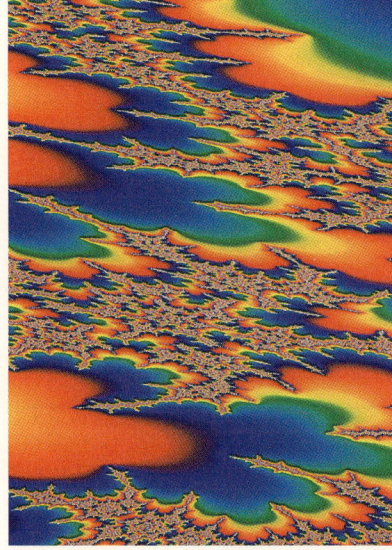

Part 3

Workplace Tools 267

Chapter 9
Computers in the Workplace: Large, Small, and Networked 269

Chapter 10
Management Information Systems: Classic Models and New Approaches 291

Chapter 11
Computer Issues in the Workplace: Security, Privacy, and Ethics 309

Chapter 12
Modern Trends in the Workplace: Expert Systems, Robotics, and Virtual Reality 337

Appendixes

Appendix A
The Programming Process: Planning the Solution A-1

Appendix B
History and Industry: The Continuing Story of the Computer Age B-1

Appendix C
Number Systems C-1

Glossary G-1
Credits CR-1
Index I-1

Detailed Table of Contents

Preface xxi

Photo Essay: The Age of Information 1
Stepping Out 2
    Forging a Computer-Based Society 2
    A Computer in Your Future 3
Computer Literacy for All 3
Everywhere You Turn 4
    The Nature of Computers 4
    Where Computers Are Used 6

Part One

Hardware Tools 13

Chapter 1
**Overview of a Computer System:
Hardware, Software, and People 15**
    Learning Objectives 15
The Big Picture 16
Hardware: Meeting the Machine 16
    Your Personal Computer Hardware 17
    Input: What Goes In 17
    The Processor and Memory: Data Manipulation 19
    Output: What Comes Out 20
    Secondary Storage 20
    The Complete Hardware System 22
    Classification of Computers 22
    Data Communications: No Need to Be There 26
Software: Telling the Machine What to Do 28
    Categories of Software 28
    Some Task-Oriented Software 30
People and Computers 32
    Computers and You, the User 32
    Computer People 33

CHAPTER REVIEW 34
    Summary and Key Terms 34
    Discussion Questions 35
    Student Study Guide 36

Margin Notes
    Computers Going Green 16
    Carrying Computers Around 20
    Help for the SAT 33

Making the Right Connections
    Coming Full Cycle 29

One Jump Ahead
    Just Don't Call It a Computer 22

Planet Internet
    What Is It All About? 39

Chapter 2
The Central Processing Unit 41
    Learning Objectives 41
    The Central Processing Unit 42
        The Control Unit 43
        The Arithmetic/Logic Unit 43
        Registers: Temporary Storage Areas 44
    Memory 45
    How the CPU Executes Program Instructions 46
    Storage Locations and Addresses: How the Control
        Unit Finds Instructions and Data 48
    Data Representation: On/Off 49
        Bits, Bytes, and Words 49
        Coding Schemes 50
    Personal Computer Chips 50
        Microprocessors 50
        Memory Components 51
    Speed and Power 53
        Computer Processing Speeds 53
        Bus Lines 53
        Cache 54
        Flash Memory 54
        RISC Technology: Less Is More 54
        Parallel Processing 55

CHAPTER REVIEW 57
    Summary and Key Terms 57
    Discussion Questions 59
    Student Study Guide 59

Margin Notes
    Chips Inside Everything 43
    A Chip to Save Your Life 53
    How Fast Is a Nanosecond? 54
    A Family of Chips 55

Making the Right Connections
    The Information Superhighway 44

One Jump Ahead
    Body Chips 52

Planet Internet
    Getting Around 63

Chapter 3
Input and Output: The User Connection 65
    Learning Objectives 65
    How Users See Input and Output 66
    Input: Getting Data from the User to the Computer 67
        Keyboard 67
        Mouse 68
        Trackball 69
        Source Data Automation: Collecting Data
            Where It Starts 70
        Magnetic-Ink Character Recognition 70
        Optical Recognition 71
        Data Collection Devices 74
        Voice Input 76
        Touch Screens 77
        Looking 77
    Output: Information for the User 78
        Computer Screen Technology 78
        Types of Screens 79
        Terminals 79
        Printers 79
        Computer Output Microfilm 82
        Voice Output 82
        Music Output 82
    Computer Graphics 83
        Business Graphics 83
        Video Graphics 84
        Computer-Aided Design/Computer-Aided
            Manufacturing 85
        Graphics Input Devices 86
        Graphics Output Devices 87

CHAPTER REVIEW 88
    Summary and Key Terms 88
    Discussion Questions 89
    Student Study Guide 90

Margin Notes
    Big Bad Bar Codes 73
    Keeping Track of Nordic Track 76
    Voice Input Pranks 78
    I Know Art When I See It 81

Making the Right Connections
    Are We Paperless Yet? 83

One Jump Ahead
    Parenthood Tryouts 87

Planet Internet
    Global Village 93

Chapter 4
Storage Devices and Stored Data: Just the Facts 95

Learning Objectives 95
The Benefits of Secondary Storage 96
Magnetic Disk Storage 96
    Diskettes 97
    Hard Disks 98
    Hard Disks in Groups 100
    How Data Is Organized on a Disk 100
Optical Disk Storage 103
    CD-ROM 103
    Multimedia 104
    Magneto-Optical 104
Magnetic Tape Storage 104
Backup Systems 107
Checking an Earlier Decision 108
Organizing and Accessing Stored Data 108
    Data: Getting Organized 108
    The File Plan: An Overview 110
    File Organization: Three Methods 111
    Disk Access to Data 113
Processing Stored Data 113
    Batch Processing 114
    Transaction Processing 114
    Batch and Transaction Processing: The Best of Both Worlds 116

CHAPTER REVIEW 118
Summary and Key Terms 118
Discussion Questions 120
Student Study Guide 120

Margin Notes
Diskette Care 97
What's in a Name? 100
Found at Last 104

Making the Right Connections
The Ultimate Connection 113

One Jump Ahead
Searching for the Killer App 107

Tips for the Macintosh Computer
Making Your Disk Space Count 102

Planet Internet
Just the FAQs, Please 123

Chapter 5
Communications: Computer Connections 125

Learning Objectives 125
Data Communications: How It All Began 126
Putting Together a Network: A First Look 126
    Getting Started 127
    Network Design Considerations 127
Data Transmission 129
    Digital and Analog Transmission 129
    Modems 130
    Asynchronous and Synchronous Transmission 131
    Simplex, Half-Duplex, and Full-Duplex Transmission 132
Communications Links 133
    Types of Communications Links 133
    Protocols 136
Network Topologies 138
Wide Area Networks 139
Local Area Networks 139
    Local Area Network Components 139
    Local Area Network Types 140
    Local Area Network Protocols 141
The Work of Networking 142
    Electronic Mail 142
    Voice Mail 143
    Facsimile Technology 144
    Teleconferencing 144
    Electronic Data Interchange 145
    Electronic Fund Transfers: Instant Banking 145
    Bulletin Boards 145
    Computer Commuting 146
    Commercial Communications Services 146
The Internet 148
    Getting Connected 149
    Getting Around 149

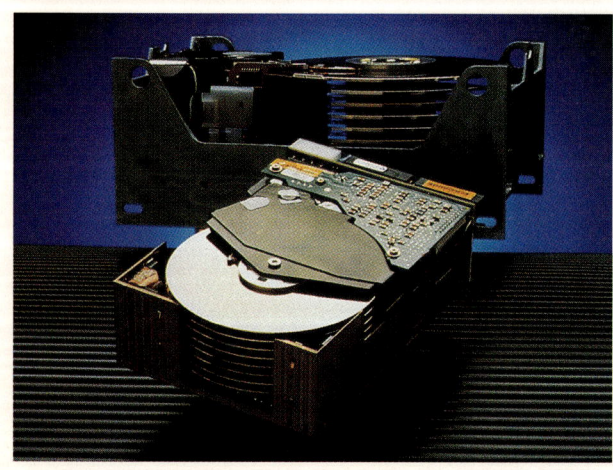

xiii
▼ Contents

    The Internet in Business 151
    So Much More 151
The Complexity of Networks 151

CHAPTER REVIEW 154
    Summary and Key Terms 154
    Discussion Questions 156
    Student Study Guide 156

Margin Notes
    No Hot Dog Lines 126
    Citizens of the Information Superhighway 127
    Bill and Craig's Excellent Adventure 136
    E-Mail Addresses of the Rich and Famous 142
    I'm So Embarrassed About My Card 151

Making the Right Connections
    Save Money, Save Paper, Save Time 129

One Jump Ahead
    2000 and Beyond 147

Tips for the Macintosh Computer
    Going Public with America Online 150

Planet Internet
    Shopping Tour 159

Part Two
Software Tools 161

Chapter 6
Programming and Languages: Telling the Computer What to Do 163
    Learning Objectives 163
Why Programming? 164
What Programmers Do 164
The Programming Process 165
    1. Defining the Problem 165
    2. Planning the Solution 165
    3. Coding the Problem 166
    4. Testing the Program 166
    5. Documenting the Program 169
Programming as a Career 169
    The Joys of the Field 169
    What It Takes 170
    Open Doors 170
Programming Languages 170
Levels of Language 171
    Machine Language 171
    Assembly Languages 171
    High-Level Languages 172

Very High-Level Languages 174
Natural Languages 175
Choosing a Language 177
Major Programming Languages 178
FORTRAN: The First High-Level Language 178
COBOL: The Language of Business 180
BASIC: For Beginners and Others 182
Pascal: The Language of Simplicity 182
Ada: Named for the Countess 184
C: A Portable Language 186
Object-Oriented Programming 186
What Is an Object? 188
Beginnings: Boats as Objects 189
Reuse 190
Activating the Object 190
Object-Oriented Languages 190
Object Technology in Business 191
Some Advice 192

CHAPTER REVIEW 194
Summary and Key Terms 194
Discussion Questions 196
Student Study Guide 196

Margin Notes
Doomsday: January 1, 2000 164
The First "Bug" Was Real 169
Keeping Up 170
Supergeeks 171
A Programming Pioneer: Grace M. Hopper 180

Making the Right Connections
Networked Children 193

One Jump Ahead
Techno Do-gooders 165

Planet Internet
Career in Gear 199

Chapter 7
Operating Systems:
Software in the Background 201
Learning Objectives 201
Operating Systems: Hidden Software 202
Operating Systems for Personal Computers:
An Overview 203
A Look at MS-DOS 205
A Brief Disk Discussion 205
Types of Files 208
Operating Environments 208
Microsoft Windows: An Overview 208
Multiprogramming/Multitasking: Comparing Mainframe Operating Systems to Windows 209
Software Applications with Windows 210
Windows 95 210
Network Operating Systems 212
Operating Systems for Large Computers:
An Overview 213
Resource Allocation 213
Sharing the Central Processing Unit 213
Sharing Memory 216
Sharing Storage Resources 218
Sharing Printing Resources 219
Service Programs 220
Generic Operating Systems 220
The UNIX Graduates 220
Is UNIX a Standard? 220
Do I Really Need to Know All This? 221

CHAPTER REVIEW 222
Summary and Key Terms 222
Discussion Questions 223
Student Study Guide 224

Margin Notes
Introducing Mr. Gates 205
All About Bob 210

Making the Right Connections
Nature Conversancy Online 219

One Jump Ahead
Invisible Computers 215

Tips for the Macintosh Computer
The Macintosh Applications Interface 204

Planet Internet
Images, Icons, and Flags 227

Chapter 8
**Systems Analysis and Design:
The Big Picture 229**
    Learning Objectives 229
The Systems Analyst 230
    The Analyst and the System 230
    The Systems Analyst as Change Agent 231
    What It Takes to Be a Systems Analyst 231
How a Systems Analyst Works: Overview of the Systems Development Life Cycle 232
Phase 1: Preliminary Investigation 234
    Problem Definition: Nature, Scope, Objectives 235
    Wrapping Up the Preliminary Investigation 237
Phase 2: Systems Analysis 237
    Data Gathering 237
    Data Analysis 239
    System Requirements 242
    Report to Management 242
Phase 3: Systems Design 243
    Preliminary Design 243
    Prototyping 244
    CASE Tools 246
    Detail Design 247
Phase 4: Systems Development 252
    Scheduling 252
    Programming 254
    Testing 254
Phase 5: Implementation 254
    Training 254
    Equipment Conversion 255
    File Conversion 256
    System Conversion 256
    Auditing 257
    Evaluation 258
    Maintenance 258
Putting It All Together: Is There a Formula? 258

CHAPTER REVIEW 260
    Summary and Key Terms 260
    Discussion Questions 261
    Student Study Guide 262

Margin Notes
    Checking the Classified Ads 231
    Some Tips for Successful Interviewing 237
    Presentations 244

Making the Right Connections
    Unknown Consequences 257

One Jump Ahead
    Community Computing 232

Case Study
    Preliminary Investigation 235
    Systems Analysis 238
    Systems Design 246
    Systems Development 251
    Implementation 255

Planet Internet
    Not Quite Perfect Yet 265

Part Three
Workplace Tools 267

Chapter 9
**Computers in the Workplace:
Large, Small, and Networked 269**
   Learning Objectives 269
Big Guys: Users of Large Computer Systems 270
Personal Computers in the Workplace 270
   Evolution of Personal Computers
      in the Workplace 270
   The Impact of Personal Computers 272
   Where Personal Computers Are Almost a
      Job Requirement 272
   The Name of the Game Is Speed 272
How Computers Change the Way We Work 273
   Groupware 274
   Computerized Meetings 276
Software at Work 277
   Communications Software 277
   Graphics at Work 278
   Decision-Making Software 279
   Storage and Retrieval Software 280
   Vertical Market Software 282
   What Every Small Business Should Know 284
What Could Possibly Be Next? 286

CHAPTER REVIEW 287
   Summary and Key Terms 287
   Discussion Questions 287
   Student Study Guide 288
Margin Notes
   The Big Payoff 270
   Computer User Groups 284
Making the Right Connections
   The Virtual Office 281
One Jump Ahead
   Kicking the Telephone Habit 273
Planet Internet
   Free and Not Free 289

Chapter 10
**Management Information Systems:
Classic Models and New Approaches 291**
   Learning Objectives 291
Classic Management Functions 292
MIS for Managers 293
The New Management Model 294
   A Flattened Pyramid 294
   The Impact of Groupware 294
   Teamwork 295
Top Managers and Computers 296
   Decision Support Systems 296
   Executive Support Systems 299
Managing Personal Computers 299
   The Personal Computer Manager 300
   Personal Computer Acquisition 301
   The Information Center 302
   Dumping Technology on Workers 302
   Do You Even Know Where Your PCs Are? 304
Leading Business into the Future 304

CHAPTER REVIEW 305
   Summary and Key Terms 305
   Discussion Questions 305
   Student Study Guide 306
Margin Notes
   Looking Intently at the Screen 299
   Troubleshooting Personal Computers 300
Making the Right Connections
   Remote Users 295
One Jump Ahead
   Your Electronic Agent 297
Planet Internet
   Resources for Living 307

Chapter 11

Computer Issues in the Workplace: Security, Privacy, and Ethics 309
    Learning Objectives 309
Intangible Issues in the Workplace 310
Computer Crime 310
    Who Is the Computer Criminal? 311
    Types and Methods of Computer Crime 312
    Discovery and Prosecution 312
Security: Playing It Safe 314
    Identification and Access: Who Goes There? 314
    When Disaster Strikes:
        What Do You Have to Lose? 315
    Disaster Recovery Plan 316
    Software Security 317
    Data Security 318
    Worms and Viruses 319
    Network Security 321
    Personal Computer Security 322
    Prepare for the Worst: Back Up Your Files 323
Privacy: Keeping Personal Information Personal 323
    Passing Your Data Around 324
    Privacy Legislation 324
    Privacy in the Workplace 326
A Matter of Ethics 326

Copying Software 327
    OK If I Copy That Software? 327
    Why Those Extra Copies? 328
    Site Licensing 328
    The Battle Continues 329

CHAPTER REVIEW 330
    Summary and Key Terms 330
    Discussion Questions 331
    Student Study Guide 332

Margin Notes
    Some "Bad Guy" Tricks 312
    Some Gentle Advice on Security 318
    I've Got a Secret 319
    Easy to Steal 323
    Own a Car? Your Name and Address Are
        Probably for Sale 324
    Copyrighted Artwork 328

Making the Right Connections
    You Have No Privacy Whatever 327

One Jump Ahead
    Just the Standard Insurance, Please 317

Planet Internet
    Internet on the Internet 335

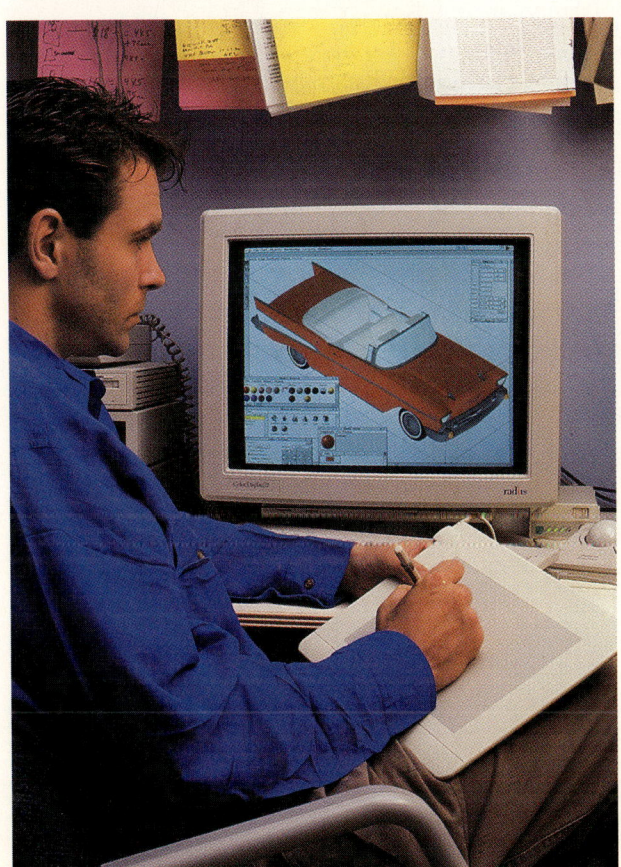

Chapter 12
Modern Trends in the Workplace: Expert Systems, Robotics, and Virtual Reality 337
Learning Objectives 337
From the Lab to the Workplace 338
The Artificial Intelligence Field 338
   Early Mishaps 338
   How Computers Learn 340
   The Artificial Intelligence Debate 340
   Brainpower: Neural Networks 340
   The Famous Turing Test 341
The Natural Language Factor 343
Expert Systems 344
   Expert Systems in Business 344
   Building an Expert System 345
   The Outlook for Expert Systems 346
Robotics 346
   Robots in the Factory 346
   Robot Vision 348
   Field Robots 349
Virtual Reality 350
   Travel Anywhere, But Stay Where You Are 350
   Getting Practical 350

CHAPTER REVIEW 352
Summary and Key Terms 352
Discussion Questions 353
Student Study Guide 353

Margin Notes
Just a Bit Garbled 338
Eliza 343

Making the Right Connections
Power and Wealth 341

One Jump Ahead
Robots Coming to Our Lives 347

Planet Internet
Eye on Business 355

Appendix A
The Programming Process: Planning the Solution A-1
Learning Objectives A-1
Flowcharts A-2
Structure Charts A-3
Pseudocode A-3
Structured Programming A-4
   Sequence A-5
   Selection A-6
   Iteration A-7
Using Flowcharts, Structure Charts, and Pseudocode A-10
   Example: Counting Salaries A-10
   Example: Checking Customer Credit Balances A-12
   Example: Determining Shift Bonus A-14
   Example: Computing Student Grades A-14
Programmers Emerge A-19
   Early Programming A-20
   Expanding the Structured Programming Concept A-20
   Modularity A-21
   Is There a Future for Structured Programming? A-21

CHAPTER REVIEW A-23
Summary and Key Terms A-23

Margin Notes
Some Pseudocode Rules A-5

Appendix B
History and Industry: The Continuing Story of the Computer Age B-1
Learning Objectives B-1
Babbage and the Countess B-2
Herman Hollerith: The Census Has Never Been the Same B-3
Watson of IBM B-3
The Start of the Modern Era B-3
The Computer Age Begins B-5
   The First Generation, 1951–1958: The Vacuum Tube B-5
   The Second Generation, 1959–1964: The Transistor B-6
   The Third Generation, 1965–1970: The Integrated Circuit B-7
   The Fourth Generation, 1971–Present: The Microprocessor B-8
   The Fifth Generation: Onward B-8
The Story of Personal Computers B-8
   I Built It in My Garage B-10
   The IBM PC Standard B-10
   The Rise of Microsoft B-10

CHAPTER REVIEW B-12
   Summary and Key Terms B-12
   Student Study Guide B-13
Margin Notes
   Watson Smart? You Bet! B-4
   The Computer Museum B-5
   Better Late Than Never B-8
Box
   The Software Entrepreneurs B-9

Appendix C
Number Systems C-1
   Learning Objectives C-1
Number Bases C-2
   Base 2: The Binary Number System C-2
   Base 8: The Octal Number System C-3
   Base 16: The Hexadecimal Number System C-3
Conversions Between Number Bases C-3
   To Base 10 from Bases 2, 8, and 16 C-4
   From Base 10 to Bases 2, 8, and 16 C-5
   To Base 2 from Bases 8 and 16 C-6
   From Base 2 to Bases 8 and 16 C-7

Glossary G-1
Credits CR-1
Index I-1

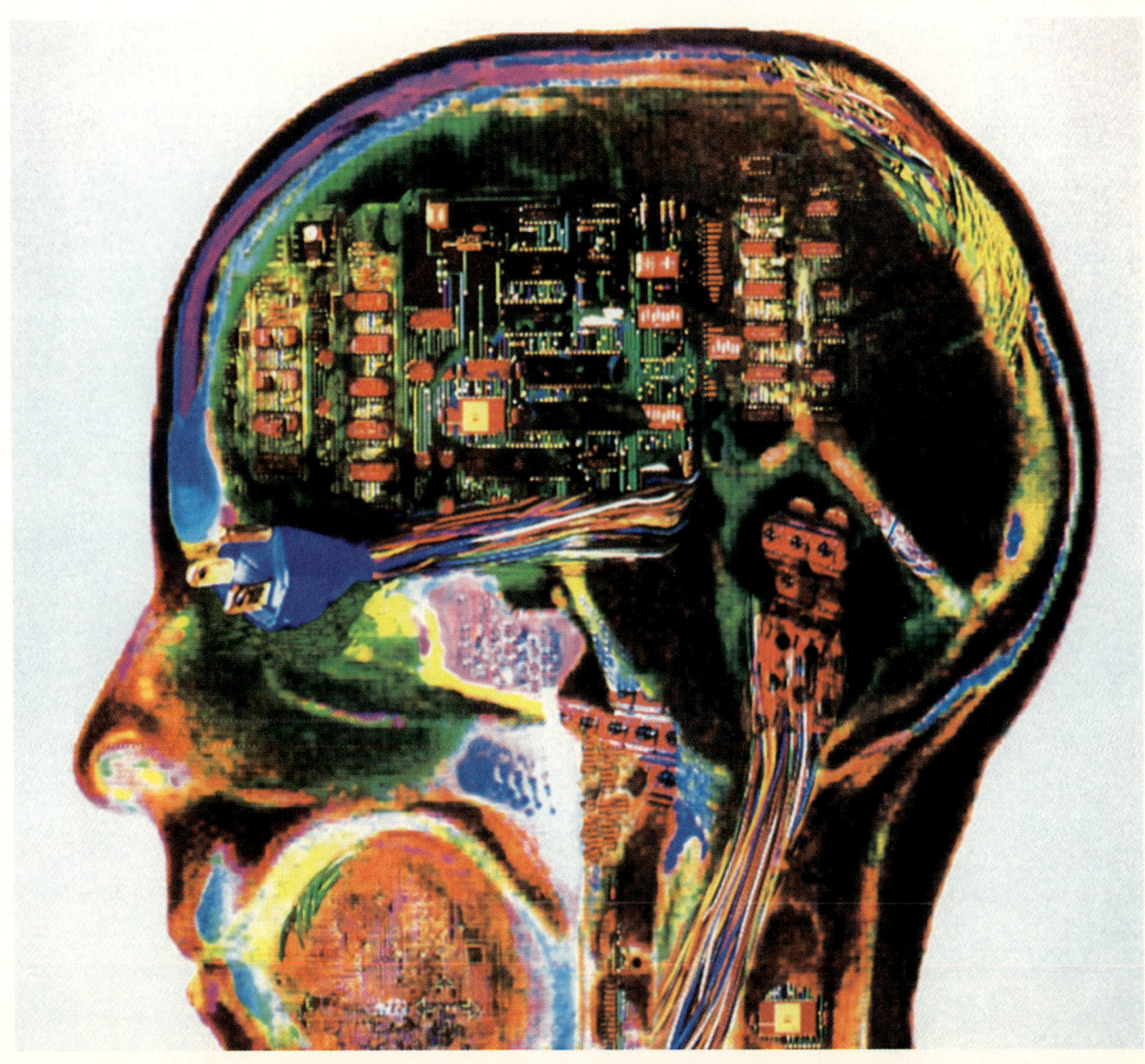

The Buyer's Guide and Galleries

Buyer's Guide
(follows page xxxii)
This eight-page section presents issues and questions to consider before buying a personal computer and software.

Gallery 1
Making Microchips
(follows page 32)
In this gallery we take a look at how silicon chips are made. We follow the process from the design stage through manufacturing, testing, and packaging the final product.

Gallery 2
Multimedia
(follows page 128)
The description of multimedia in Chapter 4 does not do justice to the excitement of multimedia. Eight pages of color screen shots—from ancient lands to baseball to art galleries—help students appreciate the sight and sound of multimedia.

Gallery 3
Computer Graphics
(follows page 256)
Rather than using paint brushes or chalk, some artists use the computer to create their artworks. The gallery features both pure art and commercial applications.

Gallery 4
Computers at Work
(follows page 320)
The computer has become an essential tool in the world of business and manufacturing. In this gallery we examine computers in a variety of workplace settings, focusing particularly on design and robotics.

Preface

The SELECT System

The Benjamin/Cummings Publishing Company is pleased to announce an innovation in publishing that will change the way you think about teaching computing concepts and microcomputer applications. We put technology to work to provide you with one convenient, affordable text that can meet the changing nature of your course. Now you can get a concise introduction to computing by the best-known authors in computer information systems, plus a customized selection of application modules for the software packages you teach. The plan is simple. Here's how it works:

The SELECT Edition of *Computers: Tools for an Information Age, Fourth Edition*

A Text with Concepts and Customized Application Coverage

You choose the SELECT Edition of *Computers: Tools for an Information Age, Fourth Edition*. Next, you choose any combination of the following modules.

Application Modules		
Application Type	Windows	DOS
Word Processing	WordPerfect 6.1 Projects for Windows	WordPerfect 6.0 Projects for DOS
	WordPerfect 6 Projects for Windows	Projects for WordPerfect 5.1
	WordPerfect 5.2 Projects for Windows	
	Microsoft Word 6 Projects for Windows	
Spreadsheets	Lotus 1-2-3 Rel. 4 Projects for Windows	Projects for Lotus 1-2-3, Rel. 2.3/2.4
	Microsoft Excel 5 Projects for Windows	Projects for Lotus 1-2-3, Rel. 2.2
	Microsoft Excel 4.0 Projects for Windows	Projects for Quattro Pro 4.0/5.0
	Projects for Excel 3.0	
	Quattro Pro 6 Projects for Windows	
	Quattro Pro 1.0/5.0 Projects for Windows	
Database	Microsoft Access 2 Projects for Windows	Projects for dBASE IV
	Paradox 5 Projects for Windows	Projects for dBASE III Plus
	Paradox Projects for Windows	Projects for Paradox 3.5

Presentation Graphics Packages		
	Microsoft PowerPoint 4 Projects for Windows	
Integrated Packages	Microsoft Works 3.0 Projects for Windows	Projects for Microsoft Works 3.0 for PCs
		Projects for Microsoft Works 2.0 for PCs
DOS/Windows Applications Modules		
	Projects for Windows 95	
	Projects for DOS 6.0 and Windows 3.1	
	Projects for DOS 5.0 and Windows 3.1	
	Projects for DOS 2.0/3.3 and Windows 3.0	
Programming Modules	Microsoft Visual Basic 3.0 Projects for Windows	Structured Basic for Beginners
		QBasic for Beginners
Also available as a stand alone text:		
Microsoft Office Projects for Windows		
This text can be bundled with any of the above titles.		

Once you place your order with your bookstore manager or your sales representative, your choice of modules will be bound together into one convenient, durable text with *Computers: Tools for an Information Age, Fourth Edition*. We will then send the bookstore your SELECT Edition. Your students will learn from a textbook that has been tailored to match the objectives of your course. Sound easy? It is.

We introduce new modules regularly, so call the SELECT Hotline at 800-854-2595 or contact your sales representative for the most current information.

Complimentary Review Copies

We have prepared the following materials for review and adoption consideration:

> The Instructor's Edition with Annotations of *Computers: Tools for an Information Age, Fourth Edition*

This edition contains the complete contents of the student text plus eight types of margin annotations to support instruction.

Custom complimentary copies of *Computers: Tools for an Information Age, Fourth Edition,* bound with your choice of SELECT modules, can be ordered upon request. Contact your Benjamin/Cummings sales representative if you wish to see a preview of a SELECT Edition.

Ordering and Pricing Information

We believe our SELECT System offers unprecedented opportunity for educators to evaluate flexible text components and build them into a customized teaching support system suitable for individual course configurations. Your Benjamin/Cummings representative will be happy to work with you and your bookstore manager to outline the ordering process, and provide pricing and delivery information. To take advantage of the SELECT System, you may also call the Benjamin/Cummings Publishing Company at 800-854-2595. This special Hotline is attended by service representatives ready to answer inquiries and to provide you with additional complimentary or desk copies.

The Concepts Portion of the SELECT Edition of *Computers: Tools for an Information Age*, Fourth Edition

For those students taking an introductory course who are just beginning to discover computers, the concepts portion of this textbook makes the process of discovery both enjoyable and educational. This introduction to computers is written in a friendly and engaging style that sparks the reader's interest.

Most introductory computer books are comprehensive in scope. *Computers: Tools for an Information Age, Fourth Edition* is too, but we want to offer more than just the basics of computing. We want to engage students and draw them into the text without detracting from the seriousness of the material. We want to present a computer book that sounds and feels like everyday living.

An important theme is connectivity. We offer two exciting new features about connecting computers and people. The first feature is a box in each chapter titled Making the Right Connections. These boxes describe people using computers to connect to the larger world in some way. The second feature, found at the end of each chapter, is called Planet Internet. This feature presents information about the Internet and its contents. Particular focus is on the World Wide Web. Chapter 5, "Communications," includes a broad discussion of the Internet.

The fourth edition retains all of the elements that have made this book a best-seller since the first edition. We have updated material and added new features. This book and its related learning materials offer students everything they need to make computers a part of their own everyday living.

New and Updated in the Fourth Edition

The entire text has been updated to reflect current technology. Topics such as the Internet and multimedia have been significantly expanded. A significant new feature is the Multimedia Gallery. Although the technology of multimedia is described in Chapter 4, "Storage Devices and Stored Data," the flavor of the multimedia phenomenon is presented in a gallery of eight color pages.

For those of you who used the third edition of this book, you can note these changes. There are substantive discussions of networking operating systems, Windows 95, and object-oriented programming. The personal computer is usually emphasized first followed, when appropriate, by information about large computers. The Buyer's Guide has been updated to indicate current offerings and buying trends. The Macintosh boxes have been rewritten. Personal Computers in Action boxes have been replaced with the more significant Making the Right Connections boxes. Also, One Jump Ahead boxes, a look to the future, have taken the place of the Perspectives boxes.

Organization of the Text

The text is divided into an introductory photo essay and three parts, followed by three appendices:

- The opening Photo Essay gives students a feeling for the exciting world of computers and shows how people can use them.
- Part 1, "Hardware Tools," explores computer hardware, including coverage of the central processing unit, input/output, storage, and communications.
- Part 2, "Software Tools," looks at software, including programming and languages, operating systems, and systems analysis and design.
- Part 3, "Workplace Tools," examines how computers are used in the work environment; management of information systems; security, privacy, and ethics; and expert systems, robotics, and virtual reality.
- Appendixes are offered on programming, the history of computing, and number systems.

Key Themes

- **Connectivity.** We place a strong emphasis on showing how people and computers connect to other people and computers through wide area networks, local area networks, information utilities, and the Internet. We show this in varied ways—through discussions, photos, and the features of Making the Right Connections and Planet Internet.
- **Focus on computers in business settings.** We provide several features that focus on the uses of computers in the business environment. Each chapter begins with a story, in which it is shown how individuals from a variety of business situations use computers on the job. Many of the photos are specifically chosen for what they convey about computers in business environments. Part 3, "Workplace Tools," is devoted to issues of current interest in business computing. Topics include large and small computers as well as networked ones; the role of the information systems manager; security, privacy, and ethics as issues in the workplace; and modern trends in the workplace such as expert systems, robotics, and virtual reality.

Special Features

- **Appealing style.** When students enjoy what they read, they remember it. The text's friendly style encourages the reader to continue and increases students' comprehension. For example, each chapter begins with an engaging story that leads the student into the material. The real-world applications included throughout the text pique student interest as well as illustrate key points from the chapter.

- **Multimedia gallery.** This color photo essay provides an in-depth look at multimedia and what it has to offer.
- **Planet Internet.** This end-of-chapter feature offers students Internet access through a web site especially developed by the publisher for this text. They will be given some background for each web site they connect to and instruction for how to access it. For those students who do not have access to the Internet, the information is complete enough in each box to learn something about it without going online.
- **Buyer's Guide.** Students and their families are making important economic decisions about the purchase of a computer for their educational, personal, and business needs. This concise eight-page guide offers students information to aid hardware and software purchases.
- **Computer Graphics gallery.** This color photo layout vividly shows the sophistication of computer graphics.
- **Making Microchips gallery and Computers at Work gallery.** These color photo layouts vividly show how microprocessors are made and how computers are commonly and not-so-commonly used in the workplace.
- **Making the Right Connections.** Each chapter includes a feature article on linking computers to people or to information. The articles range from reaching the Nature Conservancy online to children in Wisconsin using the Internet to talk with pen pals in all 50 states in the United States.
- **One Jump Ahead.** To give students a glimpse of the new directions computer technology is taking, each chapter provides a brief essay that focuses on issues and trends in the world of computing. Examples include using the computer to research and buy consumer goods and how robots may do our vacuuming and yard work in the near future.
- **Chapter Review and Student Study Guide.** To encourage students to review concepts and to confirm their comprehension of the material (as well as not having to buy another book), each chapter now concludes with a study guide. The Chapter Review provides an end-of-chapter summary of core concepts and key terms, followed by Discussion Questions, and a Student Study Guide that includes multiple-choice, true/false, and completion questions. The answers to all questions are provided at the end of the study guide.
- **Margin notes.** To further engage the student, margin notes are carefully placed throughout the text. The margin notes extend the text material by providing additional information and highlighting interesting applications of computers.
- **The Macintosh computer** is highlighted in three chapters through brief discussions of Macintosh procedures. Topics covered include the wise use of disk space, America Online, and GUI interface.

In-Text Learning Aids

Each chapter includes the following pedagogical support.

- At the beginning of each chapter, **learning objectives** provide key concepts for students.
- **Key terms** are boldfaced throughout the text.
- A **Chapter Review** offers a summary of core concepts and boldfaced key terms.

- **Discussion Questions** encourage the students to take what has been presented in the chapter and discuss it more thoroughly.
- The **Student Study Guide** gives students three types of questions (multiple choice, true/false, and completion) that they can answer to check their comprehension of essential concepts. All answers are provided for the student at the end of the chapter.
- An extensive **Glossary** and comprehensive **Index** are included.

The Application Modules in the SELECT Edition of *Computers: Tools for an Information Age,* Fourth Edition

Learning software applications is easy when students practice skills in the context of problem solving. By completing these modules, students gain realistic preparation for their future careers when software skills will be an important component of their jobs. These modules are designed to teach Windows, DOS, and popular software application packages for personal computers in an introductory computer literacy/microapplications course. The modules are intended for the first-time computer user with basic typing skills.

Each module is written by an experienced author and instructor and follows a consistent, pedagogically sound format. The modules begin with an overview of basic concepts for each software application—concepts such as starting the program, getting help, and an explanation of the conventions the modules use. Then, within the context of six or more increasingly challenging projects, students learn problem-solving techniques that enhance and reinforce comprehension of the specific software applications package.

Application Modules: The Philosophy of the Project Approach

The projects are the core of the student's learning process. They motivate the student by offering both general-interest and business-related examples. Each project title identifies a functional task in which specific commands are mastered. Students gain an appreciation of both the conceptual and keystroke levels of a software application. The modules are intended for the first-time computer user but contain selected advanced topics for the more experienced user.

To develop these project-based application modules we went one step beyond the traditional publishing model and added a consulting team of developmental editors. The team, Evelyn Spire, Nancy Canning, Shelly Langman, and Rebecca Johnson, extensively analyzed the most effective teaching philosophies to create the project approach. They carefully prepared a logical presentation of the application software. They combined their business acumen with the academic perspective of the authors and the instructors who reviewed the material. They scrutinized the modules from the standpoint of pedagogical consistency, tone, level, and writing style. They gave considerable thought to the conventions used and chose those that were easiest to use in the laboratory environment. The end result is a set of projects that will stimulate learning and prepare students for their careers.

Application Modules: Pedagogy and Learning Aids

- **Learning objectives** define in practical terms what the student will be able to do after completing each project.
- Six to eight **projects,** increasingly challenging in nature, teach students important concepts and commands in a real-world context.
- **Case Studies** lay out problems or situations that students will address in the projects. The **Designing the Solution** section analyzes each Case Study and then helps students develop problem-solving strategies.
- **Numbered steps** guide students through the projects. A computer icon cues students when to begin working on the computer.
- **Screen captures and margin figures** visually reinforce key concepts and help students check their work.
- **Key terms** appear in boldfaced italics throughout the text.
- Special features, such as **Tips, Reminders, Quick Fixes,** and **Cautions,** highlight specific material in a project that is particularly helpful, important, or pertinent to what students are learning in the project. **Exit Points** indicate places where students can save a file and temporarily stop work.
- **The Next Step** extends the material presented in the project and makes students think about other applications for the skills.
- Each project concludes with a **Summary** and a list of **Key Terms and Operations**.
- Each project also includes **Study Questions** (multiple choice, short answer, and discussion) that may be used as a self-test or as a homework assignment. End-of-project **Review Exercises** present hands-on tasks with abbreviated instructions to build on skills learned in the project.
- **Assignments,** found at the end of each project, draw on the skills introduced in the projects and require synthesis, integration, analysis, and critical thinking to complete.
- Each module has its own **Operations Reference,** extensive **Glossary of Key Terms,** and an **Index**.

SELECT Authors

Hans-Peter Appelt	Corning Community College
William J. Belisle	
Gary R. Brent	Scottsdale Community College
James R. Elam	Scottsdale Community College
J. Patrick Fenton	West Valley College
James A. Folts	Oregon State University
Marianne B. Fox	Butler University
Ahmer S. Karim	University of San Diego
Marcy Kittner	University of Tampa

Philip A. Koneman	Colorado Christian University
Pauline Johnson	
Tony Lima	California State University, Hayward
Lawrence C. Metzelaar	Purdue University at Indianapolis/IUPUI
James T. Perry	University of San Diego
Eugene J. Rathswohl	University of San Diego
Carl A. Scharpf	University of Southern California
Paul B. Thurrott	
Jane Whittenhall	Corning Community College

Supplements for the SELECT Edition of *Computers: Tools for an Information Age, Fourth Edition*

Instructors can take advantage of the complete instructional package that has been developed to support *Computers: Tools for an Information Age, Fourth Edition*. We also have designed individual instructor's manuals with transparency masters and a test bank that support the modules. The complete list of supplements for both the text and the modules follows.

- **Instructor's Edition with Annotations,** by Jim Johnson with H. L. Capron. This special edition contains annotations for lecture preparation and includes supplementary material not found in the Instructor's Guide. The annotations include chapter outlines for lecture, lecture objectives, Group Work notes, Critical Thinking Questions, Infobits, Global Perspective notes, Learn by Doing notes, test bank references, transparency references, and key terms. An icon for using the CD-ROM *How Multimedia Computers Work* (HMMCW) is also included.
- **Instructor's CD-ROM package.** This is an interactive instructional support package that includes book art and animations to illustrate complex concepts. The package allows instructors to customize lecture notes from chapter outlines. All material from the annotated Instructor's Edition and the Test Bank and Instructor's Guide is included. It can also be used to create quizzes and study tutorials. This is free to instructors upon adoption of the text.
- **Additional CD-ROM packages.** Benjamin/Cummings is pleased to offer three CD-ROM packages to adopters of the text: *How Computers Work*, by Warner New Media; *How Multimedia Computers Work*, by Mindscape; and *Computer Works*, by Software Marketing Corporation. Please contact your Benjamin/Cummings sales representative for details about these offers.
- **Test Bank and Instructor's Guide,** by H. L. Capron (288 pages). There are four types of questions: multiple choice, true/false, matching, and completion. Each question is referenced to the text by page number, and the answers are provided. The test bank is available both as hard copy and in a computerized format for the IBM PC (and compatibles) and Macintosh computers.
  The Instructor's Guide supports the concepts portion of your customized package. For each chapter you will find learning objectives, a chapter overview, a detailed lecture outline, and a list of key words. It

- also includes reference guides to the CD-ROM packages and the videotapes.
- **Color Transparency Acetates.** The 50 full-color transparency acetates include photos and artwork taken directly from the text.
- **Instructor Newsletter:** *BC Link*. Benjamin/Cummings is now offering a free instructor-oriented newsletter for teaching introductory computing. This useful resource includes articles on the use of computer technology in education, teaching strategies, and a selection designed for use with students in the classroom (and is available online at the BC web site).
- **Videotapes.** Benjamin/Cummings makes available to qualified adopters free videotapes from our library of commercially produced tapes. Use this valuable resource to enhance your lectures on concepts presented in the text. Your Benjamin/Cummings sales representative has details about this offer.
- **Instructor's Manuals for the modules.** Each module has a corresponding Instructor's Manual with a test bank and transparency masters. For each project in the module, the Instructor's Manual includes Expanded Student Objectives, Answers to Study Questions, and Additional Assessment Techniques. Two tests are included that contain multiple choice, true/false, and fill-in-the-blank questions that are referenced to pages in the module. Answers to the tests are also provided. Transparency masters illustrate 25 to 30 key concepts and screen captures from the module.
- **Instructor's Data Disk for the modules.** The Instructor's Data Disk contains student data files, answers to selected Review Exercises and Assignments, and the test files from the Instructor's Manual, in ASCII format.

Special Note to the Student

We welcome your reactions to this book. It is written to open up the world of computing for you. Expanding your knowledge will increase your confidence and prepare you for a life that will be influenced by computers. Your comments and questions are important to us. Write to the author in care of Computer Information Systems Editor, Benjamin/Cummings Publishing Company, 2725 Sand Hill Road, Menlo Park, CA 94025-7092. All letters with a return address will be answered by the author.

Acknowledgments

Many people contributed to the success of this project. Although a single sentence hardly suffices, we would like to thank some of the key people: A consulting team of developmental editors was instrumental in crafting the quality, consistency, and accuracy of the SELECT application modules. Developmental Editor Sue Ewing executed a multi-faceted role with enthusiasm and ingenuity. Jean Lake, as production editor, skillfully coordinated the efforts of many people, keeping the book on the accelerated schedule that contributes to its currency. Kelli d'Angona was a creative and tenacious photo researcher who tracked down outstanding pictures. Designer Mark Ong presented superb design solutions to endlessly changing thorny problems. Assistant Editor MaryLynne Wrye provided able assistance, on matters large and small, on a daily basis. Sponsoring Editor Maureen

Allaire's vision placed the project on the right track, and her steadying hand kept it there.

Reviewers and consultants have provided valuable contributions that improved the quality of the book. Their names are listed in the following section, and we wish to express our sincere gratitude to them.

Computers: Tools for an Information Age, Fourth Edition Reviewers

Leslie A. Blide	University of Wisconsin, Milwaukee
Fred Bounds	DeKalb College
Gary R. Brent	Scottsdale Community College
Catherine J. Brotherton	Riverside Community College
M. Edward Brunjes	San Diego Miramar College
Anne DeFrance	Montana State University
Diane G. Drott	Mississippi State University
Clinton P. Fuelling	Ball State University
Carol A. H. Hall	Louisiana State University, Shreveport
Jim Johnson	Valencia Community College
Vickie F. McCullough	Palomar College
Patrick R. Mattson	St. Cloud State University
Diana L. Miller	Flathead Valley Community College
Guy S. Mills	Benjamin/Cummings Publishing Company
Marty Murray	Portland Community College
Robert S. Plantz	Sonoma State University
Cindy Pryke	Commonwealth College
Steve St. John	Rogers State College
Sharon Underwood	Livingston University

SELECT Reviewers

Joseph Aieta	Babson College
Roger Anderson	College of Lake County
Tom Ashby	Oklahoma City Community College
Bob Barber	Lane Community College
Ronald Burgher	Metropolitan Community College
Terry Byrd	Auburn University
Robert Caruso	Santa Rosa Junior College
Paul Chase	Becker College
Robert Chi	California State University, Long Beach
Robert Clark	Buffalo State University
Martin Crossland	Southwest Missouri State University
Jill Davis	State University of New York at Stony Brook

Fredia Dillard	Samford University
Peter Drexel	Plymouth State College
Ralph Duffy	North Seattle Community College
David Egle	University of Texas, Pan American
Raymond Folse	Nicholls State University
Jonathan Frank	Suffolk University
Patrick Gilbert	University of Hawaii
Maureen Greenbaum	Union County College
Sally Ann Hanson	Mercer County Community College
Sunil Hazari	East Carolina University
Bruce Herniter	University of Hartford
Lisa Jackson	Henderson Community College
Cynthia Kachik	Santa Fe Community College
Harold Kollmeier	Franklin Pierce College
Bennett Kramer	Massasoit Community College
Charles Lake	Faulkner State Junior College
Ron Leake	Johnson County Community College
Dennis Lynch	Elgin Community College
Randy Marak	Hill College
Charles Mattox, Jr.	St. Mary's University
Jim McCullough	Porter and Chester Institute
Vickie McCullough	Palomar College
Gail Miles	Lenoir-Rhyne College
Linda Wise Miller	University of Idaho
Lawrence Molloy	Oakland Community College
Carolyn Monroe	Baylor University
Steve Moore	University of South Florida
Uday Murthy	IUPUI
Anthony Nowakowski	Buffalo State College
Gloria Oman	Portland State University
John Passafiume	Clemson University
Tonia Queen	Brevard Community College
Leonard Presby	William Paterson College
Louis Pryor	Garland County Community College
Michael Reilly	University of Denver
Dick Ricketts	Lane Community College
Dennis Santomauro	Kean College of New Jersey
Pamela Schmidt	Oakton Community College
Gary Schubert	Alderson-Broaddus College
Jennifer Sedelmeyer	Broome Community College
Patricia Smith	Temple Junior College
T. Michael Smith	Austin Community College

Cynthia Thompson	Carl Sandburg College
Marion Tucker	Northern Oklahoma College
JoAnn Weatherwax	Saddleback College
Melinda White	Santa Fe Community College
David Whitney	San Francisco State University
James Wood	Tri-County Technical College
Minnie Yen	University of Alaska, Anchorage
Allen Zilbert	Long Island University

Buyer's Guide

How to Buy Your Own Personal Computer

We cannot choose your new computer system for you any more than we might select a new car for you. But we can tell you to look for or avoid various features. We do not mean that we can lead you to a particular brand and model—so many new products are introduced every month that doing so would be impossible. If you are just starting out, however, we can help you define your needs and ask the right questions.

Where Do You Start?

Maybe you have already done some thinking and have decided that owning your own personal computer offers advantages. Now what? You can start by talking to other personal computer owners about how they got started and how to avoid pitfalls. Or you can read some computer magazines, especially ones with evaluations and ratings, to get a feel for what is available. Next find several dealers. Most dealers are listed in the yellow pages of the phone book, and many advertise in the business section of the local newspaper. Visit several dealers. Don't be afraid to ask questions. You are considering a major purchase, so plan to shop around.

Finally, you may consider buying a computer system by direct mail. You can find advertisements in any computer magazine. Call the listed 800 number and ask them to send you a free brochure.

Analyze Your Needs and Budget

Begin with a wants/needs analysis. Why do you want a computer? Be realistic: Will you use it mostly for games or for business applications? People use personal computers for a variety of reasons. At some point you will have to establish a budget ceiling. After you have examined your needs, you can relay them to the sellers, who will help you select the best hardware-software combination for your budget.

An Early Consideration

Although many brands of computers are available, the business standard is an IBM or IBM-compatible machine. If you will be using your computer for business applications and, in particular, if you need to exchange files with others in a business environment, consider sticking with the standard. However, the Apple Macintosh is an attractive alternative. The Macintosh is noted for ease of use, especially for beginners.

What to Look for in Hardware

The basic personal computer system consists of a central processing unit (CPU) and memory, a monitor (screen), a keyboard and a mouse, a storage device—probably a 3½-inch diskette drive and a hard disk drive—and a printer. Unless you know someone who can help you out with technical expertise, the best advice is to look for a packaged system—that is, one in which the above components (with the exception of the printer) are assembled and packaged by the same manufacturer. This gives you some assurance that the various components will work together.

Computer Housing

Sometimes called the computer case, or simply "the box," the housing holds the electronics and has external receptors called ports to which the monitor, printer, and other devices are connected. In an arrangement called all-in-one, the monitor and housing are a single unit. More common is the traditional desktop case, in which the monitor sits on top of the housing. A third option is the mini-tower, whereby the case sits on end, usually on the floor.

Central Processing Unit

If you plan to purchase an IBM or compatible machine, many software packages run most efficiently on computers using a Pentium microprocessor. A less expensive alternative is the 80486—also called the 486—microprocessor. Many 486 microprocessors are upgradable to the more powerful Pentium chip. A microprocessor's speed is expressed in megahertz (MHz), usually from 25 to 120 MHz and up. A 486 microprocessor, for example, may be advertised at 66 MHz. The higher the number the faster—and more expensive—the microprocessor.

Shortcut

It is probably worth your while to peruse the various hardware possibilities laid out in this guide. But many people just want a quick answer to the what-should-I-buy question. Assuming that you need the IBM standard, many experts recommend a package along these lines: Pentium or higher microprocessor, 8MB, 16MB, or more RAM, 15-inch or larger SVGA color monitor, keyboard, mouse, one 3½ inch diskette drive, a hard drive with several hundred megabytes of storage, and a fax modem. There would be less agreement on a printer, but a laser printer is often recommended. Less necessary but highly desirable is the inclusion of a multimedia hardware package: CD-ROM drive, sound card, and speakers.

Phone It In

On the whole, these direct-mail dealers sell quality hardware at good prices. Call these toll-free numbers to get free catalogs and, if you wish, eventually place an order:

1-800-348-8342
Dell Computer
1-800-846-2000
Gateway 2000
1-800-426-2968 IBM
1-800-888-5858
Compaq Computer
1-800-554-7172
Zeos International

Memory

Memory, or RAM, is measured in bytes, with each byte representing a character of data. The amount of memory you need in your computer is determined by the amount of memory required by the applications programs (like word processing or spreadsheets) that you want to use. The minimum memory threshold keeps rising, as software makers produce sophisticated products that run efficiently only with ever larger amounts of memory. We recommend at least 8 megabytes (8MB) of memory, but prefer 16 megabytes (16MB) or even more. Most machines have expandable memory, so you can add more later if you need it.

Monitor

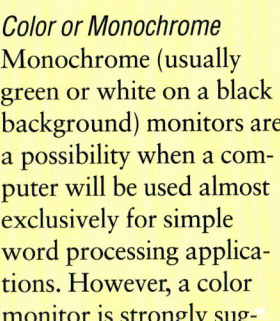

Sometimes called a video display screen, the monitor is a very important part of your computer system—you will spend all your computer time looking at it.

Color or Monochrome

Monochrome (usually green or white on a black background) monitors are a possibility when a computer will be used almost exclusively for simple word processing applications. However, a color monitor is strongly suggested. Most software, even for business use, makes impressive use of color. You will certainly want color if you want to create graphics on your screen or if you plan to run entertainment programs on your computer.

Screen Size

Monitors usually have a screen display of between 12 and 17 inches, measured diagonally. Generally, a larger screen provides a display that is easier to read, so most monitors sold today have at least 15-inch screens.

Screen Readability

Be sure to compare the readability of different monitors. First, make certain that the screen is bright and has minimum flicker. Glare is another major consideration. Harsh lighting nearby can cause glare to bounce off the screen, and some screens seem more susceptible to glare than others.

Questions to Ask the Salesperson at the Store

- ❏ Can I expand the capabilities of the computer later?
- ❏ Whom do I call if I have problems configuring the machine at home?
- ❏ Does the store offer or recommend classes on how to use this computer and software?
- ❏ What kind of warranty comes with the computer?
- ❏ Does the store or manufacturer offer a maintenance contract with the computer?

Buyer's Guide

A key factor affecting screen quality is resolution, a measure of the number of dots, or pixels, that can appear on the screen. The higher the resolution—that is, the more dots—the more solid the text characters appear. For graphics, more pixels means sharper images. Color monitors most commonly available are, in ascending order of good resolution, Video Graphics Array (VGA) and Super VGA (SVGA).

Ergonomic Considerations
Can the monitor swivel and tilt? If so, this will eliminate your need to sit in one position for a long period. The ability to adjust the position of the monitor becomes an important consideration when several users share the same computer, particularly people of different sizes, such as parents and children. Another possibility is the purchase of add-on equipment that allows you to reposition the monitor. Furthermore, if you expect to type for long periods of time, you

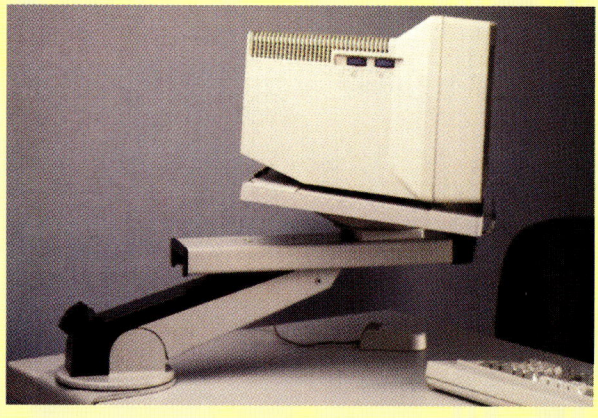

would be wise to buy a wrist pad to support your hands and wrists.

Input Devices

There are many input devices. We will mention only the two critical ones here: a keyboard and a mouse.

Keyboard
Keyboards vary in quality. To find what suits you best, sit down in the store and type. Consider how the keys feel. You may be surprised by the real differences in the feel of keyboards. Make sure the keys are not cramped together; you will find that your typing is error-prone if your fingers are constantly overlapping more than one key.

A detachable keyboard—one that you can hold on your lap, for example—is the norm. You can move a detachable keyboard around to suit your comfort. This feature becomes indispensable when a computer is used by people of different sizes.

Most keyboards follow the standard QWERTY layout of typewriter keyboards. Many have a separate numeric keypad. In addition, most keyboards have separate function keys, which simplify applications software commands.

Assess the color and layout of the keyboard. Ideally, keys should be gray with a matte finish. The dull finish reduces glare.

Mouse
A mouse is a device that you roll on a tabletop to move the cursor on the screen to make selections. A mouse was originally considered a convenient option, but now many applications software packages and even operating systems are designed to be used with a mouse. A mouse has become a necessity.

Secondary Storage

You will need disk drives to read software into your computer and to store software and data that you wish to keep.

Diskette Drive
Most personal computer software today comes on diskettes, so you need a diskette drive to accept the software. Furthermore, many users keep backup copies of their software and data files on diskette.

Most computer systems today come with a 3½-inch diskette drive; some also offer a 5¼-inch diskette drive as an option. If you have no need to be compatible with 5¼-inch diskettes, either from your old computer system or from someone else's computer, you probably do not need a 5¼-inch diskette drive.

Hard Disk Drive
Although more expensive than a diskette drive, a hard disk drive is fast and reliable and holds more data. A hard disk drive is a standard requirement. Modern software comes on a set of several diskettes or on optical disk; it would be unwieldy to load these each time the software is used. Instead, the software is stored on the hard drive, where it is conveniently accessed from that point forward.

Most computer systems offer a built-in hard disk drive, with variable storage capacity—the more storage, the higher the price. Storage capacity usually is measured in terms of millions of bytes—characters— of data. Keep in mind that software, as well as your data files, will be stored on the hard disk; even a word processing program can fill up 20 million bytes. A hard disk ususally holds several hundred megabytes or perhaps gigabytes.

Printers

A printer is probably the most expensive peripheral equipment you will buy. Although some inexpensive models are available, you will find that those costing $400 and up are the most useful. When choosing a printer, consider speed, quality, and cost.

Until recently, the dot-matrix printer was the standard for everyday printing. A dot-matrix printer forms each character with a series of closely spaced dots. But dot-matrix printers are being phased out in favor of affordable ink-jet and laser printers; each

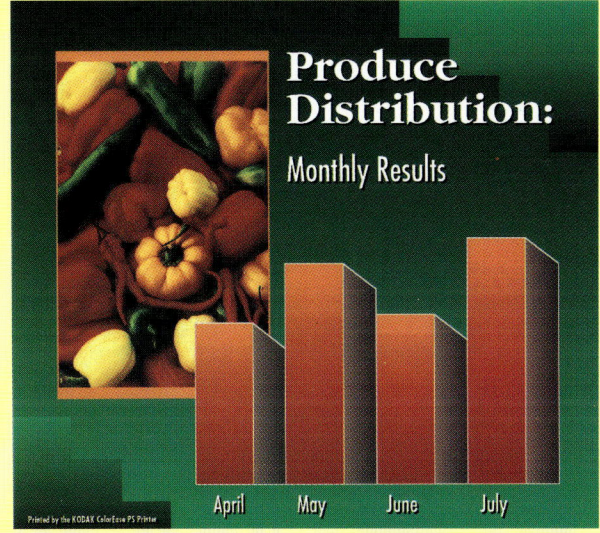

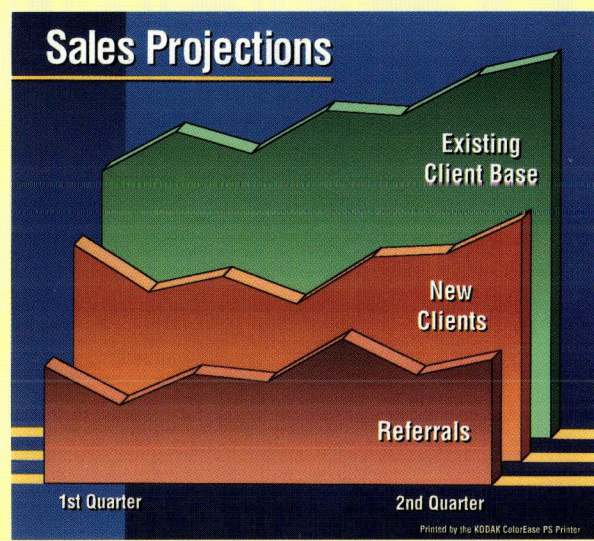

Buyer's Guide

type produces high-quality output and is much quieter than a dot-matrix printer.

Ink-jet printers, in which ink is propelled onto the paper by a battery of tiny nozzles, can produce excellent text and graphics. In fact, the quality of ink-jet printers approaches that of laser printers. The further attractions of low cost and quiet operation has made the ink-jet printer a current favorite among buyers.

Laser printers, which use technology similar to copying machines, are the top-of-the-line printers for quality and speed. The price of a low-end laser printer is now within the budget of most users. Laser printers are particularly favored by desktop publishers to produce text and graphics on the same page. For years, standard laser printers have printed text and graphics at 300 dots per inch, a resolution that produced crisp, professional documents. Now, laser printers produce output at 600 dots per inch, giving graphic images a sharpness that rivals photographs. However, this rich resolution may be of little value to a buyer who plans to produce mostly text.

Although a few color printers are available for less than $500, most are priced much higher. Even at a high price, color printers are not perfect. The rich color seen on the computer screen is not necessarily the color that will appear on the printed output. Furthermore, color printers often have high operating costs for staples such as special coated paper and color ink cartridges. Still, color printers, once prohibitively expensive and slow, are approaching affordable prices and speeds.

Portability

Do you plan to use your computer in one place, or will you be moving it around? Portable computers have found a significant niche in the market, mainly because they are packaged to travel easily. A laptop computer is lightweight (often under 8 pounds) and small enough to fit in a briefcase.

Generally, you should look for the same hardware components as you would consider in a desktop computer: a fast microprocessor, plenty of memory, a clear screen, and diskette and hard drives. You will have to make some compromises on input devices. The keyboard will probably be attached and the keys more cramped than a standard keyboard. Also, traveling users often do not have a handy surface for rolling a mouse, so the laptop may come with a trackball, in the center or on the side.

Other Hardware Options

There are a great many hardware variations; we will mention a few here. Note that, although we are describing the hardware, these devices may come with accompanying software, which must be installed according to directions before the hardware can be used.

Communications Connections
If you want to connect your computer via telephone lines to electronic bulletin boards, mainframe computers, the Internet, or information utilities such as America Online, or if you want to send and receive electronic mail, you need a modem. This device converts outgoing computer data into signals that can be transmitted over telephone lines, and does the reverse for incoming data. The Hayes family of products has become the industry standard; most new modems claim some degree of Hayes compatibility.

You may choose an external modem that can be used with different computers. But most buyers prefer an out-of-sight internal modem that fits inside the computer. Furthermore, most people choose a fax modem, which has the dual purpose of modem and fax. Using a fax modem, you can receive a fax and then print it out, or send a fax if it originated in your computer (using, for example, word processing software) or was scanned into your computer. Many new computers come equipped with a fax modem.

Other Input Devices
If you are interested in action games, you may wish to acquire a joystick, which looks similar to the stick shift on a car. A joystick allows you to manipulate a cursor on the screen. A scanner is useful if you need to store pictures and typed documents in your computer. Scanners are frequently purchased by people who want to use their computers for desktop publishing. Finally, you can purchase voice input hardware, which is basically a microphone.

Multimedia Access
A fast-growing area is multimedia: sophisticated software that offers text, sound, photos, graphics, and even movie clips. To take advantage of multimedia, which is presented on optical disks, you need a CD-ROM disk drive. Furthermore, you will probably want to invest in a sound card, to be installed in your computer, and a set of speakers. Computers tagged as "multimedia computers" will include the CD-ROM drive, a sound card, and—probably—built-in speakers.

Surge Protectors
These devices protect against the electrical ups and downs that can affect the operation of your computer. Some of the more expensive models provide up to 10 minutes of full power to your computer if the electric power in your home or office is knocked out. This gives you time to save your work on disk (so that the work won't be lost if the power fails) or to print out a report you need immediately.

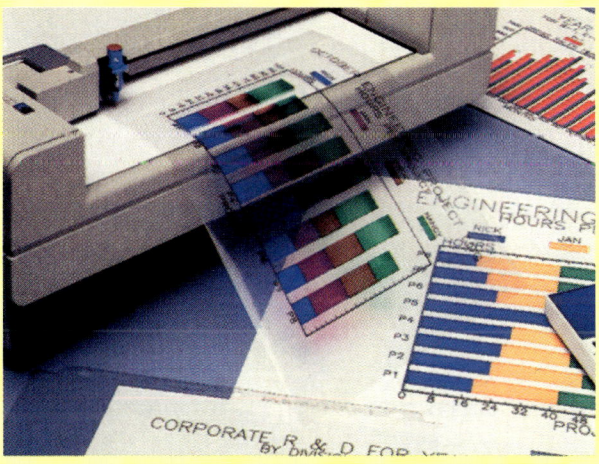

Plotters
These output devices draw hard-copy graphics: maps, bar charts, engineering drawings, overhead transparencies, and even two- or three-dimensional illustrations. Plotters often come with a set of six pens in different colors. Most users do not need a plotter.

What to Look for in Software

The first software decision is made by the choice of an IBM-compatible or Macintosh computer: You will use the operating system software that matches that machine. In the case of an IBM-compatible machine, the operating system may be MS-DOS or, more likely, Microsoft Windows. Since so much software is being written for the Windows environment, we recommend that you make Windows part of your purchase. Most new computers come bundled with certain software, particularly Windows, already installed.

Hardware Requirements for Software

Identify the type of hardware required before you buy software. Under the heading "System Requirements" right on the software package, a list will typically include a particular kind of computer and operating system, and a certain amount of memory and hard disk space.

> **System Requirements**
>
> Make sure your hardware is compatible with the requirements of the software you are buying. You can find the requirements by reading the fine print on the software package. Here is a typical blurb from a software package: Requires an IBM or compatible PC with one diskette drive, a hard drive, Windows95 or higher, a minimum of 8MB RAM, and 20MB available hard disk storage.

Brand Names

In general, publishers of well-known software offer better support than lesser-known companies. Support may be in the form of tutorials, classes by the vendor or others, and the all-important hot-line assistance. In addition, makers of brand-name software usually offer superior documentation and upgrades to new and better versions of the product.

Where To Buy Software

Not very long ago, computer users bought their software at small specialty stores where they hoped they could understand the esoteric language of the sales staff. In contrast, in enormous stores, buyers now pile software packages into their shopping carts like so many cans of soup. The choices of software vendors has expanded considerably.

Now That You Have It, Will You Be Able to Use It?

Once the proud moment has come and your computer system is at home or in the office with you, what do you do with it?

Documentation

Computer systems today come with extensive documentation, the written manuals that accompany hardware and software. Usually, a simple brochure with detailed drawings will help you plug everything together. The installation procedure, however, is often largely (and conveniently) on a diskette. The same brochure that helps you assemble the hardware will guide you to the software on the diskette. Using the software on the diskette, the computer configures itself, mostly without any assistance from you.

Software documentation usually includes a user's guide, a reference manual for the various commands available with the software. Many software packages also include a workbook of some sort to help you train yourself. Software tutorials are also common, and useful for the novice and experienced user alike. Software tutorials usually come on a separate diskette, which guide you as you work through sample problems using the software.

Training

Can you teach yourself? In addition to the documentation supplied with your computer, numerous books and magazines offer help and answer readers' questions. Consult these sources. Other sources are classes offered by computer stores and local colleges. These hands-on sessions may be the most effective learning method of all.

Maintenance Contract

Finally, when purchasing a computer, you may wish to consider a maintenance contract, which should cover labor and parts and possibly advice on a telephone hot-line. Such contracts vary in comprehensiveness. Some cover on-site repairs; others require you to pack up the computer and mail it in. Another option is that the replacement part, say, a new monitor, is sent to you and then you return the old monitor in the same packaging.

Computer Superstores
The superstores, such as CompUSA, sell a broad variety of computer hardware and software. Although their primary advantage is a vast inventory, they also offer on-site technical support.

Warehouse Stores
Often billed as clubs, such as Wal-Mart's SAM's, these giant store sell all manner of merchandise, including computer software.

Mass Merchandisers
Stores such as Sears sell software along with their other various merchandise.

Software-Only Stores
These stores, such as Egghead Software, offer a wide selection of software. Furthermore, in marked contrast to the larger stores, these stores are staffed with people who are familiar with the software.

Computer Dealers
These smaller retail stores, such as MicroAge or CompuAdd, sell hardware systems and the software that runs on them. Such a store usually has a well-informed staff and may be your best bet for in-depth consulting.

Mail Order
Users who know what they want can get it conveniently and reasonably through the mail. Once an initial contact is made, probably from a magazine advertisement, the mail-order house will send catalogs of software regularly.

Introduction

Photo Essay

The Age of Information

The dawn of a new age—the Information Age—glows before us with the promise of new ways of thinking, living, and working. The amount of information in the world is said to be doubling every six to seven years. Can we keep up? We can but not without an understanding of how computers work and the ability to control them for our own purposes.

STEPPING OUT
  Forging a Computer-Based Society
  A Computer in Your Future
COMPUTER LITERACY FOR ALL
EVERYWHERE YOU TURN
  The Nature of Computers
  Where Computers Are Used

▶ Stepping Out

Your first steps toward joining the Information Age include understanding how we got to where we are today. Perhaps you recall from history books how the Industrial Age took its place in our world. In less than 100 years, human society changed on a massive scale. To live between 1890 and 1920, for instance, was to live with the dizzying introduction of electricity, telephones, radio, automobiles, and airplanes. Compared to the Industrial Age, the Information Age is evolving much more rapidly. It is likely to continue to evolve well into the twenty-first century.

Forging a Computer-Based Society

Traditional economics courses define the cornerstones of an economy as land, labor, and capital. Today we can add a fourth key economic element: information. As we evolve from an industrial to an information society, our jobs are changing from physical to mental labor. Just as people moved physically from farms to factories in the Industrial Age, so today people are shifting muscle power to brain power in a new, computer-based society (Figure 1).

Figure 1 Workers and their computers.
(top) This engineer uses her computer to design computer circuitry. (bottom, left) These businessmen use their laptop computers during their carpool ride to work. (bottom, right) A school librarian uses the computer to track books.

Photo Essay ▼ The Age of Information

You are making your move, too, taking your first steps by signing up for this computer class and reading this book. But should you go further and get your own computer? We look next at some of the reasons why you might.

A Computer in Your Future

The television ads appeal: Computers are a must for work and play. "If you can point, you can use a computer," the ads proclaim. Small enough to sit on a desk top, personal computers have been hustled like encyclopedias: "For the price of a bicycle, you can help your child's future...."

The growth of the personal computer market has been exponential, perhaps in spite of the ads. Computers have moved into every nook and cranny of our daily lives. In our homes we use various forms of the new technology not only as playthings but also for keeping track of bank accounts; turning on lawn sprinklers or the morning coffee; monitoring inside temperature and humidity; and teaching math, reading, and other skills to children. Almost any career you map out for your future will involve a computer in some way. Clearly, the computer user no longer has to be a scientist in a laboratory somewhere. We are all computer users (Figure 2).

Computer Literacy for All

Why are you studying about computers? In addition to curiosity (and perhaps a course requirement), you probably recognize that it will not be easy to get through the rest of your life without knowing about computers. We offer a three-pronged definition of computer literacy:

- **Awareness.** As you study about computers, you will become aware of their importance, their versatility, and their pervasiveness in our society.

Figure 2 Computers in the home. Users find a variety of ways to use computers in and near home.

Figure 3 Computer-aided design. These engineers clearly have designs on a new and very round car.

- **Knowledge.** You will learn what computers are and how they work. This requires learning some technical jargon, but do not worry—no one expects you to become a computer expert.
- **Interaction.** There is no better way to understand computers than through interacting with one. So being computer literate also means being able to use a computer for some simple applications.

Note that no part of this definition suggests that you must be able to create the instructions that tell a computer what to do. That would be akin to saying that anyone who plans to drive a car must first become an auto mechanic. Someone else can write the instructions for the computer; you simply use the instructions to get your work done.

For example, an accountant might use a computer to prepare a report, a farmer to check on market prices, and a teenager to play a video game. We cannot guarantee that these people are computer literate, but they have at least grasped the "hands-on" component of the definition—they can interact with a computer.

 Everywhere You Turn

Everywhere you turn nowadays computers are at work—in stores, cars, homes, offices, hospitals, banks, theaters, and even coffee shops. To understand their prevalence in society, we must first look at the traits that make computers so useful.

The Nature of Computers

There are three features that are fundamental characteristics of every computer. These three features have by-products that are just as important. The three fundamental characteristics are

- **Speed.** Computers provide the processing speed essential to our fast-paced society. The quick service we have come to expect—for bank withdrawals, stock quotes, telephone calls, and travel reservations, to name a few—is made possible by computers. Businesses depend on the speedy processing computers provide to handle high-volume activities such as balancing ledgers and designing products (Figure 3).

Figure 4 Computers in the office. Decision-makers rely on the computer to analyze high-volume data.

- **Reliability.** Computers are extremely reliable. Of course, you might not think this from some of the stories you may have seen in the press about "computer errors." What is seldom brought out in those stories is the human side of the mistakes. True, sometimes computers do fail, but most errors supposedly made by computers are really human errors.
- **Storage capability.** Computer systems can store tremendous amounts of data, which can be located and retrieved efficiently. The capability to store volumes of data is especially important in an information age.

These three—speed, reliability, and storage capability—have by-products as follows:

- **Productivity.** Computers can increase productivity, especially where dangerous, boring, or routine tasks are involved. Jobs like punching holes in metal or monitoring water levels can be more capably performed by computers. When computers move into business offices, managers expect increased productivity as workers learn to use computers to do their jobs better and faster.
- **Decision making.** To make essential business and governmental decisions, managers need to take into account financial, geographical, and logistical factors. Using problem-solving techniques originally developed by humans, the computer helps decision-makers sort things out, analyze the implications, and make better choices (Figure 4).
- **Cost reduction.** Finally, because it improves productivity and aids decision making, the computer helps us hold down the costs of labor, energy, and paperwork. As a result, computers help reduce the costs of goods and services in our economy.

With so many wonderful traits, it is easy to see why computers have moved so quickly into every facet of our lives. Next we look at some of the ways we use them to make the workday more productive and our personal lives more rewarding.

Where Computers Are Used

Computers can do just about anything imaginable, but they really excel in certain areas. This section lists some of the principal areas of computer use.

- **Graphics.** Computer graphics help us examine change. In the field of medicine, brain-scan computers produce color-enhanced maps to help diagnose mental illness. Business people make bar graphs and pie charts from tedious figures to convey information with far more impact than numbers alone can do. Architects use computer-animated graphics to experiment with possible exteriors, to give clients a visual walk-through of proposed buildings, and to subject buildings to hypothetical earthquakes. Finally, a new kind of artist has emerged who uses computers to create landscapes, television logos, and still lifes (Figure 5).

Figure 5 Computer-generated art. Computers give artists a new medium for their creativity.

- **Retailing.** Products from meats to magazines are packaged with zebra-striped bar codes that can be read by computer scanners at supermarket checkout stands to determine prices and help manage inventory. Computers operate behind the scenes, too. Consider your copy of this book, for instance. From printer to warehouse to bookstore, its movement was tracked with the help of computers and the bar code on the back cover (Figure 6).
- **Energy.** Energy companies use computers to locate oil, coal, natural gas, and uranium. Electric companies use computers to monitor vast power networks. In addition, meter readers use handheld computers to record how much energy is used each month in homes and businesses.
- **Transportation.** Computers are used to help run rapid transit systems, load containerships, track railroad cars across the country, safeguard airport takeoffs and landings, monitor air traffic, and schedule travel (Figure 7). They are also used in cars to monitor fluid levels, temperatures, and electrical systems.

Figure 7 Airline reservations. This reservations assistant engages the power of the computer to sift through the myriad air travel possibilities.

- **Paperwork.** In some ways the computer contributes to paper use by adding to the amount of junk mail you find in your mailbox. However, in many ways it cuts down on paper handling. Using a computer, for example, you might type several drafts of a term paper before printing anything. Computerized bookkeeping, record keeping, and document sending have also made paperwork more efficient.

- **Law enforcement.** Computers have long been used to track motor vehicle tags. Recent innovations include national fingerprint files (which, once in place, solved several years-old cases), a national file on mode-of-operation of serial killers, and the computer modeling of DNA, which can be used to match traces from an alleged criminal's body, such as blood at the crime scene (Figure 8).
- **Money.** Computers speed up record keeping and allow banks to offer same-day services and even do-it-yourself banking over the phone. Computers have helped fuel the cashless economy, enabling the widespread use of credit cards and instantaneous credit checks by banks, department stores, and other retailers. Some oil companies even use credit-card activated, self-service gasoline pumps.

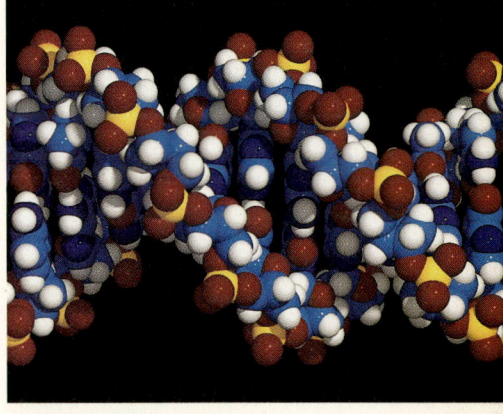

Figure 8 DNA. Computer-generated DNA.

Figure 6 Retailing. These women use the computer to help track their fabric offerings.

Photo Essay ▼ The Age of Information

Figure 9   Computers down on the farm. This farmer uses his laptop computer to enter crop data from the field.

- **Agriculture.** Farmers use small computers—purchased for less than the price of a tractor—to help with billing, crop information, cost per acre, feed combinations, and market price checks. Cattle ranchers can also use computers for information about livestock breeding and performance. In addition, computers can give people the option of working at their homes instead of in city offices. The result could lessen the isolation of country living and perhaps even stem the movement of youth from farms to cities (Figure 9).
- **Government.** The largest single user of computers is the federal government. Among other tasks, the federal government uses computers to forecast weather (Figure 10), to manage parks, to process immigrants, to produce Social Security benefit checks, and—of course—to collect taxes. State and local governments also use computers routinely.
- **Education.** Computers have been used behind the scenes for years in colleges and school districts for record keeping and accounting. Many colleges have eliminated long registration lines by using computerized, touch-tone telephone registration. Most schools in the U.S. have computers available for use in the classroom, and some colleges require entering freshmen to bring their own. Educators who once considered computers novelties in the classroom now look at them as necessities (Figure 11).

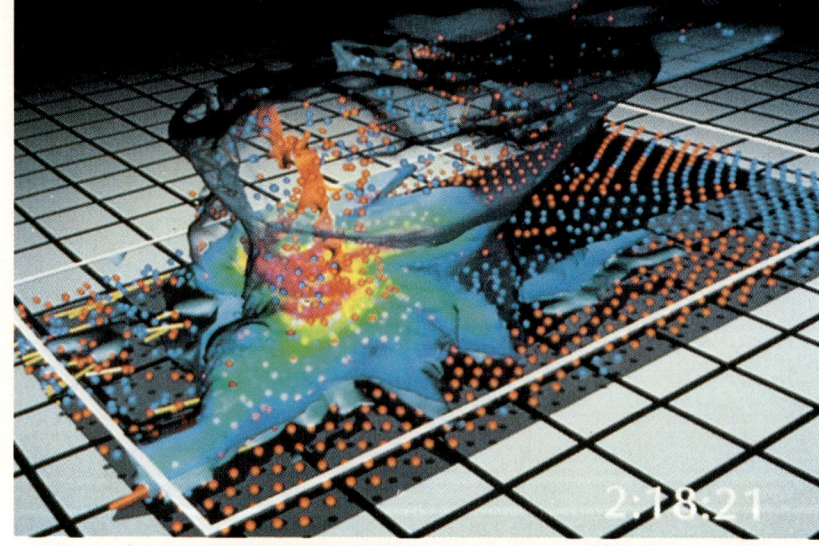

Figure 10   The approaching storm. When will the storm get here? To improve the science of weather forecasting, researchers program various weather conditions into a computerized global weather model that is useful to meteorologists. In this graphic, different colors represent different water densities in the atmosphere.

Figure 11   Computers in education. Computers are placed in all levels of learning, from pre-school through graduate school.

- **The home.** Are you willing to welcome a computer into your home? Many people already have, often justifying it as an educational tool for their children. In fact, a potent selling point is that kids who use computers tend to watch less television. But that is only the beginning. Personal computers are being used at home to keep records, write letters, prepare budgets, draw pictures, prepare newsletters, connect with others, and play games (Figure 12). The more adventurous use computers to control heating and air conditioning, answer telephone calls, safeguard the house during vacations, and so on.

Figure 12 Sophisticated game-playing. This flight simulator game is quite sophisticated, so a novice will need practice to avoid a nosedive. The cockpit shown here is realistic and responds to your commands.

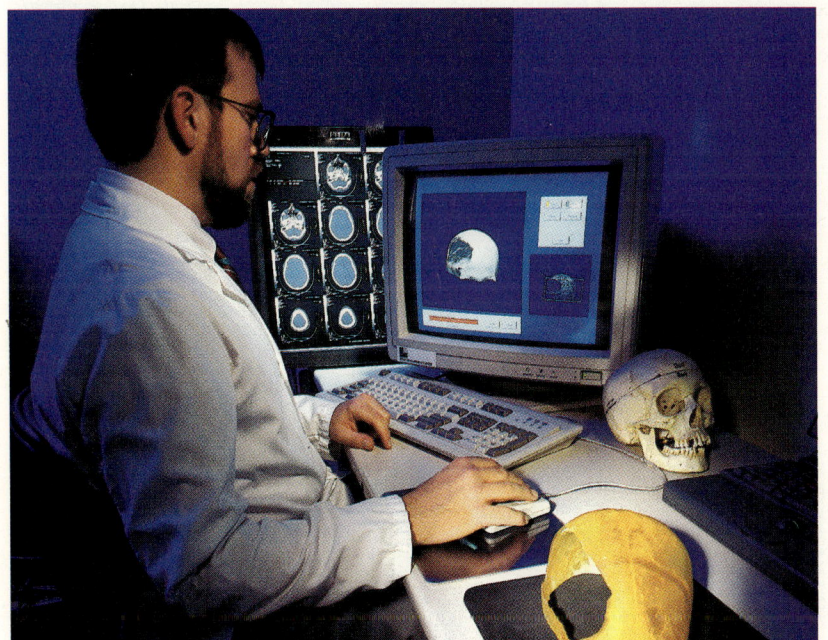

Figure 13 Medicine. This surgeon is using the computer to help plan an operation.

- **Health and medicine.** Today computers can be the difference between life and death in hospitals and medical centers. Computers help monitor the gravely ill in intensive care units and provide cross-sectional views of the body through ultrasound pictures. Physicians can also use computers to assist in diagnoses (Figure 13). In fact, computers have been shown to diagnose heart attacks correctly more frequently than physicians. If you are one of the thousands who suffer one miserable cold after another, you will be happy to know that computers have been able to map, in exquisite atomic detail, the structure of the human cold virus—the first step toward a cure for the common cold (Figure 14).

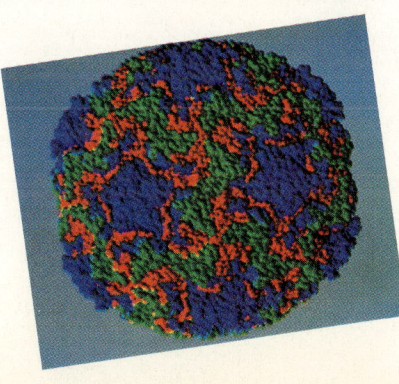

Figure 14 Cold virus. This computer-produced model of the cold virus named HRV 14 raises hopes that a cure for the common cold may be possible after all. With the aid of a computer, the final set of calculations for the model took one month to complete. Researchers estimate that without the computer the calculations might have required ten years of manual effort.

Photo Essay ▼ The Age of Information

- **Robotics.** Computers have paved the way for robots to take over many of the jobs that place human life at risk. These robots are performing tasks too unpleasant, too dangerous, or too critical for humans. Examples include robots that enter areas considered dangerous because of terrorist threats, and robots that open packages believed to contain bombs. Robots can also be used for delicate jobs such as picking fruit, not to mention handling eggs and bananas (Figure 15).
- **The sciences.** Scientific researchers have long benefited from the high-speed capabilities of computers. Computers can simulate environments (Figure 16), emulate physical characteristics, and allow scientists to provide proofs in a cost-effective manner. Consider also the beleaguered mouse, long the mainstay of scientific research. Many mice—and other animals—have been spared since computers have taken over their research roles. Space program scientists have long used onboard computers to relay data back to earth.

Figure 15 Robots. (left) This robot can gently grip a part to be added to an assembly. (right) This factory robot welds under direction of the computer.

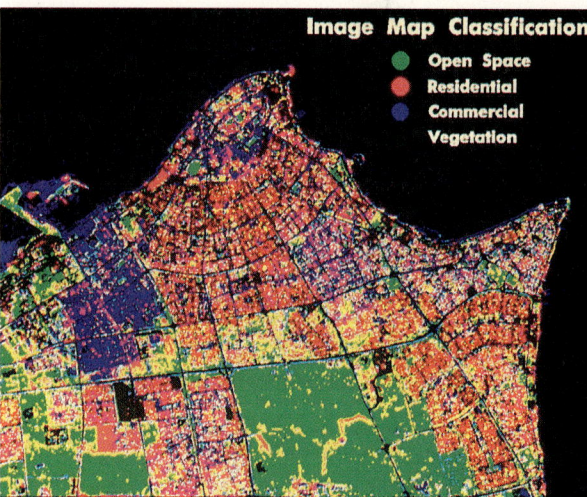

Figure 16 The environment via computer. Computers can model any landscape for environmental scrutiny. Shown here is a city color-coded to show various land uses.

Photo Essay ▼ The Age of Information

- **Training.** It is much cheaper to teach aspiring pilots to fly in computerized training "cockpits," or simulators, than in real airplanes. Novice railroad engineers can also be given the experience of running a train with the help of a computerized device. Training simulations are relatively inexpensive and always available on a one-to-one basis, making for very personal learning.
- **The human connection.** Are computers cold and impersonal? The disabled do not think so. Can the disabled walk again? Some can, with the help of computers. Can dancers and athletes improve their performance? Maybe they can, by using computers to monitor their movements. Can we learn more about our ethnic backgrounds and our cultural history with the aid of computers? Indeed we can.

▼ ▲ ▼

Computers are all around us. You have been exposed to computer hype, computer advertisements, and newspaper headlines about computers. You have interacted with computers in your everyday life—at the grocery store, your school, the library, and more. You know more than you think you do. The beginnings of computer literacy are already apparent.

We have written this book with two kinds of readers in mind. If you are contemplating a computer-related career, you will find a solid discussion of technology, computer applications, and various jobs associated with computers. Even if you are not interested in the technical side of computers, however, most careers involve computers in some way; this book will provide you with a foundation in computer literacy.

PART I

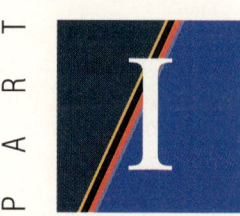

Hardware Tools

After her first year in college, Anita Jefferson got a summer job in the resort town of Friday Harbor. She waited tables for both the noon and evening shifts. Her wages were supplemented nicely by generous tips from carefree tourists. An accounting major, Anita would have preferred a job in a business office, but at least her summer income would make a significant dent in her upcoming tuition.

As it turned out, Anita's summer was more valuable than she expected. The second day on the job she learned that a colleague had signed up for a morning computer class at the local branch of a community college. Since it was a beginning class and did not interfere with her work schedule, Anita signed up too.

Anita's previous computer experiences were limited to math drills in elementary school and playing games on her mother's home computer. She knew that, as an accountant, she would certainly use a computer. In fact, two of her fall classes, according to the schedule, required computer lab time. Although somewhat apprehensive, Anita was hoping to get a head start on computers.

The course took parallel paths, teaching fundamentals about computer hardware and software in a lecture format and offering hands-on computer experience. By the end of the summer, Anita had a good grasp of computer basics and could perform such tasks as preparing memos on the computer. But her greatest reward was learning how to use spreadsheets, which let her enter, revise, and print numerical data in rows and columns. She recognized that budgets, ledgers, inventories, and other keystones of accounting would all be maintained using spreadsheets on a computer. She definitely got a head start.

Chapter 1

Overview of a Computer System
Hardware, Software, and People

LEARNING OBJECTIVES

- Know the basic components of a computer system: input, processing, output, and storage
- Become acquainted with some common input, output, and storage media
- Be able to distinguish data from information
- Become familiar with the various classifications of computers
- Appreciate the significance of networking and data communications
- Understand the categories of software
- Understand, in a general way, the nature of various types of task-oriented software
- Be able to distinguish computer users from computer professionals

THE BIG PICTURE
HARDWARE: MEETING THE MACHINE
 Your Personal Computer Hardware
 Input: What Goes In
 The Processor and Memory: Data Manipulation
 Output: What Comes Out
 Secondary Storage
 The Complete Hardware System
 Classification of Computers
 Data Communications: No Need to Be There
SOFTWARE: TELLING THE MACHINE WHAT TO DO
 Categories of Software
 Some Task-Oriented Software
PEOPLE AND COMPUTERS
 Computers and You, the User
 Computer People

Computers Going Green

Ten million personal computers a year are being sent to the scrap heap. The reasons are clear. Ever-emerging technology encourages current users to buy the latest model. Furthermore, since new computers are reasonably priced, few new buyers want an obsolete second-hand "bargain."

In addition to being a landfill headache, computer castoffs waste precious resources. Manufacturers, in environmental parlance, are "going green," producing computers made of recyclable plastic. The plastics from computer carcasses can be reheated and reused four or five times. Once exhausted, the plastic can be used as filler for other purposes.

 ## The Big Picture

A computer system has three main components: hardware, software, and people. The equipment associated with a computer system is called **hardware**. A set of instructions called **software** tells the hardware what to do. People, however, are the most important component of a computer system—people use the power of the computer for some purpose.

Software is also referred to as programs. To be more specific, a **program** is a set of step-by-step instructions that directs the computer to do the tasks you want it to do and produce the results you want. A **computer programmer** is a person who writes programs. Most of us do not write programs—we use programs written by someone else. This means we are **users**—people who purchase and use computer software. In business, users are often called **end-users** because they are at the end of the "computer line," actually making use of the computer's information.

In this chapter we will examine both hardware and software. We will also devote a separate section to computers and people. As the title of this chapter indicates, what follows is an overview, a look at the "big picture" of a computer system. Thus, many of the terms introduced in this chapter are defined only briefly here. In subsequent chapters we will discuss the various parts of a computer system in greater detail.

 ## Hardware: Meeting the Machine

What is a computer? A six-year-old called a computer "radio, movies, and television combined!" A ten-year-old described a computer as "a television set you can talk to." The ten-year-old's definition is closer but still does not recognize the computer as a machine that has the power to make changes.

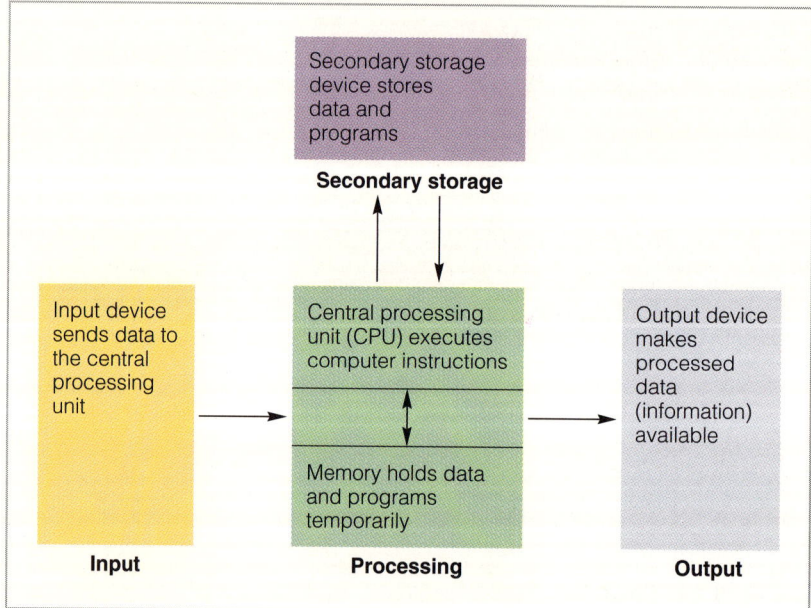

Figure 1-1 Four primary components of a computer system. To function, a computer system requires input, processing, output, and storage.

Chapter One ▼ Overview of a Computer System

A **computer** is a machine that can be programmed to accept data (*input*), process it into useful information (*output*), and store it away (in a *secondary storage* device) for safekeeping or later reuse. The *processing* of input to output is directed by the software but performed by the hardware, which we will examine in this section.

To function, a computer system requires four main aspects of data handling: input, processing, output, and storage (Figure 1-1). The hardware responsible for these four areas operates as follows:

- *Input devices* accept data in a form that the computer can use; they then send the data to the processing unit.
- The *processor,* more formally known as the *central processing unit* (*CPU*), has the electronic circuitry that manipulates input data into the information people want. The central processing unit actually executes computer instructions.
- *Output devices* show people the processed data—information—in a form they can use easily.
- *Storage* usually means *secondary storage*, which consists of secondary storage devices, such as diskettes, which can store data and programs outside the computer itself. These devices supplement *memory,* which, as we will see, can hold data and programs only temporarily.

Now let us consider the equipment related to these four aspects of data handling in terms of what you would find on a personal computer.

Your Personal Computer Hardware

Let us look at the hardware in terms a personal computer. Suppose you want to do word processing on a personal computer, using the hardware shown in Figure 1-2. Word processing software allows you to input data such as an essay, save it, revise and re-save it, and print it whenever you wish. The *input* device, in this case, is a keyboard, which you use to type in, or key in, the original essay and any changes you want to make to it. All computers, large and small, must have a *central processing unit,* so yours does, too—it is within the personal computer housing. The central processing unit uses the word processing software to accept the data you input through the keyboard. Processed data from your personal computer is usually *output* in two forms: on a screen and by a printer. As you key in the essay on the keyboard, it appears on the screen in front of you. After you examine the essay on the screen, make changes, and determine that it is acceptable, you can print the essay on the printer. Your *secondary storage device* in this case is a diskette, a magnetic medium that stores the essay until it is needed again.

Now we will take a general tour of the hardware needed for input, processing, output, and storage. These same components make up all computer systems, whether small, medium, or large. In this discussion we will try to emphasize the types of hardware you are likely to have seen in your own environment. These topics will be covered in detail in Chapters 2, 3, and 4.

Input: What Goes In

Input is the data that you put into the computer system for processing. Here are some common ways of feeding input data into the system:

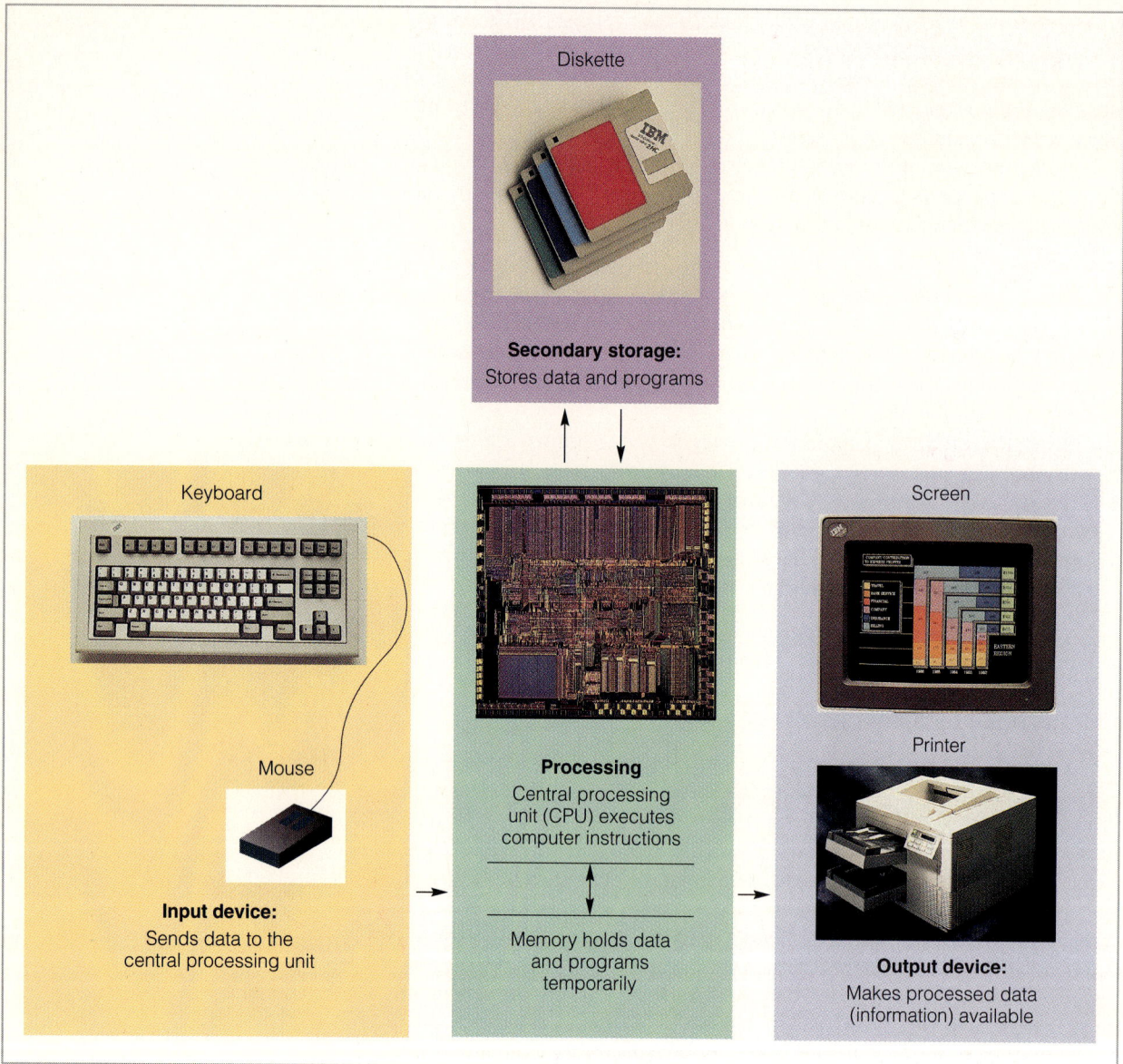

Figure 1-2 A personal computer system. In this personal computer, the input device is a keyboard or a mouse. The input device feeds data to the central processing unit, which is inside the computer housing. The two output devices in this example are the screen and the printer. The secondary storage device is a 3½-inch disk.

- *Typing* on a **keyboard** (Figure 1-3a). Computer keyboards operate in much the same way as electric typewriter keyboards. The computer responds to what you enter; that is, it "echoes" what you type by displaying it on the screen in front of you.
- *Pointing* with a **mouse** (Figure 1-3a). A mouse is a device that is moved by hand over a flat surface. As the ball on its underside rotates, the mouse movement causes corresponding movement of a pointer on the computer screen. Pressing buttons on the mouse lets you invoke commands.
- *Scanning* with a **wand reader** or **bar code reader** (Figure 1-3b). These devices, which you have seen in retail stores, use laser beams to read

(a) (b)

Figure 1-3 Input devices. (a) The keyboard is the most widely used input device, though the mouse has become increasingly popular. Movement of the mouse on a flat surface causes corresponding movement on the screen. (b) The bar code on this package of green beans is scanned into the computer.

special letters, numbers, or symbols such as the zebra-striped bar codes on many products.

An input device may be part of a terminal. A **terminal** includes an input device, a television-like screen display, and some connection to a large computer.

You can imput data to a computer in many other interesting ways, including writing, speaking, pointing, or even by just looking at the data. We will examine all these in detail in Chapter 3.

The Processor and Memory: Data Manipulation

In a computer the **processor** is the center of activity. The processor, as we noted, is also called the **central processing unit (CPU)**. The central processing unit consists of electronic circuits that interpret and execute program instructions, as well as communicate with the input, output, and storage devices.

It is the central processing unit that actually transforms data into information. **Data** is the raw material to be processed by a computer. Such material can be letters, numbers, or facts—such as grades in a class, baseball batting averages, or light and dark areas in a photograph. Processed data becomes **information**—data that is organized, meaningful, and useful. In school, for instance, an instructor could enter various student grades (data), which can be processed to produce final grades and perhaps a class average (information). Data that is perhaps uninteresting on its own may become very interesting once it is converted to information. The raw facts (data) about your finances, such as a paycheck or a donation to charity or a medical bill may not be captivating individually, but together, these and other acts can be processed to produce the refund or amount you owe on your income tax return (information).

Computer **memory,** also known as **primary storage,** is closely associated with the central processing unit but separate from it. Memory holds the data after it is input to the system and before it is processed; also, memory holds the data after it has been processed but before it has been released to

Carrying Computers Around

Is there a computer in your purse or pocket? Possibly. Not a full-blown model with keyboard and mouse, of course, but some item operated by a microprocessor—an internal chip. The most likely item is a calculator.

Now check your wrist: Some watches include a chip that performs functions once thought to be beyond the province of something so small. The Timex watch shown here can retrieve data from a personal computer. A sensor in the watch reads scheduling data from the computer, and the watch then displays the data in the digital time window. Note that the watch shown here displays WE 2-15 9:00 AM DEPT MTG, meaning that a department meeting is scheduled on Wednesday, the 15th of February at 9:00 am.

Anything else computeresque on your person? There may be a magnetic strip on a credit card or student ID card, possibly a bar code on some cosmetic item or snack bar, and probably magnetic ink characters on your personal checks.

the output device. In addition, memory holds the programs (computer instructions) needed by the central processing unit.

Output: What Comes Out

Output—the result produced by the central processing unit—is, of course, a computer's whole reason for being. Output is usable information—that is, raw input data that has been processed by the computer into information. The most common forms of output are words, numbers, and graphics. Word output, for example, may be the letters and memos prepared by office people using word processing software. Other workers may be more interested in numbers, such as those found in formulas, schedules, and budgets. In many cases numbers can be understood more easily when output in the form of charts and graphics.

The most common output devices are computer screens and printers. **Screens** can vary in their forms of display, producing text, numbers, symbols, art, photographs, and even video—in full color (Figure 1-4a). **Printers** produce printed reports as instructed by a computer program (Figure 1-4b). Many printers, particularly those associated with personal computers, can print in color.

You can produce output from a computer in other ways, including film and voice output. We will examine all output methods in detail in Chapter 3.

Secondary Storage

Secondary storage provides additional storage separate from memory. Secondary storage has several advantages. For instance, it would be unwise for a college registrar to try to keep the grades of all the students in the college in the computer's memory; if this were done, the computer would probably not have room to store anything else. Also, memory holds data and programs only temporarily—hence the need for secondary storage.

The two most common secondary storage mediums are magnetic disk and magnetic tape. A **magnetic disk** can be a diskette or a hard disk. A **diskette** may look like a small stereo record, usually 3½ inches in diameter or, in some cases 5¼ inches (Figure 1-5a). **Hard disks** usually have more storage capacity than diskettes and also offer faster access to the data they

(a)   (b)

Figure 1-4 Output devices. Screens and printers are two types of output devices. (a) This screen can display text or the colorful graphics shown here. (b) This laser printer is used to produce high-quality output.

Chapter One ▼ Overview of a Computer System

Figure 1-5 Secondary storage devices. (a) A 3½-inch diskette is being inserted into a disk drive. (b) Disk packs provide large computer systems with unlimited storage space. (c) Optical disks can hold enormous amounts of data: text, music, graphics—even video and movies. (d) Magnetic tape is used primarily as backup storage.

hold. With large computer systems, hard disks are often contained in disk packs.

Disk data is read by **disk drives**. Personal computer disk drives read diskettes; most personal computers have hard disk drives also. On large computer systems, disk packs may be removed from the drives (Figure 1-5b), permitting the use of interchangeable packs and practically unlimited storage capacity.

The most recent disk storage technology is the **optical disk,** which uses a laser beam to store large volumes of data relatively inexpensively (Figure 1-5c).

Magnetic tape, which comes on a reel or cartridge, is similar to tape that is played on a tape recorder. Magnetic tape reels are mounted on **tape drives** when the data on them needs to be read by the computer system or

ONE JUMP AHEAD

Just Don't Call It a Computer

I call it my do-everything machine. It answers the phone and takes messages. It understands and responds when I speak to it. It sends and receives both mail and faxes. It shows movies and documentaries. It draws pictures. It places ads in the newspaper and gets magazine articles from the library. It monitors my stock investments and keeps my photo collection in order. It pays the bills, orders the groceries, and turns on the heat and lights. And, of course, it is so simple to use—no harder than my microwave oven—that I think of it as my office appliance.

Farfetched? Not at all. Many personal computers can do most of these things already. Futurists predict a time, in the not-too-distant future, when do-everything machines will be as widespread as televisions are today. Consumers are still looking, however, for that elusive ease of use. A computer is not yet as easy to use as a microwave oven.

when new data is to be written on the tape (Figure 1-5d). Magnetic tape is usually used for backup purposes—for "data insurance"—because tape is inexpensive.

We will study storage media in Chapter 4.

The Complete Hardware System

The hardware devices attached to the computer are called peripheral equipment. **Peripheral equipment** includes all input, output, and secondary storage devices. In the case of personal computers, some of the input, output, and storage devices are built into the same physical unit. In many personal computers, the CPU and disk drive are all contained in the same housing; the keyboard, mouse, and screen are separate.

In larger computer systems, however, the input, processing, output, and storage functions may be in separate rooms, separate buildings, or even separate countries. For example, data may be input on terminals at a branch bank and then transmitted to the central processing unit at the headquarters bank. The information produced by the central processing unit may then be transmitted to the international offices, where it is printed out. Meanwhile, disks with stored data may be kept in bank headquarters and duplicate data kept on disk or tape in a warehouse across town for safekeeping.

Although the equipment may vary widely, from the simplest computer to the most powerful, by and large the four elements of a computer system remain the same: input, processing, output, and storage. Now let us look at the way computers have been traditionally classified.

Classification of Computers

Computers come in sizes from tiny to monstrous, in both appearance and power. The size of a computer that a person or an organization needs

Figure 1-6 Computer classifications. (a) This Cray supercomputer has been nicknamed Bubbles because of its bubbling, shimmering coolant liquids. (b) Shown here is the Control Data 7600 mainframe computer. Despite the sterile look of this photo, it does show that a mainframe has many components. (c) The VAX, a popular minicomputer made by Digital Equipment Corporation (DEC). (d) Individuals use personal computers in both the home and the office.

depends on the computing requirements. Clearly, the National Weather Service, keeping watch on the weather fronts of many continents, has requirements different from those of a car dealer's service department that is trying to keep track of its parts inventory. And the requirements of both of them are different from the needs of a salesperson using a small laptop computer to record client orders on a sales trip.

Supercomputers
The mightiest computers—and, of course, the most expensive—are known as supercomputers (Figure 1-6a). **Supercomputers** process *billions* of instructions per second. Most people do not have a direct need for the speed and power of a supercomputer. In fact, for many years supercomputer customers were an exclusive group: agencies of the federal government. The federal government uses supercomputers for tasks that require mammoth data manipulation, such as worldwide weather forecasting and weapons research. But now supercomputers are moving toward the main-

Figure 1-7 Supercomputer graphics. These graphic images, called fractals, are formed by using the computer to repeat geometric shapes with color, size, and angle variations. (a) Note the basic triangle and circle elements on which the fractals are built. (b) Here, the artist makes slight adjustments to make the fractals appear in motion.

stream, for activities as varied as stock analysis, automobile design, special effects for movies, and even sophisticated artworks (Figure 1-7).

Mainframes

In the jargon of the computer trade, large computers are called mainframes (Figure 1-6b). **Mainframes** are capable of processing data at very high speeds—millions of instructions per second—and have access to billions of characters of data. The price of these large systems can vary from several hundred thousand to many millions of dollars. With that kind of price tag, you will not buy a mainframe for just any purpose. Their principal use is for processing vast amounts of data quickly, so some of the obvious customers are banks, insurance companies, and manufacturers. But this list is not all-inclusive; other types of customers are large mail-order houses, airlines with sophisticated reservation systems, government accounting services, aerospace companies doing complex aircraft design, and the like.

Minicomputers

The next step down from mainframe computers, in terms of speed and storage capacity, are minicomputers (Figure 1-6c). When **minicomputers** first appeared on the market, their lower price fell within the range of many small businesses, greatly expanding the potential computer market. The term **supermini** has been coined to describe minis at the top of the size-price scale. Minicomputers are widely used by retail businesses, colleges, and state and city agencies. However, the minicomputer market has diminished somewhat as buyers moved toward less expensive but increasingly powerful personal computers.

Personal Computers

Most often called **personal computers,** or just **PCs,** these desktop computers are also known as **microcomputers,** or sometimes home computers (Figure 1-6d). For many years, the computer industry was on a quest for the next biggest computer. The search was always for more power and greater capacity. Prognosticators who timidly suggested a niche for a smaller computer were subject to ridicule by people who, as it turned out, could not have been more wrong. Today, for a few hundred dollars, anyone can own a personal computer. (Most people, however, are more likely to choose a computer that costs a few *thousand* dollars.) **Supermicros,** the upper-end machines used by workers such as engineers, financial traders, and graphics designers, are small enough to fit on a desk top but approach the power of a mainframe.

Laptop Computers

A computer that fits in a briefcase? A computer that weighs less than a newborn baby? A computer you do not have to plug in? A computer to use on your lap on an airplane? Yes, to all these questions. **Laptop computers,** also called **notebook computers,** are wonderfully portable and functional, and popular with travelers who need a computer that can go with them (Figure 1-8). Most laptops accept diskettes, so it is easy to move data from one computer to another. Laptops are not as inexpensive as their size might suggest; many carry a price tag equivalent to a full-size personal computer for business.

Getting Smaller Still

Using a pen-like stylus, **pen-based computers** accept handwritten input directly on a screen (Figure 1-9). Users of the handheld pen-based computers, also called **personal digital assistants (PDA),** are mainly people in companies who want to automate the work of their clipboard-carrying workers, such as parcel delivery drivers and meter readers. Other potential users are workers who cannot easily use a laptop computer because they are on their feet all day: nurses, sales representatives, real estate agents, and insurance adjusters.

Figure 1-8 Laptop computers. All these users, at work or play, find it convenient to use laptop computers.

Figure 1-9 Pen-based computers. Workers on the job sometimes prefer lightweight pen-based computers, which will accept handwritten input.

Data Communications: No Need to Be There

Originally, a computer user kept all the computer hardware in one place; that is, it was **centralized** in one room. Anyone wanting computer access had to go to where the computer was located. Although this is still sometimes the case, most computer systems are **decentralized.** That is, the computer itself and some storage devices may be in one place, but the devices to access the computer—terminals or even other computers—are scattered among the users. These devices are usually connected to the computer by telephone lines. For instance, the computer and storage that has the information on your checking account may be located in bank headquarters, but the terminals are located in branch banks all over town so a teller in any branch can find out what your balance is. The subject of decentralization is intimately tied to **data communications**, the process of exchanging data over communications facilities, such as the telephone.

In many systems processing is decentralized as well—the computers and storage devices are in dispersed locations. This arrangement is known as **distributed data processing** (ddp) because the processing is distributed among the different locations. There are several ways to configure the hardware; one common arrangement is to place smaller computers in local

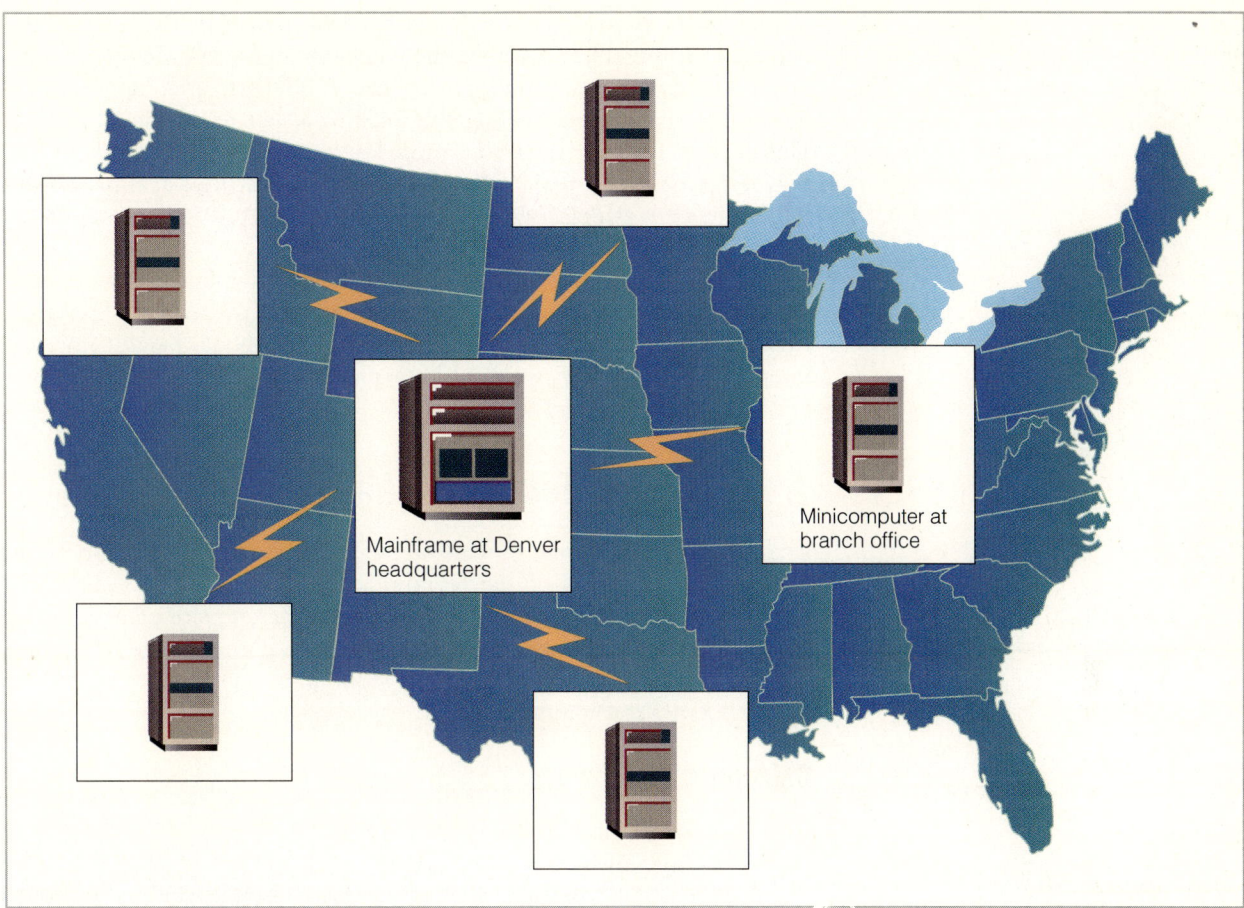

Figure 1-10   Distributed data processing system. Branch offices of an insurance company have their own computers for local processing, but they can tie into the mainframe computer in the headquarters office in Denver.

offices but still do some processing on a larger computer at the headquarters office. For example, an insurance company headquartered in Denver with branches throughout the country might process payments and claims through minicomputers or personal computers in local offices. However, summary data could be sent regularly by each office for processing by the mainframe computer in Denver (Figure 1-10).

Many organizations find that their needs are best served by a **network**, a computer system that uses communications equipment to connect computers and their resources. In one type of network, a **local area network (LAN)**, personal computers in an office are hooked together so that users can communicate with each other. Users can operate their personal computers independently or in cooperation with other computers—minis or mainframes—to exchange data and share resources. When smaller computers are connected to larger computers, the result is often referred to as a **micro-to-mainframe** link. This concept has revolutionized the way many businesses operate. Users are able to obtain data directly from the mainframe computer and immediately analyze it on their own personal computers with their own software. People have quick access to more information, which leads to better decision making.

Individuals have joined the trend to "connectivity" by connecting their personal computers, usually via the telephone lines, to other computers. From their own homes, users can connect to all sorts of computer-based services, such as getting stock quotes, making airline reservations, or shopping for videotapes. By far the most important service for individuals, however, is **electronic mail,** or **e-mail,** which lets individuals send messages via computer. A popular way to network with other users is the **Internet,** a set of networks that connects users worldwide.

Data communications and networking are so significant that we have devoted Chapter 5 to these topics. Further, we offer the feature *Making the Right Connections* in each chapter throughout the book. Finally, since the Internet is so important, we are devoting a page at the end of every chapter to a feature called *Planet Internet*.

 ## Software: Telling the Machine What to Do

In the past, when people thought about computers, they thought about machines. The tapping on the keyboard, the clacking of the printers, the rumble of whirling disk drives, the changing flashes of color on a computer screen—these are the attention-getters. However, it is really the software—the planned, step-by-step instructions required to turn data into information—that makes a computer useful.

Categories of Software

Generally speaking, software can be categorized as system software or applications software. A subset of **system software** is an **operating system,** the underlying software found on all computers. **Applications software,** software that is *applied,* can be used to solve a particular problem or to perform a particular task. Applications software may be either custom or packaged. Many large organizations pay programmers to write **custom software,** software that is specifically tailored to their needs. The average person is most likely to deal with **packaged software,** also called **commercial software**—the software that is literally packaged in a container of

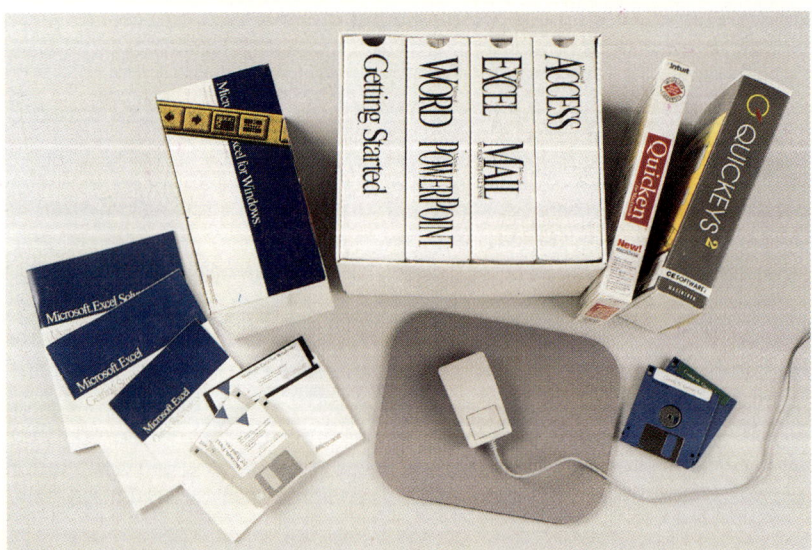

Figure 1-11 Packaged software. Each of the colorful software packages shown here includes one or more diskettes containing the software needed to run the program and an instruction manual, or documentation, describing how to use the software.

MAKING THE RIGHT CONNECTIONS

Coming Full Cycle

Could it be that we are back where we started? The original idea behind the personal computer was that it was, indeed, *personal*. On your own. Your computer, your data, your private entity. Individual users had truly broken away from the large computers that were located elsewhere and used for heavy-duty computer tasks.

Personal computer pioneers smile at that notion now. Personal computer users, whether at the office or in the home, are busily connecting their computers to everything they can find. In fact, the power of a personal computer is coming to be defined in terms of to what it can be connected. And what about those big computers, the ones personal computer users left behind? Personal computer users now gleefully connect to the big computer to access massive databases, to route messages, and much more.

Business users once were willing to isolate themselves on personal computers rather than be an insignificant cog in the large computer system. Home users, on the other hand, were willing to try a personal computer partly because it was so isolated and thus less threatening. Now, with networking, users have maintained their independence but also have expanded their horizons. They have the best of both worlds.

---

some sort, usually a box or folder, and is sold in stores or catalogs. Packaged software for personal computers often comes in a box as colorful as a board game. Inside the box you will find one or more diskettes holding the software and one or more instruction manuals, also referred to as **documentation** (Figure 1-11). To use the software, you begin by inserting the diskette in the disk drive. Then, depending on the hardware and software, you either type specified instructions on the keyboard or give a command with the click of a mouse; the software begins to run on the computer. Note, however, that complex software comes on several diskettes and usually requires a setup process before use. Furthermore, most users transfer new software to their hard disk drive.

A great assortment of software is available to help you with a variety of tasks—writing papers, preparing budgets, drawing graphs, playing games, and so much more. This wonderful array of software is what makes computers so useful.

Most personal computer software is planned to be user friendly. The term **user friendly** has become a cliché, but it still conveys meaning: It usually means that the software is supposed to be easy—perhaps even intuitive—for a beginner to use or that the software can be used with a minimum of training. Even so, such software may seem overwhelming at first. Although software is usually generalized enough to be marketed to a broad audience, it is possible to set up the features of the software to match a particular user's needs.

Some Task-Oriented Software

Most users, whether at home or in business, are drawn to task-oriented software, sometimes called productivity software, that can make their work faster and their lives easier. The collective set of business tasks is limited, and the number of general paths toward performing these tasks is

EXPENSES	JANUARY	FEBRUARY	MARCH	APRIL	TOTAL
RENT	425.00	425.00	425.00	425.00	1700.00
PHONE	22.50	31.25	17.00	35.75	106.50
CLOTHES	110.00	135.00	156.00	91.00	492.00
FOOD	280.00	250.00	250.00	300.00	1080.00
HEAT	80.00	50.00	24.00	95.00	249.00
ELECTRICITY	35.75	40.50	45.00	36.50	157.75
WATER	10.00	11.00	11.00	10.50	42.50
CAR INSURANCE	75.00	75.00	75.00	75.00	300.00
ENTERTAINMENT	150.00	125.00	140.00	175.00	590.00
TOTAL	1188.25	1142.75	1143.00	1243.75	4717.75

(a)

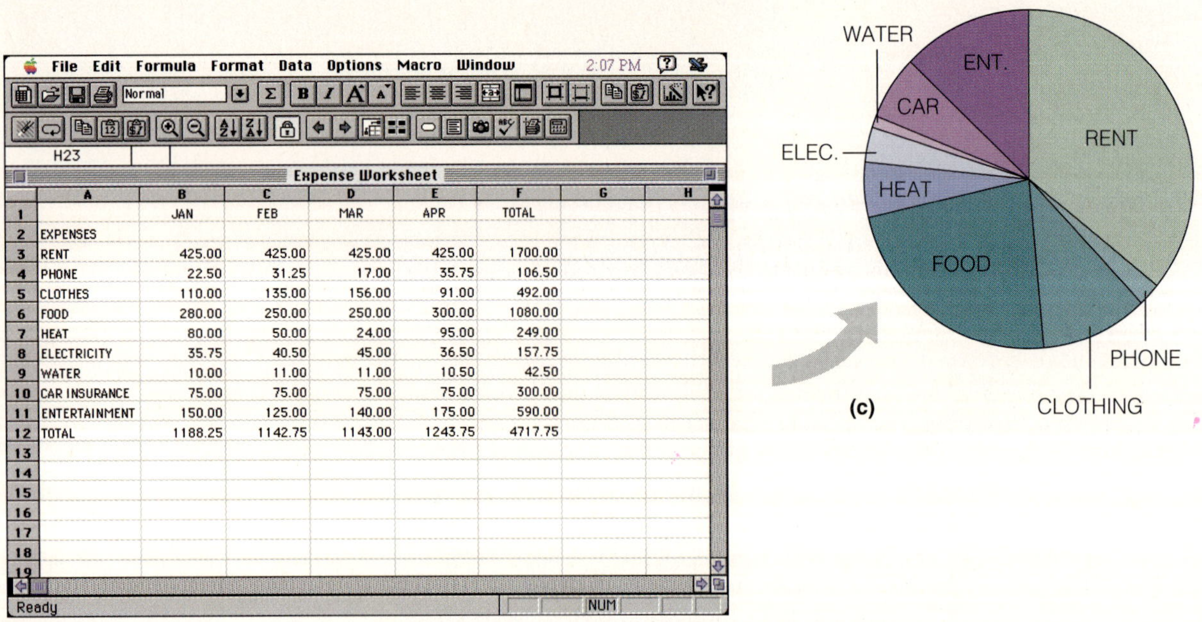

(b)

(c)

Figure 1-12 A simple expense spreadsheet. (a) This paper-and-pencil expense sheet is a typical spreadsheet of rows and columns. You have to do the calculations to fill in the totals. (b) This screen shows the same information on a computer spreadsheet program, which does the calculations for you. (c) The spreadsheet program can also present the expenses graphically in the form of a pie chart.

limited, too. Thus, the tasks and the software solutions fall, for the most part, into just a few categories, which can be found in most business environments. These major categories are word processing (including desktop publishing), spreadsheets, database management, graphics, and communications. We will present a brief description of each category here.

Word Processing/Desktop Publishing

The most widely used personal computer software is **word processing** software. This software lets you create, edit, format, store, and print text and graphics in one document. In this definition it is the three words in the middle—*edit*, *format*, and *store*— that reveal the difference between word processing and plain typing. Since you can store the memo or document you type on disk, you can retrieve it another time, change it, reprint it, or do whatever you like with it. You can see what a great time-saver word processing can be: Unchanged parts of the stored document do not need to be retyped; the whole revised document can be reprinted as if new.

As the number of features in word processing packages has grown, word processing has crossed the border into desktop publishing territory. **Desktop publishing** packages are usually better than word processing packages at meeting high-level publishing needs, especially when it comes to typesetting and color reproduction. Many magazines and newspapers today rely on desktop publishing software. Businesses use it to produce professional-looking newsletters, reports, and brochures—both to improve internal communication and to make a better impression on the outside world. We will introduce you to word processing/desktop publishing in more depth in Chapter 13.

Electronic Spreadsheets

Spreadsheets, made up of columns and rows, have been used as business tools for centuries (Figure 1-12). A manual spreadsheet can be tedious to prepare and, when there are changes, a considerable amount of calculation may need to be redone. An **electronic spreadsheet** is still a spreadsheet, but the computer does the work. In particular, spreadsheet software automatically recalculates the results when a number is changed. This capability lets business people try different combinations of numbers and obtain the results quickly. This ability to ask "What if . . . ?" helps business people make better, faster decisions. We address spreadsheets in detail in Chapter 14.

Database Management

Software used for **database management**—the management of a collection of interrelated facts—handles data in several ways. The software can store data, update it, manipulate it, report it in a variety of views, and print it in as many forms. By the time the data is in the reporting stage—given to a user in a useful form—it has become information. A concert promoter, for example, can store and change data about upcoming concert dates, seating, ticket prices, and sales. After this is done, the promoter can use the software to retrieve information, such as the number of tickets sold in each price range or the percentage of tickets sold the day before the concert. Database software can be useful for anyone who must keep track of a large number of facts. We discuss database management in Chapter 15.

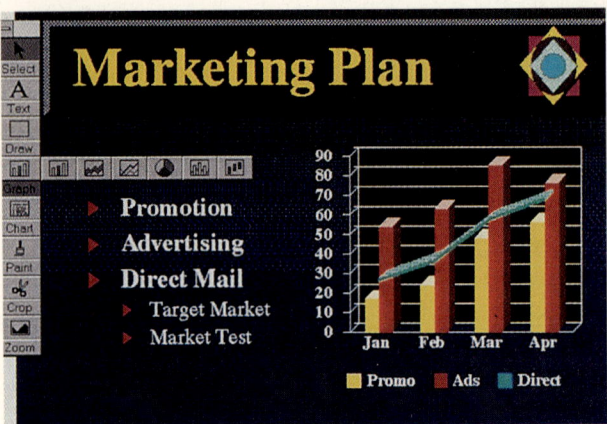

Figure 1-13 **Business graphics.** Colorful computer-generated graphics can help people compare data and spot trends.

Graphics

It might seem wasteful to show **graphics** to business people when standard computer printouts are readily available. However, graphics, maps, and charts can help people compare data and spot trends more easily, and make decisions more quickly (Figure 1-13). In addition, visual information is usually more compelling than a page of numbers. We will take a closer look at business graphics in Chapter 14.

Communications

We have already described communications in a general way. From the viewpoint of a worker with a personal computer at home, **communications** means—in simple terms—that he or she can hook a phone up to the computer and communicate with the computer at the office, or get at data stored in someone else's computer in another location.

 ## People and Computers

We have talked about hardware, software, and data, but the most important element in a computer system is people. Anyone nervous about a takeover by computers will be relieved to know that computers will never amount to much without people—the people who help make the system work and the people for whom the work is done.

Computers and You, the User

As we noted earlier, computer users have come to be called just *users*, a nickname that has persisted for years. Whereas once computer users were an elite breed—high-powered scientists, research-and-development engineers, government planners—today the population of users has broadened considerably. This expansion is due partly to user-friendly software for both work and personal use and partly to the availability of small, low-cost, personal computers. There is a strong possibility that all of us will be computer users, even if our levels of sophistication vary.

Users in a business environment may access computers and needed data in a variety of ways, including using personal computers, using personal

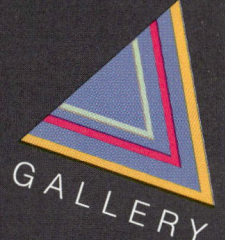

MAKING MICROCHIPS

Computer power in the hands of the people—we take it for granted now, but not so long ago computers existed only in enormous rooms behind locked doors. The revolution that changed all that was ignited by chips of silicon smaller than your fingernail: microchips.

Silicon is one of the most common elements on earth, but there is nothing commonplace about designing, manufacturing, testing, and packaging the microprocessors that are made from silicon. In this gallery we will explore the key elements in the process by which those marvels of miniaturization—microchips—are made.

The Idea Behind the Microchip

Microchips form the lightning-quick "brain" of a computer. These devices, though complex, work on a very simple principle: They "know" when electric current is on and when it is off. They can process information because it is coded as a series of on-off electric signals. Before the invention of microchips, these signals were controlled by thousands of separate devices laboriously wired together to form a single circuit. Because thousands of circuits can be embedded on a microchip, a microchip is often called an integrated circuit.

Silicon is a semiconductor—it conducts electricity only "semi" well. This does not sound like such an admirable trait, but the beauty of silicon is that it can be doped, or treated, with different materials to make it conduct electricity well or not at all. By doping various areas of a silicon chip differently, pathways can be set up for electricity to follow. The pathways consist of grooves that are etched into layers placed over silicon substrate. The silicon is doped so the pathways conduct electricity. The surrounding areas do not conduct electricity at all.

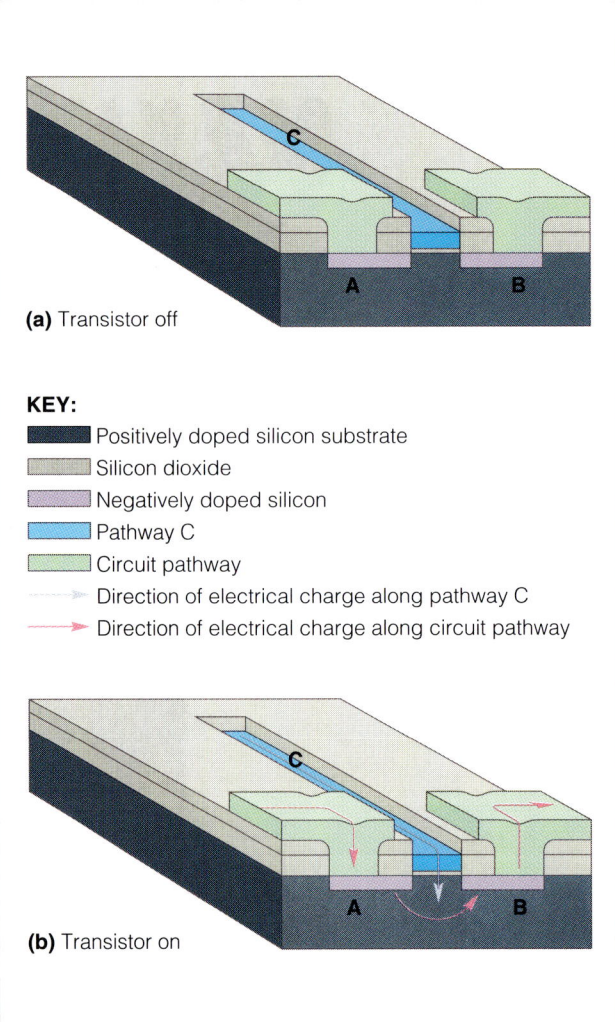

(a) Transistor off

KEY:
- Positively doped silicon substrate
- Silicon dioxide
- Negatively doped silicon
- Pathway C
- Circuit pathway
- Direction of electrical charge along pathway C
- Direction of electrical charge along circuit pathway

(b) Transistor on

1. This simplified illustration shows the layers and grooves within a transistor, one of thousands of circuit components on a single chip. Pathway C controls the flow of electricity through the circuit. (a) When no electric charge is added to pathway C, electricity cannot flow along the circuit pathway from area A to area B. Thus, the transistor is "off." (b) A charge added to pathway C temporarily allows electricity to travel from area A to area B. Now the transistor is "on," and electricity can continue to other components in the circuit. The control of electricity here and elsewhere in the chip makes it possible for the computer to process information coded as "on-off" electric signals.

Preparing the Design

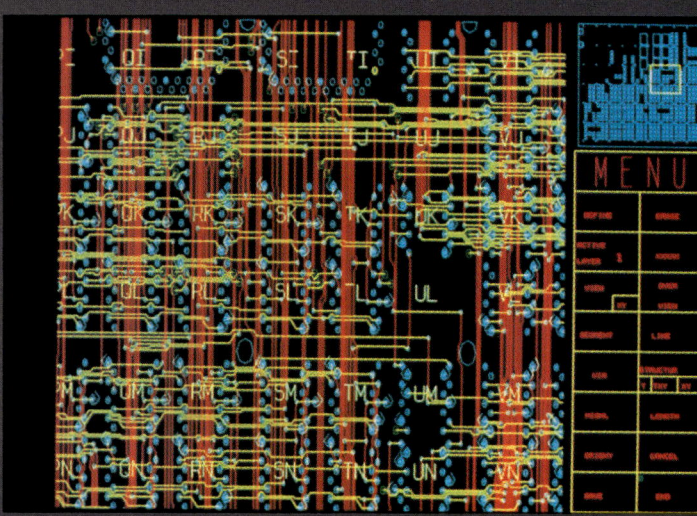

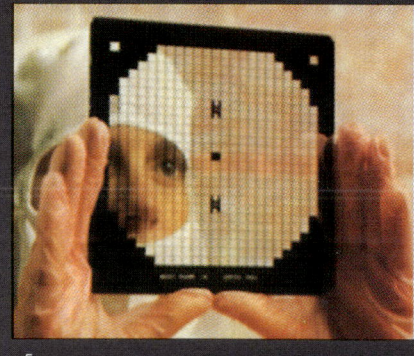

Try to imagine figuring out a way to place thousands of circuit components next to each other so that all the layers and grooves line up, and electricity flows through the whole integrated circuit the way it is supposed to. That is the job of chip designers. Essentially, they are trying to put together a gigantic multilayered jigsaw puzzle.

The circuit design of a typical chip requires over a year's work by a team of designers. Computers assist in the complex task of mapping out the most efficient pathways for each circuit layer.

2. By drawing with an electronic pen on a digitizing tablet, a designer can arrange and modify circuit patterns and display them on a screen. Superimposing the color-coded circuit layers allows the designer to evaluate the relationships between them. The computer allows the designer to electronically store and retrieve previously designed circuit patterns.

3. Here the designer has used computer graphics software to display a screen image of the circuit design. 4. The computer system can also provide a printed version of any or all parts of the design. This large-scale printout allows the design team to discuss and modify the entire chip design.

5. The final design of each circuit layer must be reduced to the size of the chip. Several hundred replicas of the chip pattern are then etched on a chemically coated glass plate called a photomask. Each photomask will be used to transfer the circuit pattern to hundreds of chips. One photomask is required for each layer of the chip. A typical design requires 7 to 12 photomasks, but more complex chips may require as many as 20.

Manufacturing the Chip

The silicon used to make computer chips is extracted from common rocks and sand. It is melted down into a form that is 99.9 percent pure silicon and then doped with chemicals to make it either electrically positive or electrically negative.

6. The molten silicon is then "grown" into cylindrical ingots in a process similar to candle dipping.
7. A diamond saw slices each ingot into circular wafers 4 or 6 inches in diameter and 4/1000 of an inch thick. The wafers are sterilized and polished to a mirror-like finish. Each wafer will eventually contain hundreds of identical chips. In the photo an engineer is holding an experimental 8-inch wafer that can comprise over 2000 chips.

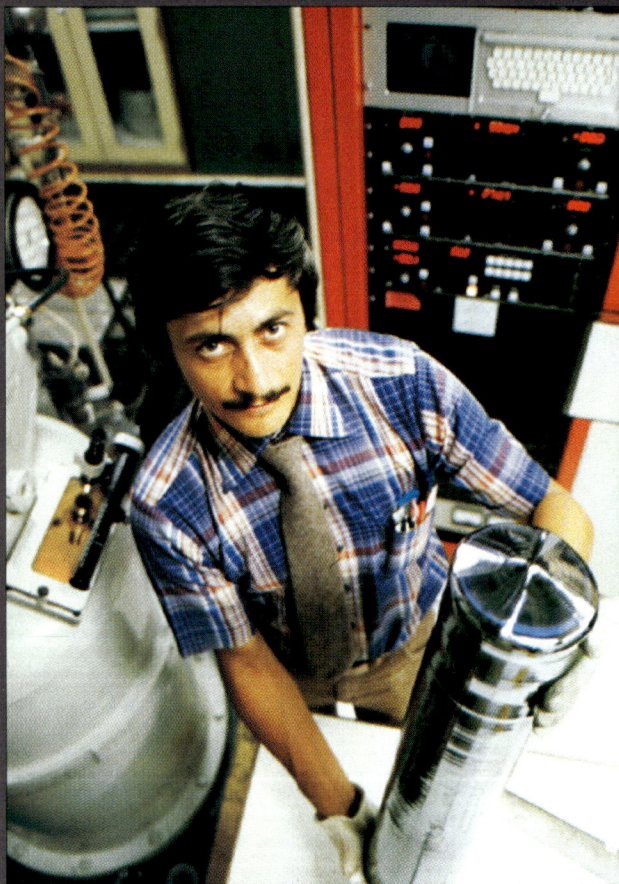

Since a single speck of dust can ruin a chip, chips are manufactured in special laboratories called clean rooms. The air in clean rooms is filtered, and workers dress in "bunny suits" to lessen the chance of chip contamination. A manufacturing lab is 100 times cleaner than a hospital operating room.

8. Chip-manufacturing processes vary, but one step is common: Electrically positive silicon wafers are placed in an open glass tube and inserted in a 1200° Celsius oxidation furnace. Oxygen reacts with the silicon, covering each wafer with a thin layer of silicon dioxide, which does not conduct electricity well. Each wafer is then coated with a gelatin-like substance called photoresist, and the first photomask pattern is placed over it. Exposure to ultraviolet light hardens the photoresist, except in the areas concealed by the dark circuit pattern on the photomask.

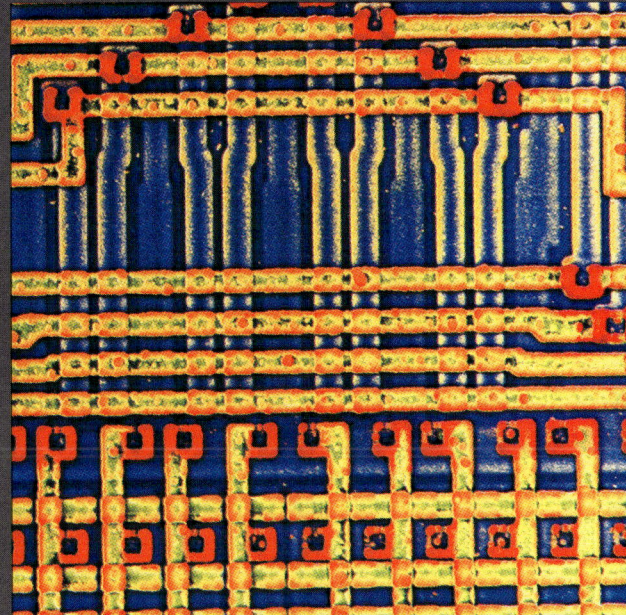

9. The wafer is then taken to a washing station in a specially lit "yellow room," where the wafer is washed in solvent to remove the soft photoresist. Next the silicon dioxide revealed by the washing is etched away by hot gases. The silicon underneath, which forms the circuit pathway, is then doped to make it electrically negative. In this way, the circuit pathway is distinguished electrically from the rest of the silicon. This process is repeated for each layer of the wafer, using a different photomask each time. In the final step aluminum is deposited to connect the circuit components and form the bonding pads to which wires will later be connected.
10. The result: a wafer with hundreds of chips.
11. Photographic lighting enhances this close-up view of a wafer with chips.
12. This close-up of a memory chip shows details of the surface architecture.

Testing the Chip

13

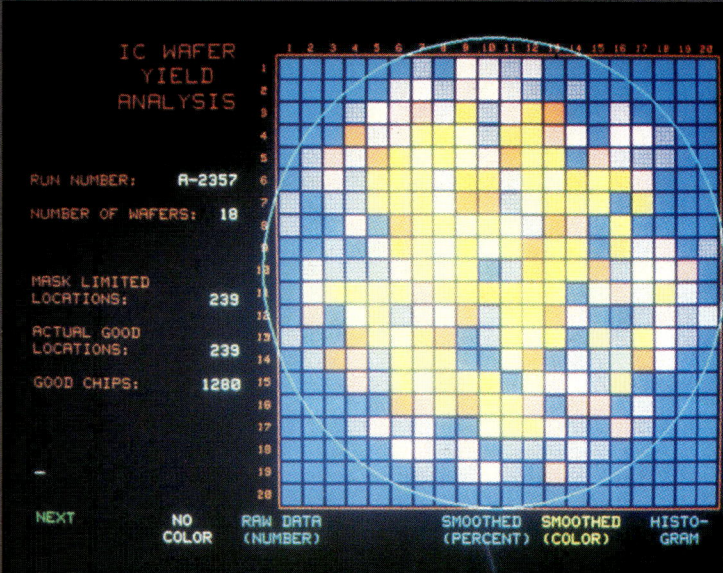

14

Although chips on a particular wafer may look identical, they do not perform identically.

13. A probe machine must perform millions of tests on each chip, to determine whether it conducts electricity in the precise way it was designed to. The needle-like probes contact the bonding pads, apply electricity, measure the results, and mark ink spots on defective chips.
14. A defect review performed by a computer finds and classifies defects in order to eliminate them from the wafer.
15. After the initial testing, a diamond saw cuts each chip from the wafer, and defective chips are discarded.

15

Packaging the Chip

16

17

18

19

Each acceptable chip is mounted on a protective package.

16. An automated wire-bonding device wires the bonding pads of the chip to the electrical leads on the package, using aluminum or gold wire thinner than a human hair.
A variety of packages are in use today.
17. Dual in-line packages have two rows of legs that are inserted into holes in a circuit board.
18. Square pin-grid array packages, which are used for chips requiring many electrical leads, look like a bed of nails. The pins are inserted into holes in a circuit board. In this photo the protective cap has been cut away, revealing the ultrafine wires connecting the chip to the package.
19. This photo shows a dual in-line package (top) compared to two surface-mount packages (bottom). Surface-mount packages do not have to be inserted in circuit board holes. Instead, a machine drops the package on the board, and a laser or infrared beam bonds the package into place. Another advantage of surface-mount packages is that they are smaller than other packages, allowing more computing power in less space.

From Chip to Computer

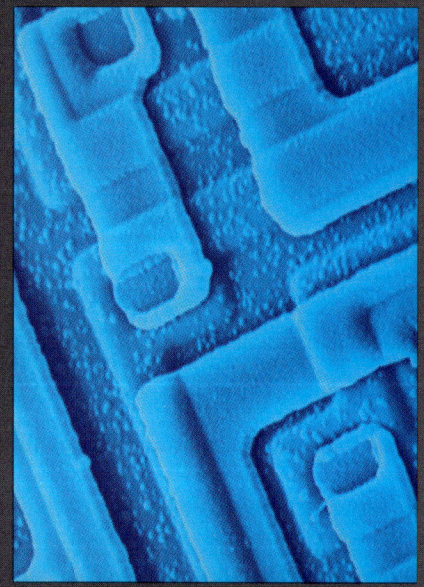

20. At the factory that manufactures computers, a robotic arm inserts a pin-grid package into holes in a circuit board. Several surface-mount packages have already been placed on the board.
21. Dual in-line packages of various sizes have been attached to this circuit board.

22. Metal lines on the board form electrical connections to the legs of the package, as shown in this color-enhanced close-up of some packages on a circuit board.

computers in a network, or using a terminal to access data on a larger computer. Data files kept on a large computer are in the hands of the computer professionals: computer people.

Computer People

Many organizations have a department called **Management Information Systems (MIS)** or **Computer Information Systems (CIS), Computing Services,** or **Information Services**. This department is made up of people responsible for the computer resources of an organization. Large organizations, such as universities, government agencies, and corporations, keep much of the institution's data in computer files: research data, engineering drawings, marketing strategy, accounts receivable, accounts payable, sales facts, manufacturing specifications, transportation plans, warehousing data, and so forth. The people who maintain the data are the same people who provide service to the users: the computer professionals. Let us touch on the essential personnel required to run large computer systems.

Data entry operators prepare data for processing, usually by keying it in a machine-readable format. **Computer operators** monitor the computer, review procedures, and keep peripheral equipment running. **Librarians** catalog the processed disks and tapes and keep them secure.

Computer programmers design, write, test, and implement the programs that process data on the computer system; they also maintain and update the programs. **Systems analysts** are knowledgeable in the programming area but have broader responsibilities. They plan and design not just individual programs but entire computer systems. Systems analysts maintain a working relationship with the users in the organization. The analysts work closely with the users to plan new systems that will meet the users' needs. The department manager, often called the **chief information officer (CIO)** must understand more than just computer technology. This person must understand the goals and operations of the entire organization. We will discuss software and systems analysis in Part 2, Chapters 6, 7, and 8.

Computer professionals also help users directly with their personal computers or terminals and the software they use. Furthermore, a professional called a **network manager** implements and maintains the network. We will discuss computer issues on the job and management issues in Chapters 9 and 10.

▼ ▲ ▼

In this chapter, we have painted the computer industry with a broad brush, touching on hardware, software, data, and people. We now move on to chapters that explain in more detail the information presented in this chapter.

Help for the SAT®
Each year thousands of students work up a full panic at the thought of taking that test of tests, the Scholastic Aptitude Test, not so fondly known as the SAT. The fortunes of many students ride on the results of the SATs. Small wonder that students seek every mode of assistance. That help has traditionally been in the form of study aids such as special texts and classes.
   Now, of course, all that has changed. The most sought-after aid is a computerized preparation course that offers at-home study on your own computer. The system includes diagnostic tests, instructions in the basics of the verbal and math SATs, and plenty of practice tests. Will these computer-tutors raise SAT scores? It looks promising.

CHAPTER REVIEW

Summary and Key Terms

- The equipment associated with a computer system is called **hardware.** The **programs,** or step-by-step instructions that run the machines, are called **software. Computer programmers** write programs for **users,** or **end-users**—people who purchase and use computer software.
- A **computer** is a machine that can be programmed to process data (input) into useful information (output). A computer system requires four main aspects of data handling—input, processing, output, and storage.
- **Input** is data to be accepted into the computer. Common input devices are the **keyboard;** a **mouse,** which translates movements of a ball on a flat surface to actions on the screen; and a **wand reader** or **bar code reader,** which uses laser beams to read special letters, numbers, or symbols such as the zebra-striped bar codes on products.
- A **terminal** includes an input device, such as a keyboard or wand reader; an output device, usually a television-like screen; and a connection to the main computer.
- The **processor,** or **central processing unit (CPU),** processes raw **data** into meaningful, useful **information.** The CPU interprets and executes program instructions and communicates with the input, output, and storage devices. **Memory,** or **primary storage,** is associated with the central processing unit but is separate from it. Memory holds the input data before processing and after processing, until the data is released to the output device.
- **Output,** which is raw data processed into usable information, is usually in the form of words, numbers, and graphics. Users can see output displayed on **screens** and use **printers** to display output on paper.
- **Secondary storage** provides additional storage space separate from memory. The most common secondary storage devices are **magnetic disks,** but **magnetic tape** also provides secondary storage. Magnetic disks are **diskettes** (usually 3½ inches in diameter, but possibly 5¼ inches) or **hard disks.** Hard disks on large systems are contained in a disk pack. Hard disks hold more data and offer faster access than a diskette. Some hard disks come in removable cartridge form. Disk data is read by **disk drives.** The most recent disk storage technology is the **optical disk,** which uses a laser beam to store large volumes of data. Magnetic tape comes on reels or in cassettes and is mainly used for backup purposes. Magnetic tape reels are mounted on **tape drives.**
- **Peripheral equipment** includes all the input, output, and secondary storage devices attached to a computer. Peripheral equipment may be built into one physical unit, as in many personal computers, or contained in separate units, as in many large computer systems.
- The most powerful and expensive computers are called **supercomputers.** Large computers called **mainframes** are used by such customers as banks, airlines, and large manufacturers to process very large amounts of data quickly. **Minicomputers,** the next step down from mainframes in terms of power and capacity, are widely used by small businesses. The largest and most expensive minicomputers are called **superminis.** Desktop computers are called **personal computers (PCs),** or **microcomputers,** or sometimes home computers. **Supermicros** combine the compactness of a desktop computer with power that almost equals that of a mainframe. **Laptop computers,** also called **notebook computers,** are small portable computers. **Pen-based computers,** also called **personal digital assistants (PDAs),** accept handwritten input directly on a screen.
- A **centralized computer system** does all processing in one location. In a **decentralized computer system,** the computer itself and some storage devices are in one place, but the devices to access the computer are somewhere else. Such a system requires **data communications**—the exchange of data over communications facilities. In a **distributed data processing (ddp)**

- system, a local office usually uses its own small computer for processing local data but is connected to a central headquarters computer for other purposes.
- Often organizations use a **network** of personal computers, which allows users to operate independently or in cooperation with other computers—exchanging data and sharing resources. Such a setup, is called a **local area network (LAN)**. Another possibility is to connect personal computers to a mainframe computer to form a **micro-to-mainframe** link, in which users can obtain data from the mainframe and analyze it on their own personal computers.
- Individuals use networking for a variety of purposes, especially **electronic mail**, or **e-mail**, often on the popular resource called the **Internet**.
- A subset of **system software** is an **operating system**, the underlying software found on all computers. **Applications software** solves a particular problem or performs a particular task. Applications software may be either custom or packaged. **Custom software** is specifically tailored to user needs. **Packaged software**, also called **commercial software**, is packaged in a container and sold in stores or catalogs.
- Software is accompanied by manuals, also called **documentation**. Software that is easy to use is considered **user friendly**.
- Task-oriented software found in most business environments includes **word processing/desktop publishing, electronic spreadsheets, database management, graphics**, and **communications**.
- An organization's computer resources department—often called **Management Information Systems (MIS)** or **Computer Information Systems (CIS), Computing Services, or Information Services**—may include **data entry operators** (who prepare data for processing), **computer operators** (who monitor and run the equipment), **librarians** (who catalog disks and tapes), **computer programmers** (who design, write, test, and implement programs), **systems analysts** (who plan and design entire systems of programs), **a chief information officer (CIO)** (who coordinates the MIS department), **and network managers** (who implement and maintain the network).

Discussion Questions

1. Consider the hardware used for input, processing, output, and storage for personal computers. If money were no object, how (generally) would you configure the hardware for your own personal computer?
2. Why do you think many companies prefer decentralized computer systems? Why might you, as an employee, prefer it?
3. Consider the task-oriented software for word processing/desktop publishing, electronic spreadsheets, database management, graphics, and communications. Which type of software (possibly more than one) would you use for each of the following tasks:
    a. Preparing an annual club report showing a comparison of the budget for last year and this year, and then including that report in an attractive monthly letter sent to members
    b. Preparing a comparison report of sales of six different products in three sales regions, and then showing the result to a group of 50 people at a sales meeting
    c. Gathering employee attendance data from managers in franchise stores in 17 locations, and then writing a memo to your boss summarizing the results
    d. Storing data as it becomes available about hotel room use—customer name, date of arrival, expected date of departure, and so forth—and later retrieving the room number for a certain customer by name or retrieving the numbers of all rooms currently available

Student Study Guide

Multiple Choice

1. The processor is an example of
   a. software
   b. hardware
   c. a program
   d. an output unit
2. Additional data and programs not being used by the processor are stored in
   a. secondary storage
   b. output units
   c. input units
   d. the CPU
3. Step-by-step instructions that run the computer are
   a. hardware
   b. CPUs
   c. documents
   d. programs
4. A computer that accepts handwritten input on a screen:
   a. minicomputer
   b. mainframe
   c. desktop computer
   d. pen-based computer
5. Desktop and personal computers are other names for
   a. microcomputers
   b. mainframes
   c. minicomputers
   d. peripheral equipment
6. The raw material to be processed by a computer is called
   a. a program
   b. software
   c. data
   d. information
7. Which is *not* a professional computer job?
   a. systems analyst
   b. user
   c. data entry operator
   d. programmer
8. A bar code reader is an example of a(n)
   a. processing device
   b. input device
   c. storage device
   d. output device
9. Computer people who design, write, test, and implement programs are
   a. programmers
   b. data entry operators
   c. computer operators
   d. systems analysts
10. Printers and screens are common forms of
    a. input units
    b. storage units
    c. output units
    d. processing units
11. The unit that transforms data into information is the
    a. CPU
    b. disk drive
    c. bar code reader
    d. wand reader
12. People who prepare data for processing are
    a. programmers
    b. printers
    c. librarians
    d. data entry operators
13. A system whereby computers and data storage are placed in dispersed locations is known as
    a. centralized processing
    b. packaged software
    c. summarizing
    d. distributed data processing
14. An example of peripheral equipment:
    a. CPU
    b. printer
    c. spreadsheet
    d. microcomputer
15. Another name for available-for-purchase software is
    a. secondary software
    b. packaged software
    c. system software
    d. peripheral software
16. Which of the following is an acronym for a computer department?
    a. MIS
    b. CPU
    c. PDA
    d. LAN
17. A device that inputs data by scanning letters and numbers is a
    a. keyboard
    b. mouse
    c. wand reader
    d. diskette
18. Another name for memory is
    a. secondary storage
    b. primary storage
    c. disk storage
    d. tape storage
19. Which is not a computer classification?
    a. maxicomputer
    b. microcomputer
    c. minicomputer
    d. mainframe
20. When all access and processing is done in one location, a computer system is said to be
    a. networked
    b. centralized
    c. distributed
    d. linked
21. An input device that translates motions of a ball rolled on a flat surface to the screen is the
    a. wand reader
    b. bar code reader
    c. keyboard
    d. mouse
22. Computer users who are not computer professionals are sometimes called
    a. librarians
    b. information officers
    c. peripheral users
    d. end-users
23. The most powerful computers are
    a. superminis
    b. supermainframes
    c. supermicros
    d. supercomputers
24. Raw data is processed by the computer into
    a. number sheets
    b. paragraphs
    c. updates
    d. information
25. Laser beam technology is used for
    a. terminals
    b. keyboards
    c. optical disk
    d. magnetic tape

True/False

1. T / F — The processor is also called the central processing unit, or CPU.
2. T / F — Secondary storage units contain the instructions and data to be used immediately by the processor.
3. T / F — A computer department may be called MIS, CIS, or IS.
4. T / F — Two secondary storage media are magnetic disk and magnetic tape.
5. T / F — A diskette holds more data than a hard disk.
6. T / F — PDAs accept handwritten data on a screen.
7. T / F — The most powerful personal computers are known as supercomputers.
8. T / F — Operating systems are a subset of system software.
9. T / F — Processed data is called information.
10. T / F — User friendly refers to a special kind of terminal.
11. T / F — Custom software is specifically tailored to user needs.
12. T / F — The people who write software are called computer operators.
13. T / F — Word processing is a type of task-oriented software.

T (F) 14. Computer hardware is always kept in one large room.
(T) F 15. These computers are arranged from least powerful to most powerful: microcomputer, mainframe, supercomputer.
T (F) 16. The Internet is an example of a peripheral device.
T (F) 17. A distributed data system is centralized in one location.
T (F) 18. A LAN is usually set up between two cities.
(T) F 19. Magnetic tape is most often used for backup purposes.
T (F) 20. Another name for memory is secondary storage.
T (F) 21. People can usually spot trends more quickly from numbers than from graphics.
T (F) 22. Desktop publishing is popular in the home but not useful for business.
(T) F 23. A spreadsheet is comprised of columns and rows.
(T) F 24. Applications software may be either custom or packaged.
T (F) 25. Another name for supermicro is personal digital assistant.

Fill-In

1. What are the four general hardware units of a computer?
   a. input
   b. output
   c. CPU (processor)
   d. storage unit

2. What kind of software presents numbers in columns and rows? spreadsheet

3. Where is data held after it is input to the system but before it is processed? memory (RAM)

4. What are magnetic tape reels mounted on when their data is to be read by the computer system? tape drives

5. What are the input, output, and secondary storage devices attached to a computer called? peripheral equipment

6. What are large computers called in the computer industry? mainframes

7. What is the term describing a system whereby computers and data storage are placed in geographically separate locations? distributed data processing

8. What is the word for raw material that is given to a computer for processing? data

9. What is the exchange of data over communication facilities called? data communications

10. Connecting personal computers to a mainframe is called: micro-mainframe link

11. What is software that is easy to use said to be? user friendly

12. What does MIS stand for? Manag. inform systems

13. What does CPU stand for? Centr. proc. unit

14. What three types of input methods are mentioned in the chapter?
    a. keyboard
    b. mouse
    c. wand reader/bar code reader

15. What is another name for a microcomputer? personal comp.

16. What kind of system does all its processing in one location? centralized

17. Using a data communications system to send messages is called: electronic mail

18. Who are the people who plan and design systems of programs? systems analysts

19. What are the planned step-by-step instructions required to turn data into information? software

20. What does LAN stand for? local area network

Answers

Multiple Choice

1. b	6. c	11. a	16. a	21. d
2. a	7. b	12. d	17. c	22. d
3. d	8. b	13. d	18. b	23. d
4. d	9. a	14. b	19. a	24. d
5. a	10. c	15. b	20. b	25. c

Part One ▼ Hardware Tools

True/False
1. T
2. F
3. T
4. T
5. F
6. T
7. F
8. T
9. T
10. F
11. T
12. F
13. T
14. F
15. T
16. F
17. F
18. F
19. T
20. F
21. F
22. F
23. T
24. T
25. F

Fill-In
1. a. input unit
   b. processor
   c. output unit
   d. storage unit
2. spreadsheet
3. memory (or primary storage)
4. tape drives
5. peripheral equipment
6. mainframes
7. distributed data processing
8. data
9. data communications
10. micro-to-mainframe link
11. user friendly
12. management information system
13. central processing unit
14. keyboard, mouse, wand reader, or bar code reader
15. personal computer
16. centralized
17. electronic mail (or e-mail)
18. systems analysts
19. software (or program)
20. local area network

PLANET INTERNET

What Is It All About?

First, just what is the Internet? The Internet, sometimes called simply "the Net," is a vast global network of computers that are connected. Technically, rather than one network, the Internet is a loosely organized collection of thousands of networks. It is accessed by students, business people, scientists, computer professionals, hobbyists, and anyone else who can tap into the Net's extraordinary resources.

It seems to me that one day I have never heard of the Internet, and then the next day I see it everywhere. That's true for a lot of people, even computer professionals. Part of the confusion is that the Net started out with a different name: Arpanet. The military created Arpanet in the 1970s to scatter their computers so that no single nuclear bomb could wipe out their computing capabilities. The network evolved and spread to other organizations, especially universities and libraries. Eventually, a more generic name was adopted: the Internet.

Why has it become so popular recently? The main reason is that it became easy to use. It is especially easy when compared to previous access methods, some of which required sophisticated technical knowledge.

OK, let's assume that it's easy enough even for me. Why should I jump in? The one answer that fits everyone is that you dare not risk being left behind. Futurists predict that networking of some kind will be as necessary to work and to living as technologies such as the telephone or computers. After that, the answer to this question depends a lot on the individual. Are you curious? Would you like to connect with people around the world? Would you like an amazing library at your fingertips? Would it amuse you just to see what other folks are up to?

Give me some for-instances. OK. Do you plan a job search in the near future? The Internet offers job boards and online help. Having a problem with your dog? Veterinary specialists answer individual questions from Net users. If you forget a birthday and thus need to send a gift quickly across the country, shopping on the Internet rivals any mall you have ever seen. Finally, consider some screens that could show up during your Internet travels. The White House image shown here signals your ability to ask questions of the president or even take a picture tour inside the White House.

Is this going to cost money? Maybe. Free Internet access is common in schools and libraries and other government organizations. Your employer may offer free access. If you want to hook up from your own personal computer, you will have to pay some sort of monthly charge and perhaps extra charges based on usage. Your local computer store could offer advice.

What's coming up later in the Internet discussions? For the most part, we'll examine various offerings on the Internet, some serious, some less so. Information about using the Internet will be tucked in here and there, so your knowledge will grow. In the next discussion of *Planet Internet,* we'll start traveling on the most popular part of the Internet, the World Wide Web. As we said, it's easy.

Mark Ong, who hoped to be a scriptwriter, planned a double major in creative writing and drama. After his first year of college, he took a summer job as an editorial assistant, where he first used word processing. He decided that it would be helpful to have a personal computer of his own when he went back to college in the fall. But he felt unsure of how to make a purchase. In fact, he felt he did not even know what questions to ask. He discussed this with an office colleague, who casually noted that any computer setup comes with the "standard stuff"—processor, keyboard, screen, disk drives—and that all he had to do was go to a computer store and pick one that fit his price range. Mark was not satisfied with this approach, especially in light of the advertisements he had seen in the local newspaper and in computer magazines.

Most advertisements displayed photos of personal computers, accompanied by cryptic descriptions of the total hardware package. A typical ad was worded this way: *Pentium chip, 100MHz, 16MB RAM, 256K Cache, 1.44MB diskette drive, 540MB hard drive.* The price for this particular machine was pretty hefty—over $2000. Mark noticed that the ads for machines with lower numbers, for example, only 75MHz, also had lower price tags. Similarly, higher numbers meant higher price tags. Although he did recognize the disk drives, he had no idea what the other items were or why the numbers mattered. Clearly, there was more to a purchasing decision than selecting a system with the "standard stuff."

Mark tore out some of the ads and marched to a nearby computer store. After asking a lot of questions, he learned that *Pentium* is a type of processor, that *MHz* stood for megahertz and is a measurement of the microprocessor's speed, that *RAM* was the computer's memory, that *cache* was a kind of handy storage place for frequently used data and software instructions, and that *MB* was an abbreviation for megabytes, a measurement of size. Most importantly, Mark learned that the number variations mattered because they were factors in determining the computer's capacity and speed.

Many buyers do select their personal computer system merely on the basis of a sales pitch and price range. Those people could argue, with some success, that they do not need to know all the computer buzzwords any more than they need to know the technical details of their television sets or sound systems. They know that they do not have to understand a computer's innards to put it to work.

But there are rewards for those who want to dig a little deeper, learn a little more. Although this chapter is not designed to help you purchase a computer (see the *Buyer's Guide* for that), it does provide some background information and gives you the foundation on which future computer knowledge can be built.

Chapter 2

The Central Processing Unit

What Goes on Inside the Computer

LEARNING OBJECTIVES

- Learn the components of the central processing unit and how they work together and interact with memory
- Understand how program instructions are executed by the computer
- Understand how data is represented in the computer
- Understand how the computer finds instructions and data
- Become acquainted with personal computer chips
- Understand the measures of computer processing speed and approaches that increase speed

THE CENTRAL PROCESSING UNIT
    The Control Unit
    The Arithmetic/Logic Unit
    Registers: Temporary Storage Areas

MEMORY

HOW THE CPU EXECUTES PROGRAM INSTRUCTIONS

STORAGE LOCATIONS AND ADDRESSES: HOW THE CONTROL UNIT FINDS INSTRUCTIONS AND DATA

DATA REPRESENTATION: ON/OFF
    Bits, Bytes, and Words
    Coding Schemes

PERSONAL COMPUTER CHIPS
    Microprocessors
    Memory Components

SPEED AND POWER
    Computer Processing Speeds
    Bus Lines
    Cache
    Flash Memory
    RISC Technology: Less Is More
    Parallel Processing

The Central Processing Unit

The computer does its primary work in a part of the machine we cannot see, a control center that converts data input to information output. This control center, called the **central processing unit (CPU),** is a highly complex, extensive set of electronic circuitry that executes stored program instructions. All computers, large and small, must have a central processing unit. As Figure 2-1 shows, the central processing unit consists of two parts: The *control unit* and the *arithmetic/logic unit*. Each part has a specific function.

Before we discuss the control unit and the arithmetic/logic unit in detail, we need to consider data storage and its relationship to the central processing unit. Computers use two types of storage: Primary storage and secondary storage. The CPU interacts closely with primary storage, or memory, referring to it for both instructions and data. For this reason this chapter will

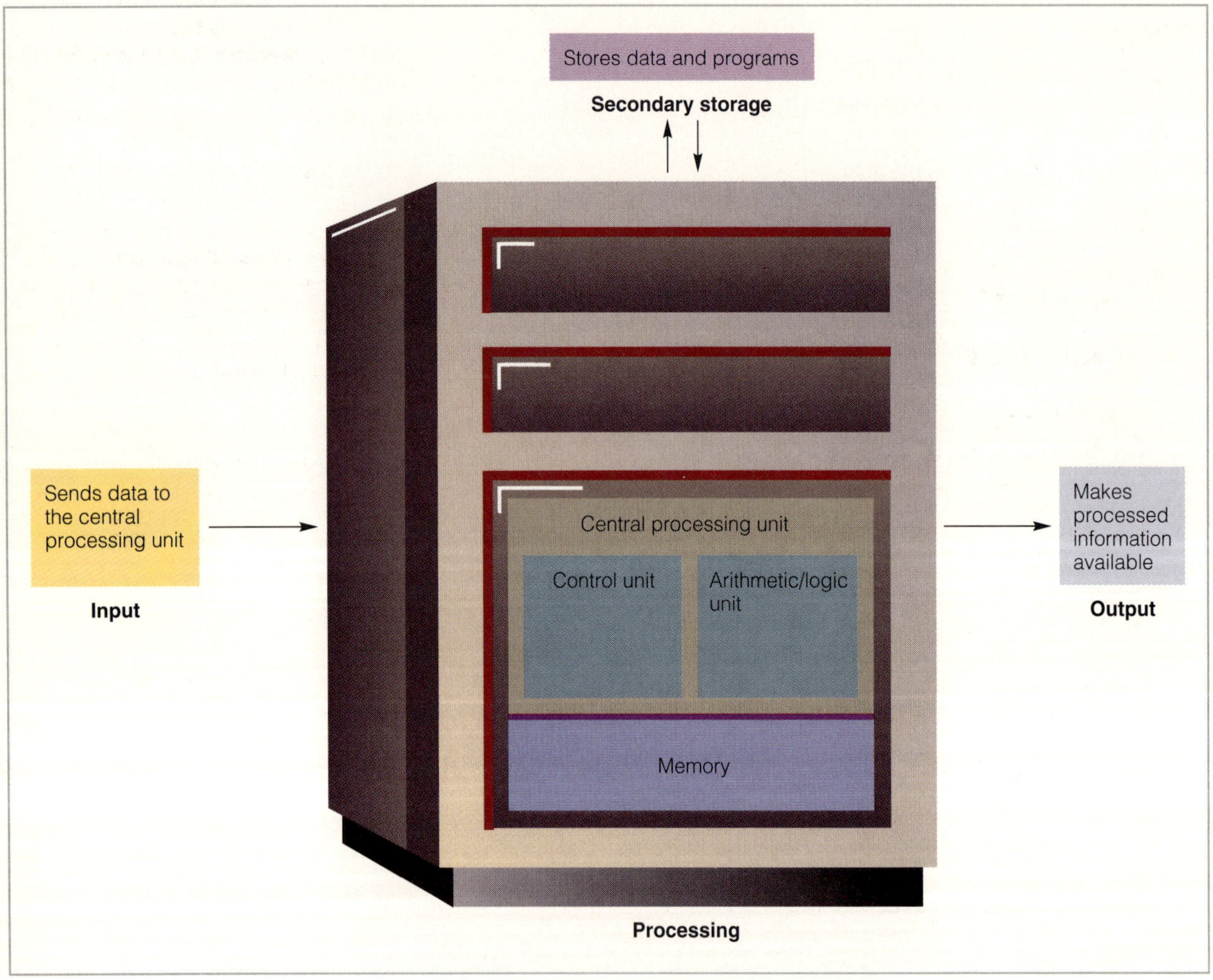

Figure 2-1 The central processing unit. The two parts of the central processing unit are the control unit and the arithmetic/logic unit. Memory holds data and instructions temporarily while the program they are part of is being executed. The CPU interacts closely with memory, referring to it for both instructions and data.

discuss memory in the context of the central processing unit. Technically, however, memory is not part of the CPU.

As Chapter 1 noted, memory holds data only temporarily, at the time the computer is executing a program. **Secondary storage** holds permanent or semipermanent data on some external magnetic or optical medium. The diskettes and CD-ROM disks that you have seen with personal computers are secondary storage devices, as are hard disks. Since the physical attributes of secondary storage devices determine the way data is organized on them, we will discuss secondary storage and data organization together in Chapter 4.

Now let us consider the components of the central processing unit.

The Control Unit

The **control unit** contains circuitry that uses electrical signals to direct the entire computer system to carry out, or execute, stored program instructions. Like an orchestra leader, the control unit does not execute program instructions; rather, it directs other parts of the system to do so. The control unit must communicate with both the arithmetic/logic unit and memory.

The Arithmetic/Logic Unit

The **arithmetic/logic unit** (ALU) contains the electronic circuitry that executes all arithmetic and logical operations.

The arithmetic/logic unit can perform four kinds of **arithmetic operations,** or mathematical calculations: addition, subtraction, multiplication, and division. As its name implies, the arithmetic/logic unit also performs logical operations. A **logical operation** is usually a comparison. The unit can compare numbers, letters, or special characters. The computer can then take action based on the result of the comparison. This is a very important capability. It is by comparing that a computer is able to tell, for instance, whether there are unfilled seats on airplanes, whether charge-card customers have exceeded their credit limits, and whether one candidate for Congress has more votes than another.

Logical operations can test for three conditions:

- **Equal-to condition.** In a test for this condition, the arithmetic/logic unit compares two values to determine if they are equal. For example: If the number of tickets sold *equals* the number of seats in the auditorium, then the concert is declared sold out.
- **Less-than condition.** To test for this condition, the computer compares values to determine if one is less than another. For example: If the number of speeding tickets on a driver's record is *less than* three, then insurance rates are $425; otherwise, the rates are $500.
- **Greater-than condition.** In this type of comparison, the computer determines if one value is greater than another. For example: If the hours a person worked this week are *greater than* 40, then multiply every extra hour by 1½ times the usual hourly wage to compute overtime pay.

A computer can simultaneously test for more than one condition. In fact, a logic unit can usually discern six logical relationships: equal to, less than, greater than, less than or equal to, greater than or equal to, and less than or greater than. Note that less than or greater than is the same as not equal to.

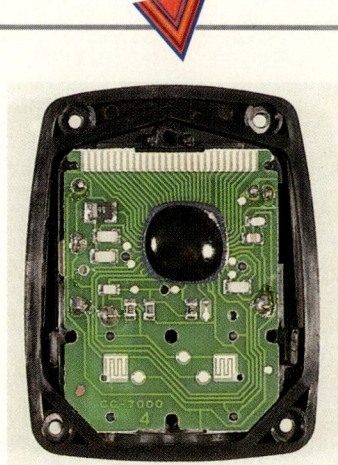

Chips Inside Everything

So popular is the tiny chip that people participate in the computer revolution every day by simple acts such as using a telephone, looking at a wristwatch, or going through a supermarket checkout line. Furthermore, chips are in cameras, blood pressure devices, microwave ovens, cars, and many other everyday devices. Homeowners can monitor heat, smoke, and security with strategically placed microprocessor chips.

Consider the bicycle odometer in the photo above. This little chip-driven device can pick up data from sensors placed on your bicycle wheels and pedals and produce the following information: current speed, average speed, maximum speed, distance, and cadence. If you were to pry the odometer open (not recommended) you would see the microprocessor revealed in the photo above.

MAKING THE RIGHT CONNECTIONS

The Information Superhighway

Just what is the information superhighway? The name is based on an analogy. Everyone understands what a highway is. Before we had the interstate highway system, just about everything you purchased was grown or built locally. It would have been pretty unusual for someone in Florida to be eating apples that were grown in Washington. But once the highway system was in place, in the 1950s, it opened up new markets. It really did not matter where goods were produced; they could be sold anywhere in the country.

Now we will do the same thing for information services. It does not matter where the information is located. A person in Oregon does not have to get on an airplane and fly to Washington, D.C., to gain access to the Library of Congress. The information is available by tapping into the information superhighway. Any information that is available any place can be accessed anywhere in the country if you are on the information superhighway. And, of course, information travels much faster than a truckload of apples from Washington to Florida.

Is the information superhighway really just another name for kinds of computer communications services I have heard about, such as America Online? Not quite. Those services are early versions of the information superhighway. The problem is that they must be accessed through a computer, and computers are not in every home. But almost every home in the United States does have a telephone and a television set, and almost everyone knows how to use them. These are the sort of components to consider for the information superhighway.

What will the people connection to the information superhighway actually look like? The truth is no one really knows. For one approach, industry is making major investments to incorporate the television system as part of the information superhighway. Your television will be like a computer screen, but it doesn't have a keyboard or anything like that. You would have some sort of remote control device, sort of like the one you use with your television now, that you would use to make selections from menus of services and information. The fact that there might be a computer embedded in that television set is pretty much immaterial to the person who is using the television set.

Others believe that it will be easier to broaden the use of computers, and make them easier to use, thus making them more palatable for the general population. This might be more realistic than making computer users out of couch potatoes. But keep in mind that computers are in relatively few homes.

When will we have an information superhighway in place? Despite all the current hype, the information superhighway is probably a ten-year, if not longer, development process. However, in the next few years there should be a significant change in government regulation for communications. There will be much greater competition, which will encourage a greater variety of services and also lower prices to consumers.

The symbols that let you define the type of comparison you want the computer to perform are called **relational operators.** The most common relational operators are the equal sign (=), the less-than symbol (<), and the greater-than symbol (>).

Registers: Temporary Storage Areas

Registers are temporary storage areas for instructions or data. They are not a part of memory; rather they are special additional storage locations that offer the advantage of speed. Registers work under the direction of the control unit to accept, hold, and transfer instructions or data and perform arithmetic or logical comparisons at high speed. The control unit uses a data storage register the way a store owner uses a cash register—as a temporary, convenient place to store what is used in transactions.

Computers usually assign special roles to certain registers, including

- An **accumulator,** which collects the result of computations.
- An **address register,** which keeps track of where a given instruction or piece of data is stored in memory. Each storage location in memory is identified by an address, just as each house on a street has an address.
- A **storage register,** which temporarily holds data taken from or about to be sent to memory.
- A **general-purpose register,** which is used for several functions—arithmetic operations, for example.

Consider registers in the context of all the means of storage discussed so far. Registers hold data *immediately* related to the operation being executed. Memory is used to store data that will be used in the *near future*. Secondary storage holds data that may be needed *later* in the same program execution or perhaps at some more remote time in the future. Now let us look at how a payroll program, for example, uses all three types of storage. Suppose the program calculates the salary of an employee. The data representing the hours worked and the data for the rate of pay are ready in their respective registers. Other data related to the salary calculation—overtime hours, bonuses, deductions, and so forth—is waiting nearby in memory. The data for other employees is available in secondary storage. As the computer finishes calculations about one employee, the data about the next employee is brought from secondary storage into memory and eventually into the registers.

Memory

Memory is also known as **primary storage, primary memory, main storage, internal storage,** and **main memory;** all these terms are used interchangeably by people in computer circles. Memory is the part of the computer that holds data and instructions for processing. Although closely associated with the central processing unit, memory is separate from it. Memory stores program instructions or data for only as long as the program they pertain to is in operation. Keeping these items in memory when the program is not running is not feasible for three reasons:

- Most types of memory only store items while the computer is turned on; data is destroyed when the machine is turned off.
- If more than one program is running at once (often the case on large computers and sometimes on small computers), a single program cannot lay exclusive claim to memory.
- There may not be room in memory to hold the processed data.

How do data and instructions get from an input device into memory? The control unit sends them. Likewise, when the time is right, the control unit sends these items from memory to the arithmetic/logic unit, where an arithmetic operation or logical operation is performed. After being processed, the information is sent to memory, where it is held until it is ready to be released to an output unit.

The chief characteristic of memory is that it allows very fast access to instructions and data, no matter where the items are within it. We will discuss the physical components of memory—memory chips—later in this chapter.

How the CPU Executes Program Instructions

Let us examine the way the central processing unit, in association with memory, executes a computer program. We will be looking at how just one instruction in the program is executed. In fact, most computers today can execute only one instruction at a time, though they execute it very quickly. Many personal computers can execute instructions in less than one-*millionth* of a second, whereas those speed demons known as supercomputers can execute instructions in less than one-*billionth* of a second.

Before an instruction can be executed, program instructions and data must be placed into memory from an input device or a secondary storage device (the process is further complicated by the fact that, as we noted earlier, the data will probably make a temporary stop in a register). As Figure 2-2 shows, once the necessary data and instruction are in memory, the central processing unit performs the following four steps for each instruction:

1. The control unit *fetches* (gets) the instruction from memory.
2. The control unit *decodes* the instruction (decides what it means) and directs that the necessary data be moved from memory to the arithmetic/logic unit. These first two steps together are called instruction time, or **I-time**.

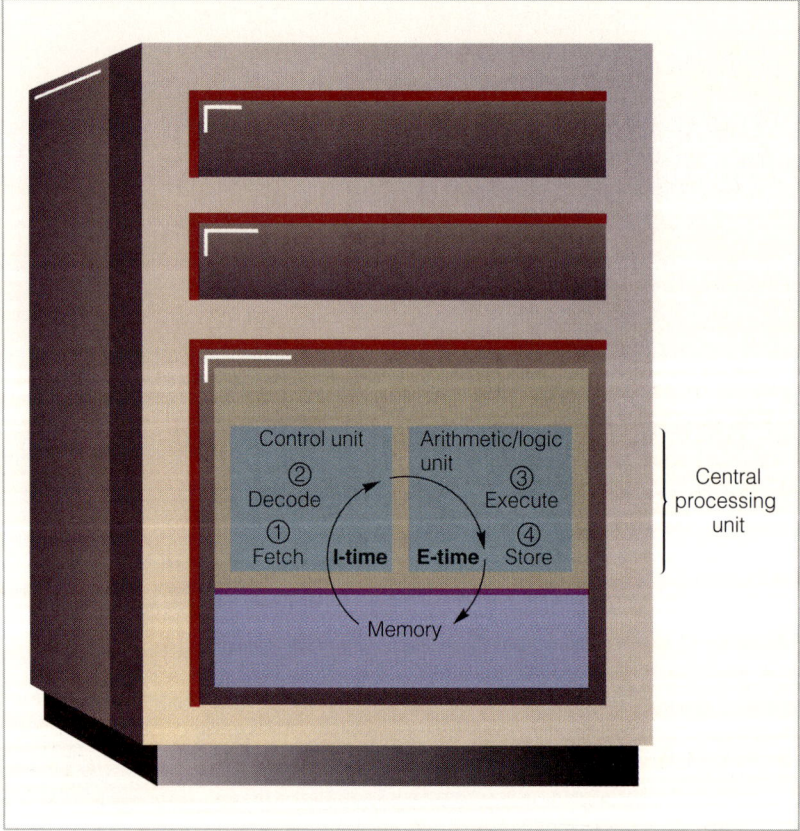

Figure 2-2 The machine cycle. Program instructions and data are brought into memory from an external device, either an input mechanism or secondary storage medium. The machine cycle executes instructions one at a time, as described in the text.

3. The arithmetic/logic unit *executes* the arithmetic or logical instruction. That is, the ALU is given control and performs the actual operation on the data.
4. The arithmetic/logic unit stores the result of this operation in memory or in a register. Steps 3 and 4 together are called execution time, or E-time.

The control unit eventually directs memory to release the result to an output device or a secondary storage device. The combination of I-time and E-time is called the **machine cycle.** Figure 2-3 shows an instruction going through the machine cycle.

Each central processing unit has an internal **clock** that produces pulses at a fixed rate to synchronize all computer operations. A single machine-cycle instruction may be made up of a substantial number of subinstructions, each of which must take at least one clock cycle. Each type of central

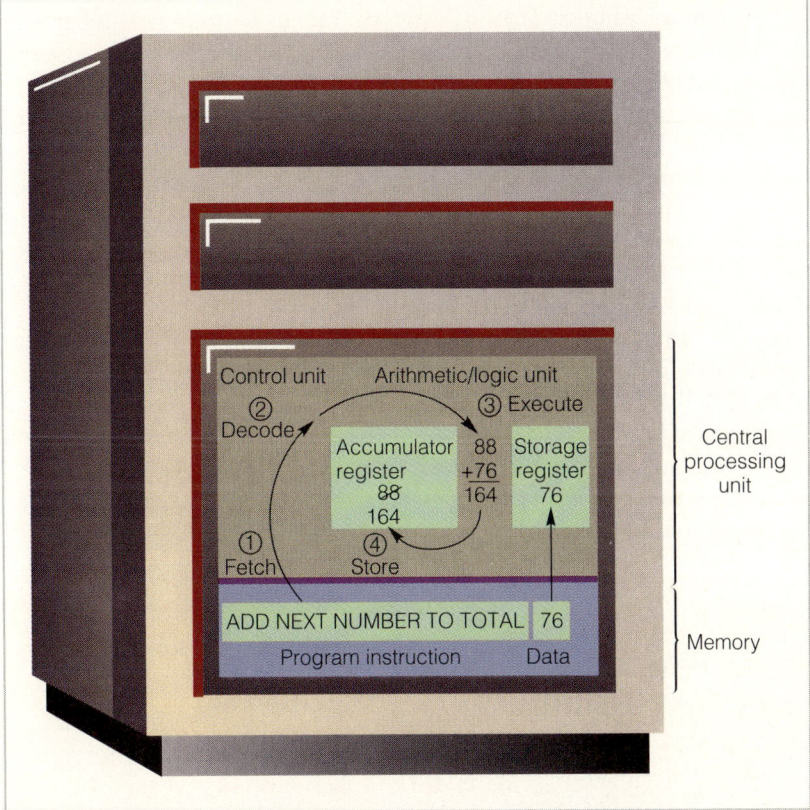

Figure 2-3 The machine cycle in action. Suppose a program must find the average of five test scores. To do this, the five scores must be totaled, then divided by 5. Begin by setting the total to 0; then add each of the five numbers, one at a time, to the total. Suppose the scores are 88, 76, 91, 83, and 87. In this figure the total has been set to 0; then 88, the first test score, has already been added to it. Now we will examine the machine cycle as it adds the next number, 76, to the total. The instruction to do so adds the next number to the total. Follow the four steps in the machine cycle. ① *Fetch:* The control unit fetches the instruction from memory. ② *Decode:* The control unit decodes the instruction. It determines that addition must take place and gives instructions for the next number (76) to be placed in a storage register for this purpose. The total so far (88) is already in an accumulator register. ③ *Execute:* The ALU does the addition, increasing the total to 164. ④ *Store:* In this case the ALU stores the new total in the accumulator register instead of memory, since more numbers still need to be added to it. When the new total (164) is placed in the accumulator register, it displaces the old total (88).

processing unit is designed to understand a specific group of instructions called the **instruction set.** Just as there are many different languages that people understand, so each different type of CPU has an instruction set it understands. Therefore, one CPU—such as the one for a Compaq personal computer—cannot understand the instruction set from another CPU—say, for a Macintosh.

Storage Locations and Addresses: How the Control Unit Finds Instructions and Data

It is one thing to have instructions and data somewhere in memory and quite another for the control unit to be able to find them. How does it do this?

The location in memory for each instruction and each piece of data is identified by an **address.** That is, each location has an address number, like the mailboxes in front of an apartment house. And, like the mailboxes, the address numbers of the locations remain the same, but the contents (instructions and data) of the locations may change. That is, new instructions or new data may be placed in the locations when the old contents no longer need to be stored in memory. Unlike a mailbox, however, a memory location can hold only a fixed amount of data; an address can hold only one number or one word.

Figure 2-4 shows how a program manipulates data in memory. A payroll program, for example, may give instructions to put the rate of pay in location 3 and the number of hours worked in location 6. To compute the employee's salary, then, instructions tell the computer to multiply the data in location 3 by the data in location 6 and move the result to location 8. The choice of locations is arbitrary—any locations that are not already spoken for can be used. Programmers using programming languages, however, do not have to worry about the actual address numbers, because each data

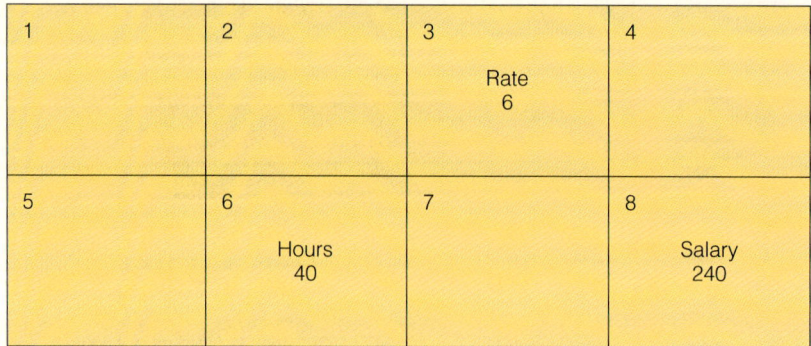

Figure 2-4 Addresses like mailboxes. The addresses of memory locations are like the identifying numbers on apartment-house mailboxes. Suppose we want to compute someone's salary as the number of hours multiplied by the rate of pay. Rate ($6) goes in memory location 3, hours (40) in location 6, and the computed salary ($6 x 40 hours, or $240) in location 8. Thus, *addresses* are 3, 6, and 8, but *contents* are $6, 40 hours, and $240. Note that the program instructions are to multiply the contents of location 3 by the contents of location 6 and move the result to location 8. (A computer language used by a programmer would use some kind of symbolic name for each location, such as Rate or Pay-Rate instead of the number 3.) The data items are the actual contents—what is stored in each location.

address is referred to by a name. The name is called a **symbolic address**. In this example, the symbolic address names are Rate, Hours, and Salary.

 ## Data Representation: On/Off

We are accustomed to thinking of computers as complex mechanisms, but the fact is that these machines basically know only two things: on and off. This two-state on/off system is called a **binary system**. Using the two states—which can be represented by electricity turned on or off—the computer can construct sophisticated ways of representing data.

Let us look at one way the two states can be used to represent data. Whereas the decimal number system has a base of 10 (with the ten digits 0, 1, 2, 3, 4, 5, 6, 7, 8, and 9), the binary system has a base of 2. This means it contains only two digits, 0 and 1, which correspond to the two states off and on. Combinations of 0s and 1s represent larger numbers (Figure 2-5).

Bits, Bytes, and Words

Each 0 or 1 in the binary system is called a **bit** (for *bi*nary digi*t*). The bit is the basic unit for storing data in computer memory—0 means off, 1 means on. Notice that since a bit is always either on or off, a bit in computer memory is always storing some kind of data.

Since single bits by themselves cannot store all the numbers, letters, and special characters (such as $ and ?) that a computer must process, the bits are put together in a group called a **byte** (pronounced "bite"). There are usually 8 bits in a byte (Figure 2-6). Each byte usually represents one **character** of data—a letter, digit, or special character.

Computer manufacturers express the capacity of memory and storage in terms of the number of bytes it can hold. The number of bytes can be expressed as **kilobytes**. *Kilo* represents 2 to the tenth power (2^{10}), or 1024. *Kilobyte* is abbreviated **KB**, or simply **K**. (Sometimes K is used casually to mean 1000, as in "I earned $30K last year.") A kilobyte is 1024 bytes. Thus, the memory of a 640K computer can store 640x1024, or 655,360 bytes. Memory capacity may also be expressed in terms of **megabytes** (1024x1024 bytes). One megabyte, abbreviated **MB**, means, roughly, one million bytes. With storage devices, manufacturers sometimes express memory amounts in terms of **gigabytes** (abbreviated **GB**)—billions of bytes. Memory in older personal computers may hold only 640K bytes; in newer machines, memory may hold anywhere from 1MB to 32MB and more. Mainframe memories can hold gigabytes.

BINARY EQUIVALENT OF DECIMAL NUMBERS 0–15	
Decimal	**Binary**
0	0000
1	0001
2	0010
3	0011
4	0100
5	0101
6	0110
7	0111
8	1000
9	1001
10	1010
11	1011
12	1100
13	1101
14	1110
15	1111

Figure 2-5 Decimal and binary equivalents. Seeing numbers from different systems side by side clarifies the patterns of progression.

Figure 2-6 Bit as light bulb. In this illustration a light bulb operates as a binary digit (bit), with off representing 0 and on representing 1. The group of eight bulbs, each of which can be on or off, represents 1 byte. Light bulbs, of course, are not used in computers, but anything that can conduct an electrical signal can be used.

A computer **word**, typically the size of a register, is defined as the number of bits that constitute a common unit of data, as defined by the computer system. The length of a word varies by computer. Generally, the larger the word, the more powerful the computer. There was a time when word size alone could classify a computer. Common word lengths are 8 bits (for very early personal computers), 16 bits (for traditional minicomputers and some personal computers), 32 bits (for full-size mainframe computers, some minicomputers, and some personal computers), and 64 bits (traditionally supercomputers). A recent microprocessor, Intel's Pentium, intended for both personal computers and larger machines, is a 64-bit chip. As you can see, the old stereotypes no longer fit very well. Note that an 8-bit machine could handle only one byte (character) at a time, whereas a 64-bit machine handles 8 bytes at a time, making its processing speed eight times faster.

Coding Schemes

As we said, a byte—a collection of bits—represents a character of data. But just what particular set of bits is equivalent to which character? In theory we could each make up our own definitions, declaring certain bit patterns to represent certain characters. Needless to say, this would be about as practical as each person speaking his or her own special language. Since we need to communicate with the computer and with each other, it is appropriate that we use a common scheme for data representation. That is, there must be agreement on which groups of bits represent which characters.

The code called **ASCII** (pronounced "AS-key"), which stands for American Standard Code for Information Interchange, uses 7 bits for each character. Since there are exactly 128 unique combinations of 7 bits, this 7-bit code can represent only characters. A more common version is ASCII-8, also called extended ASCII, which uses 8 bits per character and can represent 256 different characters. For example, the letter *A* is represented by 01000001. The ASCII representation has been adopted as a standard by the U.S. government and is found in a variety of computers, particularly minicomputers and microcomputers. Figure 2-7 shows part of the ASCII-8 code.

▶ Personal Computer Chips

The chips discussed here would be attached to the **motherboard,** the flat board within the personal computer housing. As we shall see, memory chips also may be attached to boards that can be inserted later to provide supplementary memory.

Microprocessors

A miniaturized central processing unit can be etched on a chip, a tiny square of silicon, hence the term *computer on a chip.* A central processing unit, or processor, on a chip is a **microprocessor** (Figure 2-8), often called a **logic chip** when it is used to control specialized devices (such as the fuel system of a car). Over the years the architecture of microprocessors has become somewhat standardized. Microprocessors usually include these key components: a control unit and an arithmetic/logic unit (the central

Character	ASCII–8
A	0100 0001
B	0100 0010
C	0100 0011
D	0100 0100
E	0100 0101
F	0100 0110
G	0100 0111
H	0100 1000
I	0100 1001
J	0100 1010
K	0100 1011
L	0100 1100
M	0100 1101
N	0100 1110
O	0100 1111
P	0101 0000
Q	0101 0001
R	0101 0010
S	0101 0011
T	0101 0100
U	0101 0101
V	0101 0110
W	0101 0111
X	0101 1000
Y	0101 1001
Z	0101 1010
0	0011 0000
1	0011 0001
2	0011 0010
3	0011 0011
4	0011 0100
5	0011 0101
6	0011 0110
7	0011 0111
8	0011 1000
9	0011 1001

(a)

Letter	ASCII–8
K	0100 1011
I	0100 1001
L	0100 1100
O	0100 1111
B	0100 0010
Y	0101 1001
T	0101 0100
E	0100 0101

(b)

Figure 2-7 The ASCII-8 code. (a) Shown are the ASCII-8 binary representations for letters and digits. This is not the complete code; there are many characters missing, such as lowercase letters and punctuation marks. The binary representation is in two columns to improve readability. (b) ASCII-8 representation for the word *KILOBYTE.*

processing unit), registers, and a clock. (Clocks are often on a separate chip in personal computers.) Notably missing is memory, which usually comes on its own chips.

How much smaller? How much cheaper? How much faster? Three decades of extraordinary advances in technology have packed increasingly greater power onto increasingly smaller chips. Engineers can now imprint as much circuitry on a single chip as filled room-size computers in the early days of computing. But are we approaching the limits of smallness? Current development efforts focus on a three-dimensional chip built in layers. Chip capacities in the future do seem almost limitless.

Memory Components

Earlier in the chapter we talked about memory and how it interfaces with the central processing unit. Now we will examine the memory components. Historically, memory components have evolved from primitive vacuum tubes to today's modern semiconductors. (For more information on the development of these components, see Appendix B, *History and Industry*.)

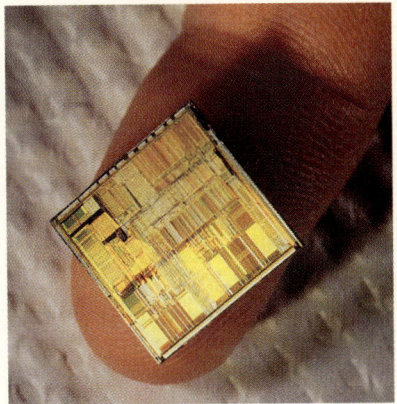

Figure 2-8 A microprocessor chip. This Pentium chip fits nicely on a fingertip.

Semiconductor Storage

Most modern computers use semiconductor storage because it has several advantages: reliability, compactness, low cost, and lower power usage. Since semiconductor memory can be mass-produced economically, the cost of memory has been considerably reduced. Chip prices have fallen and risen and fallen again, based on a variety of economic and political factors, but they remain a bargain. Semiconductor storage has one major disadvantage: It is **volatile**—that is, semiconductor storage requires continuous electric current to represent data. If the current is interrupted, the data is lost.

Semiconductor storage is made up of thousands of very small circuits—pathways for electric currents—on a silicon chip. A chip is described as **monolithic** because the circuits on a single chip compose an inseparable unit of storage. Each circuit etched on a chip can be in one of two states: either conducting an electric current or not—on or off. The two states can be used to represent the binary digits 1 and 0. As we noted earlier, these digits can be combined to represent characters, thus making the memory chip a storage bin for data and instructions.

One important type of semiconductor design is called **complementary metal oxide semiconductor (CMOS)**. This design is noted for using little electricity. This makes it especially useful for computers requiring low power consumption, such as portable computers.

RAM and ROM

Random-access memory (RAM) keeps the instructions and data for whatever programs you happen to be using at the moment. The data can be accessed in an easy and speedy manner. RAM is usually volatile; as noted above, this means that its contents are lost once the power is shut off. RAM can be erased or written over at will by the computer software.

The more RAM in your computer, the larger the programs you can run. In recent years the amount of RAM storage in a personal computer has increased dramatically. An early personal computer, for example, was advertised with "a full 4K RAM." Now 8MB or even 16MB RAM is com-

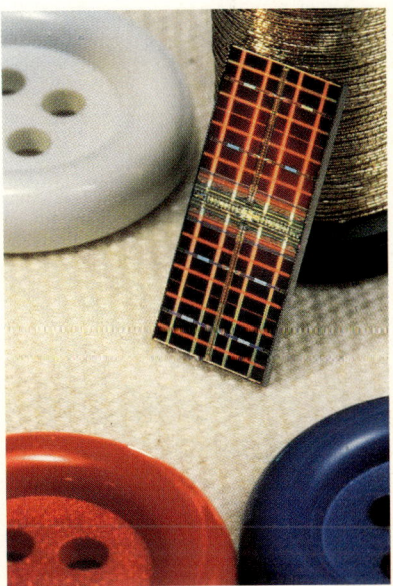

Figure 2-9 DRAM chip. This DRAM chip is smaller than a button.

ONE JUMP AHEAD

Body Chips

In the next decade we can have miniature computers inside us to monitor, and even regulate, our blood pressure, heart rate, and cholesterol. Such a chip would include a microprocessor, sensors, and a wireless radio frequency device that would permit accurate readouts of vital statistics. All this would happen, of course, without drawing any blood or attaching any external devices to the body.

Since we are already familiar with the notion of an internal pacemaker for the heart, including a chip or two may not seem all that astonishing. But this is just the beginning. Experts who specialize in the miniaturization of microprocessors foresee, within 20 years, implanted chips that can correct our ability to interact with the world. Imagine a chip that corrects hearing loss. Once implanted, the chip is invisible, unlike a hearing aid. A more common implant would be a chip to correct visual signals. No more glasses!

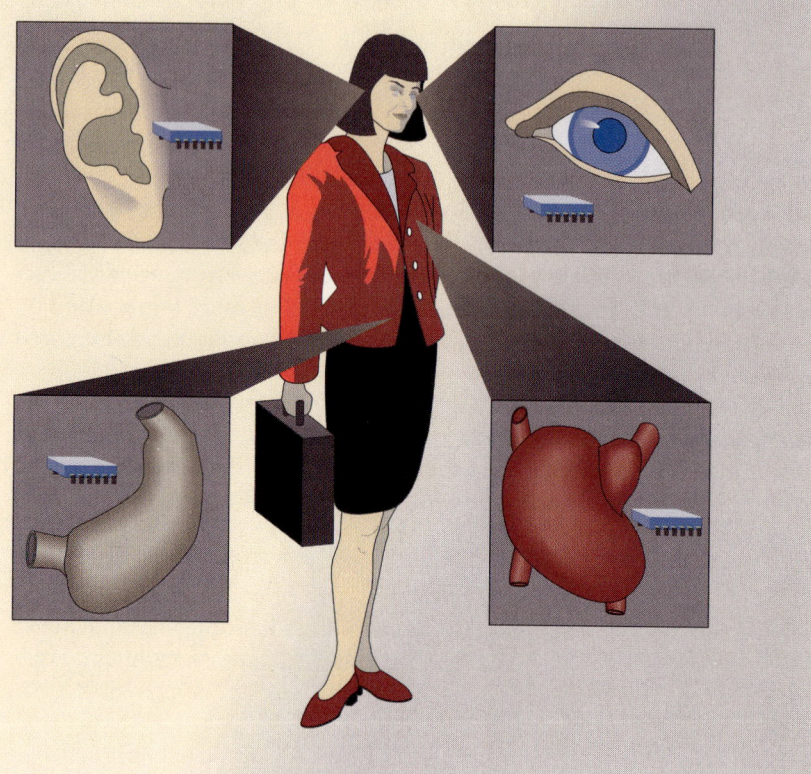

mon. More memory has become a necessity because sophisticated personal computer software requires significant amounts of memory. You can augment your personal computer's RAM by buying extra memory chips to install in your memory board or by purchasing a **single in-line memory module (SIMM)**, a board that contains memory chips. The SIMM board plugs into the computer's main circuit board, which is more convenient than attaching individual chips. In general, the more memory your computer has, the more (and bigger) tasks the computer can do.

RAM is often divided into two types: static RAM (**SRAM**) and dynamic RAM (**DRAM**). DRAM must be constantly refreshed (recharged) by the central processing unit or it will lose its contents—hence the name dynamic. Although SRAM is faster, DRAM is used in most personal computer memory because of its size and cost advantages (Figure 2-9).

Read-only memory (ROM) contains programs and data that are permanently recorded into this type of memory at the factory; they can be read and used, but they cannot be changed by the user. For example, a personal computer probably has a program for calculating square roots in ROM. ROM is nonvolatile—its contents do not disappear when the power is turned off.

Using specialized tools called **ROM burners,** the instructions within some ROM chips can be changed. These chips are known as **PROM** chips, or **programmable read-only memory** chips. There are other variations on ROM chips, depending on the methods used to alter them. The business of programming and altering ROM chips is the province of the computer engineer.

 ## Speed and Power

The characteristic of speed is universally associated with computers. Power is a derivative of speed, as well as other factors such as memory size. What makes a computer fast? Or, more to the point, what makes one computer faster than another? Several factors are involved, including microprocessor speed, bus line size, and the availability of cache. A user who is concerned about speed will want to address all of these. More sophisticated approaches to speed include flash memory, RISC computers, and parallel processing. We will discuss each approach in turn.

Computer Processing Speeds

Although all computers are fast, there is a wide diversity of computer speeds. The execution of an instruction on a very slow computer may be measured in less than a **millisecond**, which is one-thousandth of a second. Most computers can execute an instruction measured in **microseconds**, one-millionth of a second. Some modern computers have reached the **nanosecond** range—one-billionth of a second. Still to be broken is the **picosecond** barrier—one-trillionth of a second.

Microprocessor speeds are usually expressed in **megahertz (MHz)**, millions of machine cycles per second. Thus, a personal computer listed at 25MHz has a processor capable of handling 25 million machine cycles per second. A top-speed personal computer will be many times faster.

Another measure of computer speed is **MIPS**, which stands for one *million instructions per second*. MIPS is often a more accurate measure than clock speed, because some computers can use each tick of the clock more efficiently than others. A third measure of speed is the **megaflop**, which stands for one *million floating-point operations per second*. It measures the ability of the computer to perform complex mathematical operations.

Bus Lines

As is so often the case, the computer term *bus* is borrowed from its common meaning—a mode of transportation. A **bus line** is a set of parallel electrical paths that internally transports data from one place to another within the computer system. The amount of data that can be carried at one time is called the bus width—the number of electrical paths. The greater the width, the more data can be carried at a time.

In general, the larger the word size or bus, the more powerful the computer. A larger bus size means

- The computer can transfer more data at a time, making the computer faster.
- The computer can reference larger numbers, allowing more memory.
- The computer can support a greater number and variety of instructions.

Microprocessors are sometimes obscurely affixed with notations that indicate their bus size. For example, a 486DX chip has a bus width of 32 bits, whereas a 486SX chip uses a 32-bit bus within the processor but only

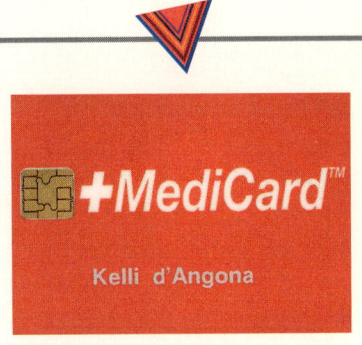

A Chip to Save Your Life

If your friend suddenly had an accident and was unconscious or incoherent, could you provide any information to an ambulance crew? Would you know her blood type, her allergies, the prescription drugs she takes? Probably not. Even family members may not have this information, or may be too distraught themselves to provide needed medical information.

Enter the MediCard, a plastic card that has an embedded chip containing all that patient information. Small computers that can read the cards are installed in ambulances and in hospital emergency rooms. This system is working successfully in some communities. The biggest hitch is making sure that people carry their cards in their wallets at all times.

How Fast Is a Nanosecond?

If one nanosecond is…	Then one second is equivalent to…
One mile	2000 trips to the moon and back
One person	Population of China and the U.S.
One minute	1900 years
One square mile	17 times the land area of the world

a 16-bit bus between the processor and memory. A buyer who cares about speed would prefer the DX chip, which carries exactly twice as much data and therefore is twice as fast as the SX chip.

Cache

A **cache** (pronounced "cash") is a relatively small amount of very fast memory designed for the specific purpose of speeding up internal transfer of data and software instructions. Think of cache as a selective memory: The data and instructions stored in cache are those that are most recently and/or most frequently used. When the processor first requests data or instructions, these must be retrieved from main memory, which is delivered at a pace that is relatively slow compared to the microprocessor. As they are retrieved, those same data/instructions are stored in cache. The next time the microprocessor needs data or instructions, it looks first in cache; if the needed items can be found there, they can be transferred at a rate that far exceeds a trip from main memory. Of course, cache is not big enough to hold everything, so the wanted data or instructions may not be there. But there is a good chance that frequently used items will be in cache. That is, since the most frequently used data and instructions are kept in a handy place, the net result is an improvement in processing speed.

Just how much cache speeds performance depends on a number of factors, including the size of the cache, the speed of the memory chips in the cache, and the software being run. Caching has become such a vital technique that some of the newer microprocessors have cache built into the processor's design.

Flash Memory

We have stated that memory is volatile—that it disappears when the power is turned off—hence the need for secondary storage to keep data on a more permanent basis. A long-standing speed problem has been the rate of accessing data from a secondary storage device such as a disk, a rate significantly slower than internal computer speeds. It seemed unimaginable that data might someday be stored on nonvolatile memory chips—nonvolatile RAM—close at hand. A breakthrough has emerged in the form of nonvolatile **flash memory**. Flash chips are currently being used in cellular phones and cockpit flight recorders, and they are replacing disks in some handheld computers.

Flash memory chips are being produced in credit-cardlike packages, which are smaller than a disk drive and require only half the power; that is why they are being used in notebook computers and the handheld personal digital assistants. Manufacturers predict that a 100-megabyte flash card will soon sell at the same price as a same-size magnetic disk drive.

Although flash memory is not yet commonplace, it seems likely that it will become a mainstream component. Since data and instructions will be ever-closer to the microprocessor, conversion to flash memory chips would have a pivotal impact on a computer's processing speed.

RISC Technology: Less Is More

It flies in the face of computer tradition: Instead of reaching for more variety, more power, more everything-for-everyone, proponents of **RISCs**—**reduced instruction set computers**—suggest that we could get by with a little less. In fact, reduced instruction set computers offer only a small subset of instructions; the absence of bells and whistles increases speed. So we have a radical back-to-basics movement in computer design.

RISC supporters say that, on conventional computers (called **CISCs,** or **complex instruction set computers**), a hefty chunk of built-in instructions—the instruction set—is rarely used. Those instructions, they note, are underused, inefficient, and impediments to performance. RISC computers, with their stripped-down instruction sets, zip through programs like racing cars—at speeds four to ten times those of CISC computers. This is heady stuff for the merchants of speed who want to attract customers by offering more speed for the money.

Parallel Processing

The ultimate speed solution is **parallel processing,** a method of using several processors at the same time (Figure 2-10). Consider the description of computer processing you have seen so far in this chapter: The processor gets an instruction from memory, acts on it, returns processed data to memory, and then repeats the process. This is conventional **serial processing.**

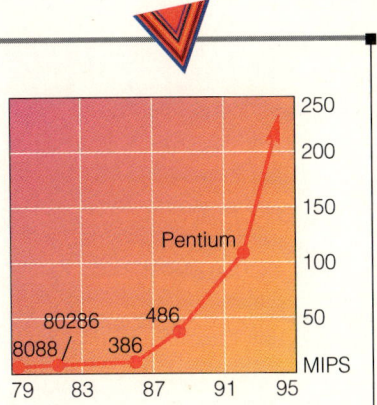

A Family of Chips

The Intel Corporation has provided personal computer makers with several generations of microprocessor chips. The first was a standard-setter: the 8088 chip used by the first IBM PC (introduced in 1981) and its many imitators. The next member of the family, the 80186 chip, was merely a transitional chip, soon replaced by the 80286 chip, which powered the IBM PC AT and, again, a slew of clones.

Intel moved to increase power and flexibility with the introduction of the 80386 chip, first brought to the market in the Compaq 386. Close on the heels of the 80386 chip was the 80486 ("the 486"), a chip whose speed and power made it popular through the early 1990s.

The next chip, trotted out in 1993, was expected to be christened the 80586. However, citing proprietary problems, Intel called it Pentium, based on the Latin root word meaning *five.* The amazing Pentium, nicknamed P5 by techies, is twice as fast as the fastest 486 chip. But the most stunning news is its word size: 64 bits.

Each new Intel chip has been increasingly powerful. In terms of MIPS, you can see from the chart that the 486 processes almost 50 million instructions per second, compared to the Pentium at about 100 million. The newest Intel chip, code named P6, will process well over 250 million instructions per second.

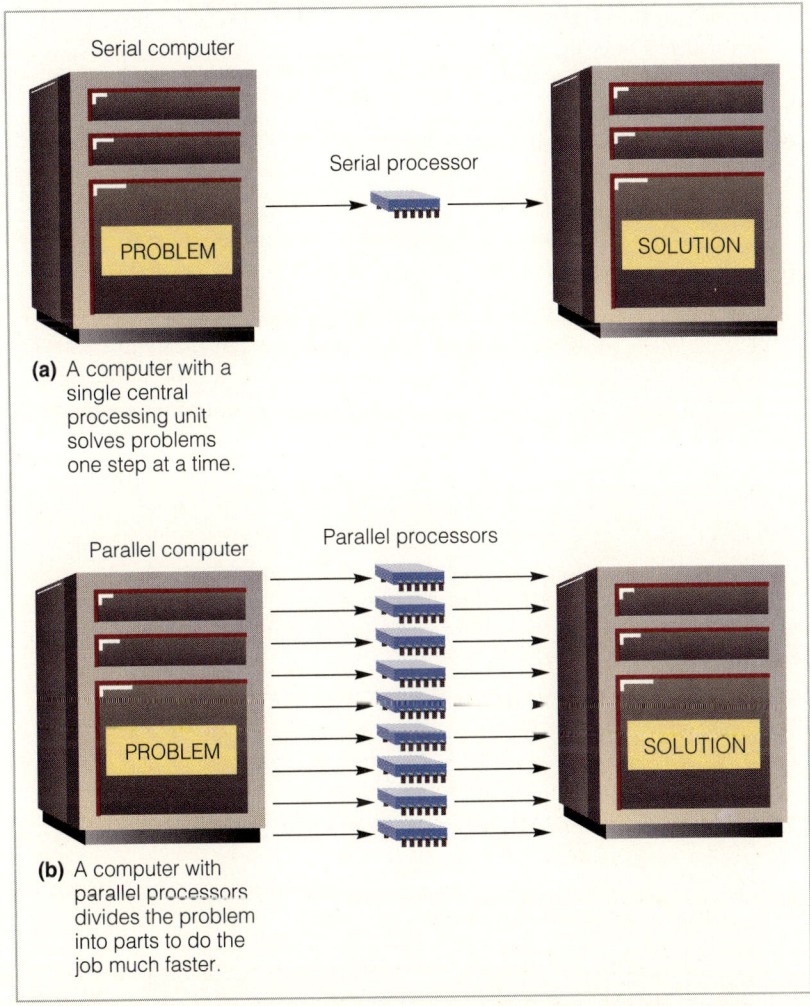

Figure 2-10 Serial vs. parallel processing. (a) Serial processing involves one processor working with one piece of data at a time. (b) Parallel processing involves a number of processors working in concert.

The problem with the conventional computer is that the single electronic pathway, the bus line, acts like a bottleneck. The computer has a one-track mind because it is restricted to handling one piece of data at a time. For many applications, such as simulating the air flow around an entire airplane in flight, this is an exceedingly inefficient procedure. A better solution? Many processors, each with its own memory unit, working at the same time: parallel processing.

A number of parallel processors are being built and sold commercially. However, do not look for parallel processing in personal computers just yet. Thus far, this technology is limited to larger computers.

▼ ▲ ▼

The future holds some exciting possibilities for computer chips. New speed breakthroughs certainly will continue. One day we may see computers that operate using light (photonics) rather than electricity (electronics) to control their operation. Light travels faster and is less likely to be disrupted by electrical interference. Also, light beams can pass through each other, alleviating some of the problems that occur in the design of electronic components, in which wires should not cross. And would you believe computers that are actually grown as biological cultures? So-called biochips may replace today's silicon chip. As research continues, so will the surprises.

Whatever the design and processing strategy of a computer, its goal is the same: to turn raw input into useful output. Input and output are the topics of the next chapter.

CHAPTER REVIEW

Summary and Key Terms

- The **central processing unit (CPU)** is an extensive, complex set of electronic circuitry that executes program instructions. It consists of two parts: a control unit and an arithmetic/logic unit.
- The central processing unit interacts closely with **primary storage,** or **memory.** Memory provides temporary storage of data while the computer is executing the program. **Secondary storage** holds the data that is permanent or semipermanent.
- The **control unit** of the central processing unit coordinates execution of the program instructions by communicating with the arithmetic/logic unit and memory—the parts of the system that actually execute the program.
- The **arithmetic/logic unit (ALU)** contains circuitry that executes the arithmetic and logical operations. The unit can perform four **arithmetic operations:** addition, subtraction, multiplication, and division. Its **logical operations** are usually making comparisons that test for three conditions: the **equal-to condition,** the **less-than condition,** and the **greater-than condition.** The computer can test for more than one condition at once, so it can discern three other conditions as well: less than or equal to, greater than or equal to, and less than or greater than.
- Symbols called **relational operators** (=, <, >) allow you to define the comparison you want the computer to perform.
- **Registers,** areas for temporary storage, quickly accept, hold, and transfer instructions or data. A register might be an **accumulator,** which collects the results of computations. An **address register** keeps track of where data is stored in memory. A **storage register** temporarily holds data taken from or about to be sent to memory. A **general-purpose register** is used for several functions.
- Registers hold data that will be processed *immediately*, and memory stores the data that will be used for operations in the *near future*. Secondary storage holds data that may be needed for operations *later.*
- **Memory** is the part of the computer that temporarily holds data and instructions before and after they are processed by the arithmetic/logic unit. Memory is also known as **primary storage, primary memory, main storage, internal storage,** and **main memory.**
- The chief characteristic of memory is that it allows very fast access to instructions and data, no matter where the items are within it.
- Most computers today can execute only one instruction at a time, though they execute it very quickly. Even a personal computer may execute instructions in less than one-*millionth* of a second, and supercomputers can execute instructions in less than one-*billionth* of a second.
- The central processing unit follows four main steps when executing an instruction: It (1) gets the instruction from memory, (2) decodes the instruction and gives instructions for the transfer of appropriate data from memory to the ALU, (3) directs the ALU to perform the actual operation on the data, and (4) directs the result of the operation to be stored in memory or a register. The first two steps are called **I-time** (instruction time), and the last two steps are called **E-time** (execution time).
- A **machine cycle** is the combination of I-time and E-time. The internal **clock** of the central processing unit produces pulses at a fixed rate to synchronize computer operations. A machine-cycle instruction may include many subinstructions, each of which must take at least one clock cycle. Each central processing unit has a set of commands it can understand. This group is called an **instruction set.**

- The location in memory for each instruction and each piece of data is identified by an **address.** Address numbers remain the same, but the contents of the locations change. The meaningful name for a memory address is called a **symbolic address.**
- Since a computer can recognize only whether electricity is on or off, data is represented by an on/off **binary system.** Two digits, 0 and 1, correspond to off and on. Combinations of 0s and 1s represent numbers, letters, or special characters.
- Each 0 or 1 in the binary system is called a **bit** (binary digit). A group of bits (usually 8 bits) is called a **byte.** Each byte usually represents one **character** of data, such as a letter, digit, or special character. Memory capacity is expressed in **kilobytes** (**KB** or **K**). One kilobyte equals 1024 bytes. A **megabyte** (**MB**), equals about one million bytes, and a **gigabyte** (**GB**), equals about one billion bytes.
- A computer **word** is the number of bits that make up a unit of data, as defined by the computer system. Common word lengths are from 8 bits to 64 bits.
- A common coding scheme for representing characters is **ASCII** (American Standard Code for Information Interchange), which uses 7-bit characters. A variation of the code, called ASCII-8, uses 8 bits per character.
- Many personal computer chips are attached to the **motherboard,** the flat board within the personal computer housing. Memory chips also may be attached to boards that can be inserted later to provide supplementary memory.
- A central processing unit, or processor, on a chip is a **microprocessor,** often called a **logic chip** when it is used to control specialized devices.
- **Semiconductor storage,** thousands of very small circuits on a silicon chip, is **volatile.** A chip is described as **monolithic** because the circuits on a single chip compose an inseparable unit of storage.
- Most modern computers use semiconductor storage because it has the advantages of reliability, compactness, low cost, and lower power usage.
- An important type of semiconductor design is **called complementary metal oxide semiconductor** (**CMOS**). This design, which uses little electricity, is especially useful for portable computers.
- **Random-access memory** (**RAM**) keeps the instructions and data for whatever programs you happen to be using at the moment. RAM is usually volatile.
- The amount of RAM storage in a personal computer has increased dramatically because larger RAM means, among other things, the ability to run larger the programs.
- A **single in-line memory module** (**SIMM**) is a plug-in board that contains memory chips.
- RAM is often divided into two types: static RAM (**SRAM**), which is faster, and dynamic RAM (**DRAM**), which is smaller and less expensive.
- **Read-only memory** (**ROM**) contains programs and data that are permanently recorded into this type of memory at the factory; they can be read and used, but they cannot be changed by the user. ROM is nonvolatile. The instructions within some ROM chips can be changed using **ROM burners;** these chips are known as **PROM** chips, or **programmable read-only memory** chips.
- Computer instruction speeds fall into various ranges, from a **millisecond,** which is one-thousandth of a second; to a **microsecond,** one-millionth of a second; to a **nanosecond,** one-billionth of a second. Still to be achieved is the **picosecond** range—one-trillionth of a second.
- Basic factors affecting computer speed are microprocessor speed, bus line size, and the availability of cache. More sophisticated factors are flash memory, RISC computers, and parallel processing.
- Microprocessor speeds are usually expressed in **megahertz** (**MHz**), millions of machine cycles per second. Another measure of computer speed is **MIPS,** which stands for one million instructions per second. MIPS is often a more accurate measure than clock speed, because some computers use each tick more efficiently than others. A third measure is the **megaflop,** which stands for one million floating-point operations per second.

- A **bus line** is an electrical path that transports data from one place to another internally within the computer system. The amount of data that can be carried at one time is called the bus width. In general, a larger word length means a more powerful computer—the computer can transfer more information at one time, can have a larger memory, and can support a greater number and variety of instructions.
- A **cache** is a relatively small amount of very fast memory that stores data and instructions that are used frequently, resulting in an improved processing speed.
- When the processor first requests data or instructions, they are retrieved from main memory, they are also stored in cache. The next time the microprocessor needs data or instructions, it looks first in cache; if the needed items can be found there, they can be transferred at a faster rate than a similar transfer from main memory.
- The emerging technology of **flash memory** will provide memory chips that are nonvolatile.
- **RISCs—reduced instruction set computers—** are fast because they use only a small subset of instructions. Conventional computers, called **CISCs**, or **complex instruction set computers**, include many instructions that are rarely used.
- Conventional **serial processing** uses a single processor and can handle just one instruction at a time. **Parallel processing** uses several processors in the same computer at the same time.
- In a conventional computer the single electronic pathway, the bus line, acts like a bottleneck; thus the computer is restricted to handling one piece of data at a time. For many applications, such as simulating the air flow around an entire airplane in flight, this is an exceedingly inefficient procedure. A better solution is parallel processing, many processors, each with its own memory unit, working at the same time.

Discussion Questions

1. Why is writing instructions for a computer more difficult than writing instructions for a person?
2. Do you think there is a continuing need to increase computer speed? Can you think of examples in which more speed would be desirable?
3. It will soon be possible to have chips implanted in our bodies to monitor or improve our physical conditions. Do you think this is a good or bad idea? Would you personally consider such a thing?

Student Study Guide

Multiple Choice

1. The complex set of electrical circuitry that executes program instructions is called the
   a. register
   b. accumulator
   c. central processing unit
   d. bus line
2. The entire computer system is coordinated by
   a. the ALU
   b. the control unit
   c. the accumulator
   d. arithmetic operators
3. A bus line consists of
   a. registers
   b. parallel data paths
   c. accumulators
   d. machine cycles
4. Equal to, less than, and greater than are examples of
   a. logical operations
   b. subtraction
   c. locations
   d. arithmetic operations
5. The primary storage unit is also known as
   a. storage registers
   b. mass storage
   c. accumulators
   d. memory
6. Data and instructions are put into primary storage by
   a. memory
   b. secondary storage
   c. the control unit
   d. the ALU
7. Registers that collect the results of computations are
   a. general-purpose
   b. storage registers
   c. main storage
   d. accumulators
8. During E-time the ALU
   a. examines the instruction
   b. executes the instruction
   c. enters the instruction
   d. elicits the instruction
9. When the control unit gets an instruction it is called
   a. E-time
   b. I-time
   c. machine time
   d. ALU time
10. When the control unit directs the ALU to perform an operation on the data, the machine cycle is involved in its
    a. first step
    b. second step
    c. third step
    d. fourth step
11. Computer operations are synchronized by
    a. the CPU clock
    b. the binary system
    c. megabytes
    d. E-time

12. Another name for primary storage is
    a. secondary storage   c. ROM
    b. binary system   d. main storage
13. Which is *not* another name for memory?
    a. primary storage   c. main storage
    b. internal storage   d. secondary storage
14. Another name for a logic chip is
    a. PROM   c. memory
    b. microprocessor   d. ROM
15. Data is represented on a computer by means of a two-state on/off system called
    a. a word   c. the binary system
    b. a byte   d. RAM
16. A letter, number, or special character is represented by a
    a. bit   c. kilobyte
    b. byte   d. megabyte
17. Memory capacity may be expressed in
    a. microseconds   c. kilobytes
    b. bits   d. cycles
18. Which is *not* a kind of register?
    a. storage   c. address
    b. accumulator   d. variable
19. A type of computer that is faster because it has fewer instructions:
    a. symbolic   c. RISC
    b. ASCII-8   d. ROM burner
20. An emerging technology that provides nonvolatile memory chips is
    a. flash memory   c. PROM
    b. CMOS   d. CISC
21. The data coding scheme that is the American standard is
    a. ASCII   c. KB
    b. SIMM   d. gigabyte
22. Tools to change PROM chips are called
    a. chip kits   c. RAM burners
    b. PROM burners   d. none of these
23. An approach to increase speed is
    a. CISC   c. parallel processing
    b. serial processing   d. CMOS
24. The shortest period of time is a
    a. millisecond   c. nanosecond
    b. picosecond   d. microsecond
25. An organic chip is called a
    a. storage chip   c. biochip
    b. microchip   d. silicon chip

True/False

T F 1. The control unit consists of the CPU and the ALU.
T F 2. Secondary storage holds data only temporarily.
T F 3. The control unit directs and coordinates the entire computer system in executing stored program instructions with electrical signals.
T F 4. A personal computer has a 16-bit word and 64 parallel data bus lines.
T F 5. The electronic circuitry that controls all arithmetic and logical operations is contained in the ALU.
T F 6. The three basic logical operations may be combined to form a total of nine commonly used operations.
T F 7. Memory allows very fast access to instructions in secondary storage.
T F 8. Registers are temporary storage areas located in memory.
T F 9. Address registers hold the addresses of locations containing data needed for an instruction.
T F 10. RISC technology uses fewer instructions than traditional computers.
T F 11. All computers except personal computers can execute more than one instruction at a time.
T F 12. The machine cycle consists of four steps, from the first step of fetching the instructions to the last step of placing the result of the operation into memory.
T F 13. The internal clock of the CPU produces pulses at a fixed rate to synchronize all computer operations.
T F 14. Cache is a small amount of secondary storage.
T F 15. Computers represent all data using a two-state on/off system called the binary system.
T F 16. A bit is commonly made up of 8 bytes.
T F 17. A kilobyte (KB) is 1024 bytes.
T F 18. A computer word is defined as the number of bits constituting a common unit of information for a specific computer system.
T F 19. Another name for memory is secondary storage.
T F 20. Serial processing is replacing parallel processing.
T F 21. Addition and subtraction are logical operations.
T F 22. Primary storage is part of the central processing unit.
T F 23. Memory is usually volatile.
T F 24. A microsecond is briefer than a millisecond.
T F 25. The ASCII coding scheme is accepted as the American standard.

Fill-In

1. What is a millionth of a second called?

   _____

2. What is the unit that consists of both the control unit and the arithmetic/logic unit?

   _____

3. Name the four kinds of arithmetic operations performed by the ALU.

   a. _____

   b. _____

    c. _____
    d. _____
 4. List five other names for memory.
    a. _____
    b. _____
    c. _____
    d. _____
    e. _____
 5. MHz is an abbreviation for: _____
 6. What is the abbreviation for memory chips that can be altered? _____
 7. What is a combination of I-time and E-time called? _____
 8. The name for the symbols =, <, and > is: _____
 9. How is the location in memory of each instruction identified? _____
 10. What is each 0 or 1 in the binary system called? _____
 11. What does MIPS stand for? _____
 12. Name a disadvantage of semiconductor storage. _____
 13. A data path to transfer data: _____
 14. What does RAM stand for? _____
 15. The flat board to which chips are attached: _____
 16. Which type of storage stores programs that will not be altered? _____
 17. Which register holds information taken from or sent to memory? _____
 18. What are the comparing operations called that are controlled by the ALU? _____
 19. What are many processors working at the same time called? _____

 20. When the control unit decodes an instruction, is the machine cycle in I-time or E-time? _____
 21. What does RISC stand for? _____
 22. What does CMOS stand for? _____
 23. Processing instructions one at a time is called: _____
 24. The two digits of the binary system: _____
 25. DRAM stands for: _____
 26. List six factors affecting computer speed:
    a. _____
    b. _____
    c. _____
    d. _____
    e. _____
    f. _____
 27. What does CPU stand for?
 28. The hardware that has the circuitry to control the computer is called: _____
 29. A comparing operation such as *greater than* is what kind of operation? _____
 30. In additon to *equal, less than,* and *greater than,* there are three other logical relationships:
    a. _____
    b. _____
    c. _____
 31. The logical operation *less than or greater than* is the equivalent of what comparison: _____
 32. Which kind of register would collect the result of a computation? _____
 33. What does CISC stand for? _____
 34. Which kind of register can be used for multiple functions? _____
 35. Which kind of register keeps track of where data or instructions are stored in memory? _____

36. Mention three reasons why it is not feasible to keep in memory the instructions and data for a program not in current operation:

    a. _____

    b. _____

    c. _____

37. When, during I-time, the control unit gets an instruction from memory, it is said to be doing what?
    _____

38. Since the circuits of a microprocessor chip compose an inseparable unit of storage, the chip is said to be:
    _____

39. A relatively small amount of fast memory used for storing recently used or frequently used instructions is called: _____

40. The measure of speed that measures the ability of a computer to perform complex mathematical operations is:
    _____

4. a. main storage
   b. internal storage
   c. primary storage
   d. primary memory
   e. main memory
5. megahertz
6. PROM
7. a machine cycle
8. relational operators
9. an address
10. a bit
11. millions of instructions per second
12. it is volatile
13. bus line
14. random access memory
15. motherboard
16. ROM
17. storage register
18. logical operations
19. parallel processing
20. I-time
21. Reduced Instruction Set Computer
22. Complementary metal oxide semiconductor
23. serial processing
24. 0 and 1
25. dynamic RAM
26. a. microprocessor speed
    b. bus line
    c. cache
    d. flash memory
    e. RISC computers
    f. parallel processing
27. Central Processsing Unit
28. control unit
29. logical operation
30. a. less than or equal to
    b. greater than or equal to
    c. less than or greater than
31. not equal to
32. accumulator register
33. complex instruction set computers
34. general-purpose register
35. address register
36. a. memory is volatile
    b. on some computers, more than one program may run at the same time
    c. insufficient memory space
37. fetch
38. monolithic
39. cache
40. megaflop

Answers

Multiple Choice
1. c	6. c	11. a	16. b	21. a
2. b	7. d	12. d	17. c	22. d
3. b	8. b	13. d	18. d	23. c
4. a	9. b	14. b	19. c	24. b
5. d	10. c	15. c	20. a	25. c

True/False
1. F	6. F	11. F	16. F	21. F
2. F	7. F	12. T	17. T	22. F
3. T	8. F	13. T	18. T	23. T
4. F	9. T	14. F	19. F	24. T
5. T	10. T	15. T	20. F	25. T

Fill-In
1. microsecond
2. central processing unit
3. a. addition, b. subtraction
   c. multiplication, d. division

PLANET INTERNET

Getting Around

Is anyone in charge here? Not exactly. Unlike a commercial product, the Internet is not owned or managed by anyone. One consequence of this is that there is no master table of contents or index for the Internet. However, several organizations have produced ordered lists that can be used as a helpful starting place.

How do I start? We said we'd do it the easy way. Briefly, you need an URL for the Web. Translation: you need a starting address (*URL*, for *Uniform Resource Locator*) to search the Net by using the *World Wide Web*, also called *WWW* or just *"the Web."*

What is the Web? The Web is a way of searching the Net that lets you move from place to place. Each place on the Web is called a *site*. Each site has a *home page*, the first page you come to on the site. The key to the Web is that each site contains embedded *links* in the text to sites that have related information; those sites probably have further links to other sites, and so on. For example, a site discussing fine dining might have links to different cities, which in turn might have links to types of ethnic restaurants, which then link to individual restaurants. This simple dining example has an end, but many links go on and on, leading a user to unanticipated places on the Net. Links are readily identifiable within the site, probably a different text color and underlined (so it will show as a link when printed), or perhaps a graphics icon (small picture). You simply click your mouse on the link and you are transported to that site.

What does an URL look like? It's often pretty messy—a long string of letters and symbols. The good news is that once you get started, you can just click on a link rather than typing more URLs. Since everything on the Net, including URLs, is subject to change, in this text we supply only the Benjamin/Cummings Publishing Company (B/C) URL, which will not change. The B/C web site, in turn, supplies up-to-date links to all the sites we mention in these discussions. Once you link to a site, its URL will show on the screen, and you can save it so you can go there directly in the future if you like. A good starting site is The Whole Internet Catalog. It has links to major topics such as Entertainment, Business, Health, and Travel. Each major category has links of its own, as do the topics at the next level, and so on.

What other starting points are there? Well, of course, you can begin anywhere with an URL of your choosing. Users often favor Planet Access and Yahoo as comprehensive starting places. Special favorites are the Cool Site of the Day (including links to prior cool sites) and The Top 25 (the busiest 25 sites for the week).

But I need to get to the Web first. Yes. To access the Web you need special software called a *browser*. You need to ask your instructor, lab personnel, or designated employee how to use the Web browser at your location. If you are using a browser purchased for your personal computer, or have access to the Internet via some online service, then these suppliers will provide instructions.

Internet Exercises

In each chapter we offer two exercises. The first exercise is called a *structured exercise* because once you are on the B/C web site we can make sure that the sites you visit remain current. No wild goose chases. Use the URL supplied here to get started. If you are feeling adventurous, try the *freeform exercise*. We start you out but make no guarantees.

1. **Structured exercise.** Go to the B/C site using URL: http://www.aw.com/bc/planet/ Link to The Whole Internet Catalog, and then link to The Top 25. From there choose three Top 25 places to link to.
2. **Freeform exercise.** At the B/C site link to Netizens, a changeable list of individuals who maintain their own sites.

▲ Input, output, and a computer system were not what Paul Yen had in mind when he thumbed through the catalog from Lands' End, a mail-order firm in Wisconsin that offers quality casual and sports clothing. Paul simply wanted to order a few knit shirts and a leather belt. Whether he knew it or not, however, Paul started the computer action rolling when he wrote these items on the order form.

The items on his order form would become input data for the computer system. In this example the input is the data related to the customer order (customer name, address, and possibly charge-card number) as well as the data for each item (catalog number, quantity, description, and price). Keyed into the Lands' End computer system as soon as it arrives, this data is placed in customer and order files to be used with files containing inventory and other related data.

What Paul provided as input can be processed into a variety of outputs, as shown in Figure 3-1. Some outputs are for individual customers, and some show information combined from several orders: warehouse orders, shipping labels (to send the shirts and belt), backorder notices, inventory reports, supply reorder reports, charge-card reports, demographic reports (showing which merchandise sells best where), and so forth. And—to keep the whole process going—Lands' End also prints out Paul's name and address on a label for the next catalog.

Chapter 3

Input and Output
The User Connection

LEARNING OBJECTIVES

- Appreciate the user relationship with computer input and output
- Understand how data is input to a computer system, and differentiate among various input equipment
- Understand the benefits of the emerging technology of voice input
- Understand how a monitor works, and know the characteristics that determine quality
- Know the different methods of computer output
- Differentiate among the different kinds of printers, especially those for personal computers
- Appreciate the large variety of input and output options available in the marketplace

HOW USERS SEE INPUT AND OUTPUT

INPUT: GETTING DATA FROM THE USER TO THE COMPUTER

  Keyboard
  Mouse
  Trackball
  Source Data Automation: Collecting Data Where It Starts
  Magnetic-Ink Character Recognition
  Optical Recognition
  Data Collection Devices
  Voice Input
  Touch Screens
  Looking

OUTPUT: INFORMATION FOR THE USER

  Computer Screen Technology
  Types of Screens
  Terminals
  Printers
  Computer Output Microfilm
  Voice Output
  Music Output

COMPUTER GRAPHICS

  Business Graphics
  Video Graphics
  Computer-Aided Design/Computer-Aided Manufacturing
  Graphics Input Devices
  Graphics Output Devices

How Users See Input and Output

The central processing unit is the unseen part of a computer system, and users are only dimly aware of it. But users are very much aware of the input and output associated with the computer. They submit input data to the computer to get processed information, the output.

Sometimes the output is an instant reaction to the input. Consider these examples:

- Zebra-striped bar codes on supermarket items provide input that permits instant retrieval of outputs—price and item name—right at the checkout counter.
- A bank teller queries the computer through the small terminal at the window by giving a customer's account number as input. The same screen immediately provides the customer's account balance as output.
- A forklift operator speaks directly to a computer through a microphone. Words like *left*, *right*, and *lift* are the actual input data. The output is the computer's instant response, which causes the forklift to operate as requested.
- A medical student studies the human body on a computer screen, inputting changes to the program to show a close-up of the leg and then to remove layers of tissue to reveal the muscles and bone underneath.

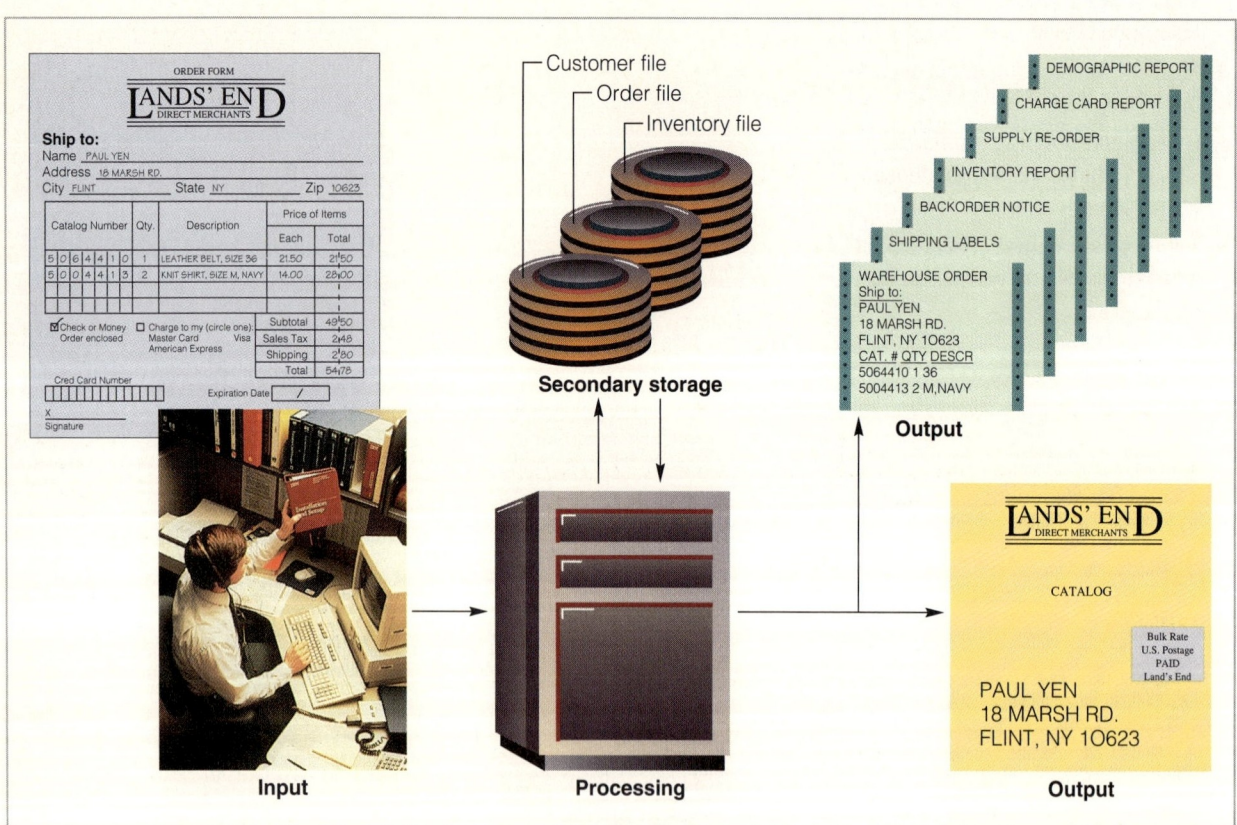

Figure 3-1 Lands' End. At this mail-order house, customer order data is input, processed, and used to produce a variety of outputs.

The screen outputs the changes, allowing the student—without donning a mask, sanitary gloves, or operating gown—to simulate surgery on the computer.

- A sales representative uses an instrument that looks like a pen to enter an order on a special pad. The handwritten characters are displayed as "typed" text and are stored in the pad, which is actually a small computer.

Input and output may sometimes be separated by time or distance or both. Here are some examples:

- Factory workers input data by punching in on a time clock as they go from task to task. The time clock is connected to a computer. The outputs are their weekly paychecks and reports for management that summarize hours per project on a quarterly basis.
- A college student writes checks. The data on the checks is used as input to the bank computer, which eventually processes the data to prepare a bank statement once a month.
- Charge-card transactions in a retail store provide input data that is processed monthly to produce customer bills.
- Water-sample data is collected at lake and river sites, keyed in at the environmental agency office, and used to produce reports that show patterns of water quality.

The examples in this section show the diversity of computer applications, but in all cases the process is the same: input–processing–output. We have already had an introduction to processing. Now, in this chapter we will examine input and output methods in detail.

Input: Getting Data from the User to the Computer

Some input data can go directly to the computer for processing. Input in this category includes bar codes, speech that enters the computer through a microphone, and data entered by means of a device that converts motions to on-screen action. Some input data, however, goes through a good deal of intermediate handling, such as when it is copied from a **source document** (jargon for the original written data) and translated to a medium that a machine can read, such as a magnetic disk. In either case the task is to gather data to be processed by the computer—sometimes called *raw data*—and convert it into some form the computer can understand.

Keyboard

A **keyboard,** which usually is similar to a typewriter, may be part of a personal computer or part of a terminal that is connected to a computer somewhere else (Figure 3-2a). Not all keyboards are traditional, however. A fast-food franchise like McDonald's, for example, uses keyboards whose keys represent items such as large fries or a Big Mac (Figure 3-2b). Even less traditional is the keyboard shown in Figure 3-2c, which is used to enter Chinese characters. Figure 3-3 shows the complete layout of a traditional keyboard.

(a)

(b)

(c)

Figure 3-2 Keyboards. (a) A traditional computer keyboard. (b) Workers at McDonald's press a key for each item ordered. The amount of the order is totaled by the computer system and then displayed on a small screen so the customer can see the amount owed. (c) Chinese characters are significantly more complicated than the letters and digits found on a standard keyboard. To enter Chinese characters into the computer system, a person uses a special keyboard. Each letter key shows the characters available by holding down other keys while typing (as you would hold down a Shift key to make capitals).

Finding Your Way Around a Keyboard

Most personal computer keyboards have three main parts: function keys, the main keyboard in the center, and numeric keys to the right. Extended keyboards, such as the keyboard shown here, have additional keys between the main keyboard and the numeric keys and status lights in the upper-right corner.

Function Keys

The function keys (highlighted in tan on the diagram) are an easy way to give certain commands to the computer. What each function key does is defined by the particular software you are using.

Main Keyboard

The main keyboard includes the familiar keys found on a typewriter keyboard (dark blue), as well as some special command keys (light blue). The command keys have different uses that depend on the software being used. Some of the most common uses are listed here.

 The Escape key, Esc, is used in different ways by different programs; often it allows you to "escape" to the previous screen of the program.

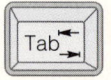 The Tab key allows you to tab across the screen and set tab stops as you would on a typewriter.

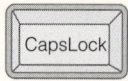 When the Caps Lock key is pressed, uppercase letters are produced. Numbers and symbols are not affected—the number or symbol shown on the bottom of a key is still produced. When the Caps Lock key is pressed, the status light under "Caps Lock" lights up.

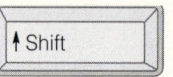 The Shift key allows you to produce uppercase letters and the upper symbols shown on the keys.

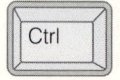 The Control key, Ctrl, is pressed in combination with other keys to initiate commands as specified by the software.

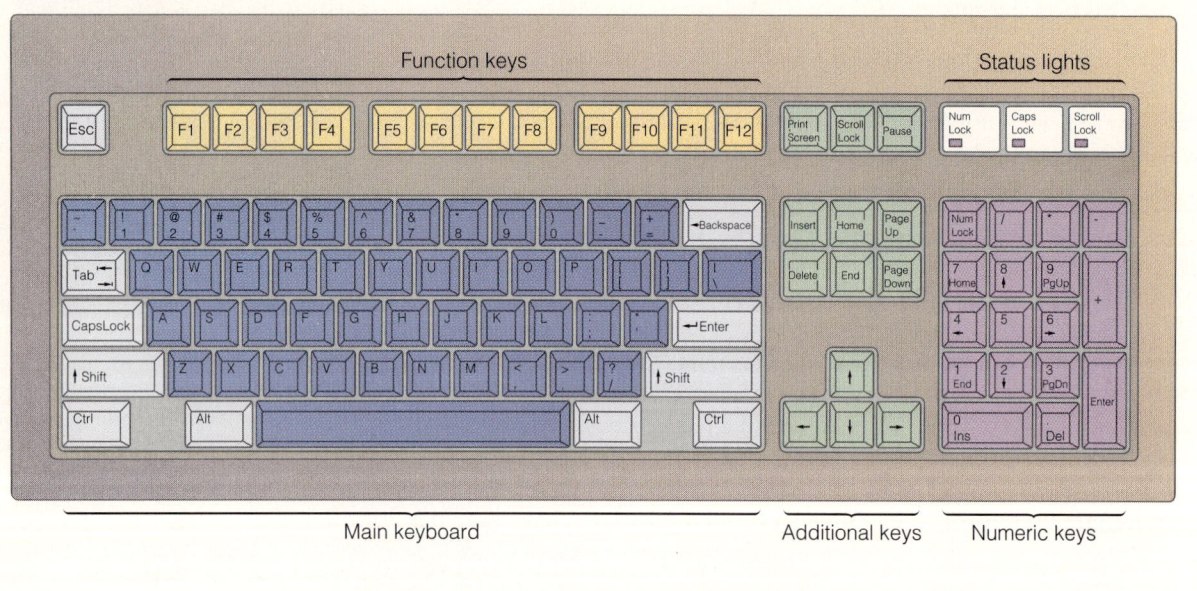

Figure 3-3 Finding your way around a keyboard.

Mouse

A **mouse** is an input device that actually looks a little bit like a mouse (Figure 3-4a and b). The mouse, which has a ball on its underside, is rolled on a flat surface, usually the desk on which the computer sits. The rolling movement causes a corresponding movement on the screen. Moving the mouse allows you to reposition the **pointer,** or **cursor,** an indicator on the screen that shows where the next interaction with the computer can take place. The cursor can also be moved by pressing various keyboard keys. You can communicate commands to the computer by pressing a button on

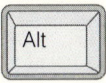 The Alternate key, Alt, is also used in combination with other keys to initiate commands.

 The Backspace key is most often used to delete a character to the left of the cursor, moving the cursor back one position. (The cursor is the flashing indicator on the screen that shows where the next character will be inserted.)

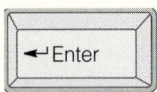 The Enter key moves the cursor to the beginning of the next line. It is used at the end of a paragraph, for instance.

Numeric Keys

The numeric keys (purple) serve one of two purposes, depending on the status of the Num Lock key. When the computer is in the Num Lock mode, these keys can be used to enter numeric data and mathematical symbols (/ for "divided by," * for "multiplied by," -, and +). In the Num Lock mode, the status light under "Num Lock" lights up. When the computer is not in the Num Lock mode, the numeric keys can be used to move the cursor and perform other functions. For example:

 In some programs the End key moves the cursor to the bottom-left corner of the screen.

 This key moves the cursor down.

 The Page Down key, PgDn, advances one full screen while the cursor stays in the same place.

 This key moves the cursor to the left.

 This key moves the cursor to the right.

 In some programs the Home key moves the cursor to the top-left corner of the screen.

 This key moves the cursor up.

 The Page Up key, PgUp, backs up to the previous screen while the cursor stays in the same place.

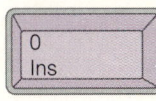 The Insert key, Ins, when toggled off, causes keyed characters to override existing characters.

 The Delete key, Del, deletes a character, space, or selected text.

Additional Keys

Extended keyboards include additional keys (green) that duplicate the cursor movement functions of the numeric keys. Users who enter a lot of numeric data can leave their computers in the Num Lock mode and use these additional keys to control the cursor.

The Arrow keys, to the left of the numeric keys, move the cursor position, just as the numeric keys 2, 4, 6, and 8 do when they are not in the Num Lock mode.

Just above the arrow keys are six keys—Insert, Delete, Home, End, Page Up, and Page Down—which duplicate functions of the numeric keys 0, decimal point (Del), 7, 1, 9, and 3.

At the top of the keyboard, to the right of the function keys, are keys that perform additional tasks. For example:

 The Print Screen key causes the current screen display to be printed. However, some programs will print a screen only when a printed page is full, so you may have to press the Print Screen key more than once.

 The Scroll Lock key causes lines of text—not the cursor—to move when cursor keys are used. When the computer is in the Scroll Lock mode, the status light under "Scroll Lock" lights up.

 The Pause key causes the screen to pause when information is appearing on the screen too fast to read.

Figure 3-3 (continued)

top of the mouse. In particular, a mouse button is often used to click on an **icon** (Figure 3-4c), a pictorial symbol on a screen; the icon represents a computer activity—a command to the computer—so clicking the icon invokes the command.

Trackball

A variation on the mouse is the **trackball.** You may have used a trackball to play a video game. The trackball is like an upside-down mouse—you roll the ball directly with your hand. The popularity of the trackball surged

 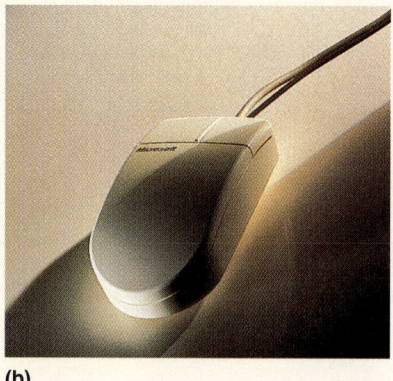 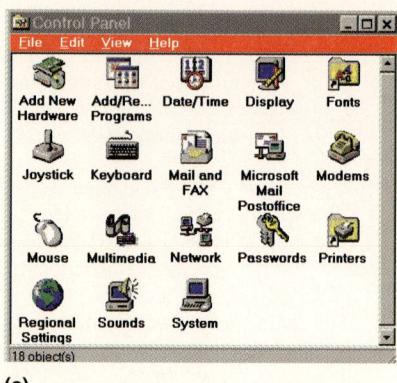

(a) (b) (c)

Figure 3-4 Mouse. (a) As the ball on the underside of the mouse moves over a smooth surface such as a desktop, the pointer on the computer screen makes a corresponding movement. (b) There are several variations on mouse design, but all are designed to fit smoothly under the hand. (c) Once the pointer is in position, a user can select an option from a list of text or icon selections, such as those shown here, by pressing a button on the mouse.

with the advent of laptop computers, when traveling users found themselves without a flat surface on which to roll the traditional mouse. Trackballs are often built in on portable computers, but they can also be used as separate input devices with standard desktop computers (Figure 3-5).

Source Data Automation: Collecting Data Where It Starts

Efficient data input means reducing the number of intermediate steps required between the origination of data and its processing. This is best accomplished by **source data automation**—the use of special equipment to collect data at the source, as a by-product of the activity that generates the data, and send it directly to the computer. Recall, for example, the supermarket bar code, which can be used to send data about the product directly to the computer. Source data automation eliminates keying, thereby reducing costs and opportunities for human-introduced mistakes. Since data about a transaction is collected when and where the transaction takes place, source data automation also improves the speed of the input operation.

For convenience we will divide this discussion into the primary areas related to source data automation: magnetic-ink character recognition, optical recognition, data collection devices, and even directly by your own voice, finger, or eye. Let us consider each of these in turn.

Magnetic-Ink Character Recognition

Abbreviated **MICR, magnetic-ink character recognition** is a method of machine-reading characters made of magnetized particles. The most common example of magnetic characters is the array of numbers across the bottom of your personal check. Figure 3-6 shows what some of these numbers and symbols represent.

Most magnetic-ink characters are preprinted on your check. If you compare a check you wrote that has been cashed and cleared by the bank with those that are still unused in your checkbook, you will note that the amount of the cashed check has been reproduced in magnetic characters in

Chapter Three ▼ Input and Output

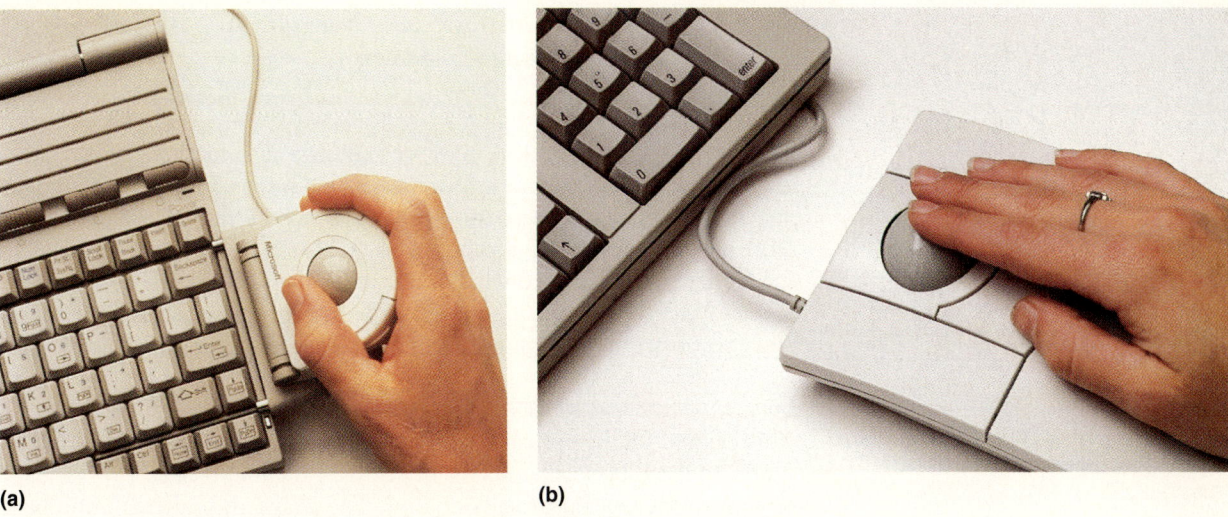

(a)   (b)

Figure 3-5 Trackball. The rotation of the ball causes a corresponding movement of the cursor on the screen. (a) Trackballs are often used with laptop computers because there may be no handy surface on which to roll a mouse. (b) Some users prefer a trackball even with their desktop computers.

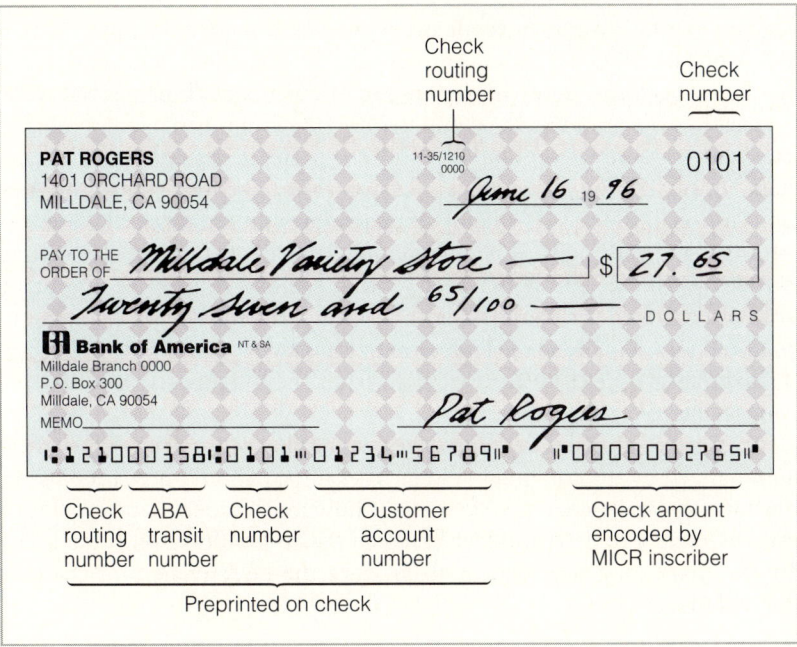

Figure 3-6 The symbols on your check. Magnetic-ink numbers and symbols run along the bottom of a check. The symbols on the left are preprinted. The MICR characters in the lower-right corner of a cashed check are entered by the bank that receives it; these numbers should correspond to the amount of the check.

the lower-right corner. These characters were added by a person at the bank by using a **MICR inscriber**.

Optical Recognition

Optical recognition systems use a light beam to scan input data to convert it into electrical signals, which are sent to the computer for processing. Optical recognition is by far the most common type of source input, appearing in a variety of ways: optical marks, optical characters, bar codes, handwritten characters, and images.

Optical Mark Recognition

Abbreviated **OMR, optical mark recognition** is sometimes called mark sensing, because a machine senses marks on a piece of paper. As a student, you may immediately recognize this approach as the technique used to score certain tests. Using a pencil, you make a mark in a specified box or space that corresponds to what you think is the answer. The answer sheet is then graded by a device that uses a light beam to recognize the marks and convert them to computer-recognizable electrical signals.

Optical Character Recognition

Abbreviated **OCR, optical character recognition** devices also use a light source to read special characters and convert them into electrical signals to be sent to the central processing unit. The characters—letters, numbers, and special symbols—can be read by both humans and machines. They are often found on sales tags on store merchandise. A standard typeface for optical characters, called **OCR-A,** has been established by the American National Standards Institute (Figure 3-7).

The handheld **wand reader** is a popular input device for reading OCR-A. There is an increasing use of wands in libraries, hospitals, and factories, as well as in retail stores. In retail stores the wand reader is connected to a **point-of-sale (POS) terminal.** This terminal is somewhat like a cash register, but it performs many more functions. When a clerk passes the wand reader over the price tag, the computer uses the input merchandise number to retrieve a description (and possibly the price, if not on the tag) of the item. A small printer produces a customer receipt that shows the item description and price. The computer calculates the subtotal, the sales tax (if any), and the total. This information is displayed on the screen and printed on the receipt; notice that both screen and printer are output, so the POS terminal is a complex machine that performs both input and output functions. Finally, some POS terminals include a device that will accept a credit card, inputting account data from the magnetic strip on a customer's charge card.

The raw purchase data becomes valuable information when it is summarized by the computer system. This information can be used by the accounting department to keep track of how much money is taken in each day, by buyers to determine what merchandise should be reordered, and by the marketing department to analyze the effectiveness of their ad campaigns.

Bar Codes

Each product on the store shelf has its own unique number, which is part of the **Universal Product Code (UPC).** This code number is represented on the product label by a pattern of vertical marks, or bars, called **bar codes.** (UPC, by the way, is an agreed-upon standard within the supermarket industry; other kinds of bar codes exist. You need only look as far as the back cover of this book to see an example of another kind of bar code.) These zebra stripes can be sensed and read by a **bar code reader,** a photoelectric device that reads the code by means of reflected light. As with the wand reader in a retail store, the bar code reader in a bookstore or grocery

Figure 3-7 Reading OCR-A typeface. This is a common typeface for optical character recognition.

store is part of a point-of-sale terminal. When you buy, say, a can of corn at the supermarket, the checker moves it past the bar code reader (Figure 3-8a). The bar code merely identifies the product to the store's computer; the code does not contain the price, which may vary. The price is stored in a file that can be accessed by the computer. (Obviously, it is easier to change the price once in the computer than to have to repeatedly restamp the price on each can of corn.) The computer automatically tells the point-of-sale terminal what the price is; a printer prints the item description and price on a paper tape for the customer. Some supermarkets are moving to self-scanning, putting the bar code reader—as well as the bagging—in the customer's hands.

Although bar codes were once found primarily in supermarkets, there are a variety of other interesting applications. Bar coding has been described as an inexpensive and remarkably reliable way to get data into a computer. It is no wonder that virtually every industry has found a niche for bar codes. In Brisbane, Australia, bar codes help the Red Cross manage their blood bank inventory (Figure 3-8b). Also consider the case of Federal Express. The management attributes a large part of the corporation's success to the bar-coding system it uses to track packages. Each package is uniquely identified by a ten-digit bar code, which is input to the computer at each point as the package travels through the system. An employee can use a computer terminal to query the location of a given shipment at any time; the sender can request a status report on a package and receive a response within 30 minutes. The figures are impressive: In regard to controlling packages, the company has an accuracy rate of better than 99 percent.

Handwritten Characters

Machines that can read handwritten characters are yet another means of reducing the number of intermediate steps between capturing data and processing it. In many instances it is preferable to write the data and immediately have it usable for processing rather than having data entry operators key it in later. However, not just any kind of handwritten scrawl will

Big Bad Bar Codes

The world's longest bar code, 375 feet long, is around the base of California's Palomar Observatory to indicate dome positioning. The world's shortest bar code is 2.8 millimeters, applied to bees' thoraxes with shellac to monitor behavior. But the most common bar codes are rapidly becoming the ones on envelopes right in your mailbox.

The post office processes letters in three tiers. At the top, mail already bar-coded by bulk mailers, such as advertisers, sails right through bar code readers. At the next level, a zip code that is typed or neatly printed is scanned and a bar code applied at the bottom of the envelope. Bar-coded letters are sorted by machine and sent on their way. The final level involves letters that must be hand-sorted, thus taking longer.

Try this experiment. Send yourself two letters, one with a neatly printed zip code, the other one barely legible. The first should arrive with a bar-code and be in your mailbox a day or two sooner than the second, especially if you have a friend mail them from another state.

(a)

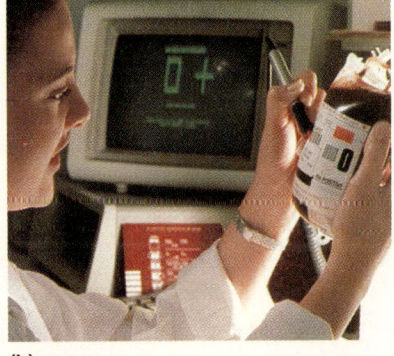
(b)

Figure 3-8 Bar codes. (a) This photoelectric bar code scanner, often seen at supermarket checkout counters, reads the product's zebra-stripe bar code. The bar code identifies the product for the store's computer, which retrieves price and description information. The price is then automatically rung up on the point-of-sale terminal. (b) The Australian Red Cross combines personal computers and handheld bar code readers to verify blood type labels.

	Good	Bad
1. Make your letters big	EWING	EWING
2. Use simple shapes	57320	57320
3. Use block printing	KENT	Kent
4. Connect lines	5BE4	5BE4
5. Close loops	9068	9068
6. Do not link characters	LOOP	LOOP

Figure 3-9 Handwritten characters. Legibility is important in making handwritten characters readable by optical recognition.

do; the rules as to the size, completeness, and legibility of the handwriting are fairly rigid (Figure 3-9).

Imaging

In a process called **imaging,** a scanner converts a drawing, a picture, or any document into computer-recognizable form by shining a light on the image and sensing the intensity of the reflection at each point of the image. **Scanners** come in both handheld and desktop models (Figure 3-10). The electronic version of the image can then be stored, probably on disk, and reproduced on screen when needed. Businesses find imaging particularly useful for documents, since they can view an exact replica of the original document at any time. If a text image is run through an optical character recognition (OCR) program, then all words and numbers can be manipulated by word processing and other software. The Internal Revenue Service, using imaging and also OCR software that can recognize characters from the image, is now scanning 17,000 tax returns per hour, a significant improvement over hand processing.

Another way to keep photos computer accessible is to have film that was shot with a conventional camera processed onto optical disk instead of prints or slides. Professional photo agencies keep thousands of images on file, ready to be leased for a fee (Figure 3-11). Typically, a couple of dozen thumbnail-size images can be displayed on the screen at one time; a particular image can be enlarged to full-screen size with a click of a mouse button.

Data Collection Devices

Another source of direct data entry is a **data collection device,** which may be located in a warehouse or factory or wherever the activity that is generating the data is located (Figure 3-12). As we noted earlier in the chapter, for example, factory employees can use a plastic card to punch job data directly into a computerized time clock. This process eliminates intermediate steps and ensures that the data will be more accurate.

Data collection devices must be sturdy, trouble-free, and easy to use because they are often located in dusty, humid, or hot or cold locations. They are used by people such as warehouse workers, packers, forklift operators, and others whose primary work is not clerical. Examples of

(a)

(b)

Figure 3-10 Scanners. Once an image has been scanned into the computer it can be stored and used again, perhaps in a document that combines text with photos. (a) As this handheld scanner is moved over a picture, the image appears on the computer screen. (b) With a desktop scanner, the image to be scanned is laid face down on the scanner, which looks something like a small copy machine. In this photo the scanner is the flat, boxlike machine on the left.

Figure 3-11 Computer photos. These images, although produced by the computer, did not originate on the computer. Instead, as photographs, they were scanned into the computer, where they are stored in electronic form on disk and can be produced on-screen on command. Alternatively, certain cameras can store photo images directly on a diskette, which can then be used with a computer.

Figure 3-12 A data collection device. Such devices are designed for use in demanding factory environments for collection of data at the source.

Keeping Track of NordicTrack

Do you have a good idea for a new exercise machine? If so, NordicTrack, based in Minneapolis, wants to hear from you. This is true. A leading provider of exercise equipment, NordicTrack has no research department of its own. It relies on outside inventors to come up with new ideas.

But once you submit an idea, would you not want to hear back soon? Therein lies the problem. NordicTrack receives about 30 documented ideas per day. To keep up with the load, they have turned to imaging. The text and any accompanying drawing from each new entry is scanned immediately into the computer, and copies are relayed to each member of the evaluating committee. With the new imaging system in place, NordicTrack now promises a ten-day turnaround time for evaluation and response.

remote data collection devices are machines for taking inventory, reading shipping labels, and recording job costs.

Voice Input

Does your computer have ears? Speaking to a computer, known as **voice input** or **speech recognition,** is another form of source input. **Speech recognition devices** accept the spoken word through a microphone and convert it into binary code (0s and 1s) that can be understood by the computer (Figure 3-13). Originally, typical users were those with "busy hands," or hands too dirty for the keyboard, or with no access to a keyboard. Such uses are changing radio frequencies in airplane cockpits, controlling inventory in an auto junkyard, reporting analysis of pathology slides viewed under a microscope, asking for stock-market quotations over the phone, inspecting items moving along an assembly line, and allowing physically disabled users to issue commands.

Most speech recognition systems are speaker dependent—that is, they must be separately trained for each individual user. The speech recognition system "learns" the voice of the user, who speaks isolated words repeatedly. The voiced words the system "knows" are then recognizable in the future.

Speech recognition systems that are limited to isolated words are called **discrete word systems,** and users must pause between words. Experts have tagged speech recognition as one of the most difficult things for a computer to do. Eventually, **continuous word systems** will be able to interpret sustained speech, so users can speak normally; so far, such systems are limited by vocabulary to a single subject, such as insurance or the weather. A key advantage of delivering input to a computer in a normal speaking pattern is ease of use. Such systems may also be propelled by the explosion of hand and wrist ailments associated with extensive computer keying. Today, software is available to let computers take dictation from people who are willing to pause . . . briefly . . . between . . . words; the best systems are quite accurate and equivalent to typing 70 words per minute.

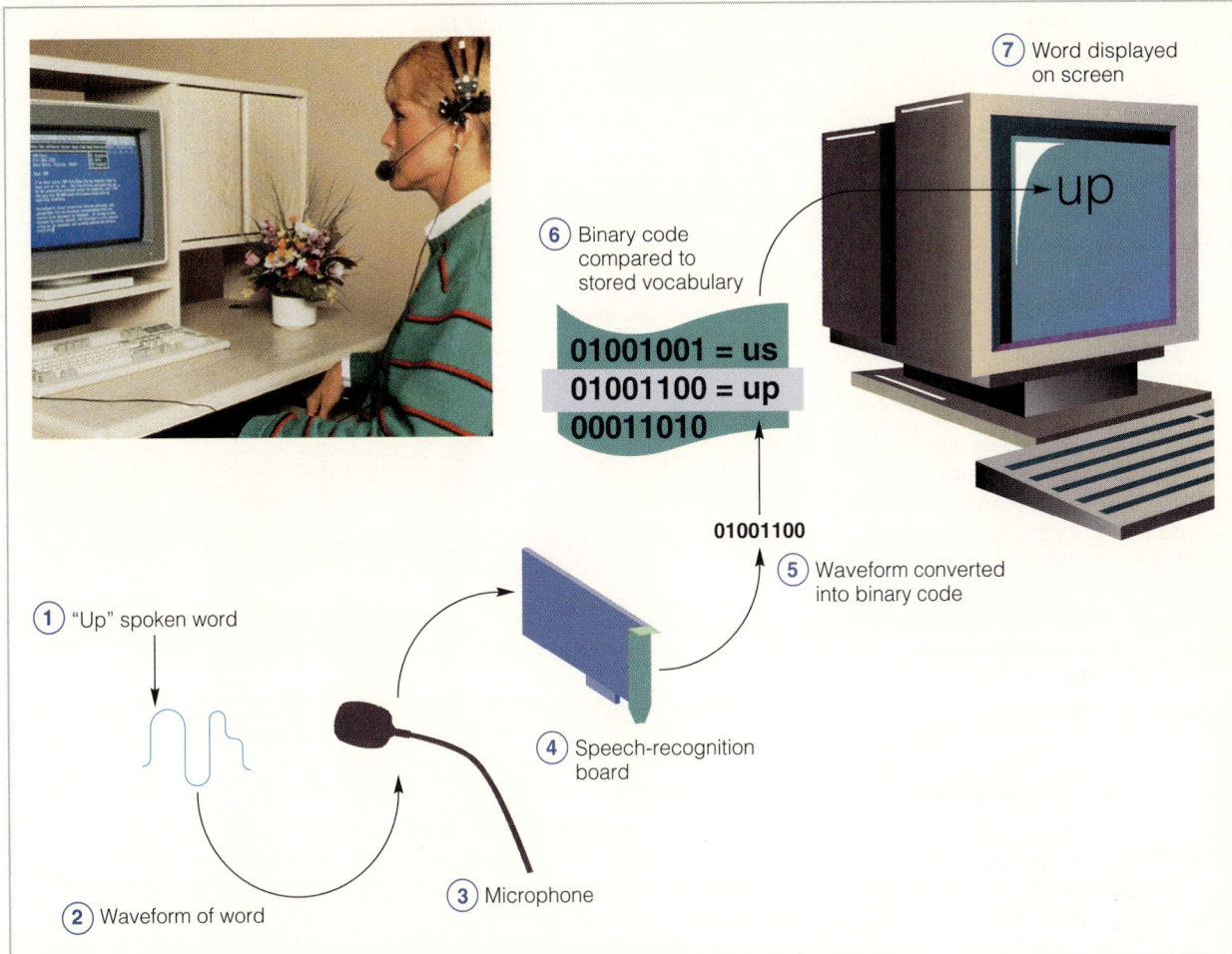

Figure 3-13 How voice input works. The user speaks into a microphone or telephone. A chip on a board inside the computer analyzes the waveform of the word and changes it to binary numbers the computer can understand. These digits are compared with the numbers in a stored vocabulary list; if a match is found, the corresponding word is displayed on the screen.

Touch Screens

One way of getting input directly from the source is to have a human simply point to a selection. The edges of the monitor of a **touch screen** emit horizontal and vertical beams of light that criss-cross the screen. When a finger touches the screen, the interrupted light beams can pinpoint the location selected on the screen. Kiosks in public places such as malls offer a variety of services via touch screens (Figure 3-14). An insurance company kiosk will let you select a policy or a government kiosk will let you order a copy of your birth certificate. Kiosks are also found in private stores. Wal-Mart, for example, uses a kiosk to let customers find needed auto parts. Many delicatessens let you point to salami on rye, among the other selections.

Looking

Delivering input to a computer by simply looking at the computer would seem to be the ultimate in capturing data at the source. The principles are

Voice Input Pranks

Voice input may become the input method of choice for brief commands. But the vision of office workers giving vocal commands to their computers is only now coming into focus. "New file" says one worker, "Italic!" cries another, while a third bellows "Add columns!" Some workers are calling the coming onslaught a new form of noise pollution.

Others see opportunities for pranksters who could prowl the office, issuing commands to other people's computers. Here are some scary possibilities: "Order new desk," "Delete file," and "Send to IRS." However, most voice recognition systems today respond only to a specific voice, limiting, for now, tricks from the office clown.

Figure 3-14 Touch screen kiosk. This kiosk is in a grocery store in Avon, Connecticut. Customers can touch the screen to make a series of choices and a suitable recipe will be printed out.

reminiscent of making a screen selection by touching the screen with a finger. Electrodes attached to the skin around the eyes respond to movement of the eye muscles, which produce tiny electric signals when they contract. The signals are read by the computer system, which determines the location on the screen where the user is looking.

Such a system is not yet in the mainstream. The first people to benefit would likely be those who, due to disabilities or busyness, cannot use their hands or voices for input.

Output: Information for the User

As we have seen, computer output often takes the form of screen or printer output. Other forms of output include voice, microfilm, and various forms of graphics output.

A computer system often is designed to produce several kinds of output. An example is a travel agency that uses a computer system. If a customer asks about airline connections to Toronto, Calgary, and Vancouver, say, the travel agent will probably make a few queries to the system and receive on-screen output indicating availability on the various flights. After the reservations have been confirmed, the agent can ask for printed output that includes the tickets, the traveler's itinerary, and the invoice. The agency may also keep customer records on microfilm. In addition, agency management may periodically receive printed reports and charts, such as monthly summaries of sales figures or pie charts of regional costs. We begin with the most common form of output, computer screens.

Computer Screen Technology

A user's first interaction with a computer screen may be the screen response to the user's input. When data is entered, it appears on the screen. Furthermore, the computer response to that data—the output—also appears on the screen. Computer screens come in many varieties, but the most common kind is the **cathode ray tube (CRT)**. Most CRT screens use a technology called **raster-scan technology**. The backing of the screen display has a phosphorous coating, which will glow whenever it is hit by a beam of electrons. But the light does not stay lit very long, so the image must be **refreshed** often. If the screen is not refreshed often enough, the fading screen image appears to flicker. A **scan rate**—the number of times the screen is refreshed—of 60 times per second is usually adequate to retain a clear screen image. As the user, you tell the computer what image you want on the screen, by typing, say, the letter *M*, and the computer sends the appropriate image to be beamed on the screen. This is essentially the same process used to produce television images.

A computer display screen that can be used for graphics is divided into dots that are called addressable, because they can be *addressed* individually by the graphics software. Each dot can be illuminated individually on the screen. Each dot is potentially a *pic*ture *el*ement, or **pixel**. The **resolution** of the screen—its clarity—is directly related to the number of pixels on the screen: The more pixels, the higher the resolution. Some computers come with built-in graphics capability. Others need a device, called a **graphics card** or **graphics adapter board**, that has to be added.

There have been several color screen standards, relating particularly to resolution. The first color display was **CGA** (color graphics adapter), which had low resolution by today's standards (320x200 pixels). This was followed by the sharper **EGA** (enhanced graphics adapter), featuring 640x350 pixels. Today, VGA and SVGA are common standards. **VGA** (video graphics array) has 640x480 pixels. **SVGA** (super VGA) offers 800x600 pixels or 1024x768 pixels, by far the superior clarity.

Types of Screens

Cathode ray tube monitors that display text and graphics are in common use today. Although most CRTs are color (Figure 3-15a), some screens are **monochrome,** meaning only one color, usually green, appears on a dark background (Figure 3-15b). Another type of screen technology is the **liquid crystal display (LCD),** a flat display often seen on watches and calculators. LCD screens are used on laptop computers (Figure 3-15c). Some LCDs are monochrome, but color screens are popular. Some laptop screens are nearing CRTs in resolution quality.

Terminals

A screen may be the monitor of a self-contained personal computer, or it may be part of a terminal that is one of many terminals attached to a large computer. A **terminal** consists of an input device, an output device, and a communications link to the main computer. Most commonly, a terminal has a keyboard for an input device and a screen for an output device, although there are many variations on this theme.

Printers

A **printer** is a device that produces printed paper output, known in the computer industry as **hard copy** because it is tangible and permanent (unlike **soft copy,** which is displayed on a screen). Some printers produce only letters and numbers, whereas others can also produce graphics.

Letters and numbers are formed by a printer either as solid characters or as dot-matrix characters. **Dot-matrix printers** create characters in the same way that individual lights in a pattern spell out words on a basketball

(a)

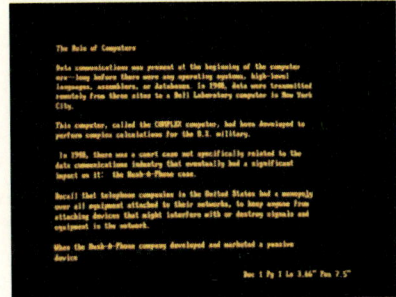
(b)

(c)

Figure 3-15 A variety of screens. (a) This high-resolution brilliance is available only on a color graphics display. (b) Monochrome screens usually feature green text on a black background. (c) Laptop computers often have color screens.

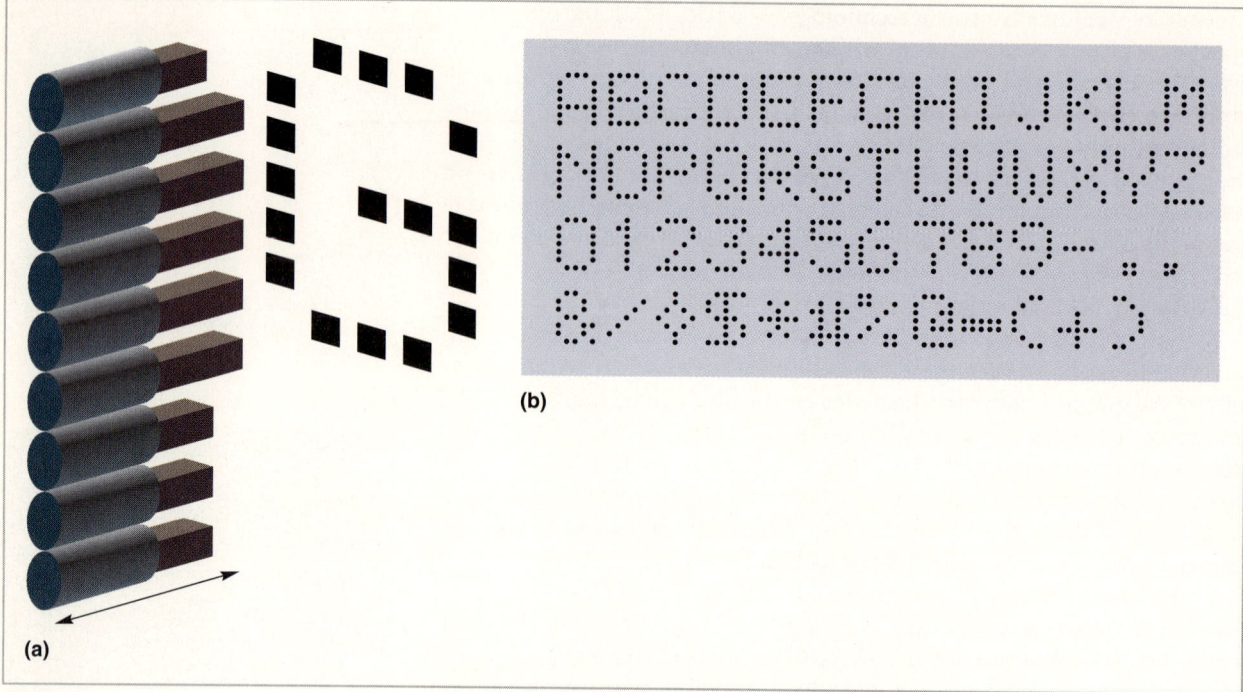

Figure 3-16 Forming dot-matrix characters. (a) The letter *G* is being printed as a 5x7 dot-matrix character. The moving matrix head has nine vertical pins, which move back and forth as necessary to form each letter. (b) Letters, numbers, and special characters are formed as 5x7 dot-matrix characters. Although not shown in this figure, dot-matrix printers can print lowercase letters, too. The two lower pins are used for the parts of lowercase letters *g, j, p,* and *y* that go below the line.

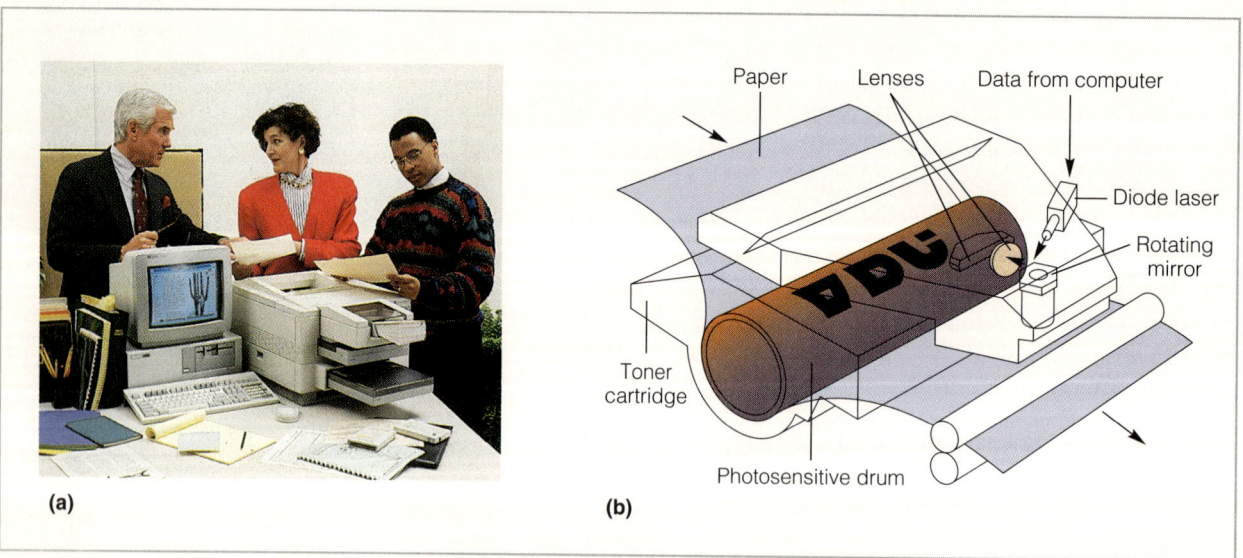

Figure 3-17 Laser printers. (a) The high-quality print and durability of the Hewlett-Packard laser printers make them best-sellers. (b) A laser printer works like a photocopy machine. Using patterns of small dots, a laser beam conveys information from the computer to a positively charged drum inside the laser printer. Wherever an image is to be printed, the laser beam is turned on, causing the drum to become neutralized. As the drum passes by a toner cartridge, toner sticks to the neutral spots on the drum. The toner is then transferred from the drum to a piece of paper. In the final printing step, heat and pressure fuse the toner to the paper. The drum is then cleaned for the next pass.

scoreboard. Dot-matrix printers construct a character by activating a matrix of pins that produce the shape of the character. Figure 3-16 shows how this works. A traditional matrix is 5x7—that is, five dots wide and seven dots high. These printers are sometimes called 9-pin printers, because they have two extra vertical dots for descenders on the lowercase letters *g*, *j*, *p*, and *y*. The 24-pin dot-matrix printer, which uses a series of overlapping dots, dominates the dot-matrix market. The more dots, the better the quality of the character produced. Some dot-matrix printers can produce color images.

There are two ways of printing an image on paper: the impact method and the nonimpact method. Let us take a closer look at the difference.

Impact Printers

The term *impact* refers to the fact that **impact printers** use some sort of physical contact with the paper to produce an image, physically striking paper, ribbon, and print hammer together. The impact may be produced by a print hammer character, like that of a typewriter key striking a ribbon against the paper, or by a print hammer hitting paper and ribbon against a character. A dot-matrix printer is one example of an impact printer. High-quality impact printers print only one character at a time.

However, users who are more concerned about high volume than high quality usually use line printers—impact printers that print an entire line at a time. Organizations that use mainframe and minicomputers usually have several line printers. Such organizations are likely to print hearty reports, perhaps relating to payroll or costs, for internal use. The volume of the report and the fact that it will not be seen by customers makes the speedy—and less expensive—line printer appropriate. One final note about impact printers: An impact printer must be used if printing a multiple-copy report so that the duplicate copies will receive the imprint.

Nonimpact Printers

A **nonimpact printer** places an image on a page without physically touching the page. The major technologies competing in the nonimpact market are laser and ink-jet. **Laser printers** use a light beam to help transfer images to paper, producing extremely high-quality results (Figure 3-17). Laser printers print a page at a time at impressive speeds. Large organizations use laser printers to produce high-volume customer-oriented reports. At the personal computer end, low-end black and white laser printers can now be purchased for a few hundred dollars. However, color laser jet printers are more expensive.

The rush to laser printers has been influenced by the trend toward desktop publishing—using a personal computer, a laser printer, and special software to make professional-looking publications, such as newsletters. We will examine desktop publishing in Chapter 13.

Ink-jet printers, by spraying ink from multiple jet nozzles, can print both black and white and in several different colors of ink to produce excellent graphics. As good as they are, color printers are not perfect. The color you see on your computer screen is not necessarily the color you will see on the printed output. Nor is it likely to be the color you would see on a four-color offset printing press. Nevertheless, with low-end printers now under $500, they may be a bargain for users who want their own color output capability.

There are many advantages to nonimpact printers over impact ones, but there are two major reasons for their growing popularity: They are

I Know Art When I See It

When you buy a computer, do you expect to see fine arts images on your screen? That is what many users enjoy on and off all day long. The images shown here, Auguste Renoir's *Girl in a Boat* and *Bouncer* from the cat series by Siri, are just two possibilities. The software that produces these images is called screen saver software, a pleasant solution to a vexing problem.

The problem is the possibility of screen burnout. Users often leave their computers unattended for periods of time, to talk on the phone, to attend a meeting, or for a number of other reasons. While they are gone the static image that remains on the screen can, on some monitors, "burn" into the screen, leaving a permanent faint shadow in the background whenever the monitor is turned on. Although modern screens are no longer subject to burn, the notion of screen savers remains as a way to personalize computers.

faster and quieter. Other advantages of nonimpact printers over conventional mechanical printers are their ability to change typefaces automatically and their ability to produce high-quality graphics.

Printers are discussed in more detail in the *Buyer's Guide*.

Computer Output Microfilm

How many warehouses would it take to store all the printed census information for this country? To save space, **computer output microfilm** (generally referred to by its abbreviation, **COM**) was developed. For COM, output takes the form of very small images on sheets or rolls of film. A microfilm record can be preserved on rolls of film (usually 35mm) or on 4x6-inch sheets of film called **microfiche**; users often call them just fiche (pronounced "fish").

The key advantage of COM is saved space. At 200 pages per microfiche, this book, for instance, could be stored on three 4x6-inch microfiche. The major disadvantage of COM is that it cannot be read without the assistance of a special reader device. COM may soon disappear, however, in favor of disk storage—when everyone has a computer to access disks.

Voice Output

We have already examined voice input in some detail. As you will see in this section, however, computers are frequently like people in the sense that they find it easier to talk than to listen. **Speech synthesis**—the process of enabling machines to talk to people—is much easier than speech recognition. "The key is in the ignition," your car says to you as you open the car door to get out. Machine voices are not real human voices. They are the product of **voice synthesizers** (also called **voice-output devices** or **audio-response units**), which convert data in main storage to vocalized sounds understandable to humans.

There are two basic approaches to getting a computer to talk. The first is **synthesis by analysis**, in which the device analyzes the input of an actual human voice speaking words, stores and processes the spoken sounds, and reproduces them as needed. The process of storing words is similar to the digitizing process we discussed earlier when considering voice input. In essence, synthesis by analysis uses the computer as a digital tape recorder.

The second approach to synthesizing speech is **synthesis by rule**, in which the device applies a complex set of linguistic rules to create artificial speech. Synthesis based on the human voice has the advantage of sounding more natural, but it is limited to the number of words stored in the computer.

Voice output has become common in such places as airline and bus terminals, banks, and brokerage houses. It is typically used when an inquiry is followed by a short reply (such as a bank balance or flight time). Many businesses have found other creative uses for voice output as it applies to the telephone. Automatic telephone voices ("Hello, this is a computer speaking...") take surveys, inform customers that catalog orders are ready to be picked up, and, perhaps, remind consumers that they have not paid their bills.

Music Output

Personal computer users have occasionally sent primitive musical messages, feeble tones that wheezed from the tiny internal speaker. Many users remain at this level, but a significant change is in progress.

MAKING THE RIGHT CONNECTIONS

Are We Paperless Yet?

In the early days of personal computing, pundits who envisioned a preference for disks and screens predicted a "paperless office." To their surprise, the opposite occurred. Given desktop publishing software and laser printers, people discovered that they could make their ideas look better on paper than ever before. The result was huge new piles of paper in every office.

It took connectivity to make the difference. Office workers, connected via computer to other workers, casually send each other electronic mail. Users on the Internet trade mail and other information.

Finally, documents prepared by computer can be sent directly to another computer without being printed first. A new type of software called document sharing software can take a document created by any software on any computer and put it in a standard format that can be understood by the receiving computer. The paperless office may be in sight.

Professional musicians lead the way, using special sound chips that simulate different instruments. A sound card, installed internally in the computer, and attached speakers complete the output environment. Now, using appropriate software, the computer can produce the sound of an orchestra or a rock band. Those of us who simply enjoy music can have a full sight/sound experience using multimedia, which we will explore in detail in the next chapter.

 ## Computer Graphics

Let us take a moment to glimpse everyone's favorite, computer graphics. Just about everyone has seen TV commercials or movies that use computer-produced animated graphics. Computer graphics can also be found in education, computer art, science, sports, and more (Figure 3-18). But perhaps their most prevalent use today is in business.

Business Graphics

It might seem wasteful to use color graphics to display what could more inexpensively be shown to managers as numbers in standard computer printouts. However, colorful graphics, maps, and charts can help managers compare data more easily, spot trends, and make decisions more quickly. Also, the use of color helps people get the picture—literally. Finally, although color graphs and charts have been used in business for

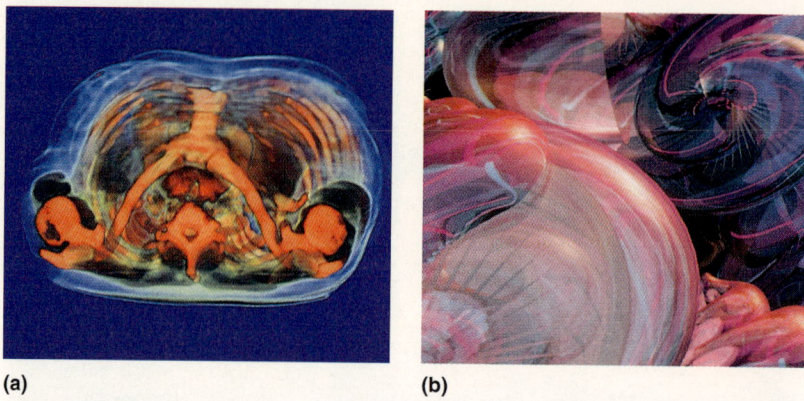

Figure 3-18 Computer graphics. Graphics are used for a variety of purposes. (a) Did you recognize this computer representation as a cross section of the chest? (b) A computer artist presents an abstraction in pink. (c) Is it a photo? No, this realistic computer graphic shows an old-fashioned diner with both a skyscraper and a mountain background.

years—usually to make presentations to higher management or outside clients—the computer allows them to be rendered quickly, before information becomes outdated. One user refers to business graphics as "computer-assisted insight."

Video Graphics

Video graphics can be as creative as an animated cartoon (Figure 3-19). Although they operate on the same principle as a moving picture or cartoon—one frame at a time in quick succession—**video graphics** are produced by computers. Video graphics have made their biggest splash on television, but many people do not realize they are watching a computer at work. The next time you watch television, skip the trip to the kitchen and pay special attention to the commercials. Unless there is a live human in the advertisement, there is a good chance that the moving objects you see, such as floating cars and bobbing electric razors, are computer output. Another fertile ground for video graphics is a television network's logo and

Figure 3-19 Video graphics. The animation in the movie *The Mask* broke new ground in video graphics. Film shots of human actors (Jim Carrey, shown here) were input to the computer where they were then manipulated by computer artists who added animation.

theme. Accompanied by music and swooshing sounds, the network symbol spins and cavorts and turns itself inside out, all with the finesse that only a computer could supply.

Computer-Aided Design/Computer-Aided Manufacturing

For more than a decade, computer graphics have also been part and parcel of a field known by the abbreviation **CAD/CAM**—short for **computer-aided design/computer-aided manufacturing**. In this area computers are used to create two- and three-dimensional pictures of everything from hand tools to tractors. CAD/CAM provides a bridge between design (planning what a product will be) and manufacturing (actually making the planned product). As a manager at Chrysler said, "Many companies have design data and manufacturing data, and the two are never the same. At Chrysler, we have only one set of data that everyone dips into." Keeping

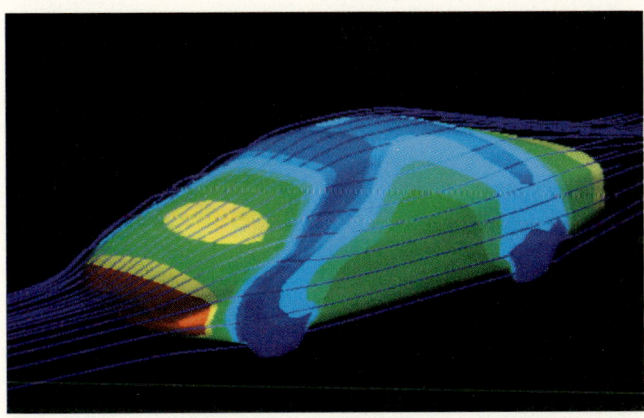 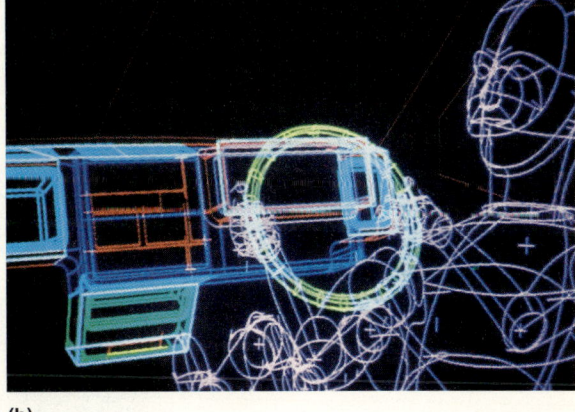

Figure 3-20 CAD/CAM. With computer-aided design and computer-aided manufacturing (CAD/CAM), the computer can keep track of all details, maintain designs of parts in storage, and combine parts electronically as required. (a) A computer-aided design wireframe of a car used to study design possibilities. (b) Engineers also use graphics to test designs relative to the consumer.

data in one place, of course, makes changes easier and encourages consistency. For an example of Chrysler's efforts, see Figure 3-20.

Graphics Input Devices

There are many ways to produce and interact with screen graphics. We have already described the mouse; the following are some other common devices that allow the user to interact with screen graphics. A **digitizing tablet** lets you create your own images (Figure 3-21). This device has a special stylus that you can use to draw or trace images, which are then converted to digital data that can be processed by the computer.

For direct interaction with your computer screen, the **light pen** is ideal. It is versatile enough to modify screen graphics or make a menu selection—that is, to choose from a list of activity choices on the screen. A light pen has a light-sensitive cell at one end. When you place the light pen against the screen, it closes a photoelectric circuit that pinpoints the spot

Figure 3-21 Digitizing tablet. This engineer is using a digitizing tablet to input his drawing to the computer.

Figure 3-22 Plotter. Designers of circuit boards, street maps, schematic diagrams, and similar applications can work in fine detail on a computer screen and then print the results on a plotter.

Chapter Three ▼ Input and Output

ONE JUMP AHEAD

Parenthood Tryouts

That cute baby may seem just a bit less cuddly when it shrieks intermittently day and—most especially—night. But this is not a real baby. This is a tryout baby, designed to give teenagers a simulated parenting experience.

The baby is aptly named Baby Think It Over. As part of a high school class, a teen takes Baby home for the weekend. Baby has an internal computer that is programmed to emit baby cries at random intervals. The cries, which can last from 5 to 35 minutes, can be stopped only by a special key attached to a wristband worn by the teen. Teens find that Baby cramps their style, both limiting their plans and interrupting their sleep.

Can toddler and teen dolls be far behind?

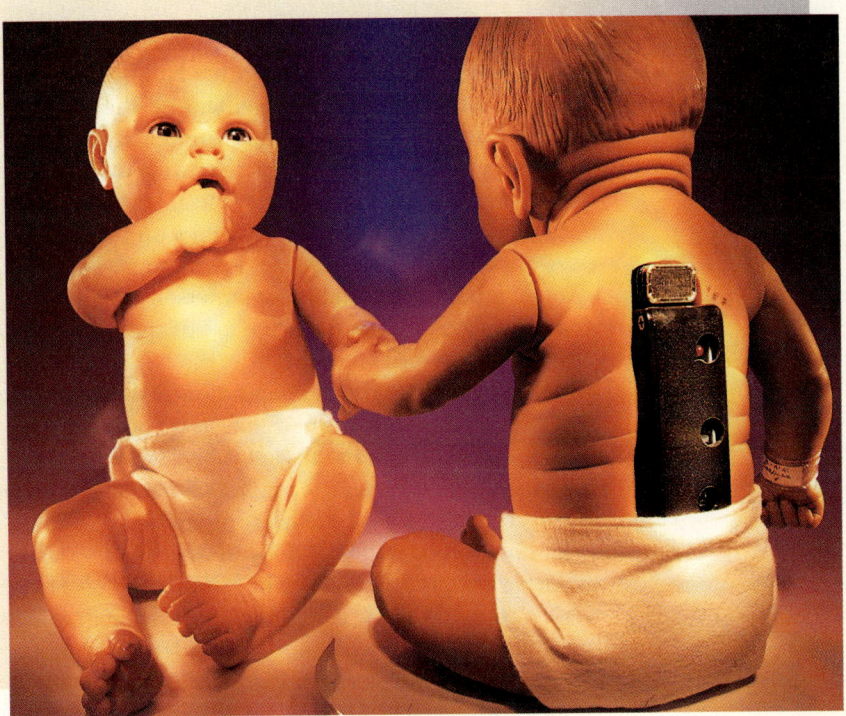

the pen is touching. This tells the computer where to enter or modify pictures or data on the screen.

Finally, a well-known graphics input device is the **joystick,** dear to the hearts of video game fans. This device allows fingertip control of figures on a CRT screen.

Graphics Output Devices

Just as there are many different ways to input graphics to the computer, there are many different ways to output graphics. Graphics are most commonly output on a screen or printed paper, as previously discussed. Another popular graphics output device is the **plotter,** which can draw hard-copy graphics output in the form of maps, bar charts, engineering drawings, and even two- or three-dimensional illustrations (Figure 3-22). Plotters often come with a set of four pens in four different colors. Most plotters also offer shading features.

New forms of computer input and output are announced regularly, often with promises of multiple benefits and new ease of use. Part of the excitement of the computer world is that these promises are usually kept, and users reap the benefits directly. Input and output just keep getting better.

CHAPTER REVIEW

Summary and Key Terms

- A **keyboard** is a common input device that may be part of a personal computer or a terminal connected to a remote computer. A **source document** is the original written input data to be keyed into the computer.
- A **mouse** is an input device with a ball on its underside, whose movement on a flat surface causes a corresponding movement on the screen. Moving the mouse (or by pressing keyboard keys) allows you to reposition the **pointer,** or **cursor,** an indicator on the screen that shows where the next interaction with the computer can take place. An **icon** is a pictorial symbol on a screen that represents a computer activity, or a command. to the computer. A computer user can interact with an icon by using a mouse.
- A **trackball** is like an upside-down mouse—the ball is rolled with the hand. Trackballs are often built in on portable computers.
- **Source data automation** is the use of special equipment to collect data at its origin and send it directly to the computer.
- **Magnetic-ink character recognition (MICR)** involves characters made of magnetized particles, such as the preprinted characters on a personal check. Some characters are preprinted, but others, such as the amount of a check, are added by a person using a **MICR inscriber.**
- **Optical recognition** uses a light beam to scan data to convert it to electric signals to send to the computer. **Optical mark recognition (OMR)** devices recognize marks on paper. **Optical character recognition (OCR)** devices read special characters, such as those on price tags. These characters are often in a standard typeface called **OCR-A.** A commonly used OCR device is the handheld **wand reader,** which is often connected to a **point-of-sale (POS) terminal** in a retail store. A **bar code reader** is a stationary photoelectric scanner used to input a **bar code,** a pattern of vertical marks that represents the **Universal Product Code (UPC)** that identifies a product. Some optical scanners can read precise handwritten characters. **Imaging** uses a **scanner** to convert a drawing or photo or document to an electronic version that can be stored and reproduced when needed. Once scanned, text documents may be processed by optical recognition software so that the text can be manipulated.
- **Data collection devices** allow direct, accurate data entry in places such as factories and warehouses.
- **Voice input,** or **speech recognition,** is the process of presenting input data to the computer through the spoken word. **Speech recognition devices** convert spoken words into a digital code that a computer can understand. The two main types of devices are **discrete word systems,** which require speakers to pause between words, and **continuous word systems,** which allow a normal rate of speaking.
- Input can be given directly to a computer via a **touch screen;** a finger touching the screen interrupts the light beams on the monitor edge, pinpointing the selected screen location.
- Input can be delivered to a computer by looking at a screen, assuming electrodes are attached to the skin near the user's eyes so that their signals can be read by the computer system.
- The most common kind of computer screen is the **cathode ray tube (CRT).** Most CRT screens use a technology called **raster-scan technology.** The backing of the screen display has a phosphorous coating, which will glow whenever it is hit by a beam of electrons. The screen image must be **refreshed** often to avoid flicker. A **scan rate**—the number of times the screen is refreshed—of 60 times per second is usually adequate to retain a clear screen image.
- A computer display screen that can be used for graphics is divided into dots that are called addressable because they can be *addressed* individually by the graphics software. Each screen dot is called a **pixel.** The more pixels, the higher the screen **resolution,** or clarity. A

computer must either come with built-in graphics capability or have a **graphics card** or **graphics adapter board** added.
- Color screen standards are the relatively low resolution **CGA** (color graphics adaptor) and **EGA** (enhanced graphics adaptor), and today's higher resolution **VGA** (video graphics array), with 640x480 pixels and **SVGA** (super VGA), with 800x600 pixels or 1024x768 pixels.
- Some computer screens are **monochrome**—the characters appear in one color, usually green, on a dark background.
- A **liquid crystal display** (**LCD**) is a type of flat screen found on laptop computers.
- A screen may be the monitor of a self-contained personal computer, or it may be part of a **terminal,** an input-output device linked to a main computer.
- **Printers** produce **hard copy,** or printed paper output. (**Soft copy** is displayed on a screen.) Some printers produce solid characters; **dot-matrix printers,** however, construct characters by producing closely spaced dots.
- Printers can be classified as being either **impact printers,** which form characters by physically striking the paper, or **nonimpact printers,** which use a noncontact printing method. Nonimpact printers, which include **laser** and **ink-jet printers,** are faster and quieter than impact printers.
- With **computer output microfilm** (**COM**), output is stored on 35mm film or 4x6-inch sheets called **microfiche.**
- Computer **speech synthesis** has been accomplished through **voice synthesizers** (also called **voice-output devices** or **audio-response units**). One approach to speech synthesis is **synthesis by analysis,** in which the computer analyzes stored tapes of spoken words. In the other approach, called **synthesis by rule,** the computer applies linguistic rules to create artificial speech.
- A personal computer user can take advantage of software that produces music and other sounds if the computer is equipped with a sound card and speakers.
- **Video graphics** are a series of computer-produced pictures.
- In **computer-aided design/computer-aided manufacturing** (**CAD/CAM**), computers are used to create two- and three-dimensional pictures of manufactured products such as hand tools and vehicles.
- Common graphics input devices include the **digitizing tablet, light pen,** and **joystick.**
- Graphics output devices include screens, printers, and **plotters.**

Discussion Questions

1. For this question use your knowledge from either reading or experience, or imagine the possibilities. What kind of input device might be convenient for these types of jobs or situations: (a) a supermarket stock clerk who takes inventory by surveying items currently on the shelf; (b) a medical assistant who must input existing printed documents to the computer; (c) an airport automated luggage tracking system; (d) a telephone worker who takes orders over the phone; (e) a restaurant in which customers place their own orders from the table; (f) an inspector at the United States Bureau of Engraving who monitors and gives a go/nogo response on printed money passing by on an assembly line; (g) a retailer who wants to move customers quickly through checkout lines, (h) a psychologist who wants to give a new client a standard test, (i) an environmental engineer who hikes woods and streams to inspect and report on effects of pollutants, (j) a small business owner who wants to keep track of employee work hours.
2. Do you think that continued research into voice input is worthwhile? In your answer discuss the practicality of current and potential uses.
3. What should a buyer consider when comparing different models of printers? If price were not a consideration, what kind of printers would you buy for your home or business personal computers? Would the kinds of software you use affect your printer decisions?

Student Study Guide

Multiple Choice

1. Computer output produced as small film images is called
   a. OCR
   b. LCD
   c. COM
   d. OMR
2. A pictorial screen symbol that represents a computer activity is called a(n)
   a. pointer
   b. icon
   c. touch screen
   d. MICR
3. Using computers to design and manufacture products is called
   a. inscribing
   b. CAD/CAM
   c. detailing
   d. imaging
4. Soft copy refers to
   a. OCR-A
   b. microfiche
   c. screen output
   d. digitizing
5. The CRT technology with the best resolution:
   a. MICR
   b. SVGA
   c. VGA
   d. LCD
6. An ink-jet printer is an example of a(n)
   a. laser printer
   b. impact printer
   c. COM printer
   d. nonimpact printer
7. Entering data into the system as a by-product of the activity that generates the data is called
   a. source data automation
   b. a discrete word system
   c. CAD/CAM
   d. MICR entry
8. The rate of screen refreshment is called
   a. pixel speed
   b. bit-map speed
   c. raster rate
   d. scan rate
9. Magnetic characters are produced on your bank checks by
   a. bar code readers
   b. mice
   c. MICR inscribers
   d. microfiche
10. "Mark sensing" is another term for
    a. MICR
    b. POS
    c. OMR
    d. VGA
11. A device used for optical-character recognition is a
    a. wand reader
    b. cursor
    c. pen
    d. MICR reader
12. OCR-A is a
    a. plotter
    b. standard typeface
    c. wand reader
    d. bar code
13. POS terminals are similar to
    a. calculators
    b. touch-tone telephones
    c. UPCs
    d. cash registers
14. A one-color screen on a black background is called
    a. monochrome
    b. blank
    c. addressable
    d. liquid crystal display
15. Voice input devices convert voice input to
    a. digital codes
    b. bar codes
    c. OCR-A
    d. optical marks
16. Imaging uses what device to input data?
    a. scanner
    b. bar code reader
    c. icon
    d. tablet
17. The cursor can be moved rolling this device on a flat surface:
    a. mouse
    b. trackball
    c. wand reader
    d. interactive tablet
18. Which input device is often attached to laptop computers:
    a. trackball
    b. graphic display
    c. inscriber
    d. wand reader
19. A display lighter and slimmer than a CRT is
    a. COM
    b. graphics card
    c. flat panel
    d. terminal
20. Computer animation is a form of
    a. LCD
    b. CAD/CAM
    c. video graphics
    d. color printer output
21. A color screen with the best resolution has the most
    a. CRT
    b. COM
    c. VGA
    d. pixels
22. The name for the screen's clarity:
    a. resolution
    b. discrete
    c. pixel
    d. LCD
23. A bar code represents the product's
    a. CGA
    b. OMR
    c. LCD
    d. UPC
24. A printer that forms characters from a series of dots is called
    a. dot-matrix
    b. MICR
    c. scanner
    d. plotter
25. Another word for pointer:
    a. monochrome
    b. pixel
    c. microfiche
    d. cursor

True/False

T F 1. The greater the number of pixels, the poorer the screen clarity.
T F 2. Printers produce hard copy.
T F 3. Video graphics are computer-produced pictures.
T F 4. Discrete word systems allow a normal rate of speaking.
T F 5. An imaged document must be input each time it is used by the computer.
T F 6. In a discrete word system, the user must pause between words.
T F 7. CRT stands for computer remote terminal.
T F 8. Optical recognition technology is based on magnetized data.
T F 9. OMR senses marks on paper.
T F 10. A wand reader can read OCR characters.
T F 11. A color screen is called monochrome.
T F 12. LCD is a type of flat screen found on laptop computers.
T F 13. LCD stands for liquid crystal display.
T F 14. COM can be read directly by a computer.
T F 15. A mouse can be clicked to invoke a command.
T F 16. The MICR process is used mainly by stores for processing purchases.

T F 17. A cursor is an indicator on a screen that shows where the next interaction will take place.
T F 18. Handwritten characters can never be used for data entry because of their general lack of uniformity.
T F 19. COM can be produced on either 35mm film or microfiche.
T F 20. The color screen standard today is CGA.
T F 21. Pixel is short for picture element.
T F 22. A screen must be refreshed to avoid flicker.
T F 23. Optical marks may be used to record test answers.
T F 24. Nonimpact printers are quieter than impact printers.
T F 25. Laser printers use a light beam to transfer images to paper.

Fill-In

1. Original written input data to be keyed into the computer: _____
2. What does LCD stand for? _____
3. The standard optical typeface: _____
4. What does POS stand for? _____
5. Another name for a pictorial screen symbol that represents a command: _____
6. MICR stands for _____
7. Using a scanner to input documents and to the computer: _____
8. What phrase is used to describe the use of special equipment to collect data at the source and send it to the computer? _____
9. What does UPC stand for? _____
10. Which input method is used mainly by banks for processing checks? _____
11. What is the general term for systems that use a light source to read data such as numbers, letters, special characters, and marks? _____
12. What method uses a light beam to sense marks on machine-readable test forms? _____

13. What is the term for the 4x6-inch sheets of film used for COM? _____
14. Which technology is more challenging, voice input or voice output? _____
15. What kind of terminal may read bar codes but is like a cash register in many ways? _____
16. What is screen output called? _____
17. What is printed computer output called? _____
18. A type of nonimpact printer that uses a light source: _____
19. An input device that reads OCR tags: _____
20. A screen that accepts input from a pointing finger: _____
21. The number of times a screen is refreshed is called its _____
22. CAD/CAM stands for _____
23. The Universal Product Code is represented on a package by a _____
24. A speech recognition system limited to isolated words is called _____
25. Screen clarity is called _____

Answers

Multiple Choice
1. c	6. d	11. a	16. a	21. d
2. b	7. a	12. b	17. a	22. a
3. b	8. d	13. d	18. a	23. d
4. c	9. c	14. a	19. c	24. a
5. b	10. c	15. a	20. c	25. d

True/False
1. F	6. T	11. F	16. F	21. T
2. T	7. F	12. T	17. T	22. T
3. F	8. F	13. T	18. F	23. T
4. F	9. T	14. F	19. T	24. T
5. F	10. T	15. T	20. F	25. T

Part One ▼ Hardware Tools

Fill-In
1. source document
2. liquid crystal display
3. OCR-A
4. point of sale
5. icon
6. magnetic-ink character recognition
7. imaging
8. source data automation
9. Universal Product Code
10. MICR
11. optical recognition
12. OMR
13. microfiche
14. voice input
15. POS terminal
16. soft copy
17. hard copy
18. laser
19. wand reader
20. touch screen
21. scan rate
22. computer-aided design/computer-aided manufacturing
23. bar code
24. discrete word system
25. resolution

PLANET INTERNET

Global Village

The Internet is big, really big. Some say that working on the Internet gives new meaning to the word infinity. One way to grasp the vastness of the Net is to start linking to sites overseas.

So far away. As intriguing as these faraway sites may be, it may take some time to access them. An option that may be available is a *mirror site*, a nearby site that has the identical offering. The WebMuseum site in Paris, for example, which offers paintings from the Louvre on your screen, urges you right away to switch (link) to one of the mirror sites on their list; California has a mirror site for the Louvre. Using a mirror site dramatically improves the speed of data access.

The CERN site. It is entirely appropriate that we mention the CERN site, a laboratory for particle physics in Geneva, Switzerland. CERN is the birthplace of the World Wide Web. The architect of the Web, Tim Berners-Lee, perceived that his work would be easier if he and his far flung colleagues could link to each other's sites. He saw the set of links from site to site as a web; hence the name. This site includes a link to the history of the relationship of CERN with the Web.

English or not English. Many foreign sites offer an English version of their site to present to visitors whose language is English; you could as easily be in Kansas. Other sites offer a choice: Would you prefer English or Swedish? Others simply launch into their native tongue. You either speak it or just go along for the pictures. It is intriguing to see the Russian alphabet come on the screen as you tour the Kremlin.

Worth the trip? You decide. A good place to begin is the World Communities site, where you can hop from country to country. Be prepared for the fact that some developing countries have relatively primitive offerings at this point; in fact, some cannot be reached at all. But many sites are fascinating. The sites are as diverse as those of the three home page logos shown here: Japan's Hottest Links, the Egypt Interactive Home Page, and The University of Bristol Philosophy Department. If you check out the Australian National University site you'll find a Mediterranean art and architecture exhibit, complete with sight and sound, that won an award for its multimedia presentation.

Internet Exercises

1. **Structured exercise.** Begin with the B/C URL http://www.aw.com/bc/planet/ and use the link supplied to go to the Canadian site.
2. **Freeform exercise.** Beginning with the World Communities site, compare the home pages you find for countries in Europe, Asia, the Pacific, and South America.

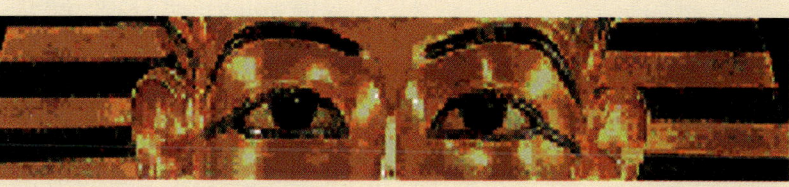

When Ken Koll began his own landscaping business, he did not hesitate about buying a computer. From everything he had read, he was convinced that the computer would be the key to a successful small business. Although Ken was not an experienced computer user, he had taken an evening course that introduced word processing, desktop publishing, and spreadsheet programs. He planned to use a computer to write letters, to keep track of customers and appointments, to plan his budget, to track billing, and even to design his own letterhead and circulars.

Ken took a little time to investigate personal computers. Working with a salesperson from a recommended outlet, Ken considered several midpriced models. Ken's main concern was the secondary storage. He knew he would need a diskette drive for transferring software and files, and he also decided that he should get a CD-ROM drive so he could use software that came on optical disk. But he had misgivings about the capacity of the hard disk.

Could he really use the 540-megabyte disk the salesperson recommended? How could he possibly come up with 540 million characters of *anything*? Ken did choose the 540MB disk, but he wondered whether he had bought more than he needed.

The choices for storage, whether for a large or small computer, are complicated. Ken actually chose quite well. We will check back with him later in the chapter to see how his choice worked out. We switch now from personal computer storage to the broader needs of a large corporation or government agency.

Chapter 4

Storage Devices and Stored Data

Just the Facts

LEARNING OBJECTIVES

- Know the benefits of secondary storage
- Understand the principal types of secondary storage: magnetic disk, optical disk, and magnetic tape
- Understand how data is stored on a disk
- Understand the storage media available for personal computers
- Understand how data is organized, accessed, and processed
- Become acquainted with three methods of file organization: sequential, direct, and indexed sequential
- Understand the difference between batch and transaction processing

THE BENEFITS OF SECONDARY STORAGE

MAGNETIC DISK STORAGE
    Diskettes
    Hard Disks
    Hard Disks in Groups
    How Data Is Organized on a Disk

OPTICAL DISK STORAGE
    CD-ROM
    Multimedia
    Magneto-Optical

MAGNETIC TAPE STORAGE

BACKUP SYSTEMS

CHECKING AN EARLIER DECISION

ORGANIZING AND ACCESSING STORED DATA
    Data: Getting Organized
    The File Plan: An Overview
    File Organization: Three Methods
    Disk Access to Data

PROCESSING STORED DATA
    Batch Processing
    Transaction Processing
    Batch and Transaction Processing: The Best of Both Worlds

The Benefits of Secondary Storage

Picture, if you can, how many filing-cabinet drawers would be required to hold the millions of files of, say, tax records kept by the Internal Revenue Service or historical employee records kept by General Motors. The record storage rooms would have to be enormous. Computers, in contrast, permit storage on tape or disk in extremely compressed form. Storage capacity is unquestionably one of the most valuable assets of the computer.

Secondary storage, sometimes called **auxiliary storage**, is storage separate from the computer itself, where you can store software and data on a semipermanent basis. Secondary storage is necessary because memory, or primary storage, can be used only temporarily. If you are sharing your computer, you must yield memory to someone else after your program runs; if you are not sharing your computer, your programs and data will disappear from memory when you turn off the computer. However, you probably want to store the data you have used or the information you have derived from processing; that is why secondary storage is needed. Furthermore, memory is limited in size, whereas secondary storage media can store as much data as necessary.

The benefits of secondary storage can be summarized as follows:

- **Space.** Organizations may store the equivalent of a roomful of data on sets of disks that take up less space than a breadbox. A simple diskette for a personal computer holds the equivalent of 500 printed pages, or one book. An optical disk can hold the equivalent of approximately 400 books.
- **Reliability.** Data in secondary storage is basically safe, since secondary storage is physically reliable. Also, it is more difficult for unscrupulous people to tamper with data on disk than data stored on paper in a file cabinet.
- **Convenience.** With the help of a computer, authorized people can locate and access data quickly.
- **Economy.** Together the three previous benefits indicate significant savings in storage costs. It is less expensive to store data on tape or disk (the principal means of secondary storage) than to buy and house filing cabinets. Data that is reliable and safe is less expensive to maintain than data subject to errors. But the greatest savings can be found in the speed and convenience of filing and retrieving data.

These benefits apply to all the various secondary storage devices but, as you will see, some devices are better than others. We begin with a look at the various storage media, including those used for personal computers, and then consider what it takes to get data organized and processed.

Magnetic Disk Storage

Diskettes and hard disks are magnetic media; that is, they are based on a technology of representing data as magnetized spots on the disk—with a magnetized spot representing a 1 bit and the absence of such a spot representing a 0 bit. Reading data from the disk means converting the magnetized data to electrical impulses that can be sent to the processor. Writing data to disk is the opposite: sending electrical impulses from the processor to be converted to magnetized spots on the disk. As Figure 4-1 shows, the

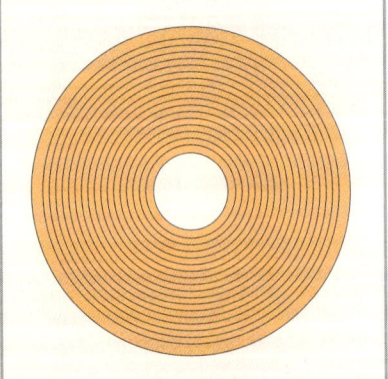

Figure 4-1 Surface of a disk. Note that each track is a closed circle. This drawing is only to illustrate the location of the tracks; you cannot actually see the tracks on the disk surface.

surface of each disk has concentric tracks on it. The number of tracks per surface varies with the particular type of disk.

Diskettes

Made of flexible mylar, a diskette can record data as magnetized spots on tracks on its surface. Diskettes became popular along with the personal computer. The older diskette, 5¼ inches in diameter, is still in use, but newer computers use the 3½-inch diskette (Figure 4-2). The 3½-inch diskette has the protection of a hard plastic jacket, a size to fit conveniently in a shirt pocket or purse, and the capacity to hold significantly more data than a 5¼-inch diskette. Diskettes offer particular advantages which, as you will see, are not readily available with hard disk:

- **Portability.** Diskettes easily transport data from one computer to another. Workers, for example, carry their files from office computer to home computer and back on a diskette instead of in a briefcase. Students use the campus computers but keep their files on their own diskettes.
- **Backup.** It is convenient to place an extra copy of a hard disk file on a diskette.

Diskette Care

Although diskettes come in a protective jacket, users must take at least ordinary care of them to protect the data contents.

- Never put anything heavy on a disk or pile disks on top of one another; disks should be stored vertically.
- Do not put disks near a magnet or anything that could generate a magnetic field, such as telephones and stereo speakers.
- Keep disks away from the sun and excessive heat, including the vent from your car heater.
- Do not move the sliding metal shutter on a 3½-inch disk or touch the disk inside; in fact, never touch the mylar surface of any diskette.
- Never attempt to clean a disk surface.
- Do not use disk labels in layers, lest the disks get too thick and get stuck in the drive.
- Never attempt to force a disk into the drive nor to remove a stuck disk; get help.
- Never remove a disk from the drive when the drive light is on; the light tells you the disk is in use, and removing it could damage the data.
- If you are using 5¼-inch disks, write on the label with a felt tip pen, and avoid paper clips and rubber bands.

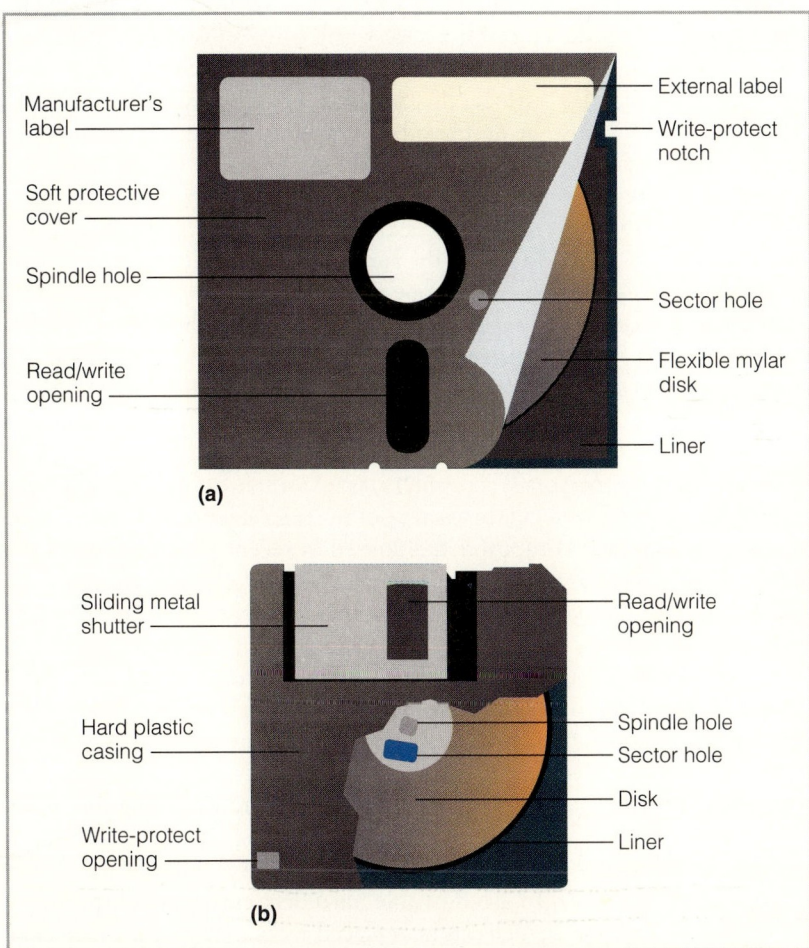

Figure 4-2 Diskettes. (a) A cutaway view of a 5¼-inch diskette. (b) A cutaway view of a 3½-inch diskette.

- **New software.** Although, for convenience, software packages are kept on hard disk, new software out of the box may come on diskettes (new software also may come on CD-ROM disks, which we will discuss shortly).

Hard Disks

A **hard disk** is a metal platter coated with magnetic oxide that can be magnetized to represent data. Hard disks come in a variety of sizes. Hard disks for mainframes and minicomputers may be as large as 14 inches in diameter (Figure 4-3). Several disks can be assembled into a **disk pack**. There are different types of disk packs, with the number of platters varying by model. Each disk in the pack has top and bottom surfaces on which to record data. Many disk devices, however, do not record data on the top of the top platter or on the bottom of the bottom platter.

A **disk drive** is a machine that allows data to be read from a disk or written on a disk. A disk pack is mounted on a disk drive that is a separate unit connected to the computer. Large computers have dozens or even hundreds of disk drives. In a disk pack all disks rotate at the same time, although only one disk is being read or written on at any one time. The mechanism for reading or writing data on a disk is an **access arm**; it moves a read/write head into position over a particular track (Figure 4-4a). The **read/write head** on the end of the access arm hovers just above the track but does not actually touch the surface. When a read/write head does accidentally touch the disk surface, this is called a **head crash** and all data is destroyed. Data can also be destroyed if a read/write head encounters even minuscule foreign matter on the disk surface (Figure 4-4b). A disk pack has a series of access arms that slip in between the disks in the pack (Figure 4-4c). Two read/write heads are on each arm, one facing up for the surface above it and one facing down for the surface below it. However, only one read/write head can operate at any one time.

In some disk drives the access arms can be retracted; then the disk pack can be removed from the drive. Most disk packs, however, combine the disks, access arms, and read/write heads in a **sealed module** called a **Winchester disk**. Winchester disk assemblies are put together in clean rooms so even microscopic dust particles do not get on the disk surface.

Hard disks for personal computers are 5¼-inch or 3½-inch disks in sealed modules and even gigabytes are not unusual (Figure 4-5). Hard disk capacity for personal computers has soared in recent years; capacities of hundreds of megabytes are common and gigabytes are not unusual. Although an individual probably cannot imagine generating enough output—letters, budgets, reports, and so forth—to fill a hard disk, software packages take up a lot of space and can make a dent rather quickly. Furthermore, graphics images and audio and video files require large file capacities. Perhaps more important than capacity, however, is the convenience of speed. Personal computer users find accessing files on a hard disk is significantly faster and thus more convenient than accessing files on a diskette.

Personal computer users, who never seem to have enough hard disk storage space, may turn to a **removable hard disk cartridge**. Once full, a removable hard disk cartridge can be replaced with a fresh one. In effect, a removable cartridge is as portable as a diskette, but the disk cartridge holds much more data. Removable units also are important to businesses concerned with security, because the units can be used during business

(a)

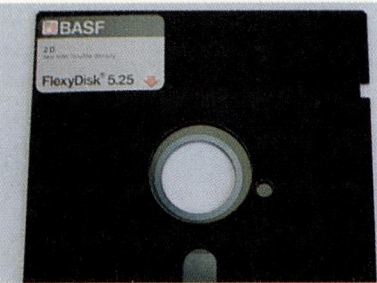

(b)

(c)

Figure 4-3 Magnetic disks. (a) Hard magnetic disks come in a variety of sizes (b) This 5¼-inch diskette is in a square protective paper jacket. (c) These 3½-inch diskettes are protected by a firm plastic exterior cover.

Chapter Four ▼ Storage Devices and Stored Data

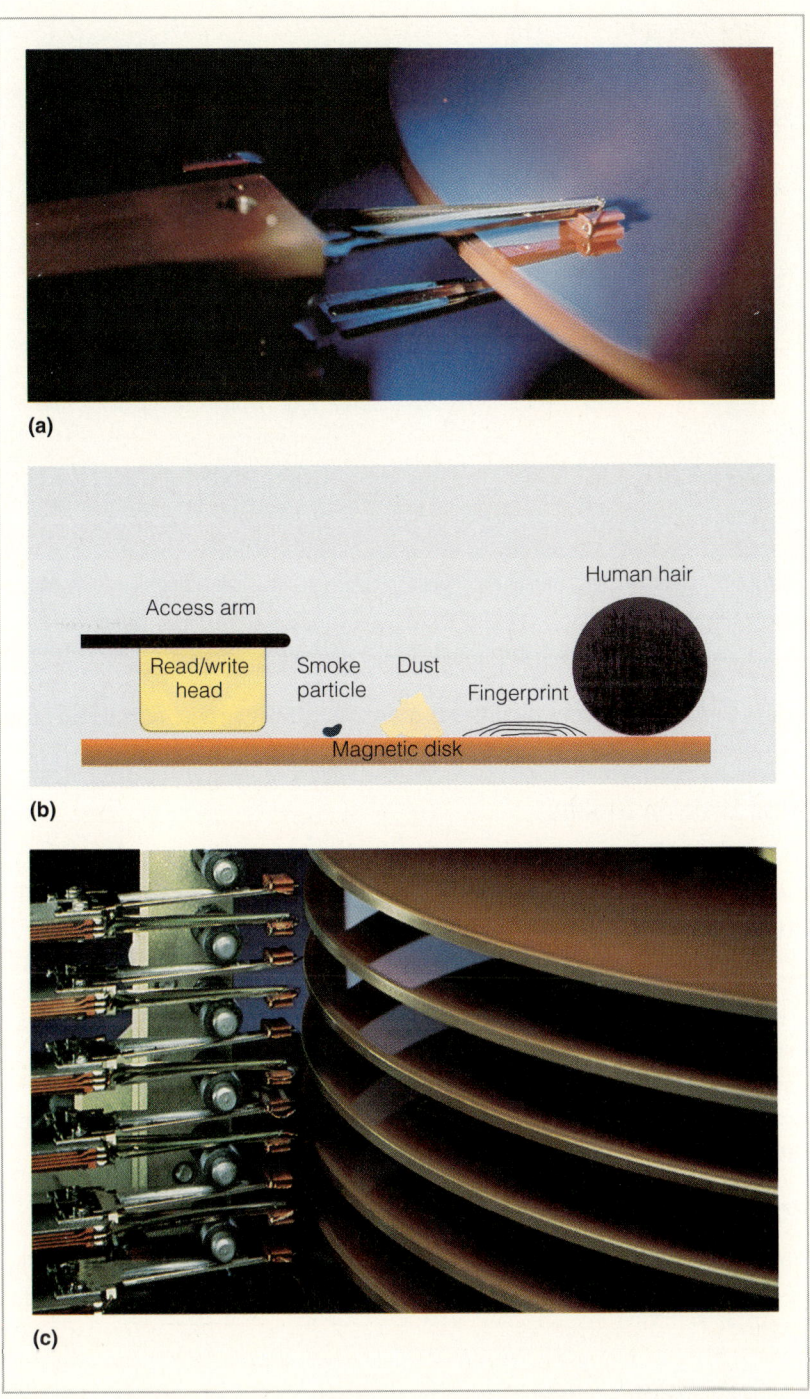

Figure 4-4 Read/write heads and access arms. (a) This photo shows a read/write head on the end of an access arm poised over a hard disk. (b) When in operation the read/write head comes very close to the surface of the disk. On a disk, particles as small as smoke, dust, a fingerprint, and a hair loom large. If the read/write head encounters one of these, data is destroyed and the disk damaged. (c) Note that there are two read/write heads on each access arm. Each arm slips between two disks in the disk pack. The access arms move simultaneously, but only one read/write head operates at any one time.

Figure 4-5 Hard disk for a personal computer. Innards of a 3½-inch hard disk with the access arm visible.

What's in a Name?

Most of us reconciled ourselves long ago to the fact that our names get passed around on computer files. This annoyance is part of the price we pay for the computer age. But what if you were *not allowed* to have your name on a computer file? This might prove to be a much bigger annoyance, as it was for the gentleman named Stephen O.

Mr. O—yes, that is a single letter for his last name—has been turned down for credit cards, insurance, a compact disk club, and even a driver's license. None of their computer systems could handle a single-character last name. The last straw was a credit bureau reference to Mr. O—Mr. "zero." Mr. O gave up and had his name legally changed to Stephen Oh. Now his name is accepted on computer files.

hours but hidden away during off hours. A disadvantage of a removable hard disk is that it takes longer to access data than a built-in hard drive.

Hard Disks in Groups

A concept of using several small disks that work together as a unit is called a **redundant array of inexpensive disks,** or simply **RAID** (Figure 4-6). The group of connected disks operates as if it were just one large disk, but it speeds up reading and writing by having multiple access paths. The data file for, say, aircraft factory tools, may be spread across several disks; thus, if the computer is used to look up tools for several workers, the computer need not read the data in turn but instead read them at the same time in parallel. Furthermore, data security is improved because if a disk fails, the disk system can reconstruct data on an extra disk; thus, computer operations can continue uninterrupted. This is significant data insurance.

How Data Is Organized on a Disk

There is more than one way of physically organizing data on a disk. The methods we will consider here are the sector method and the cylinder method.

The Sector Method

In the **sector method** each track is divided into sectors that hold a specific number of characters (Figure 4-7a). Data on the track is accessed by referring to the surface number, track number, and sector number where the data is stored. The sector method is used for diskettes as well as disk packs.

Zone Recording

The fact that a disk is circular presents a problem: The distances around the tracks on the outside of the disk are greater than that of the tracks on the inside. A given amount of data that takes up 1 inch of a track on the

Chapter Four ▼ Storage Devices and Stored Data

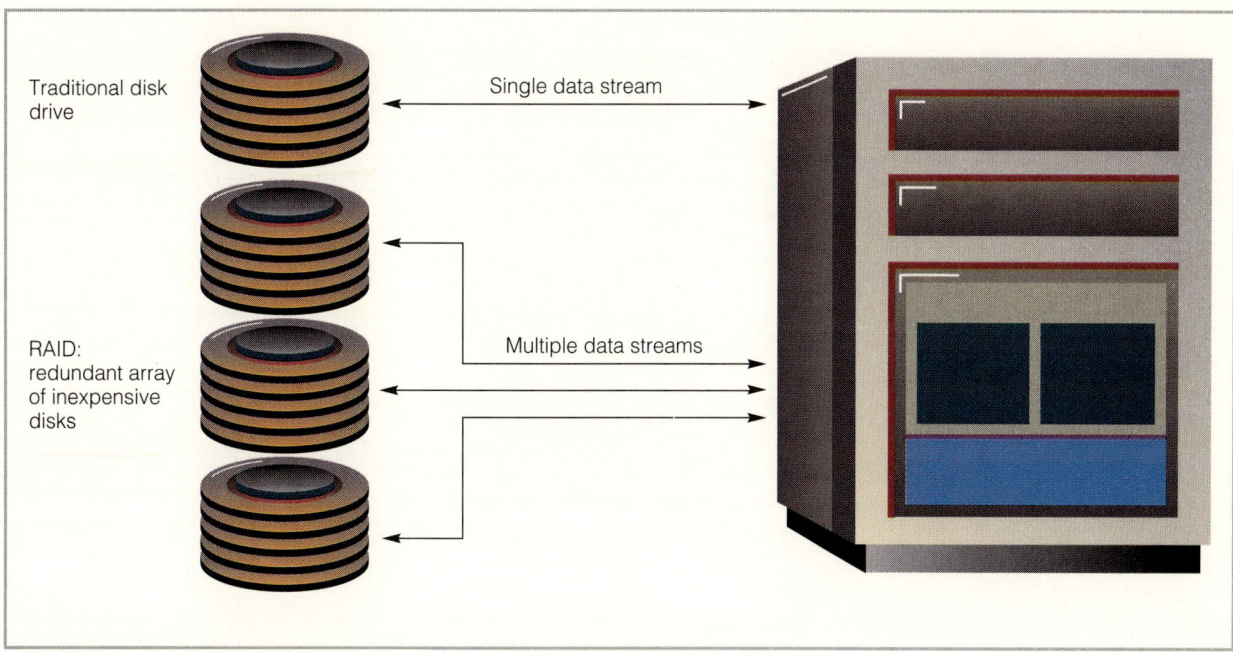

Figure 4-6   RAID. This illustration compares a traditional hard disk, which has a single path for data to travel back and forth to the central processing unit, with a redundant array of inexpensive disks (RAID), which has multiple access paths.

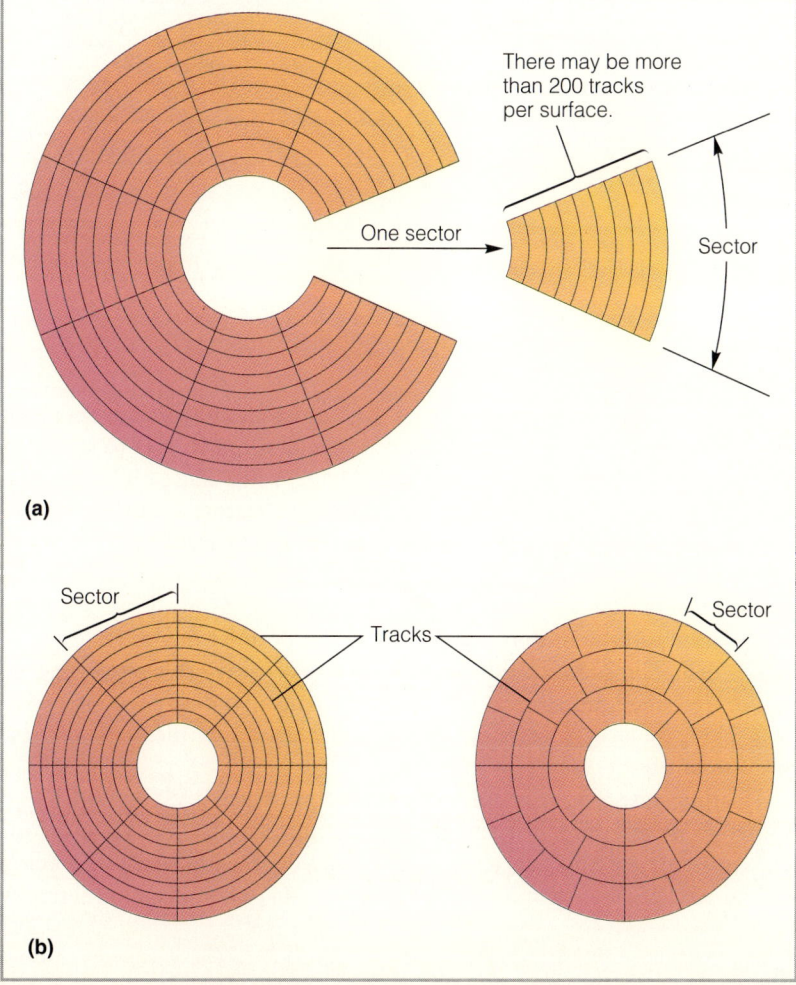

Figure 4-7   Sectors and zone recording. (a) When data is organized by sector, the address is the surface, track, and sector where the data is stored. (b) If a disk is divided into traditional sectors, as shown here on the left, each track has the same number of sectors. Sectors near the outside of the disk are wider, but they hold the same amount of data as sectors near the inside. If the disk is divided into recording zones, as shown on the right, the tracks near the outside have more sectors than the tracks near the inside. Each sector holds the same amount of data, but since the outer zones have more sectors, the disk as a whole holds more data than the disk on the left.

Tips for the Macintosh® Computer

Making Your Disk Space Count

Sooner or later it happens to almost everyone. No matter how inexhaustible we imagined our hard disks to be, the day arrives when we run out of storage space. One reason for this is that new applications use more memory than ever before. Luckily, storage devices are getting less expensive, and hard disks provide more megabytes per dollar as the years go by. The 80-megabyte disk that seemed so boundless in capacity a few years ago is now considered minimal, as users are opting for larger hard disks when they purchase their computers.

The increasing demand for disk space is likely to continue as storage prices decline. If you want your computer to be the powerful machine it was intended to be, you need to pay attention to your disk's capacity.

One way of increasing your storage potential is to invest in a compression utility program. SuperDoubler, developed by Symantec Corporation, compresses files to as little as half their size. When the files need to be accessed, SuperDoubler decompresses them back into their normal format.

A removable hard drive can increase the capacity of your system's memory indefinitely with disks that have 40 MB or 80 MB of memory.

SuperDoubler's enormous popularity is largely due to its ability to compress your files "in the background," so that you never have to go through the steps of compressing and decompressing them. SuperDoubler can nearly double your available disk space, and the process is invisible to you, the user. The only problem with this option is that the activities of compression and decompression require some of your processor's time and can slow the performance of your machine a little.

As an alternative, you should consider investing in a larger disk. If you already have an internal hard drive in your Mac and you don't want to replace it, look into an external drive, which resides in a case of its own and sits next to or under your Mac. If you anticipate that your storage needs will continue to grow, you might consider buying a removable hard drive, for which you can purchase indefinitely many 40 MB or 80 MB disks that pop in and out of the drive, just like 3½-inch diskettes. If you work heavily with sounds, images, or desktop publications, a removable drive can be of great value to you. Another advantage of a removable drive is that you can easily port huge amounts of data to other machines that have the same kind of drive.

If you do buy a new hard disk, make sure to invest in plenty of extra capacity. The trend of programs and data files getting larger is one that is likely to continue, so planning now will ensure you have enough disk space to last years; now, most users are buying machines with hundreds of megabytes for hard drive space.

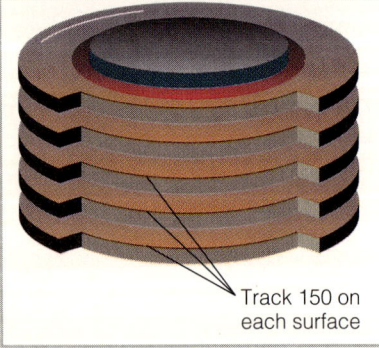

Figure 4-8 Cylinder data organization. To visualize the cylinder form of organization, imagine that a cylinder such as a tin can were dropped straight down through all the disks in the disk pack. Within cylinder 150, the track surfaces are vertically aligned, and numbered vertically from top to bottom.

inside of a disk might be spread over several inches on a track near the outside of a disk. This means that the tracks on the outside are not storing data as efficiently.

Zone recording involves dividing a disk into zones to take advantage of the storage available on all tracks, by assigning more sectors to tracks in outer zones than to those in inner zones (Figure 4-7b). Since each sector on the disk holds the same amount of data, more sectors mean more data storage than if all tracks had the same number of sectors.

The Cylinder Method

A way to organize data on a disk pack is the **cylinder method**, shown in Figure 4-8. The organization in this case is vertical. The purpose is to reduce the time it takes to move the access arms of a disk pack into position. Once the access arms are in position, they are in the same vertical position on all disk surfaces.

To appreciate this, suppose you had an empty disk pack on which you wished to record data. You might be tempted to record the data horizontally—to start with the first surface, fill track 000, then fill track 001, track 002, and so on, and then move to the second surface and again fill tracks

000, 001, 002, and so forth. Each new track and new surface, however, would require movement of the access arms, a relatively slow mechanical process.

Recording the data vertically, on the other hand, substantially reduces access arm movement. The data is recorded on the tracks that can be accessed by one positioning of the access arms—that is, on one **cylinder.** To visualize cylinder organization, pretend a cylindrically shaped item, such as a tin can, were figuratively dropped straight down through all the disks in the disk pack. All the tracks thus encountered, in the same position on each disk surface, comprise a cylinder. The cylinder method, then, means all tracks of a certain cylinder on a disk pack are lined up one beneath the other, and all the vertical tracks of one cylinder are accessible by the read/write heads with one positioning of the access arms mechanism. Tracks within a cylinder are numbered according to this vertical perspective: A 20-surface disk pack contains cylinder tracks numbered 0 through 19, top to bottom.

Figure 4-9 An optical disk. Optical disks store data using laser beam technology.

 ## Optical Disk Storage

The explosive growth in storage needs has driven the computer industry to provide cheaper, more compact, and more versatile storage devices with greater capacity. This demanding shopping list is a description of the **optical disk** (Figure 4-9). The technology works like this: A laser hits a layer of metallic material spread over the surface of a disk. When data is being entered, heat from the laser produces tiny spots on the disk surface. To read the data, the laser scans the disk, and a lens picks up different light reflections from the various spots.

Optical storage technology is categorized according to its read/write capability. **Read-only media** are recorded on by the manufacturer and can be read from but not written to by the user. Such a disk cannot, obviously, be used for your files, but manufacturers can use it to supply software. Applications software packages sometimes include a dozen diskettes or more; all these could fit on one optical disk with plenty of room to spare.

Write-once, read-many media, also called **WORM media,** may be written to once. Once filled, a WORM disk becomes a read-only media. A WORM disk is nonerasable. For applications demanding secure storage of original versions of valuable documents or data, the primary advantage of nonerasability is clear: Once they are recorded, no one can erase or modify them.

CD-ROM

A variation on optical technology is the **CD-ROM,** for **compact disk read-only memory.** CD-ROM has a major advantage over other optical disk designs: The disk format is identical to that of *audio* compact disks, so the same dust-free manufacturing plants that are now stamping out digital versions of Mozart or Mary Chapin Carpenter can easily convert to producing anything from software to an encyclopedia. Furthermore, CD-ROM storage is gargantuan—up to 660 megabytes per disk, the equivalent of over 400 3½-inch diskettes.

Keep in mind that a CD-ROM disk cannot be used in your personal computer's diskette drive; you must have a CD-ROM drive installed on your computer. Today, even some laptop computers have CD-ROM

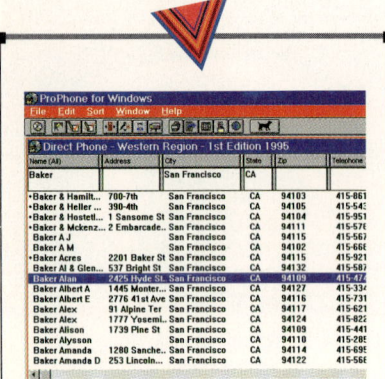

Found at Last

Stories of reunited families are so compelling that tearful airport embraces are often featured in newspapers and on television. Few people fit that scenario, but many people have lost track of someone they would like to find again—a childhood friend, a school chum, a neighbor, a colleague.

Until recently, the best recourse was to hire a private detective, an avenue few can afford. Another approach is directory assistance, but that also gets expensive quickly and, if a few years have passed since your last contact, would probably be unsuccessful.

CD-ROM to the rescue. For a modest investment (under $50) you can purchase software that includes all residential white page phone listings in the United States. Begin your search in the most likely states, and you may find your party in literally seconds. If your friend has moved across the country, it may take a little longer. You may encounter barriers, such as name changes and unlisted phones numbers. But a little creativity may lead you to a person who knows the whereabouts of the person for whom you are searching.

drives. If you have a CD-ROM drive, you could be on your way to one of the computer industry's great adventures: multimedia.

Multimedia

Software described as **multimedia** typically presents information with text, illustrations, photos, narration, music, animation, and film clips (Figure 4-10). Until the optical disk, placing this much data on a disk was impossible. However, the high volume capacity of optical disks means that the kinds of data that take up huge amounts of storage space—photographs, film clips, music—can now be accommodated.

To use multimedia software, you must have the proper hardware. In addition to the aforementioned CD-ROM drive, you also need a sound card (installed internally) and speakers, which may rest externally on either side of the computer or be built into the computer housing. Special software accompanies the drive and sound card. See the multimedia gallery for more information and photos of multimedia offerings.

Magneto-Optical

A hybrid disk, called **magneto-optical (MO),** combines the best features of magnetic and optical disk technologies. A magneto-optical disk has the high-volume capacity of an optical disk but can be written over as a magnetic disk. The disk surface is coated with plastic and embedded with magnetically sensitive metallic crystals. To write data, a laser beam melts a tiny spot on the plastic surface and a magnet aligns the crystals before the plastic cools. The crystals are aligned so that some reflect light and others do not. When the data is later read by a laser beam, only the crystals that reflect light are picked up.

 ## Magnetic Tape Storage

We saved magnetic tape storage for last because it has taken a subordinate role in storage technology. **Magnetic tape** looks like the tape used in music cassettes—plastic tape with a magnetic coating. As in other magnetic media, data is stored as extremely small magnetic spots. Tapes come in a number of forms, including ½-inch-wide tape wound on a reel, ¼-inch-wide tape in data cartridges and cassettes, and tapes that look like ordinary music cassettes but are designed to store data instead of music (Figure 4-11). The amount of data on a tape is expressed in terms of **density,** which is the number of **characters per inch (cpi)** or **bytes per inch (bpi)** that can be stored on the tape.

The highest-capacity tape is the **digital audio tape,** or **DAT,** which uses a different method of recording data. Using a method called **helical scan recording,** DAT wraps around a rotating read/write head that spins vertically as it moves. This places the data in diagonal bands that run across the tape rather than down its length. This method produces high density and faster access to data.

Figure 4-12a shows a **magnetic tape unit** that might be used on a minicomputer or mainframe. (A unit that uses cartridges or cassettes would be much smaller.) The tape unit reads and writes data using a **read/write head** (Figure 4-12b). When the computer is writing on the tape, the **erase head** first erases any data previously recorded on the medium.

Figure 4-10 Multimedia applications. (a) A multimedia package that tours the London Gallery of Art offers narrated commentary on hundreds of paintings. Shown here is Seurat's *Bathers at Asnieres.* (b) A very tuneful package indeed, the screens from the package Musical Instruments exhort users to click icons to play an instrument in a variety of ways. Here, the French horn can be ripped (played rapidly) or hand stopped (moving hand to change pitch). (c) The wonderful Dinosaurs offering shows dinosaurs of every size and shape, complete with ominous dinosaur snorting. Shown here are baryonyx fishing. (d) This screen photo is taken from a multimedia offering called From Alice to Ocean, which beautifully chronicles a young woman's journey, with her camels, across the Australian outback.

Two reels are used, a **supply reel** and a **take-up reel.** The supply reel, which has the tape with data on it or on which data will be recorded, is the reel that is changed. The take-up reel always stays with the magnetic tape unit. Many cartridges and cassettes have the supply and take-up reels built into the same case.

Tape now has a limited role because disk has proved the superior storage medium. Disk data is quite reliable, especially within a sealed module. Furthermore, as we will see, disk data can be accessed directly, as opposed to data on tape, which can be accessed only by passing by all the data ahead of it on the tape. Consequently, the primary role of tape today is as an inexpensive backup medium.

Figure 4-11 Magnetic tape. Magnetic tape comes in several forms, including (a) the ½-inch tape reels, the workhorse of the large computer system. Other forms of magnetic tape are (b) cartridges, cassettes, and digital audio tape, which uses helical scan recording technology.

(a) (b)

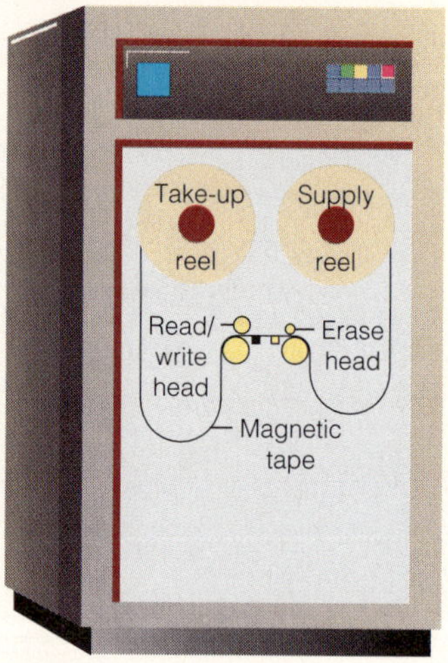

Figure 4-12 Magnetic tape units. Tapes are always protected by glass from outside dust and dirt. (a) Magnetic tape on reels is run on these traditional tape drives. (b) This diagram highlights the read/write head and the erase head found in magnetic tape units. (c and d) These modern tape drives, called "stackers," accept several cassette tapes, each with its own supply and take-up reels. Files can flow from one cassette to another.

ONE JUMP AHEAD

Searching for the Killer App

It is a slang term—killer app—used casually in the computer industry. Killer app is short for *killer application,* the software so important, so useful, so integral to living that most everyone would want to—*have to*—buy it.

There have been killer apps before. Perhaps the most famous is spreadsheet software, which made business people realize that the personal computer was not a toy. They *had to* have the small computer with the useful software; this killer app propelled the personal computer into businesses everywhere. Killer apps to a greater or lesser degree are word processing, desktop publishing, database management, graphics, and data communications.

The talk these days is about a multimedia killer app. What unique application might convince users that they simply must have a CD-ROM disk drive and the killer app software? So far, multimedia applications have emphasized entertainment and information. As dazzling as they may be, these applications are not yet compelling to a mass audience. Partnerships among software, communications, and cable providers are most likely to produce the can't-live-without application. Thus far, multimedia apps are filler, not killer.

 ## Backup Systems

Although a hard disk is an extremely reliable device, a hard disk drive is subject to electromechanical failures that cause loss of data. Furthermore, data files, particularly those accessed by several users, are subject to errors introduced by users. There is also the possibility of errors introduced by software. With any method of data storage, a **backup system**—a way of storing data in more than one place to protect it from damage and errors—is vital. As we have already noted, magnetic tape is used primarily for backup purposes. For personal computer users, an easy and inexpensive way to back up a hard disk file is to simply copy it to a diskette whenever it is updated. But this is not practical for a system with many files or many users.

Personal computer users have the option of purchasing their own tape backup system, to be used on a regular basis for copying all data from hard disk to a high-capacity tape (Figure 4-13). Data thus saved can be restored to the hard disk later if needed. A key advantage of a tape backup system is that it can copy the entire hard disk in minutes, saving you the trouble of swapping diskettes in and out of the machine.

Figure 4-13 Tape backup systems. These tape backup units—one for external and the other for internal use—permit backup of part or all of a hard disk for security or archival purposes.

▶ Checking an Earlier Decision

We left Ken Koll with his personal computer equipped with a 540MB hard disk. Now that we have considered all kinds of storage media, we can take a more informed look at Ken's decision. The capacity of the hard disk was just fine for Ken's original business purposes. The hard disk also accommodated his initial forays into multimedia, a reference work and two games (the multimedia offerings used optical disk but also required hard disk space). But he soon began branching out in other directions, using various types of software: He planned bids, analyzed financial data, and even produced some complex graphics. The software for these activities took up a surprising amount of space on his hard drive. But the clincher was multimedia: When Ken installed an encyclopedia, an art gallery, and several historical offerings, the disk became crowded.

A rule of thumb among computer professionals is to estimate disk needs generously and then double that amount. But estimating future needs is rarely easy. Many users, therefore, make later adjustments. Ken added a removable hard disk cartridge to accommodate his expanding storage needs. To quote Ken, "I just couldn't envision how I could use all that disk space. Now I can imagine even the extra disk filling up."

▶ Organizing and Accessing Stored Data

As users of computer systems, we just offer data as we are instructed to do, such as punching in our identification code at an automated teller machine or perhaps filling out a form with our name and address. But data cannot be dumped helter-skelter into a computer. Some computer professional—probably a programmer or systems analyst—had to plan how data from users will be received, organized, and stored, and also in what manner data will be processed by the computer. First consider how data is organized for processing.

Data: Getting Organized

To be processed by the computer, raw data is organized into characters, fields, records, files, and databases. We will start with the smallest element, the character.

- A **character** is a letter, digit, or special character (such as $, ?, or *).
- A **field** contains a set of related characters. For example, suppose a health club is making address labels for a mailing. For each person it might have a member-number field, a name field, a street-address field, a city field, a state field, a zip-code field, and a phone-number field.
- A **record** is a collection of related fields. Thus, on the health club mailing list, one person's member number, name, address, city, state, zip code, and phone number constitute a record.
- A **file** is a collection of related records. All the member records for the health club compose a file. Figure 4-14 shows how the health club data might look.
- A **database** is a collection of interrelated files stored together with minimum redundancy. Specific data items can be retrieved for various applications. For instance, if the health club is opening a new outlet, it can

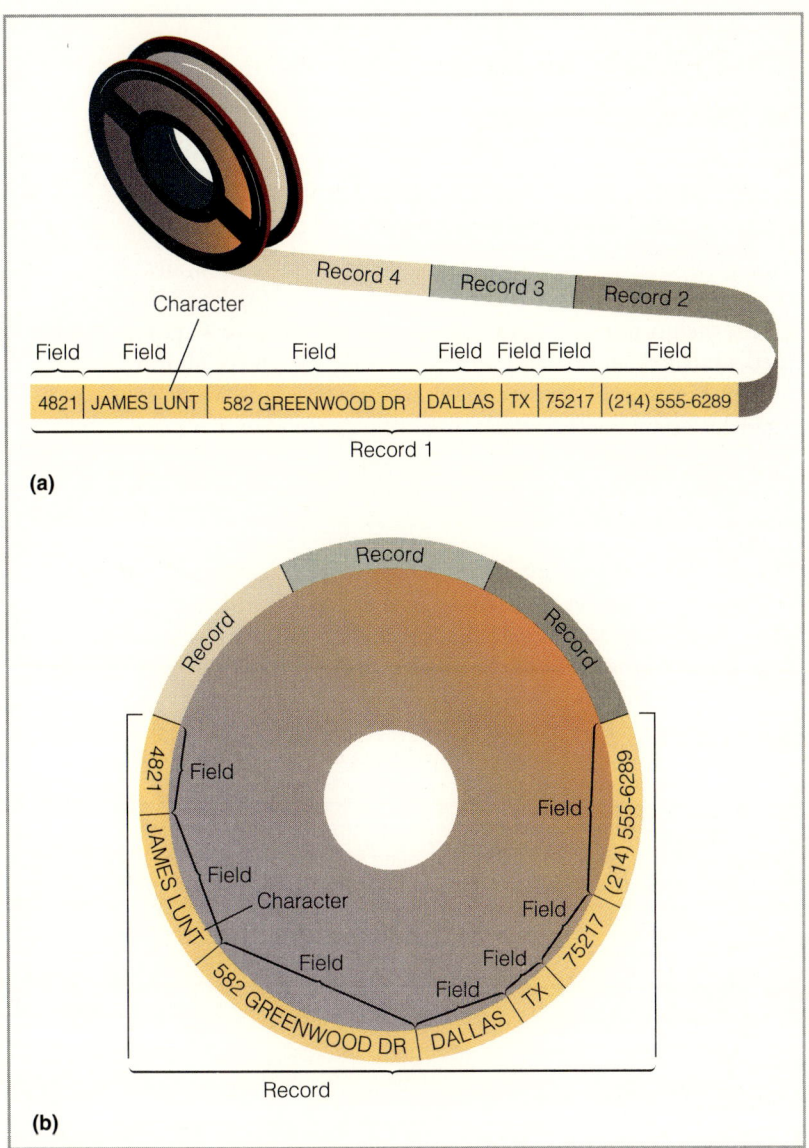

Figure 4-14 How data is organized. Whether stored on tape or on disk, data is organized into characters, fields, records, and files. A file is a collection of related records. These drawings represent (a) magnetic tape and (b) magnetic disk.

pull out the names of people with zip codes near the new club, and send them an announcement. The concept of a database is complicated; we will return to it in detail in Chapter 15.

A field of particular interest is the field called the **key**, a unique identifier for a record. It might seem, at first, that name—of a person, say, or a product—would be a good key; however, since people and products may have the same name, a name field is not a choice for a key. When a file is first computerized, existing description fields are seldom used as keys. Although a file describing people might use a Social Security number as the key, it is more likely that a new field will be developed that can be assigned unique values, such as customer number or product number.

In addition to organizing the expected data, a plan must be made to access the data on files.

The File Plan: An Overview

Now that we have a general idea of how data is organized, we need to consider what way would be appropriate to place data on a storage medium—tape or disk. Consider this chain: (1) It is the application—payroll, airline reservations, inventory control, whatever—that determines the way the data must be accessed by users. (2) Once an access method has been determined, then it follows that there are certain ways the data must be organized so that the needed access is workable. (3) The organization method, in turn, limits what storage medium may be used. We will discuss both organization and access in detail, but let us begin with an appreciation of application demands.

Consider these application examples to see how an access decision might be made.

1. A department store offers its customers charge accounts. When a customer makes a purchase, a sales clerk needs to be able to check the validity of the customer's account while the customer is waiting. The clerk needs immediate access to the individual customer record in the account file.
2. A major oil company supplies its charge customers with credit cards, which it considers sufficient proof for purchase. The charge slips collected by gas stations are forwarded to the oil company, which processes them in order of account number. Unlike the retail example above, the company does not need access to any one record at a specific time but merely needs access to all customer charge records when it is time to prepare bills.
3. A city power and light company employee accepts reports of burned-out streetlights from residents over the phone. Using a key made up of unique address components, the clerk immediately finds the record for the offending streetlight and prints out a one-page report that is routed to repair units within 24 hours. To produce such quick service based on an individual streetlight, the employee needs to be able to access the individual streetlight record.
4. Reports of airline flight attendant next-month schedules are computer-produced monthly and delivered to the attendants' home-base mailboxes. The schedules are put together from information based on flight records, whose entire file can be accessed monthly at the convenience of the airline and the computer-use plan.

As you can see, the question of access seems to come down to whether a particular record is needed right away, as it was in examples 1 and 3. As we will see, this immediate need for a particular record means access must be *direct*. It follows that the organization must also be direct, or at least *indexed*, and that the storage medium must be disk. Furthermore, the type of processing, a related topic, must be *transaction processing*. There are other possibilities, but the critical distinction is whether or not access to an individual record is needed. We now examine all these topics in detail. Although, as we have just seen, organization type is determined by the type of access required, the file must be organized before it can be accessed, so we begin there.

File Organization: Three Methods

There are three major methods of storing files of data in secondary storage:

- Sequential file organization, in which records are organized in a particular order
- Direct file organization, in which records are not organized in any special order
- Indexed file organization, in which records are organized sequentially, but indexes built into the file allow a record to be accessed either sequentially or directly

Sequential File Organization

Sequential file processing means records are in order according to a key field. As noted earlier, a people file will be in order by a key that uniquely identifies each person, such as Social Security number or customer number. If a particular record in a sequential file is wanted, then all the prior records in the file must be read before reaching the desired record. Tape storage is limited to sequential file organization. Disk storage may be sequential, but records on disk can also be accessed directly.

Direct File Organization

Direct file processing, or **direct access,** allows the computer to go directly to the desired record by using a record key; the computer does not have to read all preceding records in the file as it does if the records are arranged sequentially. Direct processing requires disk storage; in fact, a disk device is called a **direct-access storage device (DASD)** because the computer can go directly to the desired record on the disk. It is this ability to access any given record instantly that has made computer systems so convenient for people in service industries—for catalog order-takers determining if a particular sweater is in stock, for example, or bank tellers checking individual bank balances. An added benefit of direct access organization is the ability to read, change, and return a record to its same place on the disk; this is called **updating in place.**

Obviously, if we have a completely blank area on the disk and can put records anywhere, then there must be some predictable system for placing a record at a disk address and then retrieving the record at a subsequent time. In other words, once the record has been placed on a disk, it must be possible to find it again. This is done by choosing a certain formula to apply to the record key, thereby deriving a number to use as the disk address. **Hashing,** or **randomizing,** is the name given to the process of applying a mathematical operation to a key to yield a number that represents the address. Even though the record keys are unique, it is possible for a hashing scheme to produce the same disk address, called a **synonym,** for two different records; such an occurrence is called a **collision.** There are various ways to recover from a collision; one way is simply to use the next available record slot on the disk.

There are many different hashing schemes; although the example in Figure 4-15 is too simple to be realistic, you can get a general idea of how the process works. An example of how direct processing works is provided in Figure 4-16.

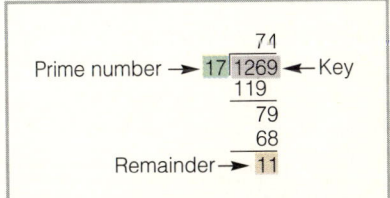

Figure 4-15 A hashing scheme. Dividing the key number 1269 by the prime number 17 yields remainder 11, which can be used to indicate the address on a disk.

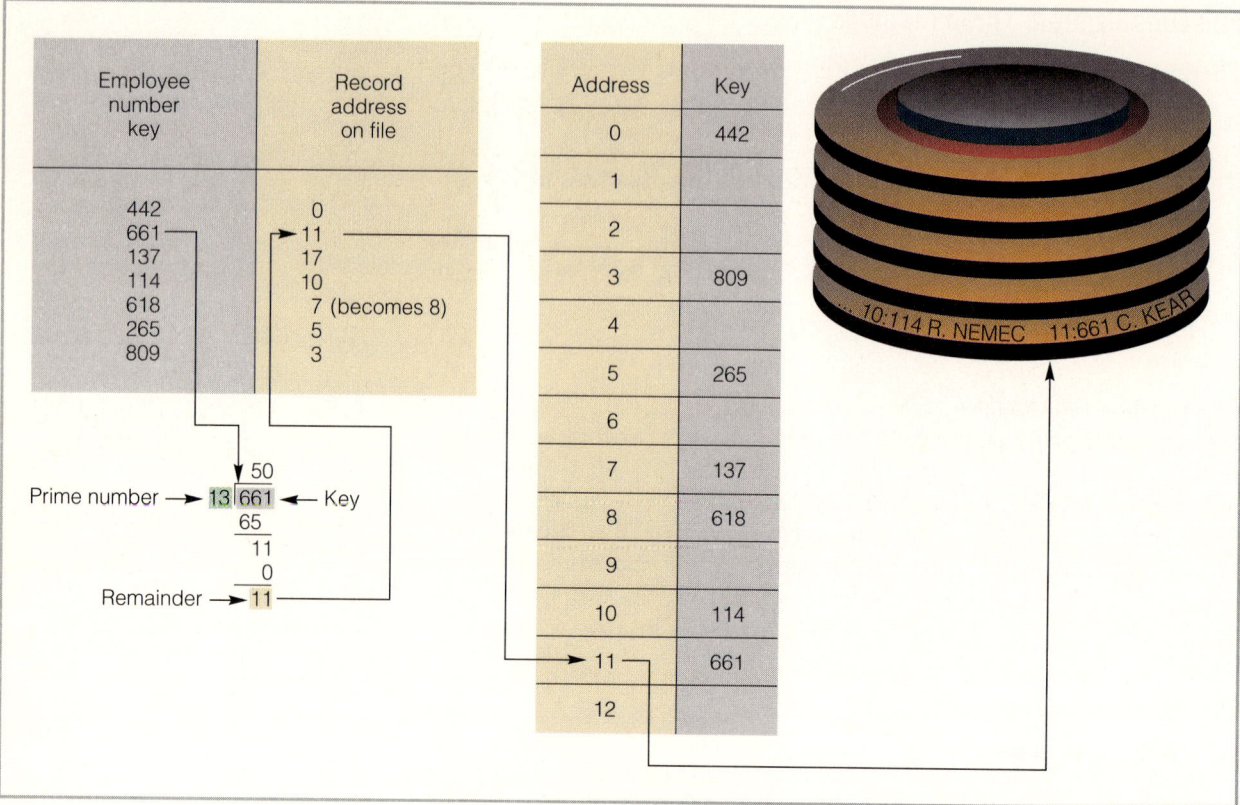

Figure 4-16 An example of direct access. Assume there are 13 addresses (0 through 12) available in the file. Dividing the key number 661, which is C. Kear's employee number, by the prime number 13 yields remainder 11. Thus, 11 is the address for key 661. However, for the key 618, dividing by 13 yields remainder 7, a synonym, since this address has already been used by the key 137, which also has a remainder of 7. Hence, the address becomes the next location—that is, 8. Note, incidentally, that keys (and therefore records) need not appear in any particular order. (The 13 record locations available are, of course, too few to hold a normal file; a small number was used to keep the example simple.)

Indexed File Organization

Indexed file processing, or **indexed processing,** is a third method of file organization, and it represents a compromise between the sequential and direct methods. It is useful in applications where a file needs to be in sequential order but, in addition, access to individual records is needed.

An indexed file works as follows: Records are stored in the file in sequential order, but the file also contains an index. The index contains entries consisting of the key to each record stored on the file and the corresponding disk address for that record. The index is like a directory, with the keys to all records listed in order. To access a record directly, the record key must be located in the index; the address associated with the key is then used to locate the record on the disk. To access the entire file of records sequentially, begin with the first record and proceed through the rest of the records.

Before we proceed with the actual processing of data, we need to pause briefly to consider the physical activity of the disk as it accesses records directly.

MAKING THE RIGHT CONNECTIONS

The Ultimate Connection

Can we really be computer-connected to planet Jupiter? Not literally. But if you can access the Internet, you can access a file with hundreds of space photos from NASA and observatories around the world. The photos can be shown right on your computer screen. The space photos have been so popular that they have been accessed by approximately 1 million users per month. The photo on the left shows the explosion on Jupiter right after it was whacked by a comet in 1994. The image on the right is a computer simulation of the earth if given a similar blow.

The comet is called the Shoemaker-Levy comet, named for the scientists who discovered it and predicted its collision course. This is the first collision of two solar bodies ever to be observed as it happened. As the comet approached Jupiter, it broke into 21 pieces, some as large as two miles in diameter, which struck the planet over the seven-day period from July 16 through July 22. Considered by scientists to be of tremendous importance, all of this was watched carefully by observatories around the world.

Disk Access to Data

Three primary factors determine **access time,** the time needed to access data directly on disk:

- **Seek time.** This is the time it takes the access arm to get into position over a particular track. Keep in mind that all the access arms move as a unit, so they are simultaneously in position over a set of tracks that comprise a cylinder.
- **Head switching.** The access arms on the access mechanism do not move separately; they move together, all at the same time. However, only one read/write head can operate at any one time. Head switching is the activation of a particular read/write head over a particular track on a particular surface. Since head switching takes place at the speed of electricity; the time it takes is negligible.
- **Rotational delay.** Once the access arm and read/write head is in position, ready to read or write data, the read/write head waits for a short period until the desired data on the track moves under it.

Once the data has been found, we must consider **data transfer,** the process of transferring data between memory and the place on the disk track—from memory to the track if you are writing, from the track to memory if you are reading.

One measure for the performance of disk drives is the average access time, which is usually measured in milliseconds (ms). Another measure is the **data transfer rate,** which tells how fast data can be transferred once it has been found. This usually will be stated in terms of megabytes of data per second.

Processing Stored Data

Now that there is a plan for accessing the files, they can be processed. There are several methods of processing data files in a computer system.

The two main methods are batch processing (processing data in groups at a more convenient later time) and transaction processing (processing data immediately, as it is received).

Batch Processing

Batch processing is a technique in which transactions are collected into groups, or batches, to be processed at a time when the computer may have few online users and thus be more accessible, usually during the night. Unlike transaction processing, which we will examine momentarily, batch processing includes no direct user interaction. Let us suppose that we are going to update the health club address-label file. The **master file,** a semi-permanent set of records, is, in this case, the list of all members of the health club and their addresses. The **transaction file** contains all changes to be made to the master file: additions (transactions to create new master records for new members), deletions (transactions with instructions to delete master records of members who have resigned from the health club), and revisions (transactions to change items such as street addresses or phone numbers in fields in the master records). Periodically, perhaps monthly or weekly, the master file is **updated** with the changes called for in the transaction file. The result is a new, up-to-date master file (Figure 4-17).

In batch processing, before a transaction file is matched against a master file, the transaction file must be sorted (usually by computer) so that all the transactions are in sequential order according to a key field. In updating the health club address-label file, the key is the member number assigned by the health club. The records on the master file are already in order by key. Once the changes on the transaction file are sorted by key, the two files can be matched and the master file updated.

During processing, the computer matches the keys from the master and transaction files, carrying out the appropriate action to add, revise, or delete. At the end of processing, a newly updated master file is created; in addition, an error report is usually printed. The error report shows actions such as an attempt to delete a nonexistent record or an attempt to add a record that already exists.

Transaction Processing

Transaction processing is a technique of processing transactions—a bank withdrawal, an address change, a credit charge—in random order, that is, in any order they occur. Transaction processing is real-time processing. **Real-time processing** means that a transaction is processed fast enough for the result to come back and be acted upon right away. For example, a teller at a bank can find out immediately what your bank balance is. For processing to be real-time, it must also be **online**—that is, the terminals must be connected directly to the computer. Transaction processing systems use disk storage because the disk drive can move directly to the desired record.

Advantages of transaction processing are immediate access to stored data (and thus immediate customer service), and immediate updating of the stored data. A salesclerk, for example, could access the computer via a terminal to verify the customer's credit and also record the sale via the computer (Figure 4-18). Later, by the way, those updated records can be batch processed to bill all customers.

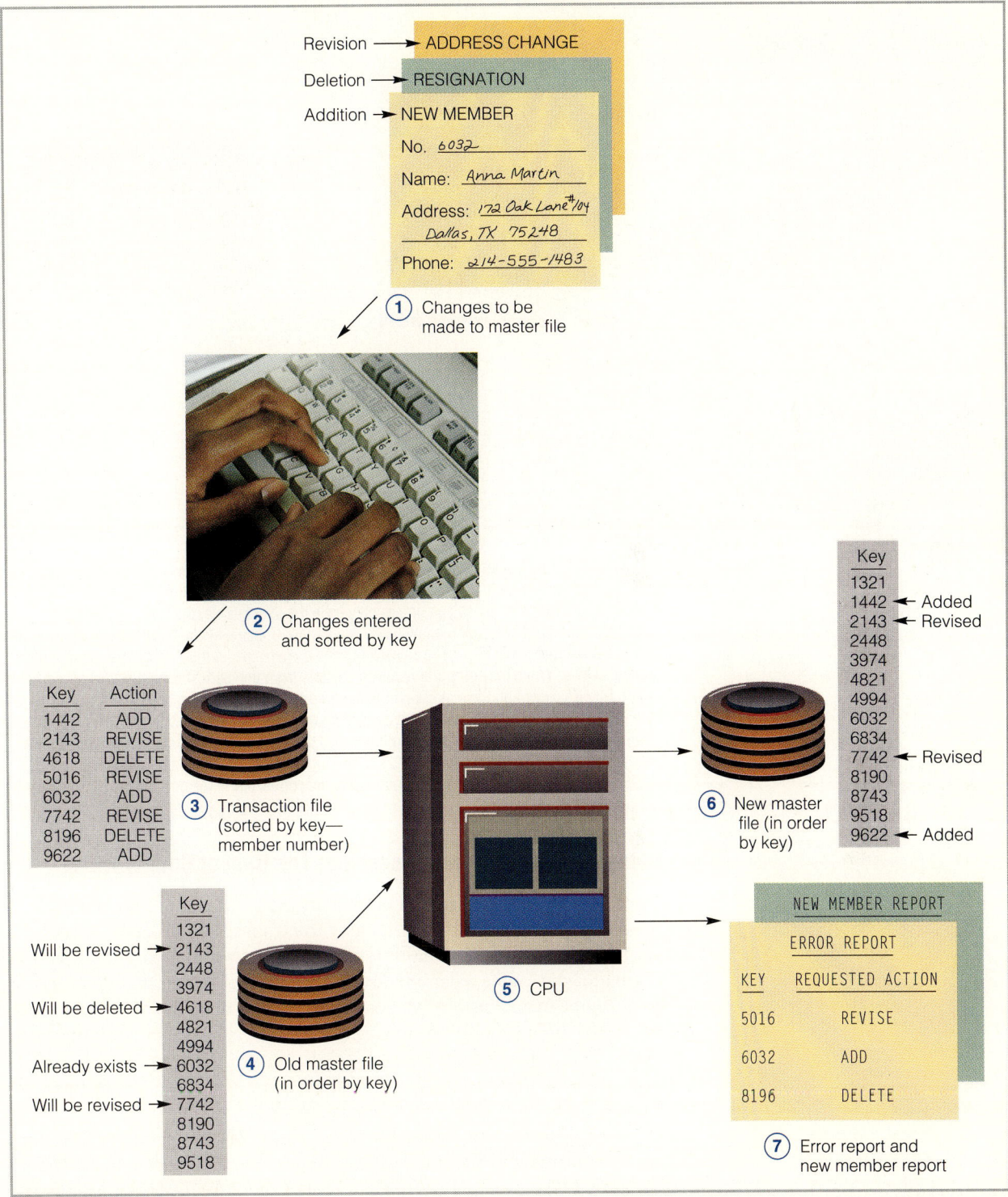

Figure 4-17 How batch processing works. The purpose of this system is to update the health club's master address-label file. The updating will be done sequentially. ① Changes to be made (additions, deletions, and revisions) are input with ② a keyboard, sorted, and sent to a disk, where it is ③ the transaction file. The transaction file contains records in sequential order, according to member number, from lowest to highest. The field used to identify the record is called the key; in this instance the key is the member number. ④ The master file is also organized by member number. ⑤ The computer matches transaction file data and master file data by member number to produce ⑥ a new master file and ⑦ an error report and a new member report. Note that, since this was a sequential update, the new master file is a completely new file, not just the old file updated in place. The error report lists member numbers in the transaction file that were not in the master file and member numbers that were included in the transaction file as additions that were already in the master file.

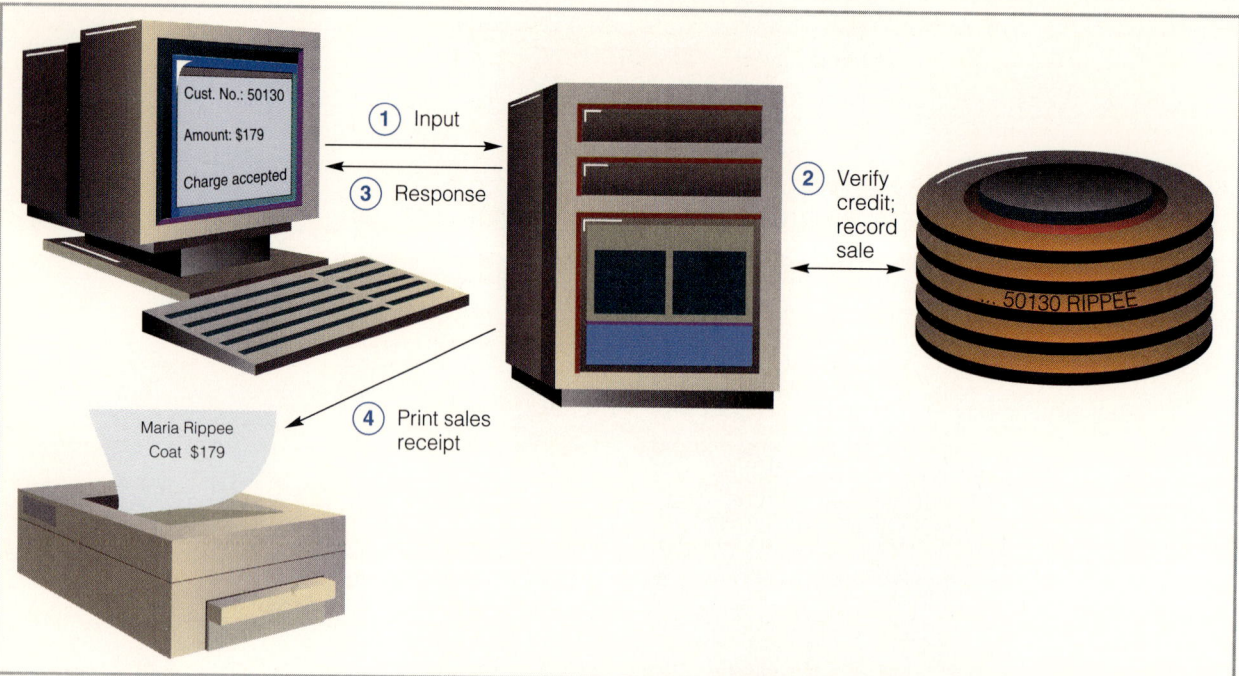

Figure 4-18 How transaction processing works. The purposes of this retail sales system are to verify that a customer's credit is good, record the credit sale on the customer's record, and produce a sales receipt. Since customers may have the same name, the file is organized by customer account number rather than by name. Here Maria Rippee, account number 50130, wishes to purchase a coat for $179. ① The salesclerk uses the terminal to input Maria's account number and the sale. ② When the computer receives the data from the clerk, it uses the account number to find Maria's record on the disk file, verify her credit, and record the sale so that she will later be billed for it. ③ The computer returns an acceptance back to the clerk's terminal. ④ The computer sends sales receipt information to the clerk's printer. All this is done within seconds while the customer is waiting. This example is necessarily simplified, but it shows a system that is real-time (immediate response), and online (directly connected to the computer).

Batch and Transaction Processing: The Best of Both Worlds

Numerous computer systems combine the best features of both methods of processing. A bank, for instance, may use transaction processing to check your balance and individually record your cash withdrawal transaction during the day at the teller window. However, the deposit that you leave in an envelope in an "instant" deposit drop may be recorded during the night by means of batch processing. Most store systems also combine both methods: A point-of-sale terminal finds the individual item price as a sale is made. But that same process captures inventory data, which may be batched and totaled to produce inventory reports.

Police license-plate checks for stolen cars work the same way. As cars are sold throughout the state, the license numbers, owners' names, and so on, are updated in the motor vehicle department's master file, usually via batch processing on a nightly basis. But when police officers see a car they suspect may be stolen, they can radio headquarters, where an operator with a terminal uses transaction processing to check the master file immediately to see if the car was reported missing.

Auto junk yards, which often are computerized big businesses, can make an individual inquiry for a record of a specific part needed by a customer waiting on the phone or in person. As parts are sold, sales records are kept to update the files nightly using batch processing.

▼ ▲ ▼

What is the future of storage? Perhaps holographic storage, which would provide gigabytes of capacity and be much faster than even the fastest hard drives. Whatever the technology, it seems likely that we will be seeing greater storage capabilities in the future to hold the huge data files for law, medicine, science, education, business, and, of course, the government.

To have access to all that data from any location, we need data communications, to which we turn in the next chapter.

CHAPTER REVIEW

Summary and Key Terms

- **Secondary storage,** sometimes called **auxiliary storage,** is storage separate from the computer itself, where you can store software and data on a semipermanent basis. Secondary storage is necessary because memory, or primary storage, can be used only temporarily.
- The benefits of secondary storage are **space, reliability, convenience,** and **economy.**
- Diskettes and hard disks are magnetic media, based on a technology of representing data as magnetized spots on the disk—with a magnetized spot representing a 1 bit and the absence of such a spot representing a 0 bit. The surface of each disk has concentric tracks on it.
- **Diskettes** are made of flexible mylar. Older diskettes are 5¼ inches in diameter, but newer computers use the 3½-inch diskette, which has a hard plastic jacket, a convenient size, greater storage capacity. Advantages of diskettes, compared to hard disks, are portability, backup, and delivery of new software to consumers.
- A **hard disk** is a metal platter coated with magnetic oxide that can be magnetized to represent data. Several disks can be assembled into a **disk pack.**
- A **disk drive** is a machine that allows data to be read from a disk or written on a disk. A disk pack is mounted on a disk drive that is a separate unit connected to the computer. The disk **access arm** moves a **read/write head** into position over a particular track, where the read/write head hovers above the track. A **head crash** occurs when a read/write head touches the disk surface and causes all data to be destroyed. Most disk packs combine the disks, access arms, and read/write heads in a **sealed module** called a **Winchester disk.**
- Hard disks for personal computers are 5¼-inch or 3½-inch disks in sealed modules. Hard disk capacity for personal computers can be hundreds of millions of bytes or gigabytes; much of this will be for storing software. Personal computer users may supplement hard disk storage with a **removable hard disk cartridge,** which, once filled, can be replaced with a fresh one.
- A concept of using several small disks that work together as a unit is called **redundant array of inexpensive disks,** or simply **RAID.**
- The **sector method** of recording data on disk divides each track into sectors that hold a specific number of characters. Data on the track is accessed by referring to the surface number, track number, and sector number where the data is stored. **Zone recording** involves dividing a disk into zones to take advantage of the storage available on all tracks, by assigning more sectors to tracks in outer zones than to those in inner zones.
- The organization of data on disk using the **cylinder method** is vertical. When the access arms are in position, they are in the same vertical position for all disk surfaces. The set of tracks that can be accessed by one positioning of the access arms is called a **cylinder.**
- **Optical disk** technology uses a laser beam to enter data as spots on the disk surface. To read the data, the laser scans the disk, and a lens picks up different light reflections from the various spots. **Read-only media** are recorded on by the manufacturer and can be read from but not written to by the user. **Write-once, read-many media,** also called **WORM media,** may be written to once. **CD-ROM,** for **compact disk read-only memory,** which has a disk format identical to that of audio compact disks, can hold up to 660 megabytes per disk.
- **Multimedia** software typically presents information with text, illustrations, photos, narration, music, animation, and film clips—possible because the high volume capacity of optical disks can accommodate photographs, film clips, and music. To use multimedia software, you must have the proper hardware: a CD-ROM drive, a sound card, and speakers.
- A hybrid disk, called **magneto-optical (MO),** has the high volume capacity of an optical disk but can be written over as a magnetic disk. It uses both a laser beam and a magnet to properly align magnetically sensitive metallic crystals.

- **Magnetic tape** stores data as extremely small magnetic spots. The amount of data on a tape is expressed in terms of **density**, which is the number of **characters per inch (cpi)** or **bytes per inch (bpi)** that can be stored on the tape. The highest-capacity tapes are **digital audio tape,** or **DAT,** which use a different method of recording data. Using a method called **helical scan recording,** the data is placed in diagonal bands that run across the tape rather than down its length.
- A **magnetic tape unit** reads and writes data using a **read/write head;** when the computer is writing on the tape, the **erase head** first erases any data previously recorded on the medium. Two reels are used, a **supply reel** that has the data tape and a **take-up reel** that stays with the magnetic tape unit.
- A **backup system** is a way of storing data in more than one place to protect it from damage and loss. Most backup systems use tape.
- A **character** is a letter, digit, or special character (such as $, ?, or *). A **field** contains a set of related characters. A **record** is a collection of related fields. A **file** is a collection of related records. A **database** is a collection of interrelated files stored together with minimum redundancy; specific data items can be retrieved for various applications.
- The application determines the way the data must be accessed by users, and it follows that there are certain ways the data must be organized so that the needed access is workable. The organization method, in turn, limits what storage medium may be used.
- **Sequential file processing** means records are in a certain order by a unique identifier field called a **key.** If a particular record in a sequential file is wanted, then all the prior records in the file must be read before reaching the desired record.
- **Direct file processing,** or **direct access,** allows the computer to go directly to the desired record by using a record key. Direct processing requires disk storage; a disk device is called a **direct-access storage device (DASD).** In addition to instant access to any record, an added benefit of direct access organization is the ability to read, change, and return a record to its same place on the disk; this is called **updating in place. Hashing,** or **randomizing,** is the name given to the process of applying a formula to a key to yield a number that represents the address. A hashing scheme may produce the same disk address, called a **synonym,** for two different records; such an occurrence is called a **collision.**
- **Indexed file processing,** or **indexed processing,** stores records in the file in sequential order, but the file also contains an index of keys; the address associated with the key is then used to locate the record on the disk.
- Three factors determine **access time,** the time needed to access data directly on disk: **seek time,** to get the access arm into position over a particular track; **head switching,** the activation of a particular read/write head over a particular track on a particular surface; and **rotational delay,** the brief wait until the desired data on the track moves under the read/write head. Once data has been found, we refer to **data transfer,** the transfer of data between memory and the place on the disk track—from memory to the track if you are writing, from the track to memory if you are reading.
- Access time is usually measured in milliseconds (ms). The **data transfer rate,** which tells how fast data can be transferred once it has been found, is usually stated in terms of megabytes of data per second.
- **Batch processing** is a technique in which transactions are collected into groups, or batches, to be processed at a time when the computer has few online users and thus is more accessible. A **master file** is a semipermanent set of records. A **transaction file,** sorted by key, contains all changes to be made to the master file: additions, deletions, and revisions. The master file is **updated** periodically with the changes called for in the transaction file.
- **Transaction processing** is a technique of processing in any order they occur. **Real-time processing** means that a transaction is processed fast enough for the result to come back and be acted upon right away. **Online** processing means that the terminals must be connected directly to the computer.

Discussion Questions

1. If you were buying a personal computer today, what would you choose for secondary storage?
2. Can you think of massive data files that a personal computer user might want to have on optical disk? Can you imagine new multimedia applications that take advantage of sound, photos, art, and perhaps video?
3. Provide your own example to illustrate how characters of data are organized into fields, records, files, and (perhaps) databases. If you wish, you may choose one of the following examples: department store data, airline reservations, or Internal Revenue Service data. Would you organize these files directly or sequentially? Would your examples use batch processing, transaction processing, or a combination of the two?

Student Study Guide

Multiple Choice

1. The density of data stored on magnetic tape is expressed as
   a. units per inch
   b. tracks per inch
   c. packs per inch
   d. bytes per inch
2. Another name for secondary storage:
   a. cylinder
   b. density
   c. auxiliary
   d. memory
3. A magnetized spot represents
   a. cpi
   b. 0 bit
   c. MB
   d. 1 bit
4. A field contains one or more
   a. characters
   b. databases
   c. records
   d. files
5. The reel that is changed on a magnetic tape unit is the
   a. take-up reel
   b. RAID
   c. supply reel
   d. record
6. A hard disk can be backed up efficiently using
   a. zoning
   b. a tape backup system
   c. a transaction file
   d. WORM
7. Relatively permanent data is contained in
   a. a field
   b. memory
   c. a transaction
   d. a master file
8. A limitation of magnetic tape as a method of storing data is that it is
   a. not reusable
   b. organized sequentially
   c. relatively expensive
   d. not portable
9. DASD refers to
   a. disk storage
   b. tape storage
   c. field
   d. sorting
10. Optical disk technology uses
    a. helical scanning
    b. DAT
    c. laser beam
    d. RAID
11. The mechanism for reading or writing data on a disk is called a(n)
    a. track
    b. WORM
    c. key
    d. access arm
12. A disk pack within a sealed data module is a
    a. backup unit
    b. diskette
    c. Winchester
    d. CD-ROM
13. The time required for the access arm to get into position over a particular track is
    a. rotational delay
    b. seek time
    c. data transfer
    d. head switching
14. A way of physically organizing data on a disk pack to minimize seek time uses
    a. sequential file
    b. the cylinder method
    c. removable hard disk cartridge
    d. Winchester technology
15. The speed with which a disk can find data being sought is called
    a. access time
    b. direct time
    c. data transfer time
    d. cylinder time
16. The disk storage that uses both a magnet and a laser beam
    a. hashing
    b. CD-ROM
    c. magneto-optical
    d. WORM
17. Which is *not* a benefit of secondary storage?
    a. convenience
    b. economy
    c. DAT
    d. space
18. Before a sequential file can be updated the transactions must first be
    a. numbered
    b. sorted
    c. labeled
    d. updated
19. Hashing, to get an address, is the process of applying a formula to a
    a. key
    b. file
    c. record
    d. character
20. Personal computer users may wish to increase their hard disk storage capacity with
    a. higher density
    b. read-only media
    c. DAT
    d. removable hard disk cartridge

21. CD-ROM has the same format as a(n)
    a. backup tape   c. DAT
    b. diskette   d. audio compact disk
22. The concept of using several small disk packs that work together as a unit is
    a. CD-ROM   c. RAID
    b. WORM   d. MO
23. Assigning more sectors to outer disk tracks is called
    a. zone recording c. randomizing
    b. data transfer   d. sectoring
24. The ability to return a changed disk record to its original location is called
    a. magneto-optical c. rotational delay
    b. multimedia   d. updating in place
25. Processing transactions in groups is called
    a. data transfer   c. head switching
    b. transaction processing d. batch processing

True/False

T F 1. Real-time processing means a transaction is processed fast enough for the result to come back and be acted upon right away.
T F 2. Processing data by groups of transactions is called batch processing.
T F 3. A 0 bit is represented on magnetic disk by a magnetized spot.
T F 4. A magnetic tape unit records data on tape but cannot retrieve it.
T F 5. A transaction file contains records to update the master file.
T F 6. WORM can be written once; then it becomes read-only.
T F 7. Rotational delay comes before seek time.
T F 8. Density is the number of characters per inch.
T F 9. The most common backup medium is CD-ROM.
T F 10. Another name for randomizing is zoning.
T F 11. Transaction processing systems are real-time systems.
T F 12. Multimedia software can include film clips.
T F 13. Winchester technology places disks, access arms, and read/write heads in a sealed module.
T F 14. Magneto-optical refers to a special type of tape that records data diagonally.
T F 15. A magnetic disk has concentric tracks.
T F 16. Access time is measured in terms of megabytes.
T F 17. The cylinder method is a means of physically organizing data on a disk pack horizontally to minimize seek time.
T F 18. Direct file organization is a combination of sequential and indexed file organization.
T F 19. Zone recording takes full advantage of the space on a disk track.
T F 20. Hashing is a process used to locate records sequentially.
T F 21. RAID uses several large disks in place of several small disks.
T F 22. A field is a set of related records.
T F 23. Auxiliary storage can be used only temporarily.
T F 24. A database is a collection of interrelated records.
T F 25 Indexed records are stored in sequential order.

Fill-In

1. Adding more sectors to the outer tracks of a disk is called _____

2. What method processes transactions in a group? _____

3. List four benefits of secondary storage.
   a. _____
   b. _____
   c. _____
   d. _____

4. The type of software that can offer photos, narration, music, and more: _____

5. DASD stands for _____

6. The type of access required by a file is determined by _____

7. Name the two reels that are used with magnetic tape.
   a. _____
   b. _____

8. If a read/write head touches a hard disk surface, this is called _____

9. What does CD-ROM stand for? _____

10. The disk that has magnetically sensitive metallic crystals embedded in the plastic coating: _____

11. The concept of using several disks together as a unit: _____

12. What is the primary advantage of optical disk technology? _____

13. Name the three kinds of components in a sealed data module.

    a. _____

    b. _____

    c. _____

14. List the three primary factors that determine the time needed to access disk data.

    a. _____

    b. _____

    c. _____

15. What are three major methods of organization for storing files of data in secondary storage?

    a. _____

    b. _____

    c. _____

16. What does sequential batch processing require before transactions can be used to update a file? _____

17. What is a unique identifier for a record called? _____

18. The smallest unit of raw data is _____

19. Organizing data vertically on a disk pack is called what method? _____

20. The two terms for the process used in direct file processing to get the numerical representation of a record's address are

    a. _____

    b. _____

Answers

Multiple Choice
1. d	6. b	11. d	16. c	21. d
2. c	7. d	12. c	17. c	22. c
3. d	8. b	13. b	18. b	23. a
4. a	9. a	14. b	19. a	24. d
5. c	10. c	15. a	20. d	25. d

True/False
1. T	6. T	11. T	16. F	21. F
2. T	7. F	12. T	17. F	22. F
3. F	8. T	13. T	18. F	23. F
4. F	9. F	14. F	19. T	24. T
5. T	10. F	15. T	20. F	25. T

Fill-In
1. zone recording
2. batch processing
3. a. space
   b. reliability
   c. convenience
   d. economy
4. multimedia
5. direct-access storage device
6. the application
7. a. supply reel
   b. take-up reel
8. head crash
9. compact disk read-only memory
10. magneto-optical
11. redundant array of inexpensive disks(RAID)
12. high capacity
13. a. disks
    b. access arms
    c. read/write heads
14. a. seek time
    b. head switching
    c. rotational delay
15. a. sequential file
    b. direct file
    c. indexed file
16. sorting transactions by key
17. key
18. character
19. cylinder method
20. a. hashing
    b. randomizing

PLANET INTERNET

Just the FAQs, Please

When people begin to learn something new, they usually have many questions. In fact, beginners on the same subject often have the same questions. Rather than answer each question individually, it makes sense to keep the most Frequently Asked Questions—FAQs—in a handy place that anyone can access. FAQs are a long-standing tradition on the Internet. We will answer a few FAQs here and also refer you to online FAQs that you can check at your convenience.

Why do URLs all begin with http? Most do, but not all. HTTP stands for HyperText Transfer Protocol, the means of communicating using, among other things, links. The html you often see at the end of an URL stands for HyperText Markup Language, a language to make pages that take advantage of multimedia—pictures, sound, and even film clips. And, while mentioning Web software, we should note that it was software called Mosaic, invented by Marc Andreessen when he was a college student, that made Web page multimedia possible. Today there are many competitive browsers, but it was Mosaic that got the Web off the ground.

How can I avoid typing URLs? The easy way, of course, is to simply click links from one site to another. But, inevitably, there will be a site to which you want to return without going through a chain of links. Most browsers offer a *hot list* where you can store your favorite sites and their URLs. Then, in another session, simply go to the hot list and click the site without typing the URL.

So far we have used just the Web. Is that all there is to the Internet? No. The Web is one way to access the Internet. The Web uses links to go from site to site, but Gopher, named for the mascot at the University of Minnesota, where it was developed, lets you "go fer" a certain file through a series of menus that zero in on your choice. For example, when a region on the world map shown here is clicked, the screen switches to a Gopher menu, from which you select a country, then a smaller geographical area such as a state, which lists all the WWW servers in that area. Another possibility is FTP, for File Transfer Protocol, which lets you transfer files from a remote computer to your computer. These and other Net access methods came before the Web and many have been incorporated into the Web. However, since non-Web sites are textual only, no graphics, the Web is becoming dominant. You can tell which access method is being used by the first part of the URL: http, ftp, and so forth.

I know there is no "top" of the Internet, but what are some good starting places? A good place to begin is Where to Start, a site that has a directory of directories. Many directories feature "What's new" so you can take off in a new direction. Also, we can't resist The Awesome List, featuring not only the Awesome but also the Truly Awesome, and World Wide Web Must-Sees.

What help sources are online? A helpful FAQ list for the beginner is the World Wide Web FAQs. Several sites offer lists of definitions, including The World Wide Web Starter Kit. A favorite for beginners is The Newbie Adoption Agency, which will match a beginner with a seasoned user.

Internet Exercises

1. **Structured exercise.** Begin with the B/C URL http://www.aw.com/bc/planet/ and link to the World Wide Web FAQs.
2. **Freeform exercise.** If you are a "newbie," we suggest that you put yourself up for "adoption."

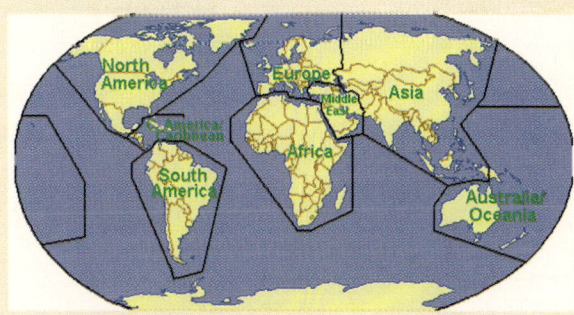

Kim Warren has heard a lot about the information superhighway. She read that the pieces are being assembled by a hodgepodge of computer and software vendors, by phone and cable companies, and by any other visionary who wants to jump aboard. Yet, she read, this project lacks a central architect and a basic blueprint. The articles go on and on, in every newspaper and magazine. In fact, a recent article noted that more has been written about the information superhighway than was written about personal computers when they first came out. But a personal computer, unlike the information superhighway, is something real that Kim can see and even buy.

Kim has a personal computer, her third. And she understands the basics about the information superhighway—that eventually information will be readily available to almost anyone from almost any place. She also knows that the information sending/receiving devices will be computers in some form and that the "highway" carrying the information will be some assemblage of communications equipment such as telephone and cable lines. Even if it is not yet a reality, Kim believes most of the predictions about the information superhighway, particularly that it will be pervasive.

Kim wants to get a head start. She has taken three approaches. First, she equipped her own computer with the hardware and software needed to communicate with other computers. Second, she signed up for the Internet, which some consider to be the early stages of the information superhighway, and has begun investigating its many options. Third, she took a two-week lunch-hour course offered by her employer to learn about the basics of communications. Kim is well-positioned at an on-ramp for the information superhighway. You can do the same by understanding the information in this chapter.

Chapter 5

Communications
Computer Connections

LEARNING OBJECTIVES

- Become acquainted with the evolution of data communications systems, from centralized to teleprocessing to distributed data processing to local area networks
- Know the basic components of a data communications system
- Know data transmission methods, including types of signals, modulation, and choices among transmission modes
- Differentiate the various kinds of communications links, and appreciate the need for protocols
- Understand network configurations
- Know local area network components, types, and protocols
- Appreciate the complexity of networking
- Become acquainted with examples of networking
- Appreciate the importance and expanse of the Internet

DATA COMMUNICATIONS: HOW IT ALL BEGAN

PUTTING TOGETHER A NETWORK: A FIRST LOOK
  Getting Started
  Network Design Considerations

DATA TRANSMISSION
  Digital and Analog Transmission
  Modems
  Asynchronous and Synchronous Transmission
  Simplex, Half-Duplex, and Full-Duplex Transmission

COMMUNICATIONS LINKS
  Types of Communications Links
  Protocols

NETWORK TOPOLOGIES

WIDE AREA NETWORKS

LOCAL AREA NETWORKS
  Local Area Network Components
  Local Area Network Types
  Local Area Network Protocols

THE WORK OF NETWORKING
  Electronic Mail
  Voice Mail
  Facsimile Technology
  Teleconferencing
  Electronic Data Interchange
  Electronic Fund Transfers
  Bulletin Boards
  Computer Commuting
  Commercial Communications Services

THE INTERNET
  Getting Connected
  Getting Around
  The Internet in Business
  So Much More

THE COMPLEXITY OF NETWORKS

125

No Hot Dog Lines

Unless they want to settle for peanuts and warm beer, baseball fans who want a snack usually have to stumble through the bleachers, wait in a long line and, worst of all, possibly miss a bases-loaded home run. But not at the Houston Astrodome, where order-taking has gone high-tech.

A hungry fan merely waves a menu to get the attention of the in-dome waiter, who approaches and then enters the order, the fan's location, and a credit card charge on a small handheld terminal. The data is sent over a wireless link to a computer that processes the food order in the kitchen. When the food is ready, a runner brings it to the fan, who has not missed a moment of the game.

Data Communications: How It All Began

Mail, telephone, TV and radio, books, newspapers, and periodicals—these are the principal ways we send and receive information, and they have not changed appreciably in a generation. However, **data communications systems**—computer systems that transmit data over communications lines such as telephone lines or cables—have been gradually evolving since the mid-1960s. Let us take a look at how they came about.

In the early days of computing, **centralized data processing** placed everything—all processing, hardware, and software—in one central location. Computer manufacturers responded to this trend by building even larger, general-purpose computers so that all departments within an organization could be serviced. Eventually, however, total centralization proved inconvenient. All input data had to be physically transported to the computer, and all processed material had to be picked up and delivered to the users. Insisting on centralized data processing was like insisting that all conversations between people occur face-to-face in one designated room.

The next logical step was **teleprocessing** systems—terminals connected to the central computer via communications lines. Teleprocessing systems permitted users to have remote access to the central computer from their terminals in other buildings and even other cities. However, even though access to the computer system was decentralized, all processing was still centralized—that is, performed by a company's one central computer.

In the 1970s businesses began to use minicomputers, which were often at a distance from the central computer. These were clearly decentralized systems because the smaller computers could do some processing on their own, yet some also had access to the central computer. This new setup was labeled **distributed data processing (DDP)**. It is similar to teleprocessing, except that it accommodates not only remote *access* but also remote *processing*. A typical application of a distributed data processing system is a business or organization with many locations—perhaps branch offices or retail outlets.

The whole picture of distributed data processing has changed dramatically with the advent of networks of personal computers. By **network** we mean a computer system that uses communications equipment to connect two or more computers and their resources. DDP systems are networks. But of particular interest in today's business world are local area networks (LANs), which are designed to share data and resources among several individual computer users in an office or building (Figure 5-1). We will examine networking in more detail in later sections of the chapter.

In the next section we will preview the components of a communications system to give you an overview of how these components work together.

Putting Together a Network: A First Look

Even though the components needed to transmit data from one computer to another seem quite basic, the business of putting together a network can be extremely complex. We begin with the initial components and then move to the list of factors that a network designer would have to consider.

Figure 5-1 Local area network. Although allocated to individual workers, the computers shown here are hooked together so that their users can communicate with each other.

Getting Started

The basic configuration—how the components are put together—is pretty straightforward, but there is a great variety of components to choose from, and the technology is ever changing. Assume that you have some data—a message—to transmit from one place to another. The basic components of a data communications system used to transmit that message are (1) a sending device, (2) a communications link, and (3) a receiving device. Suppose, for example, that you work at a sports store. You might want to send a message to the warehouse to inquire about a Wilson tennis racquet, an item you need for a customer. In this case the sending device is your computer terminal at the store, the communications channel is the phone line, and the receiving device is the computer at the warehouse. As you will see later, however, there are many other possibilities.

There is another often-needed component that must be mentioned in this basic configuration, as you can see in Figure 5-2. This component is a modem, which is usually needed to convert computer data to signals that can be carried by the communications channel and vice versa. We will discuss modems in detail shortly.

Large computer systems may have additional components. At the computer end, data may travel through a communications control unit called a **front-end processor**, which is actually a computer in itself. Its purpose is to relieve the central computer of some of the communications tasks and thus free it for processing applications programs. In addition, a front-end processor usually performs error detection and recovery functions.

Network Design Considerations

The task of network design is a complex one, usually requiring the services of a professional specifically trained in that capacity. Although we cannot learn how to design a network in this brief chapter, we can ask some questions that help us appreciate what the designer must contemplate. Here, in the vernacular, is a list of questions that might occur to a customer consid-

Citizens of the Information Superhighway

When computer guru Esther Dyson was appointed to a 27-member committee to give the government advice about the emerging information superhighway, she wondered who she should be representing. After some thought, she decided that, first and foremost, she is a citizen. As such, she wrote a column in *Computerworld,* a national newspaper, giving her Internet address and asking for advice.

The number of e-mail letters she got in response surprised her; apparently a lot of people care about this subject. The highlights of advice from her fellow citizens are as follows: the government should keep hands off as much as possible and leave developments to the free market, access should be universal but not totally free (discouraging waste), and privacy is important, especially for conversations and financial transactions.

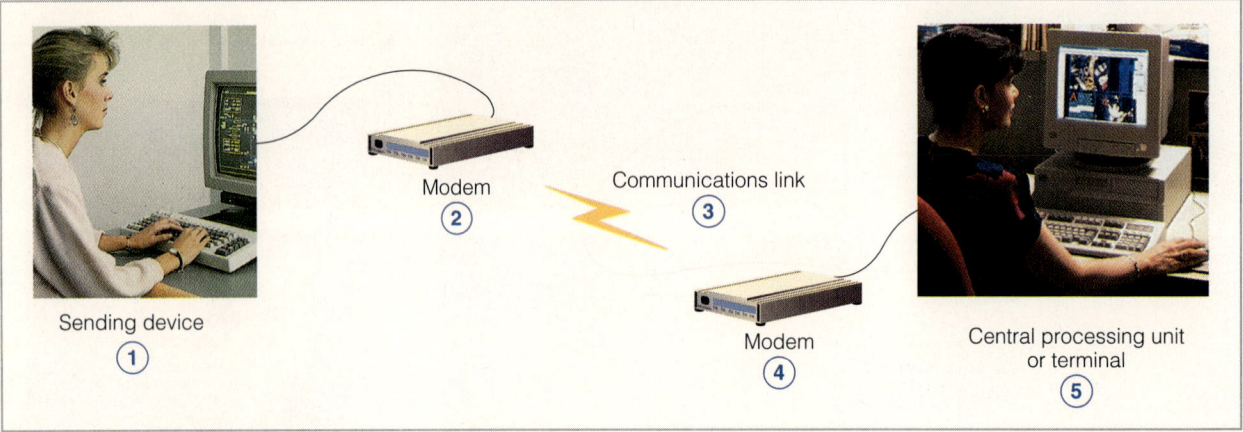

Figure 5-2 Communications system components. Data originated from ① a sending device is ② converted by a modem to data that can be carried over ③ a communications link and ④ reconverted by a modem at the receiving end before ⑤ being received by the destination computer.

ering installing a network—and also will provide hints of what is to come in the chapter.

Question: I've heard that modems send data at different speeds. Does that matter?

Answer: Yes. The faster, the better. Generally, faster means less expensive too. That also goes for synchronous transmission, which, as we will see, permits data to be sent quickly in big clumps.

Question: Am I limited to communicating via the telephone system?

Answer: Not at all. There are all kinds of communications media, with varying degrees of speed, reliability, and cost. There are trade-offs. A lot depends on distance, too—you wouldn't choose a satellite, for example, to send a message to the office next door.

Question: So the geographical area of the network is a factor?

Answer: Definitely. In fact, network types are described by how far-flung they are: *a wide area network* might span the nation or even the globe, but a *local area network* would probably be in an office or possibly just a department. The computers vary, too; generally, smaller networks can be served by small computers.

Question: Can I just cable the computers together and start sending data?

Answer: You must decide on some sort of plan. There are various standard ways, called *topologies*, to physically lay out the computers and other elements of a network. Also available are standard software packages, which provide a sort of language, called a *protocol*, for data communications.

Question: I know one of the advantages of networking is sharing disk files. Where are the files kept? And can any user get any file?

Answer: The files are usually kept with a particular computer, one that is more powerful than the other computers on the network. Access depends on the network setup. In some arrangements, for example, a user might be sent a whole file, but in others the user would be sent only the particular records needed to fulfill a request. The latter is called *client/server*, a popular alternative we will examine.

Question: This is getting complicated.

Answer: Yes.

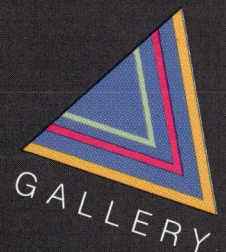

GALLERY

MULTIMEDIA
THE NEW SIGHT AND SOUND

Whether or not you are interested in architecture, you may be captivated by the multimedia images of works by Frank Lloyd Wright, called "America's architect." Wright said that the site and the building are one, as he demonstrates here with a house built over a waterfall. Working clockwise from the waterfall, the colorful archway is in a private home, the chandelier is in a synagogue, and, in a work that astonished the architectural world, this indoor room is in a business research building.

The Multimedia Story

Multimedia is different. For example, have you ever thought that you could see a film clip from Gone with the Wind on your computer screen? One could argue that such treats are already available on videocassette, but the computer version provides an added dimension for this and other movies: reviews by critics, photographs of movie stars, lists of Academy Awards, and much more.

Although an "ordinary" personal computer is certainly adequate for most personal and business uses, a computer equipped for multimedia offers a greater variety of information. Multimedia software typically presents information with text, illustrations, photos, narration, music, animation, and film clips.

How Is This Possible?

The key to multimedia is the high-volume capacity of optical disks. One CD-ROM disk can hold approximately 500 times as much data as an ordinary data diskette. This capacity accommodates the kinds of data that take up huge amounts of storage space, such as photographs, film clips, and music.

From a hardware standpoint, a multimedia computer must be equipped with a CD-ROM drive to read the disks. Also needed are a sound card (installed internally) and speakers, which normally rest on either side of the computer. Special software may accompany the drive and sound card.

A multimedia computer looks much like an ordinary personal computer on the surface. The main difference that can be seen is the speakers on either side.

To Buy or Not To Buy

Should your next computer be a multimedia personal computer? Absolutely. There is no doubt that multimedia applications will be the standard very shortly. Furthermore, if you get multimedia components pre-installed, it is the computer maker's job to see that everything works together properly.

But suppose you already own a personal computer and do not plan to get a new one any time soon. Should you upgrade your current computer with a multimedia kit? This is a tougher question. A multimedia kit usually includes a CD-ROM drive, a sound card, speakers, required software, and several application diskettes. Installing the required hardware and software may not be an easy task. Unless you are fairly skilled, or the hardware is labeled "plug and play," you should make sure you have support from a vendor or some other source before you begin this task. Once a multimedia system is set up, you can take advantage of the growing list of multimedia software packages.

The Coming Deluge

If you take a moment to peruse the racks of multimedia software in your local store, you can see that most of the current offerings come under the categories of entertainment or education — or pos-

sibly both. You can study *and hear* works by Stravinsky or Schubert. You can explore the planets or the ocean bottom through film clips and narrations by experts. You can be "elected" to Congress, after which you tour the Capitol, decorate your office, hire staff, and vote on issues. You can study the Japanese language, seeing the symbols and hearing the intonation. You can buy multimedia versions of reference books, magazines, children's books, and entire novels.

But this is just the beginning. Businesses are already moving to this high-capacity environment for street atlases, national phone directories, and sales catalogs. Coming offerings will include every kind of standard business application — all tricked out with fancy animation, photos, and sound. Educators will be able to draw upon the new sight and sound for everything from human anatomy to time travel. And just imagine the library of the future, consisting of not only the printed word but also photos, film, animation, and sound recordings

These scenes are from a multimedia package that provides a narrated tour of the London Gallery of Art. The scene above, by Titian, depicts the ancient myth of *Bacchus and Ariadne*. The scene below, by Renoir, is called *Boating on the Seine*.

People who wish to upgrade their existing computers can buy a multimedia kit, which typically contains a CD-ROM drive, a sound card, speakers, and several diskettes.

3

GALLERY

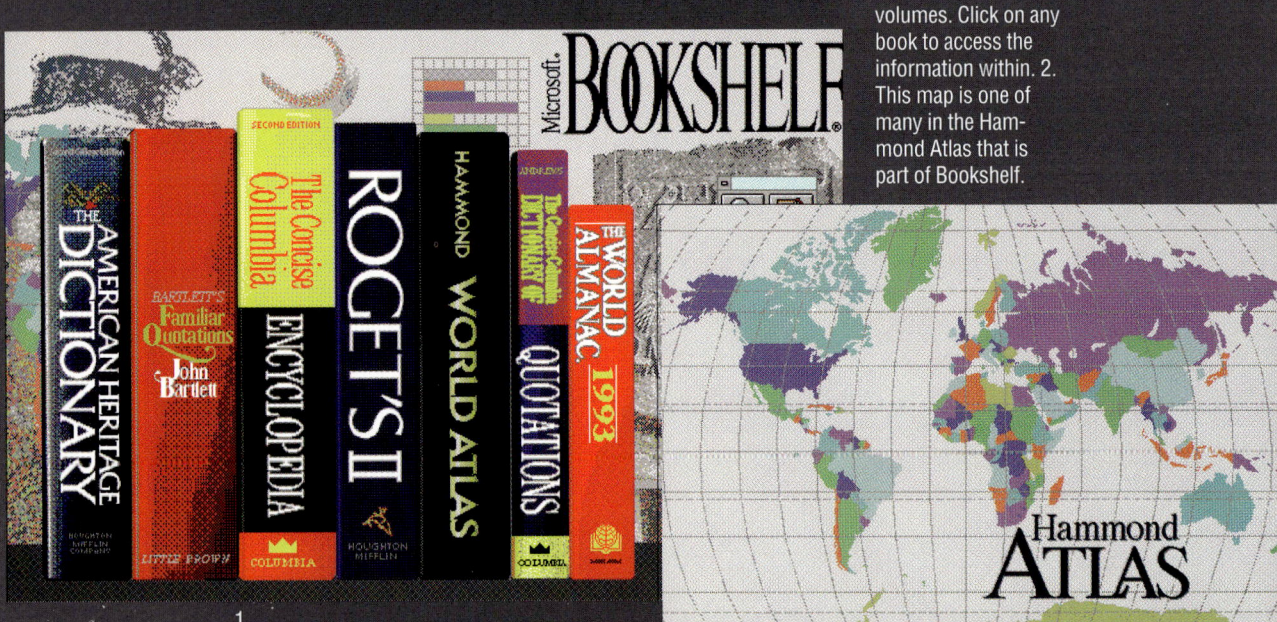

1. The Microsoft Bookshelf software opens with this screen shot of actual volumes. Click on any book to access the information within. 2. This map is one of many in the Hammond Atlas that is part of Bookshelf.

A multimedia presentation called Renoir, Cezanne, Matisse, and Dr. Barnes offers guided tours of Dr. Barnes collection of those artists and more. Shown here are 3. artist Paul Cezanne's village of *Gardanne*, 4. Henri Matisse's painting called *Etretat, the Sea*, and 5. Pierre-Auguste Renoir's *Girl with a Hat*.

Several comprehensive musical packages are offered through multimedia, including Franz Schubert's *Trout Quintet*. Although the emphasis is on listening, the package includes 6. a portrait of Schubert, 7. a picture of a manuscript with his glasses, and 8. a "trout game," an audio matching game.

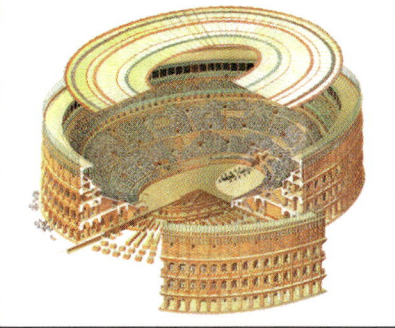
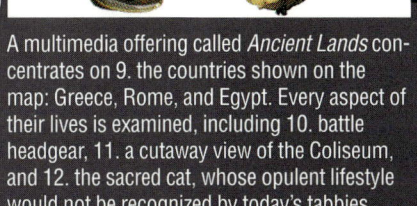

A multimedia offering called *Ancient Lands* concentrates on 9. the countries shown on the map: Greece, Rome, and Egypt. Every aspect of their lives is examined, including 10. battle headgear, 11. a cutaway view of the Coliseum, and 12. the sacred cat, whose opulent lifestyle would not be recognized by today's tabbies.

The package called *Cinemania* provides users with everything they want to know about movies and movie stars. This package specializes in film clips of action scenes from movies as diverse as *Amadeus* and *Star Wars*, and even has Gene Kelly performing *Singing in the Rain*. Photos are available of any star, including 13. Brad Pitt, 14. Whoopie Goldberg, 15. Tom Hanks, and 16. Meryl Streep.

17. This screen for Mrs. Doubtfire shows a photo scene and, to the left, icons indicating (clockwise from top left) rating, color, academy award nomination, availability on cassette, and (in the center) length of 125 minutes. 18. This image is from the screen for The Lion King.

19

All multimedia packages offer several methods of exploring the information offered, including a menu of some sort. Here, 19. shows a menu that greets users of a package called *In the Company of Whales*. To the accompaniment of whale singing and much splashing, a user is presented with a variety of shots, including 20. a killer whale breaching the water.

20

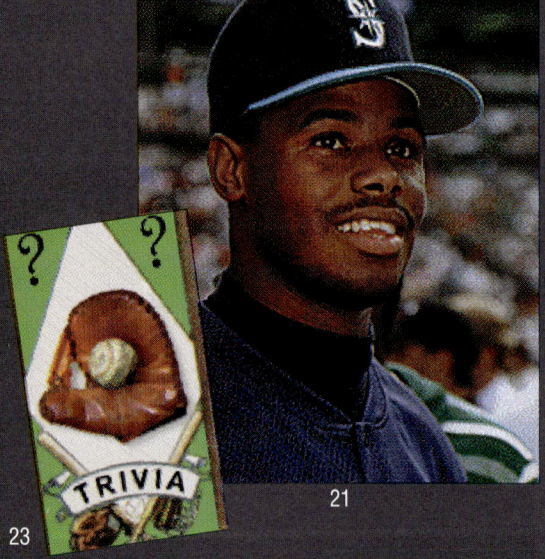

21

22

Baseball? Yes, *Complete Baseball* is the subject of this all-encompassing offering of statistics, photos, history, and even videos of baseball highlights. Shown here are superstars 21. Ken Griffey, Jr. of the 1990s and 22. Roberto Clemente of the 1960s, along with 23. the clickable screen trivia button and 24. an early baseball manual.

23

24

25

25. What better way to understand a Civil War battle than to see it unfold before your eyes — on the screen. The *Gettysburg* multimedia package begins with a screen of the empty green terrain that gradually changes as the military forces maneuver and change locations. A southern accented narration keeps you posted minute by minute.

26

28

27

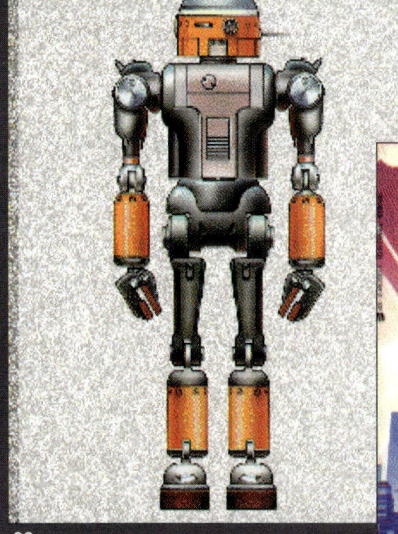

An offering called *Dangerous Creatures* lets you see and hear them in their own environments. The narrators make a point of noting that such photography is not for wimps. Shown here are 26. the not-yet-dangerous baby grizzly bear, 27. a very dangerous jaguar, and 28. an alligator looking for lunch.

The Ultimate Robot is just that — a source for everything anyone would want to know about robots, including fact and fiction. A highlight is 29. a hands-on opportunity to build your own on-screen robot. 30. Shown here is the cover for one of Isaac Asimov's famous robot fiction books.

29

30

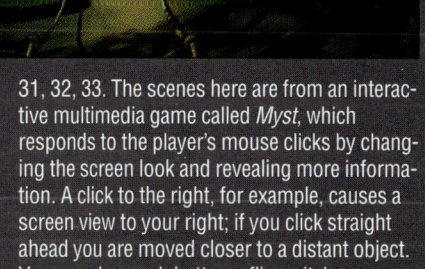

33

32

31, 32, 33. The scenes here are from an interactive multimedia game called *Myst*, which responds to the player's mouse clicks by changing the screen look and revealing more information. A click to the right, for example, causes a screen view to your right; if you click straight ahead you are moved closer to a distant object. You can also push buttons, flip switches, open doors, and creep down winding staircases.

31

7

34. This browse screen is one of several menu presentations for Microsoft Encarta, a multimedia encyclopedia that features photos, illustrations, narration, and animation.
35. Screens typically show a photo accompanied by text, as shown here for the image orthicon tube. Other screen photos of interest, clockwise from the top, arc 36. integral calculus, 37. a dry cell battery, 38. a marmoset, 39. altocumulus clouds, 40. the Mississippi Queen paddleboat, 41. Mount Rushmore, 42. The University of Virginia, 43. the international flag alphabet, and 44. pilot Amelia Earhart.

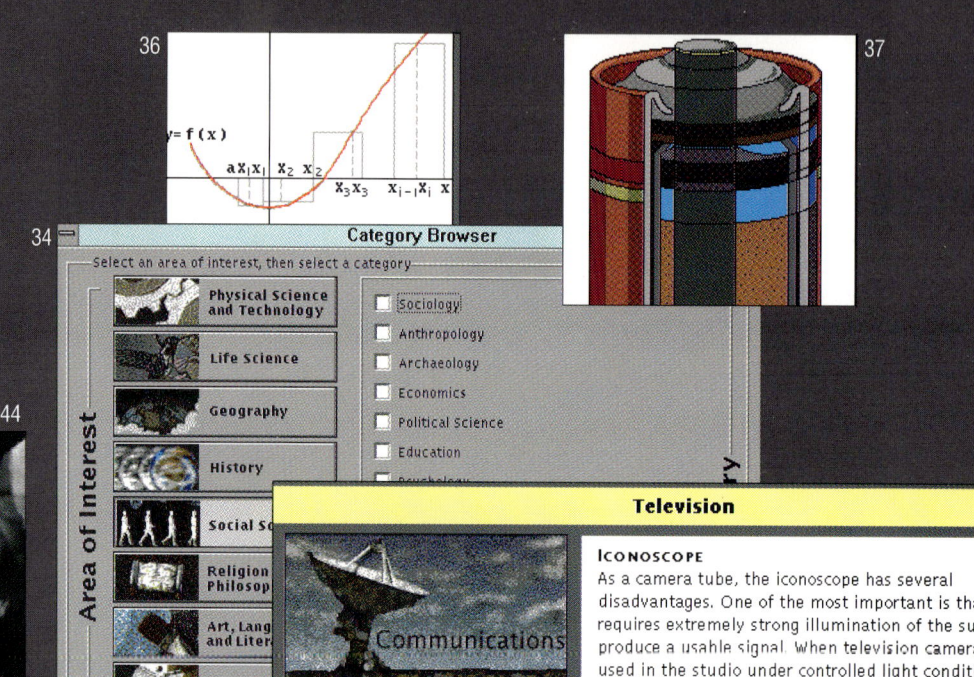

MAKING THE RIGHT CONNECTIONS

Save Money, Save Paper, Save Time

Time magazine, that is. If you are a subscriber to America Online, you can skip the magazine subscription fee and read *Time* right on your computer screen. Time Online, as the service is called, makes the entire weekly magazine available online. Users can download any article they choose and even save the article on their own disk. A further advantage is that Time Online is available a day or so before you would receive the paper version in your mailbox.

Other magazines are moving to on-line access via various services. Current entries include *National Geographic, Scientific American, Consumer Reports, People, Smithsonian Magazine, Cycle World,* and even the daily *Congressional Record.*

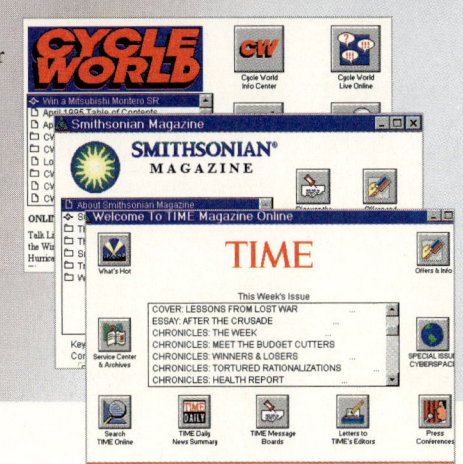

After we have a look at these and other related topics, we will present an example of a complex network or, rather, a set of networks. You need not understand all the details, but you will have an appreciation for the effort required to put together a network. Let us see how the components of a communications system work together, beginning with how data is transmitted.

 ## Data Transmission

A terminal or computer produces digital signals, which are simply the presence or absence of an electric pulse. The state of being on or off represents the binary number 1 or 0, respectively. Some communications lines accept digital transmission directly, and the trend in the communications industry is toward digital signals. However, most telephone lines through which these digital signals are sent were originally built for voice transmission, and voice transmission requires analog signals. We will look at these two types of transmission and then study modems, which translate between them.

Digital and Analog Transmission

Digital transmission sends data as distinct pulses, either on or off, in much the same way that data travels through the computer. However, most communications media are not digital. Communications devices such as telephone lines, coaxial cables, and microwave circuits are already in place for voice transmission. The easiest choice for most users is to piggyback on one of these. Thus, the most common communications devices all use **analog transmission,** a continuous electric signal in the form of a wave.

To be sent over analog lines, a digital signal must first be converted to an analog form. It is converted by altering an analog signal, called a **carrier wave,** which has alterable characteristics (Figure 5-3a). One such characteristic is the **amplitude,** or height of the wave, which can be increased to represent the binary number 1 (Figure 5-3b). Another char-

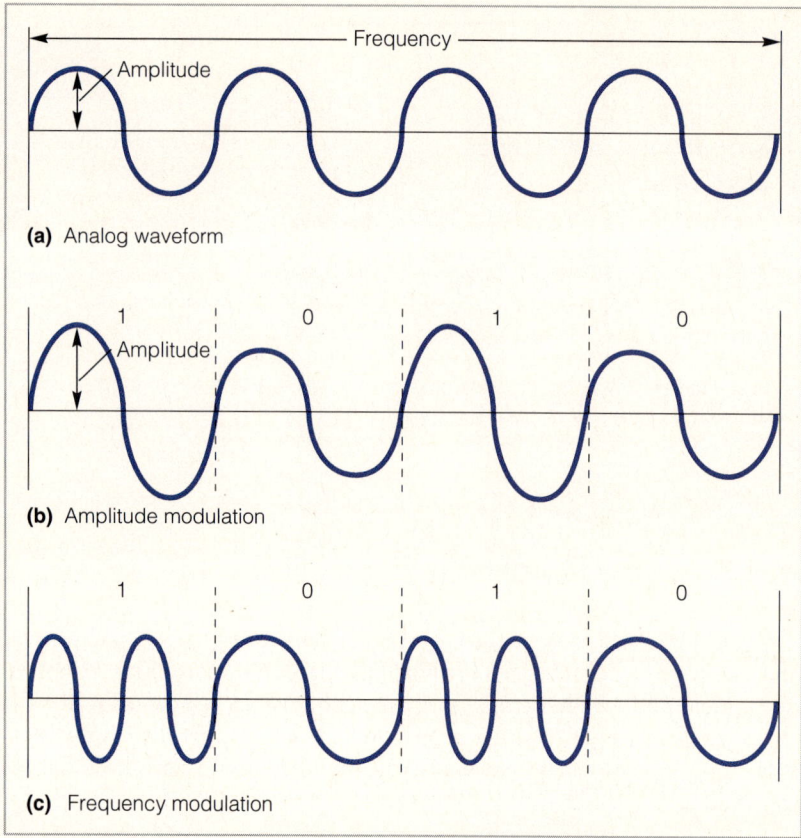

Figure 5-3 Analog signals. (a) An analog carrier wave moves up and down in a continuous cycle. (b) The analog waveform can be converted to digital form through amplitude modulation. As shown, the wave height is increased to represent a 1 or left the same to represent a zero. (c) In frequency modulation the amplitude of the wave stays the same but the frequency increases to indicate a 1 or stays the same to indicate a zero.

acteristic that can be altered is the **frequency,** or number of times a wave repeats during a specific time interval; frequency can be increased to represent a 1 (Figure 5-3c).

Conversion from digital to analog signals is called **modulation,** and the reverse process—reconstructing the original digital message at the other end of the transmission—is called **demodulation.** (You probably know amplitude and frequency modulation by their abbreviations, AM and FM, the methods used for radio transmission.) An extra device is needed to make the conversions: a modem.

Modems

A **modem** is a device that converts a digital signal to an analog signal and vice versa (Figure 5-4). Modem is short for *modulate/dem*odulate.

Types of Modems

Modems vary in the way they connect to the telephone line. There are two main types: acoustic coupler modems and direct-connect modems. **Acoustic coupler modems** include a cradle to hold the telephone handset. Most modems today, however, are directly connected to the phone system.

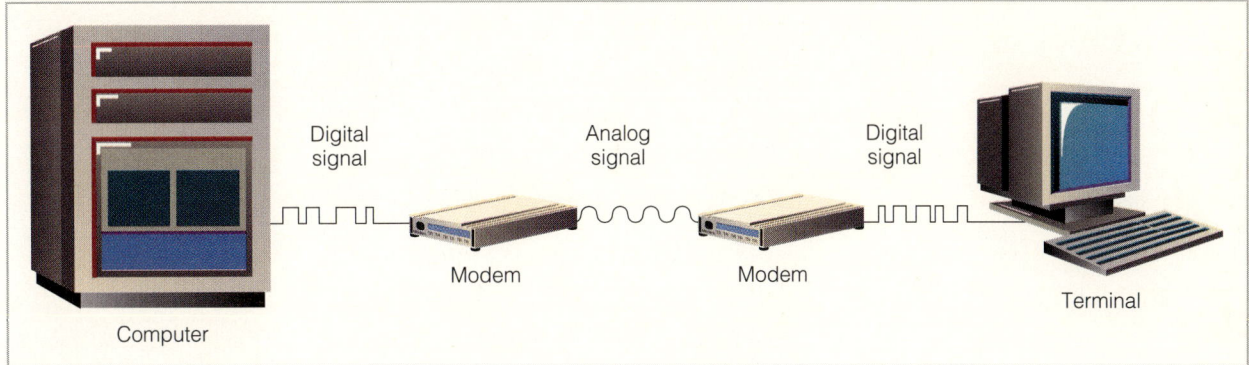

Figure 5-4 Modems. Modems convert—modulate—digital data signals to analog signals for traveling over communications links, then reverse the process—demodulate—at the other end.

A **direct-connect modem** is directly connected to the telephone line by means of a telephone jack. An **external modem** is separate from the computer (Figure 5-5). Its main advantage is that it can be used with a variety of computers. If you buy a new personal computer, for example, you can probably keep the same external modem. For those personal computer users who regard an external modem as one more item taking up desk space, new modem-on-a-chip designs have produced a modem that is so small you will hardly notice it. For a modem that is out of sight—literally—an **internal modem** board can be inserted into the computer by the user; it might even come installed in a personal computer as standard equipment. Hayes is the brand name that is the standard for modems; most modems are **Hayes compatible**. As we will discuss shortly, most modems today also have fax capability.

Figure 5-5   An external modem. This is a Hayes modem, the brand that has set the standard for modems.

Modem Data Speeds

Users who connect their computers via communications services may pay charges based on the time the computers are connected. Thus, there is a strong incentive to transmit as quickly as possible. The old standard modem speeds of 1200, 2400, and 9600 bits per second (bps) have now been superseded by modems that transmit an astonishing 14,400 or 28,800 bps. The speedsters using these rates are usually corporations sending data from one office to another. If you buy a modem today, you will probably find only modems with the 14,400 bps or higher for sale. However, transmission from one modem to another can be no faster than the speed of the slower modem, and many services to which you may be connected still receive and transmit at the lower rates. Still, there is no harm having the faster rate in anticipation of the day when all services will also transmit at that rate. Note the transmission time comparisons in Table 5-1.

Asynchronous and Synchronous Transmission

Sending data off to a far destination works only if the receiving device is ready to accept it. By *ready* we mean more than just available; the receiving device must be able to keep in step with the sending device. Two techniques commonly used to keep the sending and receiving units dancing to the same tune are asynchronous and synchronous transmission.

When **asynchronous transmission** (also called **start/stop transmission**) is used, a special start signal is transmitted at the beginning of each group

Table 5-1	Data transfer rates compared
Data Transfer rate (bps)	Time to transmit a 20-page single-spaced report
1,200	10 min.
2,400	5 min.
9,600	1.25 min.
14,400	50 sec.
28,800	25 sec.

of message bits—a group is usually just a single character. Likewise, a stop signal is sent at the end of the group of message bits (Figure 5-6a). When the receiving device gets the start signal, it sets up a timing mechanism to accept the group of message bits.

Synchronous transmission is a little trickier because characters are transmitted together in a continuous stream (Figure 5-6b). There are no call-to-action signals for each character. Instead, the sending and receiving devices are synchronized by having their internal clocks put in time with each other by a bit pattern transmitted at the beginning of the message. Furthermore, error check bits are transmitted at the end of each message to make sure all characters were received properly. Synchronous transmission equipment is more complex and more expensive but, without all the start/stop bits, transmission is much faster.

Simplex, Half-Duplex, and Full-Duplex Transmission

Data transmission can be characterized as simplex, half duplex, or full duplex, depending on permissible directions of traffic flow (Figure 5-7). **Simplex transmission** sends data in one direction only; everyday examples are television broadcasting and arrival/departure screens at airports. **Half-duplex transmission** allows transmission in either direction, but only one way at a time. An analogy is talk on a CB radio. In a bank a teller using

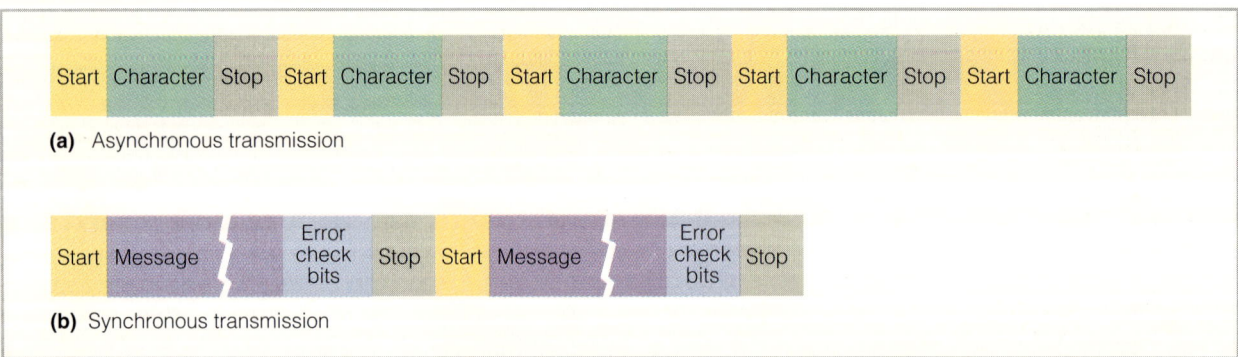

Figure 5-6 **Asynchronous and synchronous transmission.** (a) Asynchronous transmission uses start/stop signals surrounding each character. (b) Page width constraints preclude showing the true amount of continuous data that can be transmitted synchronously between start and stop characters. As contrasted with asynchronous transmission, which has one start/stop set per character, synchronous transmission can send many characters, even many messages, between one start/stop set. Note that synchronous transmission requires a set of error check bits to make sure all characters were received properly.

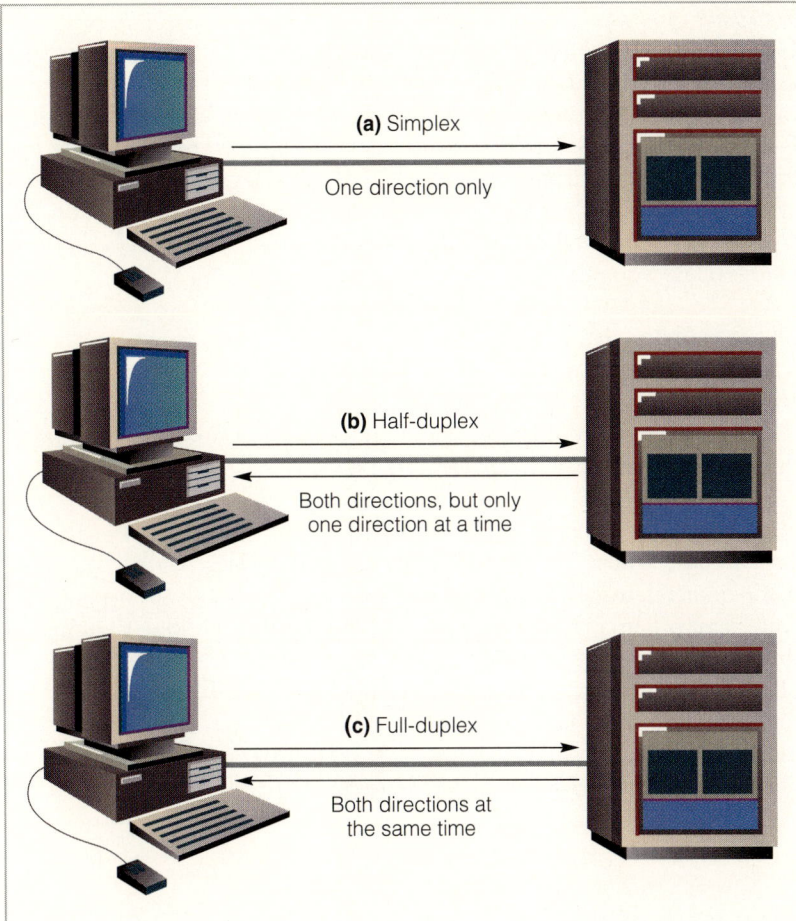

Figure 5-7 Transmission directions. (a) Seldom-used simplex transmission sends data in one direction only. (b) Half-duplex transmission can send data in either direction, but only one way at a time. (c) Full-duplex transmission can send data in both directions at once.

half-duplex transmission can send the data about a deposit and, after it is received, the computer can send a confirmation reply. **Full-duplex transmission** allows transmission in both directions at once. An analogy is a telephone conversation in which, good manners aside, both parties can talk at the same time.

We have discussed data transmission at some length. Now it is time to turn to the actual media that transmit the data.

Communications Links

The cost for linking widely scattered computers is substantial, so it is worthwhile to examine the communications options. Telephone lines are the most convenient communications channel because an extensive system is already in place, but there are many other options. A communications **link** is the physical medium used for transmission.

Types of Communications Links

There are several kinds of communications links. Some may be familiar to you already.

Wire pairs

One of the most common communications media is the **wire pair,** also known as the **twisted pair.** Wire pairs are wires twisted together to form a cable, which is then insulated (Figure 5-8a). Wire pairs are inexpensive. Further, they are often used because they had already been installed in a building for other purposes or because they are already in use in telephone systems. However, they are susceptible to electrical interference, or noise. **Noise** is anything that causes distortion in the signal when it is received. High-voltage equipment and even the sun can be sources of noise.

Coaxial Cables

Known for sending a strong signal, a **coaxial cable** is a single conductor wire within a shielded enclosure (Figure 5-8b). Bundles of cables can be laid underground or undersea. These cables can transmit data much faster than wire pairs and are less prone to noise.

Fiber Optics

Traditionally, most phone lines transmitted data electrically over wires made of metal, usually copper. These metal wires had to be protected from water and other corrosive substances. **Fiber optics** technology eliminates this requirement (Figure 5-8c, d). Instead of using electricity to send data, fiber optics uses light. The cables are made of glass fibers, each thinner than a human hair, that can guide light beams for miles. Fiber optics transmits data faster than some technologies, yet the materials are substantially lighter and less expensive than wire cables. It can also send and receive a

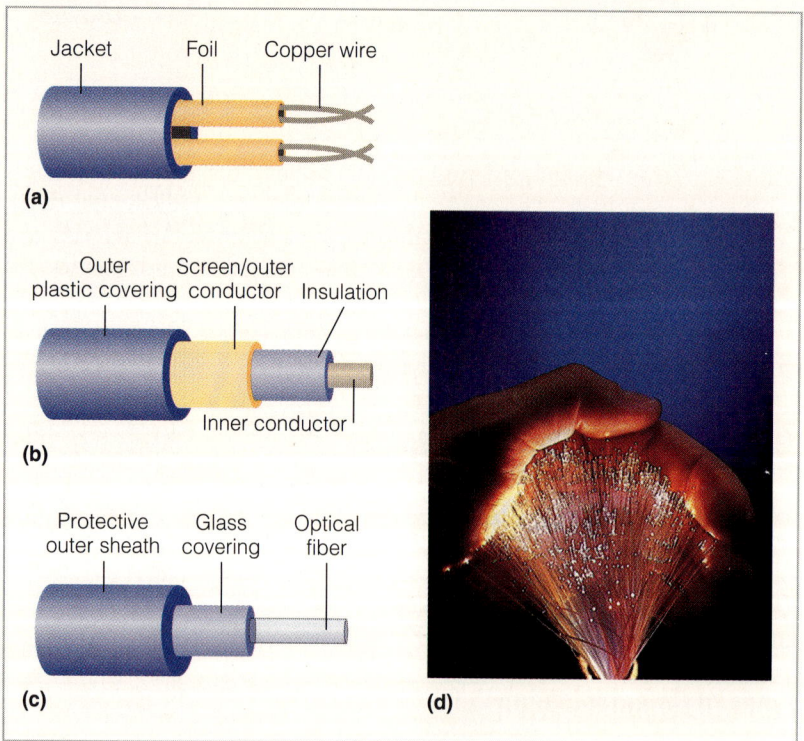

Figure 5-8 Communications links. (a) Wire pairs are pairs of wires twisted together to form a cable, which is then insulated. (b) A coaxial cable is a single conductor wire surrounded by insulation. (c) Fiber optics consists of hair-like glass fibers that carry voice, television, and data signals. (d) This photo shows light emitted from a handful of fiber optic cables.

wider assortment of data frequencies at one time. The range of frequencies that a device can handle is known as its **bandwidth;** bandwidth is a measure of the capacity of the link. The broad bandwidth of fiber optics translates into promising multimedia possibilities, since fiber optics is well suited for handling all types of data—voice, pictures, music, and video—at the same time.

Microwave Transmission

Another popular medium is **microwave transmission,** which uses what is called line-of-sight transmission of data signals through the atmosphere (Figure 5-9a). Since these signals cannot bend around the curvature of the earth, relay stations—often antennas in high places such as the tops of mountains and buildings—are positioned at points approximately 30 miles apart to continue the transmission. Microwave transmission offers speed, cost-effectiveness, and ease of implementation. Unfortunately, in major metropolitan areas tall buildings may interfere with microwave transmission.

Satellite Transmission

The basic components of **satellite transmission** are **earth stations,** which send and receive signals, and a satellite component called a transponder. The **transponder** receives the transmission from an earth station, amplifies the signal, changes the frequency, and retransmits the data to a receiving earth station (Figure 5-9b). (The frequency is changed so that the weaker incoming signals will not be impaired by the stronger outgoing signals.) This entire process takes a matter of a few seconds.

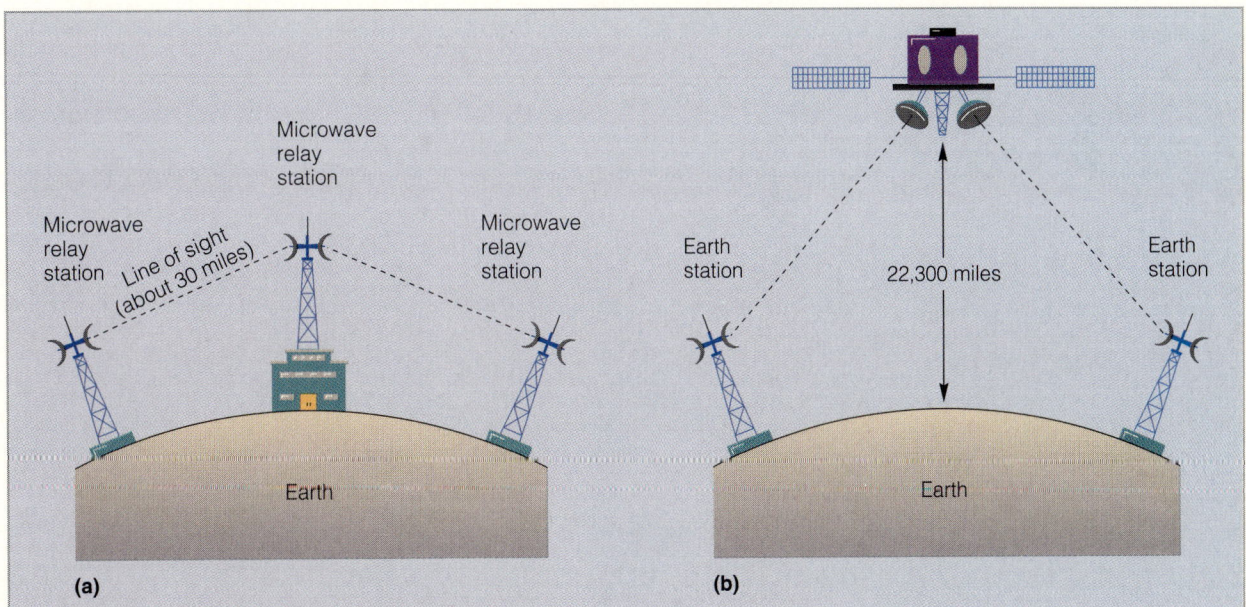

Figure 5-9 Microwave and satellite transmission. (a) To relay microwave signals, dish-shaped antennas such as these are often located atop buildings, towers, and mountains. Microwave signals can follow a line-of-sight path only, so stations must relay this signal at regular intervals to avoid interference from the curvature of the earth. (b) In satellite transmission, a satellite acts as a relay station and can transmit data signals from one earth station to another. A signal is sent from an earth station to the relay satellite, which changes the signal frequency before transmitting it to the next earth station.

Bill and Craig's Excellent Adventure

What does a fellow do when he is already a billionaire but looking for something new and exciting? He hooks up with another billionaire and plans a dream in the sky.

Bill Gates, founder of Microsoft, Inc., and Craig McCaw, founder of McCaw Cellular Communications, began a new company. Teledesic Corporation, founded in 1994, will build and launch 840 small communications satellites by the year 2001.

The founders expect their satellites to be the access road to the information superhighway in rural and underdeveloped countries of the world. They envision a day when, for example, a doctor in a field hospital in Africa could get help from images transmitted via satellite from a medical specialist in Massachusetts.

If a signal must travel thousands of miles, satellites are usually part of the link. A message being sent around the world probably travels by cable or some other physical link only as far as the nearest satellite earth transmission station (Figure 5-10). From there it is beamed to a satellite, which sends it back to earth to another transmission station near the data destination. Communications satellites are launched into space where they are suspended about 22,300 miles above the earth. Why 22,300 miles? That is where satellites reach geosynchronous orbit—the orbit that allows them to remain positioned over the same spot on the earth.

Mixing and Matching

A network system is not limited to one kind of link and, in fact, often works in various combinations, especially over long distances. An office worker who needs data from a company computer on the opposite coast will most likely use wire pairs in the phone lines, followed by microwave and satellite transmission (Figure 5-11). Astonishingly, the trip across the country and back, with a brief stop to pick up the data, may take only seconds.

Protocols

A **protocol** is a set of rules for the exchange of data between a terminal and a computer or between two computers. A protocol is embedded in the network software. Think of protocol as a sort of pre-communication to make sure everything is in order before a message or data is sent. Protocols are handled by software related to the network, so that users need only worry about their own data.

Protocol Communications

Two devices must be able to ask each other questions (Are you ready to receive a message? Did you get my last message? Is there trouble at your end?) and to keep each other informed (I am sending data now). In

Figure 5-10 Satellite dishes. These dishes, shown here in the New Mexico sunset, send data to and receive data from satellites in space.

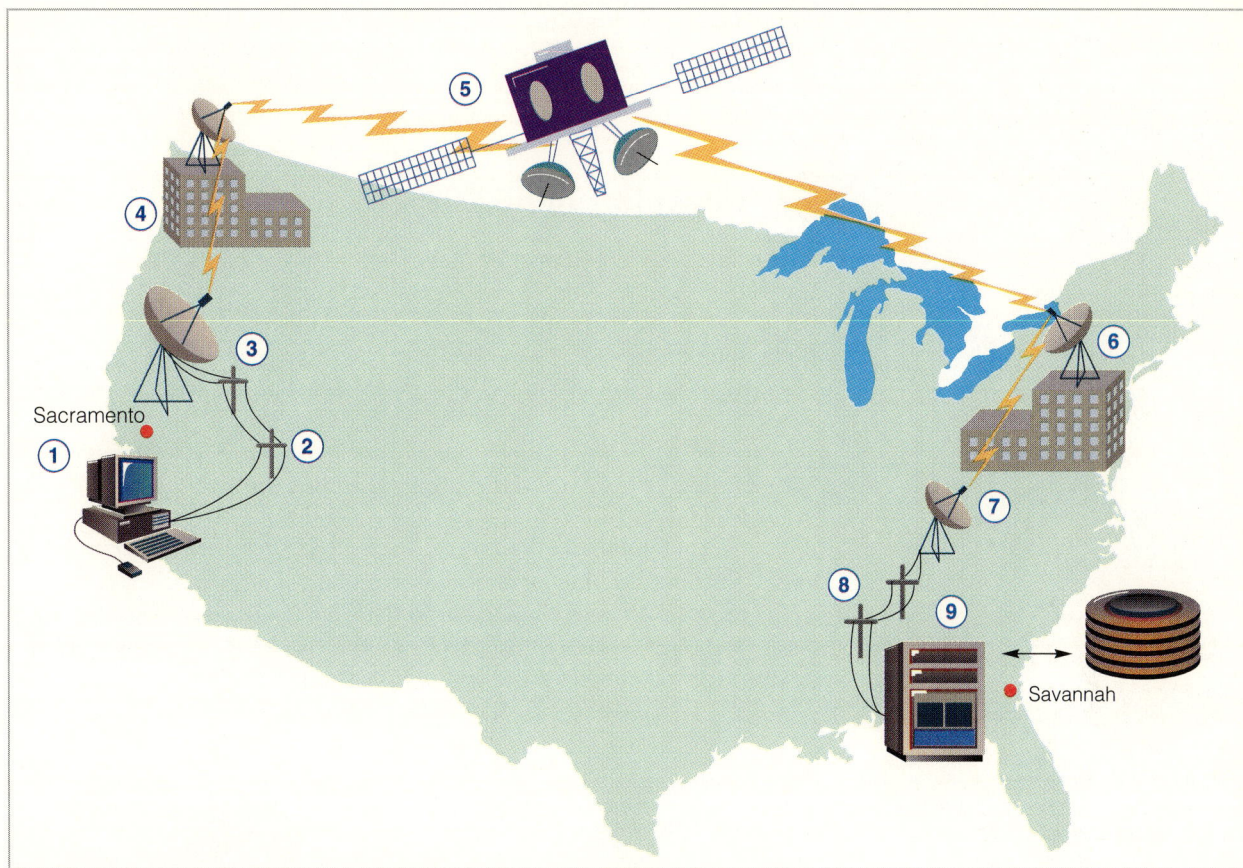

Figure 5-11 A variety of communications links. Say an accountant working in the Sacramento office needs certain tax records from the headquarters computer files in Savannah. One possibility for the route of the user request and the response is as follows. ① The office worker makes the request for the records, which ② goes out over the local phone system to ③ a nearby microwave station, which transmits the request to ④ the nearest earth satellite transmission station, where ⑤ it is relayed to a satellite in space, which relays it back to earth ⑥ to an earth satellite station near Savannah, where it is sent to ⑦ a microwave station and then ⑧ via the phone lines to ⑨ the headquarters computer. Once the tax records are retrieved from the Savannah computer files, the whole process is reversed as the requested records are sent back to Sacramento.

addition, the two devices must agree on how data is to be transferred, including data transmission speed and duplex setting. But this must be done in a formal way. When communication is desired among computers from different vendors (or even different models from the same vendor), the software development can be a nightmare because different vendors use different protocols. Standards would help.

Setting Standards
Standards are important in the computer industry; it saves money if we can all coordinate effectively. Nowhere is this more obvious than in data communications systems, where many components must "come together." But it is hard to get people to agree to a standard.

Communications standards exist, however, and are constantly evolving and being updated for new communications forms. Standards provide a framework for how data is transmitted. The International Standards Organization (ISO), based in Geneva, Switzerland, has defined a set of communications protocols called the **Open Systems Interconnection (OSI)** model. (Yes, that is ISO giving us OSI.) The OSI model has been endorsed by the United Nations. As we will discuss shortly, particular types of protocols are used for local area networks.

 ## Network Topologies

As we have noted, a network is a computer system that uses communications equipment to connect computers. They can be connected in different ways. The physical layout of a network is called a **topology**. There are three common topologies: star, ring, and bus networks. In describing a network topology, we often refer to a **node**, which is a computer on a network.

A **star network** has a hub computer that is responsible for managing the network (Figure 5-12a). All messages are routed through the central computer, which acts as a traffic cop to prevent collisions. Any connection failure between a node and the hub will not affect the overall system. However, if the hub computer fails, the network fails.

A **ring network** links all nodes together in a circular chain (Figure 5-12b). Data messages travel in only one direction around the ring. Any data that passes by is examined by the node to see if it is the addressee; if not, the

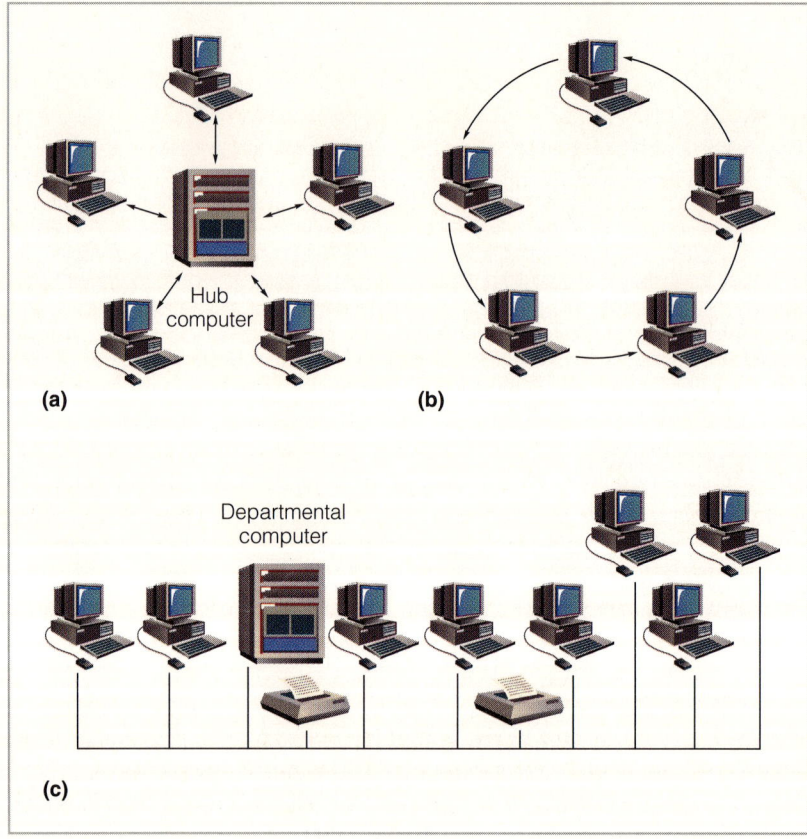

Figure 5-12 Topologies. (a) The star network topology has a central computer that runs the network. (b) The ring network topology connects computers in a circular fashion. (c) The bus network topology connects all nodes in a line and can preserve the network if one computer fails.

data is passed on to the next node in the ring. Since data travels in only one direction, there is no danger of data collision. However, if one node fails, then the entire network fails.

A **bus network** has a single line to which all the network nodes are attached (Figure 5-12c). Computers on the network transmit data in the hope that it will not collide with data transmitted by other nodes; if this happens, the sending node simply tries again. Nodes can be attached to or detached from the network without affecting the network. Furthermore, if one node fails, it does not affect the rest of the network.

Wide Area Networks

There are different kinds of networks. We begin with the geographically largest, a wide area network.

A **wide area network (WAN)** is a network of geographically distant computers and terminals. In business, a personal computer sending data any significant distance is probably sending it to a minicomputer or mainframe computer. Since these larger computers are designed to be accessed by terminals, a personal computer can communicate with a minicomputer or mainframe only if the personal computer emulates, or imitates, a terminal. This is accomplished by using **terminal emulation software** on the personal computer. The larger computer then considers the personal computer or workstation as just another user input/output communications device—a terminal.

The larger computer to which the terminal or personal computer is attached is called the **host computer.** If a personal computer is being used as a terminal, **file transfer software** permits users to download data files from the host or upload data files to the host. To **download** a file means to retrieve it from another computer and to send it to the computer of the user who requested the file. To **upload** a file, a user sends a file to another computer.

Local Area Networks

A **local area network (LAN)** is a collection of computers, usually personal computers, that share hardware, software, and data. In simple terms, LANs hook personal computers together through communications media so that each personal computer can share the resources of the others. As the name implies, LANs cover short distances, usually one office or building or a group of buildings that are close together.

Local Area Network Components

LANs do not use the telephone network. Networks that are LANs are made up of a standard set of components.

- All networks need some system for interconnection. In some LANs the nodes are connected by a shared **network cable.** Low-cost LANs are connected with twisted wire pairs, but many LANs use coaxial cable or fiber optic cable, which are both more expensive and faster. Some local area networks, however, are **wireless**, using infrared or radio wave transmissions instead of cables. Wireless networks are easy to set up and reconfigure, since there are no cables to connect or disconnect, but

they have slower transmission rates and limit the distance between nodes.
- A **network-interface card,** sometimes called a **NIC,** connects each computer to the wiring to the network. A NIC is a circuit board that fits in one of the computer's internal expansion slots.
- Similar networks can be connected by a **bridge,** which recognizes the messages on a network and passes on those addressed to nodes in other networks. For example, a fabric designer whose computer is part of a department LAN for a textile manufacturer could send cost data, via a bridge, to someone in the accounting department whose computer is part of another company LAN, one used for financial matters.
- A **gateway** is a collection of hardware and software resources that lets a node communicate with a computer on another dissimilar network. A gateway, for example, could connect an attorney on a local area network to a legal service offered through a wide area network.

Local Area Network Types

Two ways to organize the resources of a LAN are client/server and peer-to-peer.

Client/Server

A **client/server** arrangement involves a **server,** which is a computer that controls the network. In particular, a server has the hard disks holding shared files and often has the highest-quality printer, which can be used by all nodes (Figure 5-13). The clients are all the other computers on the network. Under the client/server arrangement, processing is usually done by the server, and only the results are sent to the node. Sometimes the server and the node share processing. For example, a server, upon request from the node, could search a database of cars in the state of Maryland and come up with a list of all Jeep Cherokees. This data could be passed on to the node computer, which could process the data further, perhaps looking for certain equipment or license plate letters. This method can be contrasted with a **file server** relationship, in which the server transmits the entire file to the node, which does all its own processing. Using the Jeep

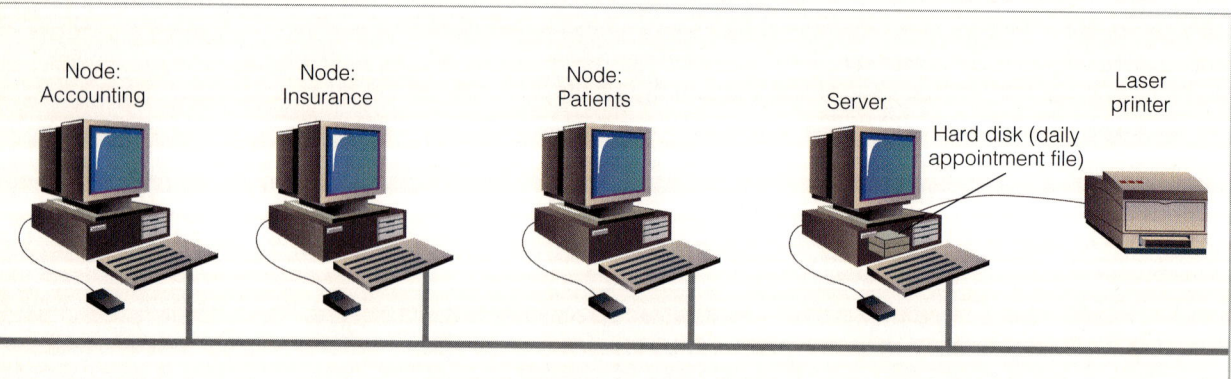

Figure 5-13   Server and peripheral hardware. In this network for a clinic with seven doctors, the daily appointment records for patients are kept on the hard disk associated with the server. Workers who, using their own computers, deal with accounting, insurance, and patient records can access the daily appointment file to update their own files.

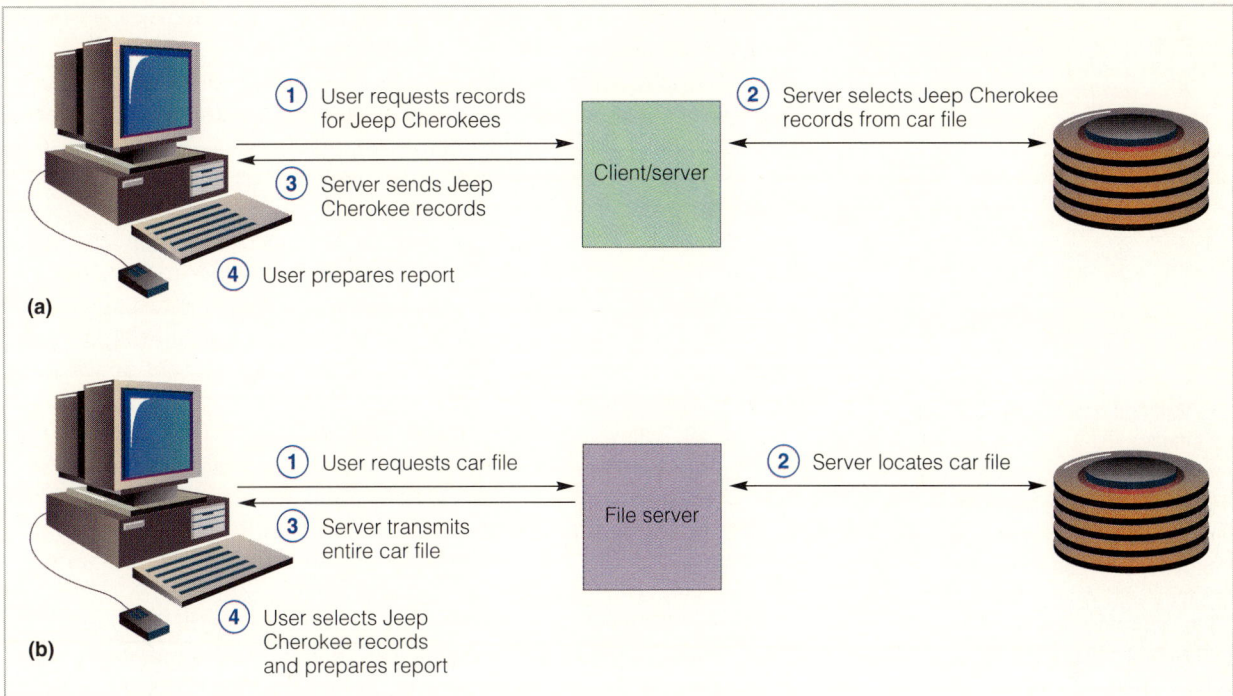

Figure 5-14 Client/server contrasted with file server. (a) In a client/server relationship, ① a user makes a request to the server to select only Jeep Cherokee records from a state car file; ② the server does so and ③ sends the records back to the user, who ④ uses those specific records to prepare a report. (b) In a file server relationship, ① a user asks for the entire state car file, which ② the server locates and then ③ transmits to the user, who then ④ selects the Jeep Cherokee records and prepares a report. The client/server setup puts most of the processing burden on the more powerful server and also significantly reduces the amount of data being transferred between server and user.

example, the entire car file would be sent to the node, instead of just the extracted Jeep Cherokee records (Figure 5-14).

Client/server has attracted a lot of attention because a well-designed system reduces the volume of data traffic on the network and allows faster response at each node. Also, since the server does most of the heavy work, less expensive computers can be used as nodes.

Peer-to-Peer

All computers in a **peer-to-peer** arrangement have equal status; no one computer is in control. With all files and peripheral devices distributed across several computers, users share each other's data and devices as needed. An example might involve a corporate building in which marketing wants its files kept on its own computer, public relations wants its files kept on its own computer, personnel wants its files kept on its own computer, and so on; all can still gain access to the other's files when needed. The main disadvantage is lack of speed—most peer-to-peer networks slow down under heavy use. Many networks are hybrids, containing elements of both client/server and peer-to-peer arrangements.

Local Area Network Protocols

We have already noted that networks must have a set of rules—protocols—to access the network to send data. Recall that a protocol is embed-

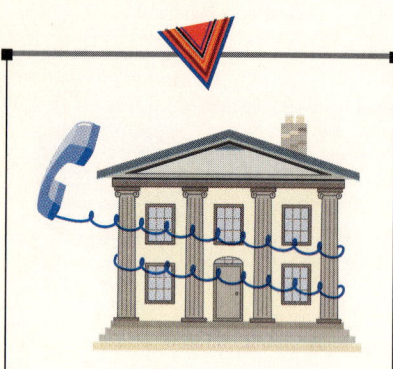

E-Mail Addresses of the Rich and Famous

The title of this note is also the title of a book, wherein you can find the e-mail addresses of journalist Tom Brokaw, Microsoft mogul Bill Gates, actor James Woods, novelist Toni Morrison, humorist Dave Barry, movie critic Roger Ebert, U.S. Senator Ted Kennedy, tycoon Ross Perot, and hundreds of others.

ded in the network software. The two most common network protocols for LANs are Ethernet and the Token Ring network.

Ethernet, currently the most common network protocol, uses a high-speed network cable. Ethernet uses a bus topology and is inexpensive and relatively simple. Since all the nodes—computers—in a LAN use the same cable to transmit and receive data, the nodes must follow a set of rules about when to communicate; otherwise, two or more nodes could transmit at the same time, causing garbled or lost messages. Operating much like a party line, before transmitting data a node "listens" to find out if the cable is in use. If the cable is in use, the node must wait. When the cable is free from other transmissions, the node can begin transmitting immediately. This transmission method is called by the fancy name of **carrier sense multiple access with collision detection,** or **CSMA/CD.**

If, by chance, two nodes transmit data at the same time, the messages collide. When a **collision** occurs a special message, lasting a fraction of a second, is sent out over the network to indicate that it is jammed. Each node stops transmitting, waits a random period of time, and then transmits again. Since the wait period for each node is random, it is unlikely that they will begin transmitting at the same time again.

A **Token Ring network,** which is closely associated with IBM, works on the concept of a ring network topology and a token—a kind of electronic signal. The method of controlling access to the shared network cable is called **token passing.** The idea is similar to the New York City subway: If you want to ride—transmit data—you must have a token. The token circulates from node to node along the ring-shaped LAN.

Only one token is available on the network. When a node on the network wishes to transmit, it first captures the token; only then can it transmit data. When the node has sent its message, it releases the token back to the network. Since only one token is circulating around the network, only one device is able to access the network at a time.

The Work of Networking

Think of the millions of telephones installed throughout the world; theoretically, you can call any one of them. Further, every one of these phones has the potential to be part of a networking system. Although we have discussed other communications media, it is still the telephone that is the basis for action for the user at home or in the office. Revolutionary changes are in full swing in both places, but particularly in the office.

The use of automation in the office is as varied as the offices themselves. As a general definition, however, **office automation** is the use of technology to help people do their jobs better and faster. Much automated office innovation is based on communications technology. We begin this section with several important office technology topics—electronic mail, voice mail, facsimile technology, teleconferencing, and electronic data interchange.

Electronic Mail

Electronic mail, or **e-mail,** is the process of sending messages directly from one computer to another. A user can send data to a colleague downstairs, a message across town to that person who is never available for phone calls, a query to the headquarters office in Switzerland, and even memos simultaneously to regional sales managers in Chicago, Raleigh, and San Antonio.

Electronic mail works, of course, only if the intended receiver has the electronic mail facility to which the sender is connected. There are several electronic mail options. One option is for a user to enlist a third-party service bureau that provides electronic mail service for its customers. Another option is to use a public data network such as the Internet.

Electronic mail users shower it with praise. It crosses time zones, can reach many people with the same message, reduces the paper flood, and does not interrupt meetings the way a ringing phone does. Furthermore, with software called **smart e-mail**, e-mail's role is being broadened to display screen messages, schedule meetings, and even file expense reports.

Voice Mail

You know all about telephone tag. Say office worker Ted Marshall calls customer Leslie Tanaka. She is not in, so Ted leaves a message with her secretary. Ted leaves for a meeting and finds a message from Leslie when he returns. And so it goes. Few of us, it seems, are sitting around waiting for the phone to ring.

Voice mail releases workers from the tyranny of the telephone. Here is how voice mail typically works from the point of view of the user. If the person being called does not answer, the caller is given instructions to dictate a message to the system. The voice mail computer system stores the message in the recipient's "voice mailbox." Later, when the recipient dials the mailbox, the system delivers the message.

Is voice mail just a fancy answering machine? They serve similar purposes, but they do not use the same storage techniques. A voice mail system translates the word of a message into digital impulses, which it then stores on disk, just as any other data. Later, the stored message is reconverted to audio form (Figure 5-15). Voice mail also may sound like a spoken version of electronic mail. There is one big difference between electronic mail and voice mail, however. To use electronic mail, you and the mail recipient must have compatible devices with a keyboard and be able

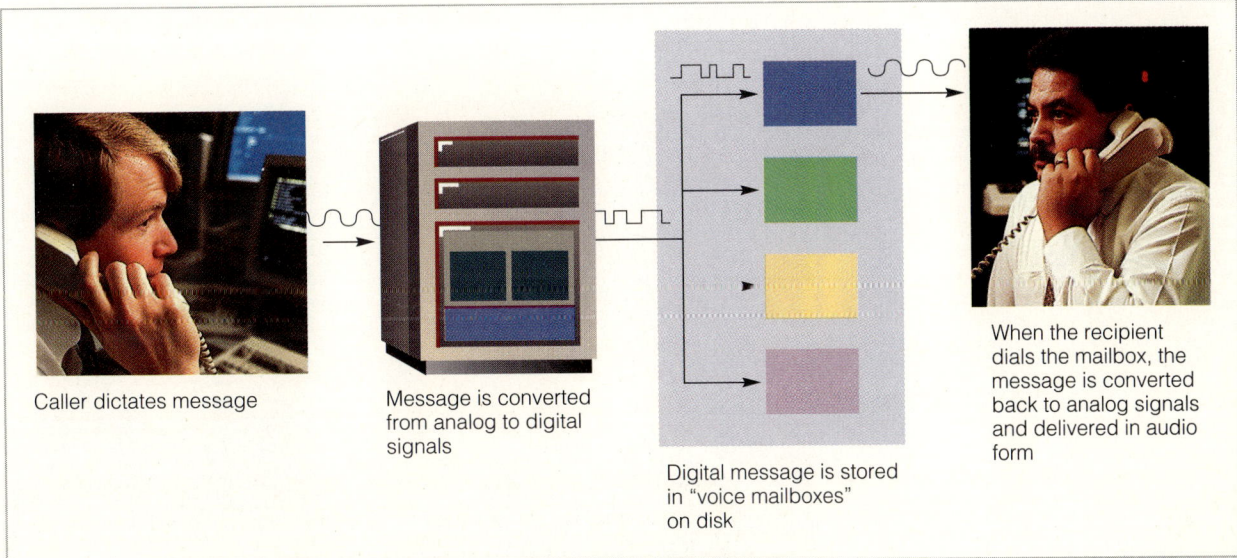

Caller dictates message

Message is converted from analog to digital signals

Digital message is stored in "voice mailboxes" on disk

When the recipient dials the mailbox, the message is converted back to analog signals and delivered in audio form

Figure 5-15 A voice mail system. The caller's message is stored in the recipient's voice mailbox on disk. Later, the recipient can check his mailbox to get the message.

Figure 5-16 Faxing it. This facsimile machine can send and receive text, drawings, and graphs long-distance.

to use them. In contrast, telephones are everywhere and everyone already knows how to use them.

Facsimile Technology

Operating something like a copy machine connected to a telephone, **facsimile technology** uses computer technology and communications links to send quality graphics, charts, text, and even signatures almost anywhere in the world. The drawing—or whatever—is placed in the facsimile machine at one end, as shown in Figure 5-16, where it is digitized. Those digits are transmitted across the miles and then reassembled at the other end to form a nearly identical version of the original picture. All this takes only minutes—or less. Facsimile is not only faster than overnight delivery services, it is less expensive. Facsimile is abbreviated **fax,** as in "I sent a fax to the Chicago office."

Personal computer users can send and receive faxes directly by means of a **fax modem,** which also performs the usual modem functions. A user can send computer-generated text and graphics. When a fax comes in, it can be reviewed on the computer screen and printed out. The only missing ingredient in this scheme is paper; if the document to be sent is available only on paper, it must be scanned into the computer first or else be sent using a separate fax machine.

Teleconferencing

An office automation development with cost-saving potential is **teleconferencing,** a method of using technology to bring people and ideas together despite geographic barriers (Figure 5-17). There are several varieties of teleconferencing, but most common today is **videoconferencing,** whose components usually include a large screen, cameras that can send live pictures, and an on-line computer system to record communication among participants. Although this setup is expensive to rent and even more expensive to own, the costs seem trivial when compared to travel expenses—airfare, lodging, meals—for in-person meetings.

Videoconferencing has some drawbacks. Some people are uncomfortable about their appearance on camera. A more serious fear is that the loss

Figure 5-17 A videoconferencing system. Geographically distant groups can hold a meeting with the help of videoconferencing. A camera transmits images of local participants for the benefit of distant viewers.

of personal contact will detract from some business functions, especially those related to sales.

Electronic Data Interchange

Businesses use a great deal of paper in transmitting orders. One method devised to cut down on paperwork is **electronic data interchange (EDI)**. EDI is a series of standard formats that allow businesses to transmit invoices, purchase orders, and the like electronically. In addition to eliminating paper-based ordering forms, EDI can help to eliminate errors in transmitting orders that result from transcription mistakes made by people. Since EDI orders go directly from one computer to another, the tedious process of filling out a form at one end and then keying it into the computer at the other end is eliminated.

Many firms use EDI to reduce paperwork and personnel costs. Some large firms, especially discounters such as Wal-Mart, require their suppliers to adopt EDI and, in fact, have direct computer hookups with their suppliers.

Figure 5-18 An automated teller machine. Users can use bank services 24 hours a day through ATMs.

Electronic Fund Transfers: Instant Banking

Using **electronic fund transfers (EFTs)**, people can pay for goods and services by having funds transferred from various accounts electronically, using computer technology. One of the most visible manifestations of EFT is the **ATM**—the **automated teller machine** that people use to obtain cash quickly (Figure 5-18). A high-volume EFT application is the disbursement of millions of Social Security payments by the government directly into the recipients' checking accounts. Unlike those sent via U.S. mail, no such payment has ever been lost.

Electronic funds transfers are not limited to transfers between institutions and individuals. Banks and other financial institutions transfer funds among themselves electronically, on both the national and international level.

Bulletin Boards

Person-to-person data communication is one of the more exhilarating ways of using your personal computer. A **bulletin board system (BBS)** uses data communications systems to link personal computers to provide public-access message systems. Most bulletin boards are formed to benefit people in a club or with a common hobby; others are linked to a particular business.

Electronic bulletin boards are similar to the bulletin boards you see in laundromats or student lounges. Somebody leaves a message, but the person who picks it up does not have to know the person who left it. To get access to someone else's computer, all you really have to know is that computer's bulletin board phone number. You can use any kind of computer, but you need a modem so you can communicate over the phone lines.

Anyone who has a personal computer can set up a bulletin board. It takes a computer (usually with a hard disk drive), a phone line, a modem, and particular (inexpensive) software. You just tell a few people about your board, start up your computer using the BBS software, and sit back and watch the messages start scrolling down your screen. But note that your computer must be left on to receive the calls.

Figure 5-19 Telecommuters. (a) This engineer has all the equipment he needs to work at home. (b) Although this worker can hardly call the beach his home office, he has indeed taken his computer along to do some work.

Computer Commuting

A logical outcome of computer networks is **telecommuting,** the substitution of communications and computers for the commute to work (Figure 5-19). That is, a telecommuter works at home on a personal computer and probably uses the computer to communicate with office colleagues or customers. Although the original idea was that people would work at home all the time, telecommuting has evolved into a mixed activity. That is, most telecommuters stay home two or three days a week and come into the office the other days. Time in the office permits the needed face-to-face communication with fellow workers and also provides a sense of participation and continuity.

The ideal telecommuting candidate is one who needs little personal contact with colleagues, has access to a quiet work space at home, is able to work with little supervision, and reports to a supervisor who manages by results rather than by surveillance or a time clock. Although only a half million workers telecommuted in 1992, almost eight million are forecasted to do so by the year 2002. The numbers surely will continue to rise.

Potential benefits of telecommuting include savings in fuel costs and commuting time, an opportunity to work at your own pace, and an opportunity for workers to work in an undisturbed environment. There are, of course, some problems. One problem associated with telecommuting is the strain on families that results when a family member works at home. A more common complaint is that at-home employees miss the interaction with co-workers at the office. At the head of the list, however, is this, from the telecommuters themselves: They work too much!

Commercial Communications Services

We have talked about specific services, but some companies offer a wide range of services. Users can connect their personal computers to commercial, consumer-oriented communications systems via telephone lines. These services—known as **information utilities**—are widely used by both home and business customers. Popular information utilities include

ONE JUMP AHEAD

2000 and Beyond

The approach of the millennium has inspired both deep thinkers and not-so-deep thinkers to contemplate our lives in the next century. The ability to access information and services on-line will be a major factor. Here are some guesses about online future tripping:

- Using a computer and online services will be akin to using a phone. Everyone will know how.
- Most shopping will be done from the home by perusing color catalogs on the computer screen and then placing orders via computer. Shopping malls will evolve to specialty restaurants, entertainment centers, and—even more—teenage hangouts.
- Since most workers will telecommute most of the time, work will be less central to people's lives. Work will become less of a place to go and more of a thing to do. Since most workers will be at home, the importance of family and community will increase.
- A telecommuting society means we will stop building skyscrapers to house office workers. However, some people may go to a "work center" just to hang out with other humans and not feel isolated.
- Large public companies will be replaced by hundreds of smaller entrepreneurial companies that survive nicely by ordering supplies online, advertising their goods and services online, and selling directly to their customers via home computers.
- People will use their computers to access information and services related to accounting, the law, and medicine. Thus, since these services will become less labor intensive, their prices to the consumer will drop significantly.
- Televised talk shows will become interactive. Your computerized TV will include a camera, so everyone who tunes in can see you as you ask a question or debate the issues.
- Eventually, people will use their in-home online computer services to bank, vote, send gifts, get advice, download entertainment, retrieve library materials, and chat with friends. Leaders will realize that special organizations must be formed to encourage people to get out of the house.

CompuServe Information Service, America Online, and Prodigy. With any utility, you must take a few minutes to install the system software on your personal computer. Then, for a fee, the world opens up to you via your computer.

These utilities each offer myriad services, including news, weather, shopping, games, educational materials, electronic mail, forums, and financial information (Figure 5-20). Generally speaking, CompuServe is of greatest value to sophisticated users and computer professionals, offering program packages, text editors, a software exchange, and a number of programming languages. America Online offers a superior, easy-to-use graphical environment, with mouse-controlled icons and overlaid screen

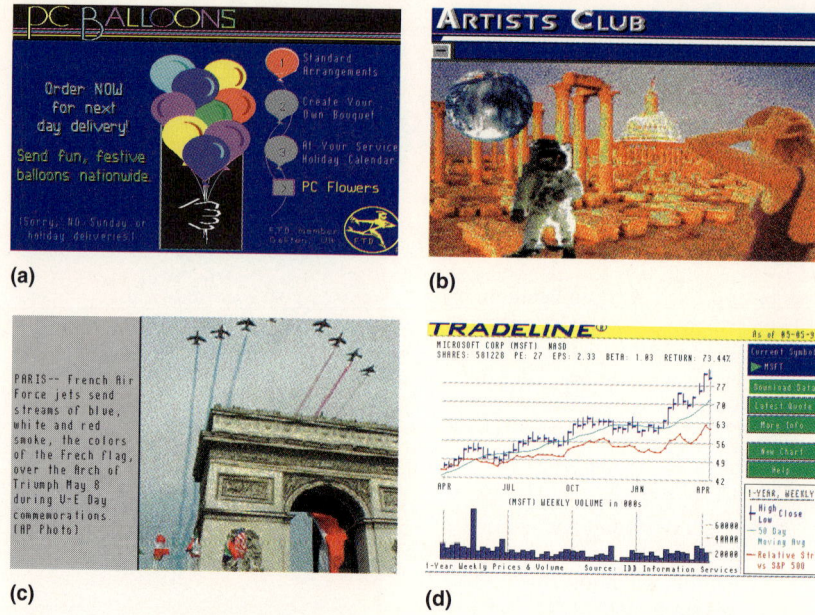

Figure 5-20 Commercial services. Computer users can use their personal computers to get information on a variety of topics through informational utilities such as America Online and Prodigy. Shown here are (a) balloon shopping screen, (b) the monthly winner in the Artists' Club, (c) a news photo of an airplane show commemorating V-E Day, (d) a chart of Microsoft's stock price.

windows. Macintosh or Windows users are familiar with this type of environment.

Charges for these services vary. Most charge a nominal sum for the initial setup software. The services offer some sort of ongoing package deal, usually a monthly fee that includes all basic services and a certain amount of connection time, with extra charges for extra time. People who live in populated areas can connect to the service at no extra phone charge through a local phone number. However, people in remote areas may have to access the service through a long distance phone number, a disadvantage that can generate a shocking phone bill.

There are many online commercial services, in addition to the ones we mentioned here. The number of users—already approaching 10 million—grows dramatically every year.

The Internet

Although the Internet could fall under the previous section on the work of networking, we choose to give it its own section because it is unique and important. The **Internet,** sometimes called simply the Net, is the largest and most far-flung network system of them all. Surprisingly, the Internet is not really a network at all but a loosely organized collection of about 25,000 networks accessed by computers worldwide. Many people are astonished to discover that no one owns the Internet; it is run by volunteers. It has no central headquarters, no centrally offered services, and no comprehensive online index to tell you what information is available.

How can all the different types of computers talk to each other? They use a standardized protocol called **Transmission Control Protocol/Internet**

Chapter Five ▼ Communications

Protocol (TCP/IP). A user must access the Internet through a computer called a **server,** which has special software that uses the Internet protocol.

Originally developed and still subsidized by the United States government, the Internet connects libraries, college campuses, research labs, and businesses. The great attraction of Internet for these users is that, once the sign-up fees are paid, there are no extra charges. Therefore, and this is a key drawing card, electronic mail is free, regardless of the amount of use. In contrast, individuals using the Internet on their own personal computers must pay ongoing monthly fees to whoever is their service provider.

Getting Connected

The Internet is available to individuals through third-party vendors, such as America Online. But the level of service may vary. Since the offered services change rapidly, along with the prices, we can only suggest that you ask before you sign up. Alternatively, individuals can install personal computer software and be billed at an hourly or monthly rate for access to the Internet.

Getting Around

Since the Internet did not begin as a commercial customer-pleasing package, it did not initially offer attractive options for finding information. The arcane commands were invoked only by a hardy and determined few. Furthermore, the vast sea of information, including news and trivia, can seem an overwhelming challenge to navigate. As both the Internet user population and the available information grew, new ways were developed to tour the Internet.

The most attractive method used to move around the Internet is called browsing. Using a program called a **browser,** you can use a mouse to point and click on screen icons to navigate the Internet, particularly the **World**

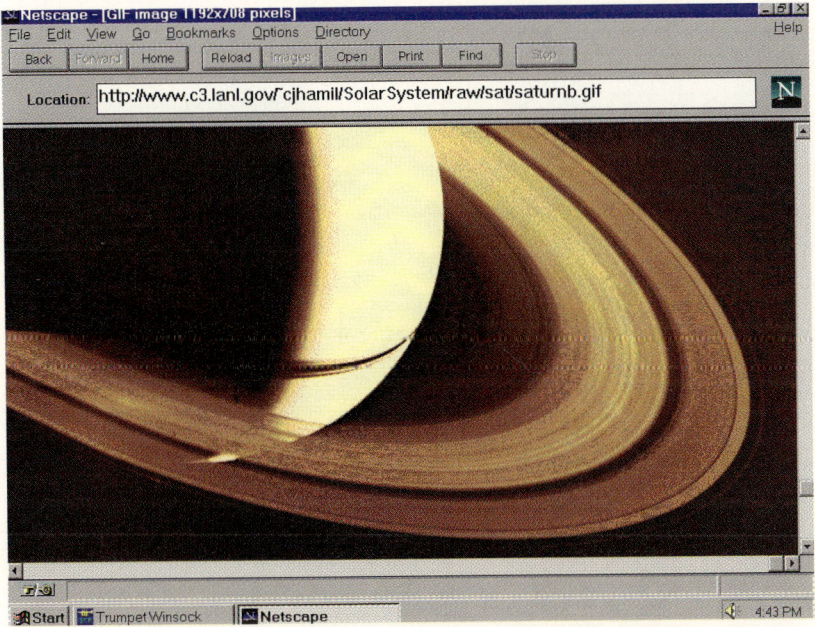

Figure 5-21 An image from the Internet. This NASA image of the rings of Saturn, found on the Internet's World Wide Web, was accessed using a browser called Netscape.

Tips for the Macintosh® Computer

Going Public with America Online

There was a time when people equated personal computers with isolation. The classic image was that of the antisocial computer nerd who spent most of his or her hours interacting with machines instead of people. That image is changing, partly because of the friendliness of modern online subscriber services—mainframe computers to which users can connect from thousands of miles away. Among the most popular of such services is America Online.

Established in 1985 by Steve Case, America Online (AOL) is now the world's fastest growing subscriber service. If you own an Apple Macintosh or DOS-based machine with Windows, getting an account on AOL is easy. You will, of course, need to have a modem connected to your computer and some related software. AOL will send you a free copy of their software, which will manage all the modem settings and connection specifications necessary to begin your membership. Once you've installed the software and provided it with a little information, (your name, address, and so forth), connecting to AOL is as simple as clicking a button.

The interface you work with once you've connected to AOL is equally friendly, in the tradition of the Macintosh graphical interface: Choices are presented to you in the form of windows and colorful icons that you activate by pointing and clicking. AOL offers many services but is organized in such a way that you can find the items you want quickly and easily.

Each window presented to you by AOL offers several choices of what to do or where to go next. For instance, when you first log on, you have the choices of reading the top news items of the day, checking any electronic mail you might have received, or going directly to the Departments section. If you choose Departments, you are given a window of choices such as *Entertainment, Lifestyles & Interests,* and *News & Finances.*

What lies at the end of this stream of choices? Possibly a display of information; for instance, you can look up a topic in Compton's Encyclopedia, or display the day's stock market activities. You can copy files from AOL to your own computer's hard drive—perhaps some computer-scanned works of art from the Smithsonian Department, or interactive educational programs, or some good games. AOL, like most subscriber services, provides massive software libraries through which you can browse.

Services like AOL are not just software storehouses, though; in fact, they're more aptly described as electronic communities. You can send mail to and receive mail from other users of the service, or even users of different services or machines. You can actively participate in a number of different forums—groups of users who exchange messages, post notices, and provide programs and documents to each other. Each forum is organized around some topic of interest, and there are many to choose from.

If you are interested in giving it a try, you can get your free software and a free trial period by consulting a major software retailer or bookstore, or by calling (800) 827-6364.

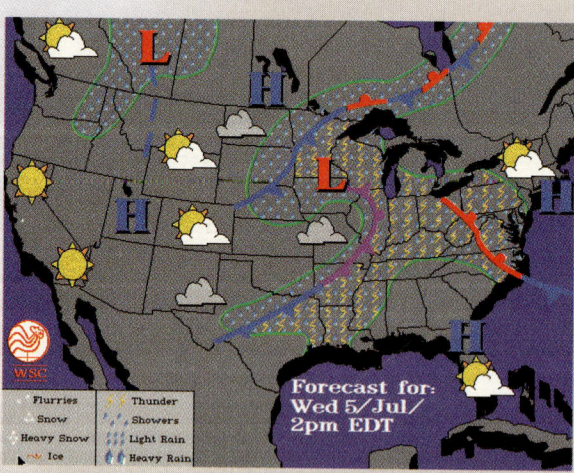

a) America Online features a wide variety of services and forums centered around different topics. You can go to a specific department just by clicking on its icon.

b) Here the user views a weather map on the screen.

Wide Web (the Web), an Internet subset of text, images, and sounds linked together to allow users to peruse related topics (Figure 5-21). It is important, by the way, for Web cruisers to have a fast modem, 14,400 bps or even 28,800 bps.

The Internet in Business

Are serious businesses interested in the Internet? You bet. General Electric, IBM, J. P. Morgan, Merrill Lynch, Motorola, and Xerox, to name a few. Some portions of the Internet, particularly those devoted to research and education, specifically ban commercial data such as advertising. However, much of the Internet is embraced by—and supported by—business. More than half of Internet networks are business related.

There are both short-term and long-term Internet benefits for business people. In the short term there is e-mail, information gathering, collaboration with strategic partners, and even direct marketing. The latter must be done with care and imagination, perhaps a "test drive" or free sample; Internet subscribers give negative feedback to ordinary online junk mail. Long term, the most compelling reason for connecting is that the Internet represents the way business is going to be transacted in the future. For now, the Internet is the nearest thing to a working prototype of the coming information superhighway.

So Much More

The subject of the Internet is so complex that it has spawned newspaper columns, magazines, dozens of books, software access packages, and special classes at libraries, campuses, and businesses around the world. In this brief section we can only glimpse the possibilities. Even so, you can appreciate the magnitude of the Internet and its emerging importance.

The future of the Internet seems limitless. Many experts predict that the Internet is destined to become the centerpiece of all online communications.

Maureen Allaire
Executive Vice President
556 S. Barrington Ave.
Los Angeles, CA 90049

Office:
(310) 476-1124
Fax: (310) 476-8080
Internet: Sorry

I'm So Embarrassed About My Card

Businesses have long offered logo-embossed cards to their employees. A few years ago such cards were updated to include a fax number. Now a new update is taking place to add an Internet address, which includes a name designation followed by @ and the company name, and a final appendage of .com for commercial enterprise:

MALLAIRE@CLAREN.COM

But the card shown here is missing an Internet address because the company, worried about employee dawdling, discourages its employees from using the Internet. This will change soon. Lack of an Internet address sends the message that the company is not serious about electronic commerce. As one employee put it, just say yes to the Internet.

 ## The Complexity of Networks

Networks can be designed in an amazing variety of ways, from a simple in-office group of three personal computers connected to a shared printer to a global spread including thousands of personal computers, minicomputers, and mainframes. The latter, of course, would not be a single network but, instead, a collection of connected networks.

You have already glimpsed the complexity of networks. Review Figure 5-11, for example, showing the variety of communications links that can be used in a nation-spanning network. Now let us consider a set of networks for a toy manufacturer (Figure 5-22).

The toy company has a bus local area network for the marketing department, consisting of six personal computers, a modem used by outside field representatives to call in for price data, a shared laser printer, shared marketing program and data files, and a server. The LAN for the design department, also a bus network, consists of three personal computers, a shared printer, shared files, and a server. Both LANs use the Ethernet protocol and have client/server relationships. The design department sometimes sends its in-progress work to the marketing representatives for their evaluation; similarly, the marketing department sends new ideas from

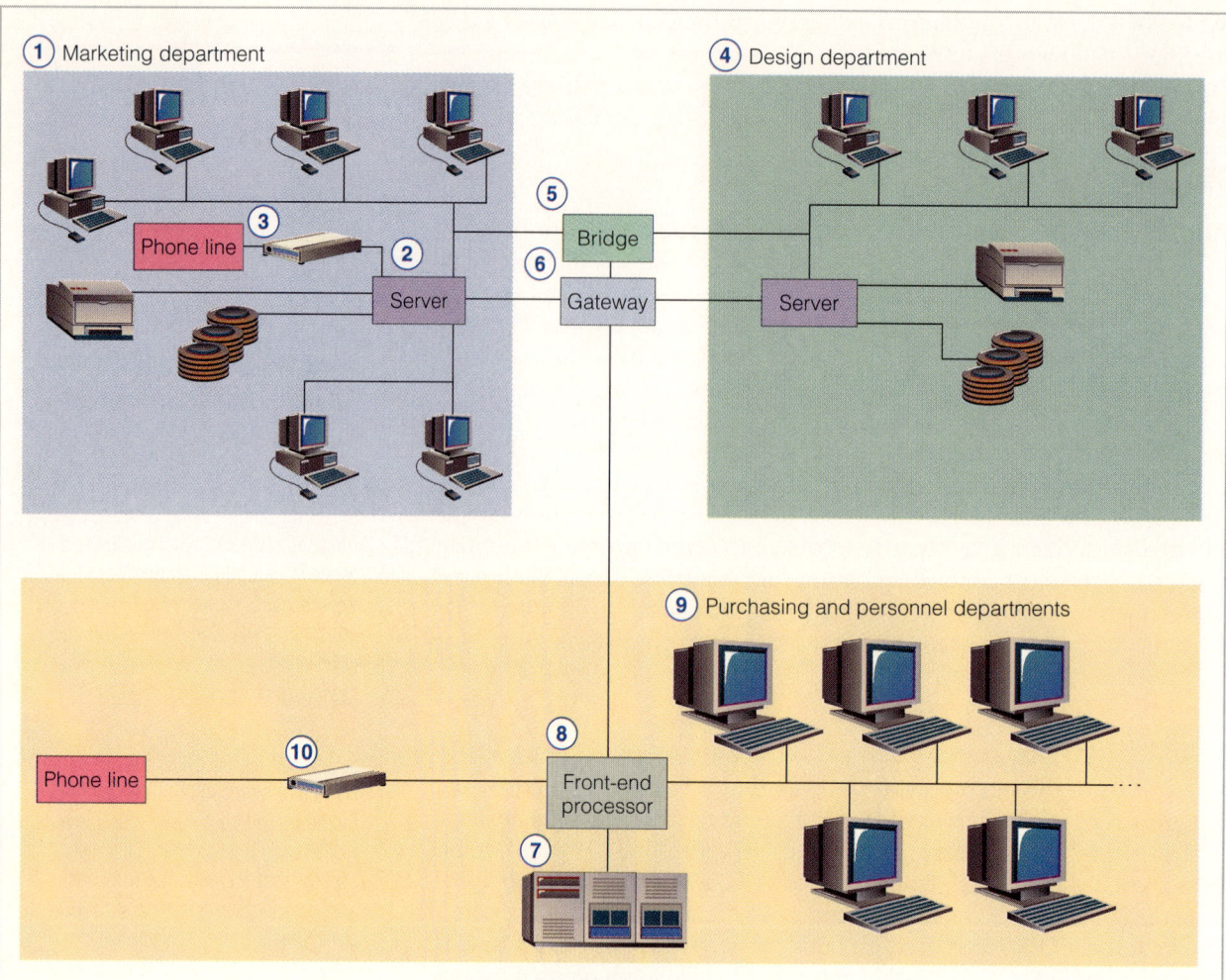

Figure 5-22 Example of a network. In this set of networks for a toy manufacturer, ① the marketing department has a bus local area network whose six personal computers use a shared printer and both program and data files stored with the ② server. Note ③ the modem that accepts outside inquiries from field representatives. ④ The design department, with just three personal computers, has a similar LAN. The two LANs can communicate via ⑤ a bridge. Either LAN, via ⑥ a gateway, can access ⑦ the mainframe computer, which uses ⑧ a front-end processor to handle communications. Users in ⑨ the purchasing and personnel departments have terminals attached directly to the mainframe computer. The mainframe computer also has ⑩ a modem that connects to the telephone lines and then, via satellite, to the mainframe at the headquarters office in another state.

the field to the design department. The two departments communicate, one LAN to another, via a bridge. It makes sense to have two separate LANs, rather than one big LAN, because the two departments need to communicate with each other only occasionally.

In addition to communicating with each other, users on each LAN, both marketing and design, occasionally need to communicate with the mainframe computer, which can be accessed through a gateway. All communications for the mainframe are handled by the front-end processor. Users in the purchasing, administrative, and personnel departments have terminals attached directly to the mainframe computer. The mainframe also has a modem that connects to telephone lines and then, via satellite, to the mainframe computer at corporate headquarters in another state.

Network factors that add to complexity but are not specifically addressed in Figure 5-22 include the electronic data interchange setups between the toy manufacturer's purchasing department and seven of its major customers, the availability of electronic mail throughout the networks, and the fact that—via a modem to an outside line—individual employees can access the Internet.

The near future in data communications is not difficult to see. The demand for services is just beginning to swell. Electronic mail already pervades the office and the campus and is beginning to reach the home. Expect instant access to all manner of information from a variety of convenient locations. Prepare to become blasé about communications services available in your own home and everywhere you go.

CHAPTER REVIEW

Summary and Key Terms

- **Data communications systems** are computer systems that transmit data over communications lines, such as telephone lines or cables.
- **Centralized data processing** places all processing, hardware, and software in one central location.
- In **teleprocessing** systems, terminals at various locations are connected by communications lines to the central computer that does the processing.
- Businesses with many locations or offices often use **distributed data processing (DDP)**, which allows both remote access and remote processing. Processing can be done by the central computer and the other computers that are hooked up to it.
- A **network** is a computer system that uses communications equipment to connect two or more computers and their resources.
- The basic components of a data communications system are a sending device, a communications link, and a receiving device.
- Data may travel to a large computer through a communications control unit called a **front-end processor,** which is actually a computer in itself. Its purpose is to relieve the central computer of some communications tasks.
- **Digital transmission** sends data as distinct on or off pulses. **Analog transmission** uses a continuous electric signal in a **carrier wave** having a particular **amplitude** and **frequency.**
- Computers produce digital signals, but most types of communications equipment use analog signals. Therefore, transmission of computer data involves altering the analog signal, or carrier wave. Digital signals are converted to analog signals by **modulation** (change) of a characteristic, such as the amplitude of the carrier wave. **Demodulation** is the reverse process; both processes are performed by a device called a **modem.**
- **Acoustic coupler modems** include a cradle to hold the telephone handset. A **direct-connect modem** is connected directly to the telephone line by means of a telephone jack. An **external modem** is not built into the computer and can therefore be used with a variety of computers. An **internal modem** is on a board that fits inside a personal computer. The Hayes brand name is the standard for modems; most modems are **Hayes compatible.**
- Two common methods of coordinating the sending and receiving units are **asynchronous transmission** and **synchronous transmission.** The asynchronous, or **start/stop,** method keeps the units in step by including special signals at the beginning and end of each group of message bits—a group is usually a character. In synchronous transmission the internal clocks of the units are put in time with each other at the beginning of the transmission, and the characters are transmitted in a continuous stream.
- **Simplex transmission** allows data to move in only one direction (either sending or receiving). **Half-duplex transmission** allows data to move in either direction but only one way at a time. With **full-duplex transmission,** data can be sent and received at the same time.
- A communications **link** is the physical medium used for transmission. Common communications links include **wire pairs** (or **twisted pairs**), **coaxial cables**, fiber optics, microwave transmission, and satellite transmission. In satellite transmission, which uses **earth stations** to send and receive signals, a **transponder** ensures that the stronger outgoing signals do not interfere with the weaker incoming ones. **Noise** is anything that causes distortion in the received signal. **Bandwidth** refers to the number of frequencies that can fit on one link at the same time, or the capacity of the link.
- A **protocol** is a set of rules for exchanging data between a terminal and a computer or between two computers. The **Open Systems Interconnection (OSI)** protocol model, developed by the International Standards Organization (ISO), has been endorsed by the United Nations.

- The physical layout of a local area network is called a **topology**. A **node** is a computer on a network. A **star network** has a central computer, the hub, that is responsible for managing the network. A **ring network** links all nodes together in a circular manner. A **bus network** has a single line, to which all the network nodes and peripheral devices are attached.
- Computers that are connected so that they can communicate among themselves are said to form a network. A **wide area network** (**WAN**) is a network of geographically distant computers and terminals. To communicate with a mainframe, a personal computer must employ **terminal emulation software**. The large computer to which a terminal or personal computer is attached is called the **host computer**. In a situation in which a personal computer or workstation is being used as a network terminal, **file transfer software** enables a user to **download** files (retrieve them from another computer and store them) and **upload** files (send files to another computer).
- A **local area network** (**LAN**) is usually a network of personal computers that share hardware, software, and data. The nodes on some LANs are connected by a shared **network cable** or by **wireless** transmission. A **network interface card** (**NIC**) may be inserted into a slot inside the computer to allow it to send and receive messages on the LAN.
- If two LANs are similar, they may send messages among their nodes by using a **bridge**. A **gateway** is a collection of hardware and software resources that connect two dissimilar networks.
- A **client/server** arrangement involves a **server**, a computer that controls the network and has the hard disks holding shared files and often has the highest-quality printer. Processing is usually done by the server, and only the results are sent to the node. A **file server** transmits the entire file to the node, which does all its own processing.
- All computers in a **peer-to-peer** arrangement have equal status; no one computer is in control. With all files and peripheral devices distributed across several computers, users share each other's data and devices as needed.
- **Ethernet** is a type of network protocol that accesses the network by first "listening" to see if the cable is free; this method is called **carrier sense multiple access with collision detection**, or **CSMA/CD**. If two nodes transmit data at the same time, this is called a **collision**. A **Token Ring network** controls access to the shared network cable by using **token passing**.
- **Office automation** is the use of technology to help people do their jobs better and faster. **Electronic mail** (**e-mail**) allows workers to transmit messages to other people's computers. **Smart e-mail** can screen messages, schedule meetings, and file expense reports. **Voice mail** lets a caller dictate a message, which is translated into digital impulses, stored, and later is delivered in audio form. **Facsimile technology** (**fax**) can transmit text, graphics, charts, and signatures. **Fax modems** for personal computers can send or receive faxes, as well as handle the usual modem functions. **Teleconferencing** is usually **videoconferencing,** in which computers are combined with cameras and large screens. **Electronic data interchange** (**EDI**) allows businesses to send common business forms electronically.
- In **electronic fund transfers** (**EFTs**), people pay for goods and services by having funds transferred from various checking and savings accounts electronically, using computer technology. The **ATM**—the **automated teller machine**—is a type of EFT.
- A **bulletin board system** (**BBS**) uses data communications to link personal computers into a public-access message system.
- **Telecommuting** means a worker works at home on a personal computer and probably uses the computer to communicate with office colleagues or customers.
- CompuServe, America Online, and Prodigy are major commercial communications services, or **information utilities**.
- The **Internet** is a loose collection of networks accessed by computers worldwide. Computers on the Internet, called **servers**, communicate using a **protocol** called **Transmission Control Protocol/Internet Protocol** (**TCP/IP**). Using a program called a browser, users can navigate the Internet, particularly the **World Wide Web** (**the Web**).

Discussion Questions

1. Assuming that your personal computer is suitably equipped, determine for what purposes you might use one or more—or all—of the following if you ran a business out of your home: e-mail, fax modem, information services such as America Online, the Internet, electronic fund transfers, and electronic data interchange. Pick your own business or choose one of the following: catering, motorcycle repair, financial services, a law office, roofing, photography, or photo research.
2. Discuss the advantages and disadvantages of teleconferencing versus face-to-face business meetings.
3. If you plan to work in an office or in a business that operates mostly in the field (such as contracting or inspecting), what kinds of data communications services do you expect to be available to the workers? Will their availability depend on the kind of company or its size? How easily do you think you could learn the various technologies available to you on the job?

Student Study Guide

Multiple Choice

1. Electronic banking:
   a. token ring   c. Mosaic
   b. EFT   d. BBS
2. Centralized processing but with access from terminals is known as
   a. a teleprocessing system   c. a ring network
      d. telecommuting
   b. DDP
3. Computer systems that use data communications equipment to connect two or more computers and their resources are
   a. host computer systems   c. networks
      d. centralized
   b. teleprocessing systems
4. When all hardware, software, storage, and processing is housed in one location it is called
   a. a time-sharing system   c. centralized processing
      d. a host computer system
   b. a DDP system
5. Transmission permitting data to move only one way at a time is
   a. half duplex   c. simplex
   b. full duplex   d. start/stop
6. The process of converting from analog to digital is called
   a. modulation   c. line switching
   b. telecommuting   d. demodulation
7. The device used with satellite transmission that ensures that strong outgoing signals do not interfere with weak incoming signals is called a
   a. microwave   c. transponder
   b. cable   d. modem
8. The most convenient communications links are
   a. coaxial cables   c. via satellites
   b. telephone lines   d. via microwaves
9. The Token Ring network controls access to the network using
   a. facsimile   c. a bus
   b. auto-disconnect   d. token passing
10. The arrangement in which most of the processing is done by the server:
    a. simplex transmission   c. electronic data interchange
       d. client/server
    b. file server
11. A computer-based system in which a telephone message is recorded in digital form and then forwarded to others is
    a. a teleconferencing   c. voice mail
    b. a bulletin board   d. telecommuting
12. One or more computers connected to a hub computer is a(n)
    a. ring network   c. node
    b. information utility   d. star network
13. A connection for similar networks:
    a. satellite   c. gateway
    b. bridge   d. fax
14. The physical layout of a LAN is called a
    a. topology   c. contention
    b. link   d. telecommunications
15. A network type in which all computers have equal status:
    a. communications link   c. peer-to-peer
       d. direct-connect
    b. WAN
16. The type of modulation that changes the height of the signal is called
    a. frequency   c. phase
    b. amplitude   d. prephase

17. A network that places all nodes on a single cable:
    a. star   c. ring
    b. switched   d. bus
18. The signals produced by a computer or terminal to be sent over phone lines must be converted to
    a. modems   c. analog signals
    b. digital signals d. microwaves
19. Each computer in a LAN is called a
    a. bus   c. node
    b. host   d. server
20. Microwave transmission, coaxial cables, and fiber optics are examples of
    a. modems   c. gateways
    b. communication   d. ring networks
       links
21. A network of geographically distant computers and terminals is called a
    a. bus   c. WAN
    b. gateway   d. LAN
22. Two dissimilar networks can be connected by a
    a. gateway   c. node
    b. bus   d. server
23. Graphics and other paperwork can be transmitted directly using which technology?
    a. CSMA/CD   c. token passing
    b. facsimile   d. bulletin board
24. Software to peruse the Internet:
    a. gateway   c. EFT
    b. browser   d. teleconferencing
25. To make a personal computer act like a terminal, which type of software must be used?
    a. fax   c. videoconferencing
    b. bridge   d. emulation

True/False

T F 1. Frequency modulation varies the position in time of a complete wave cycle.
T F 2. Local area networks are designed to share data and resources among several computers in the same geographical location.
T F 3. A WAN is usually limited to one office building.
T F 4. The Open System Interconnection protocol has been approved by the United Nations.
T F 5. Teleprocessing allows a user to make queries of a computer 1000 miles away.
T F 6. An internal modem is normally used with a variety of computers.
T F 7. A modem can be used for either modulation or demodulation.
T F 8. Start/stop transmission transmits characters in a stream.
T F 9. Satellites use line-of-sight transmission.
T F 10. Full-duplex transmission allows transmission in both directions at once.
T F 11. Fiber optics are a cheaper form of communication than wire cables.
T F 12. Another name for the information superhighway is America Online.
T F 13. Another name for file server is peer-to-peer.
T F 14. A digital signal can be altered by frequency modulation.
T F 15. Synchronous transmission is also called start/stop transmission.
T F 16. Interactions among networked computers must use a protocol.
T F 17. Smart e-mail can screen messages.
T F 18. Ethernet and Token Ring are identical protocols.
T F 19. A ring network has no central host computer.
T F 20. A file server usually transmits the entire requested file to the user.
T F 21. A gateway connects two similar computers.
T F 22. A bus network uses a central computer as the server.
T F 23. Fax boards can be inserted inside computers.
T F 24. Ethernet systems "listen" to see if the network is free before transmitting data.
T F 25. Telecommuting is a kind of information utility.

Fill-In

1. What is the term for computer systems that transmit data over telephone lines? _____
2. What does OSI stand for? _____
3. Which kind of signal do most telephone lines require? _____
4. What device converts a digital signal to an analog signal or vice versa? _____
5. What are America Online and CompuServe examples of? _____
6. Distortion in the received signal: _____
7. The number of frequencies that can fit on a link at one time: _____
8. What is the general term for the use of technology in the office? _____
9. The name of the extra computer often used by large computers to perform communications functions: _____
10. What kind of network links distant computers and terminals? _____

11. The physical layout of a network: _____

12. A type of server that delivers the entire file to a user node: _____

13. How does a Token Ring network control access to the cable? _____

14. If a modem meets the current standard it is said to be _____

15. NIC stands for _____

16. What is the term for computer networks that share resources in a limited geographical location? _____

17. To move files from the user computer to another computer: _____

18. Personal computers and other hardware attached to a LAN are called _____

19. A network in which all computers have equal status and share resources: _____

20. What does BBS stand for? _____

21. What hardware device connects two dissimilar networks? _____

22. The protocol that uses CSMA/CD: _____

23. To communicate with a larger computer, a personal computer must use what kind of software? _____

24. Changing a signal from digital to analog is called _____

25. Another name for the start/stop method of data transmission: _____

Answers

Multiple Choice

1. b	6. d	11. c	16. b	21. c
2. a	7. c	12. d	17. d	22. a
3. c	8. b	13. b	18. c	23. b
4. c	9. d	14. a	19. c	24. b
5. a	10. d	15. c	20. b	25. d

True/False

1. F	6. F	11. T	16. T	21. F
2. T	7. T	12. F	17. T	22. F
3. F	8. F	13. F	18. F	23. T
4. T	9. F	14. F	19. T	24. T
5. T	10. T	15. F	20. T	25. F

Fill-In

1. data communications systems
2. open systems interconnection
3. analog
4. a modem
5. information utilities
6. noise
7. bandwidth
8. office automation
9. front-end processor
10. wide area network
11. topology
12. file server
13. token passing
14. Hayes compatible
15. network interface card
16. local area network
17. upload
18. nodes
19. peer-to-peer
20. bulletin board systems
21. gateway
22. Ethernet
23. terminal emulation software
24. modulation
25. asynchronous

PLANET INTERNET

Shopping Tour

Shopping conveniences have existed since catalogs were invented. Convenience is at a high point today because computer shopping offers goods and services handily bundled together. The variety of available goods is stunning. What would you like to buy? Fishing tackle? An ergonomic chair? A gift box of fruit? Mexican pottery? (The images shown here are from the home pages of the stores.) All these products are available from companies that sell their wares on the Internet. As for the service, with a few clicks of a mouse and taps on the keyboard an order can be on its way to you.

So many stores. The term *electronic mall* refers to a group of Internet stores that rivals physical malls in size and variety. A good place to start is the Hall of Malls, which lists dozens of malls, each of which has many stores. Many shoppers find that, in a rather short time, they have stumbled on favorite stores. We offer these from our list: Net Sweats and Tees, The Plastic Princess (Barbie!), and the Chocolate Lover's Page. You can use the B/C URL to check these out, but you will soon make your own list.

Finding what you want. It is all well and good to know the names and the URLs of several shopping sites. And, although more convenient than traipsing around by car and foot, you are still faced with the prospect of searching for what you want, store by Internet store. What if there was a better way? What if you could just say what you want and have the computer search through the stores for you? One possibility is a *search engine*, a site that allows you to key in a request and then returns locations that you can click on. The only problem with search engines is that they are often busy. If you get a busy response, you can simply click your browser's Reload button to try again. If you repeatedly get a busy response, switch to a different search engine. We offer several search engines at the B/C site so you have a choice. Search engines, by the way, are not just for shopping; they can come in handy any time you need to find something specific on the Net.

What if I don't want to buy anything? No problem. It's just like old-fashioned window shopping, where you are welcome to look to your heart's content. Of course, as do retailers in real stores, they hope you will see something you want to buy.

Charge it? Standards for secure transmission of transactions through the Internet are still evolving. For this reason, some people prefer to do their shopping on the Net but place the actual order by telephone or some other secure means.

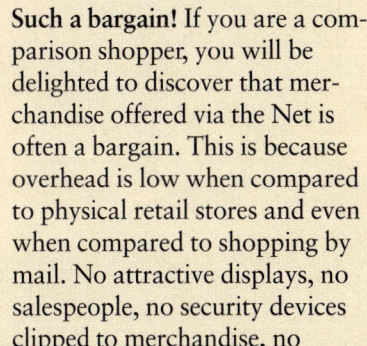

Such a bargain! If you are a comparison shopper, you will be delighted to discover that merchandise offered via the Net is often a bargain. This is because overhead is low when compared to physical retail stores and even when compared to shopping by mail. No attractive displays, no salespeople, no security devices clipped to merchandise, no printed catalogs, and possibly even no advertising. And, of course, a successful business has a potential worldwide audience and can thus purchase in high volume and pass the savings on to consumers.

Internet Exercises

1. **Structured exercise.** Begin with the B/C URL http://www.aw.com/bc/planet/ and go to one of the search engine sites. Choose an object to purchase, and see what stores the search turns up.
2. **Freeform exercise.** This is obvious: Head right down the Hall of Malls.

PART II

Software Tools

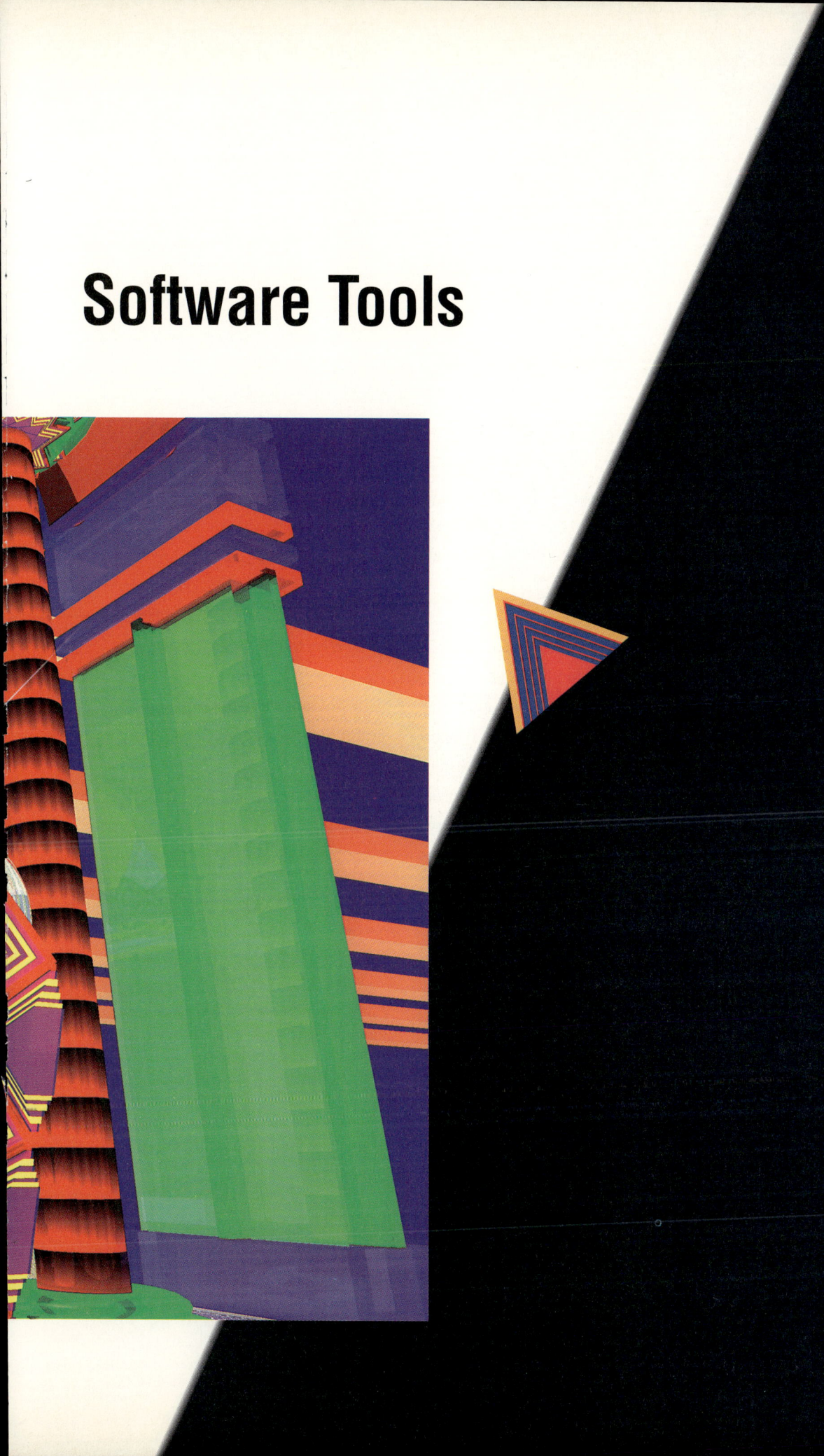

Bob Fergusen and Sean O'Connor met as freshmen at Ohio State University in an introductory business class and became fast friends when they discovered they were both majoring in accounting. The next semester they signed up for an introductory computer course to get a good foundation in computer technology and, in particular, to become proficient in the use of spreadsheets.

After studying what programmers do, however, Bob recalled his rusty BASIC programming from high school and decided to take a closer look. He signed up for a learn-at-your-own-pace lab course in BASIC programming. From there he gravitated to the computer science department, where he studied more languages and took a variety of theoretical courses. Bob eventually decided on a computer science major but minored in accounting. Sean was not as taken with computers, particularly with the details of programming, and remained an accounting major. After graduation, each found a job in his own field.

Their paths crossed again by chance seven years later when they began attending an evening M.B.A. program at a private university. Both accounting skills and computer skills were needed for various projects in the program. Bob and Sean were able to help each other out, each with his own expertise. In particular, Sean came to appreciate the care and precision needed to write a computer program.

Chapter 6

Programming and Languages
Telling the Computer What to Do

LEARNING OBJECTIVES

- Understand what programmers do and do not do
- Learn how programmers define a problem, plan the solution, and then code, test, and document the program
- Appreciate the possibilities of a programming career
- Know the levels of programming languages—machine, assembly, high level, very high level, and natural
- Become acquainted with some major programming languages
- Understand the concepts of object-oriented programming

WHY PROGRAMMING?
WHAT PROGRAMMERS DO
THE PROGRAMMING PROCESS
    1. Defining the Problem
    2. Planning the Solution
    3. Coding the Program
    4. Testing the Program
    5. Documenting the Program
PROGRAMMING AS A CAREER
    The Joys of the Field
    What It Takes, Open Doors
PROGRAMMING LANGUAGES
LEVELS OF LANGUAGE
    Machine, Assembly Languages
    High-Level Languages
    Very High-Level Languages
    Natural Languages
CHOOSING A LANGUAGE
MAJOR PROGRAMMING LANGUAGES
    FORTRAN
    COBOL: The Language of Business
    BASIC: For Beginners and Others
    Pascal: The Language of Simplicity
    Ada: Named for the Countess
    C: A Portable Language
OBJECT-ORIENTED PROGRAMMING
    What Is an Object?
    Beginnings: Boats as Objects
    Reuse, Activating the Object
    Object-Oriented Languages
    Object Technology in Business
SOME ADVICE

Doomsday: January 1, 2000

Tick, tick, tick. Some computer professionals are predicting a doomsday scenario when the clock passes midnight and we plunge into a new century—the year 2000. The problem is the date, specifically the year, which has been recorded in millions of computer files in a two-digit format: MM/DD/YY. The year1990 was recorded as 90, 1995 as 95, and so forth. The year 2000 would be recorded as 00.

This seeming bit of trivia will cause serious problems after the year 2000 in such time-related computer program activities as determining age and computing interest. Suppose, for example, your birth year is 1978. In 1999 the computer can subtract 78 from 99 to reveal your age correctly as 21. In the year 2000, however, a similar subtraction, 78 from 00, would present your age as –78.

The solution is to change all years to their original four digits. As straightforward as this sounds, experts estimate the cost of programming to adjust all systems for the year 2000 at approximately $50 *billion*. An even bigger problem, however, is the amount of time it will take to change data in existing files and to identify and change date-sensitive code in existing programs. Some organizations, it is feared, do not recognize this and will simply be out of time when the millennium arrives.

 ## Why Programming?

You may already have used software, perhaps for word processing or spreadsheets, to solve problems. Perhaps now you are curious to learn how programmers write software. As we noted earlier, a **program** is a set of step-by-step instructions that directs the computer to do the tasks you want it to do and produce the results you want.

There are at least three good reasons for learning programming:

- Programming helps you understand computers. The computer is only a tool. If you learn how to write simple programs, you will gain more knowledge about how a computer works.
- Writing a few simple programs increases your confidence level. Many people find great personal satisfaction in creating a set of instructions that solve a problem.
- Learning programming lets you find out quickly whether you like programming and whether you have the analytical turn of mind programmers need. Even if you decide that programming is not for you, understanding the process certainly will increase your appreciation of what programmers and computers can do.

A set of rules that provides a way of telling a computer what operations to perform is called a **programming language.** There is not, however, just one programming language; there are many. In this chapter you will learn about controlling a computer through the process of programming. You may even discover that you might want to become a programmer.

An important point before we proceed: You will not be a programmer when you finish reading this chapter or even when you finish reading the final chapter. Programming proficiency takes practice and training beyond the scope of this book. However, you will become acquainted with how programmers develop solutions to a variety of problems.

 ## What Programmers Do

In general, the programmer's job is to convert problem solutions into instructions for the computer. That is, the programmer prepares the instructions of a computer program and runs those instructions on the computer, tests the program to see if it is working properly, and makes corrections to the program. The programmer also writes a report on the program. These activities are all done for the purpose of helping a user fill a need, such as paying employees, billing customers, or admitting students to college.

The programming activities just described could be done, perhaps, as solo activities, but a programmer typically interacts with a variety of people. For example, if a program is part of a system of several programs, the programmer coordinates with other programmers to make sure that the programs fit together well. If you were a programmer, you might also have coordination meetings with users, managers, systems analysts, and with peers who evaluate your work—just as you evaluate theirs.

Let us turn to the programming process.

ONE JUMP AHEAD

Techno Do-gooders

Want to be a hero? How about a high-tech hero? Computer professionals have only begun to flex their power to do good. One particularly satisfying application of computer technology is helping people walk again.

Computer programmers work hand-in-hand with exercise physiologists to design computer-assisted walking systems for quadriplegics with spinal injuries and for stroke victims. Very thin stainless steel wires are inserted into a patient's lower body muscles. The system is programmed to stimulate muscles electrically so that certain muscle groups contract in sequence, enabling users to stand, walk, and climb stairs again.

When such systems began in the early 1980s, a patient's movements were jerky and uncertain. Improvements have been made by activating a greater number of muscles, providing the user with a smoother gait. The ultimate goal is to make locomotion as smooth and as cosmetically appealing as possible.

The Programming Process

Developing a program involves steps similar to any problem-solving task. There are five main ingredients in the programming process:

1. Defining the problem
2. Planning the solution
3. Coding the program
4. Testing the program
5. Documenting the program

Let us discuss each of these in turn.

1. Defining the Problem

Suppose that, as a programmer, you are contacted because your services are needed. You meet with users from the client organization to analyze the problem, or you meet with a systems analyst who outlines the project. Specifically, the task of defining the problem consists of identifying what it is you know (input—given data), and what it is you want to obtain (output—the result). Eventually, you produce a written agreement that, among other things, specifies the kind of input, processing, and output required. This is not a simple process. It is closely related to the process of systems analysis, which will be discussed in Chapter 8.

2. Planning the Solution

Two common ways of planning the solution to a problem are to draw a flowchart and to write pseudocode, or possibly both. Essentially, a **flow-**

chart is a pictorial representation of a step-by-step solution to a problem. It consists of arrows representing the direction the program takes and boxes and other symbols representing actions. It is a map of what your program is going to do and how it is going to do it. The American National Standards Institute (ANSI) has developed a standard set of flowchart symbols. Figure 6-1 shows the symbols and how they might be used in a simple flowchart of a common everyday act—preparing a letter for mailing. As a practical matter, few programmers use flowcharting in their work, but flowcharting retains its value as a visual learning tool.

Pseudocode is an English-like nonstandard language that lets you state your solution with more precision than you can in plain English but with less precision than is required when using a formal programming language. Pseudocode permits you to focus on the program logic without having to be concerned just yet about the precise syntax of a particular programming language. However, pseudocode is not executable on the computer. We will illustrate these later in this chapter, when we focus on language examples. (A detailed look at flowcharting and pseudocode, in addition to the approach called structured programming, is offered in Appendix A.)

3. Coding the Program

As the programmer, your next step is to code the program—that is, to express your solution in a programming language. You will translate the logic from the flowchart or pseudocode—or some other tool—to a programming language. As we have already noted, a programming language is a set of rules that provides a way of instructing the computer what operations to perform. There are many programming languages: BASIC, COBOL, Pascal, FORTRAN, and C are some examples. You may find yourself working with one or more of these. We will discuss the different types of languages in detail later in this chapter.

Although programming languages operate grammatically, somewhat like the English language, they are much more precise. To get your program to work, you have to follow exactly the rules—the **syntax**—of the language you are using. Of course, using the language correctly is no guarantee that your program will work, any more than speaking grammatically correct English means you know what you are talking about. The point is that correct use of the language is the required first step. Then your coded program must be keyed, probably using a terminal or personal computer, in a form the computer can understand.

One more note here: Programmers usually use a **text editor**, which is somewhat like a word processing program, to create a file that contains the program. However, as a beginner, you will probably want to write your program code on paper first.

4. Testing the Program

Some experts insist that a well-designed program can be written correctly the first time. In fact, they assert that there are mathematical ways to prove that a program is correct. However, the imperfections of the world are still with us, so most programmers get used to the idea that their newly written programs probably have a few errors. This is a bit discouraging at first, since programmers tend to be precise, careful, detail-oriented people who take pride in their work. Still, there are many opportunities to introduce

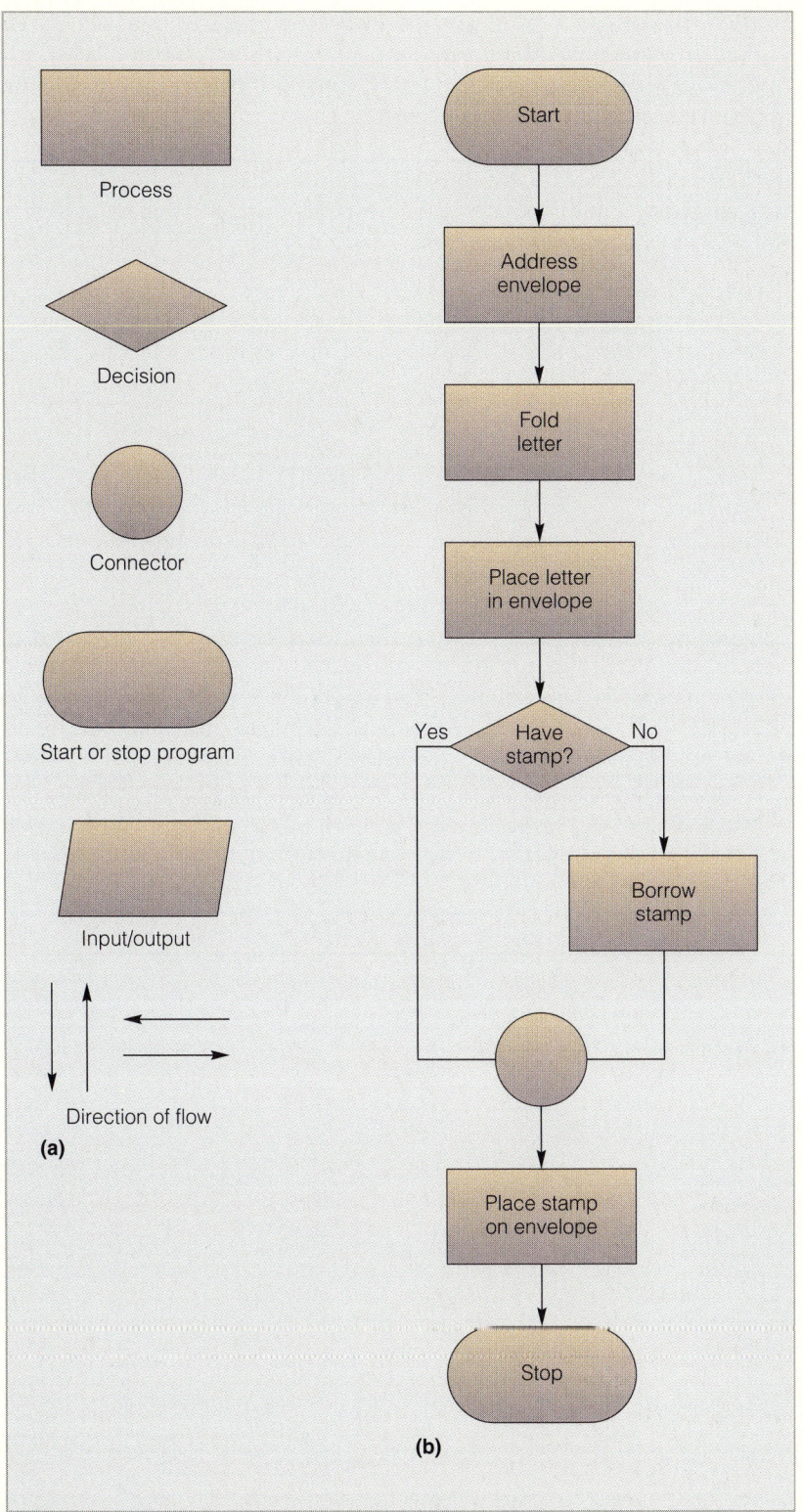

Figure 6-1 Flowchart symbols and a simple flowchart. (a) The ANSI standard flowchart symbols. (b) A flowchart shows how the standard symbols might be used to prepare a letter for mailing. There can be as many flowcharts to represent the task as there are ways of mailing a letter.

mistakes into programs, and you, just as those who have gone before you, will probably find several of them.

Eventually, after coding the program, you must prepare to test it on the computer. This step involves these phases:

- **Desk-checking.** This phase, similar to proofreading, is sometimes avoided by the programmer who is looking for a shortcut and is eager to run the program on the computer once it is written. However, with careful desk-checking you may discover several errors and possibly save yourself time in the long run. In desk-checking you simply sit down and mentally trace, or check, the logic of the program to attempt to ensure that it is error-free and workable. Many organizations take this phase a step further with a **walkthrough**, a process in which a group of programmers—your peers—review your program and offer suggestions in a collegial way.

- **Translating.** A **translator** is a program that (1) checks the syntax of your program to make sure the programming language was used correctly, giving you all the syntax-error messages, called **diagnostics,** and (2) then translates your program into a form the computer can understand. A by-product of the process is that the translator tells you if you have improperly used the programming language in some way. These types of mistakes are called **syntax errors.** The translator produces descriptive error messages. For instance, if in FORTRAN you mistakenly write N = 2*(I + J))—which has two closing parentheses instead of one—you will get a message that says, "UNMATCHED PARENTHESES." (Different translators may provide different wording for error messages.) Programs are most commonly translated by a compiler. A **compiler** translates your entire program at one time. As shown in Figure 6-2, the

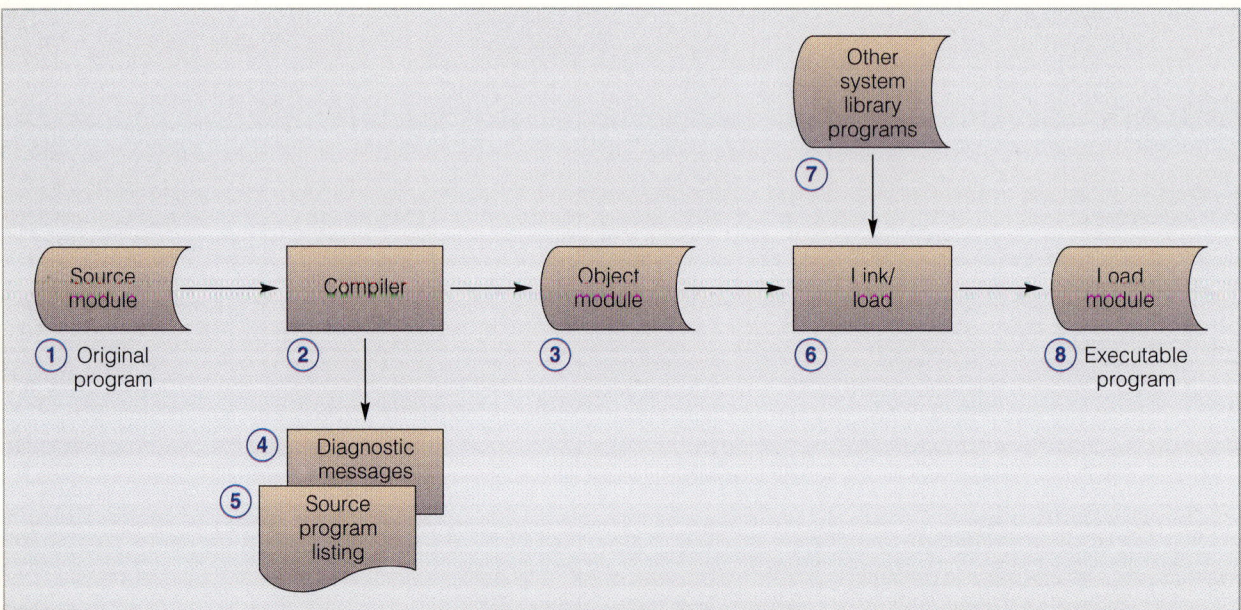

Figure 6-2 Preparing a program for execution. An original program, ① the source module, is translated by ② the compiler into ③ an object module, which represents the program in a form the machine can understand. The compiler may produce ④ diagnostic messages, indicating syntax errors. ⑤ A listing of the source program may also be output from the compiler. After the program successfully compiles, the object module is linked in the ⑥ link/load phase with ⑦ system library programs as needed, and the result is a ⑧ load module, or executable program.

translation involves your original program, called a **source module,** which is transformed by a compiler into an **object module.** Prewritten programs from a system library may be added during the **link/load phase,** which results in a **load module.** The load module can then be executed by the computer.

- **Debugging.** A term used extensively in programming, debugging means detecting, locating, and correcting bugs (mistakes), usually by running the program. These bugs are **logic errors,** such as telling a computer to repeat an operation but not telling it how to stop repeating. In this phase you run the program using test data that you devise. You must plan the test data carefully to make sure you test every part of the program.

5. Documenting the Program

Documenting is an ongoing, necessary process, although, as many programmers are, you may be eager to pursue more exciting computer-centered activities. **Documentation** is a written detailed description of the programming cycle and specific facts about the program. Typical program documentation materials include the origin and nature of the problem, a brief narrative description of the program, logic tools such as flowcharts and pseudocode, data-record descriptions, program listings, and testing results. Comments in the program itself are also considered an essential part of documentation. Many programmers document as they code. In a broader sense, program documentation can be part of the documentation for an entire system, as you will learn in Chapter 8, which discusses systems analysis and design.

The wise programmer continues to document the program throughout its design, development, and testing. Documentation is needed to supplement human memory and to help organize program planning. Also, documentation is critical to communicate with others who have an interest in the program, especially other programmers who may be part of a programming team. And, since turnover is high in the computer industry, written documentation is needed so that those who come after you can make any necessary modifications in the program or track down any errors that you missed.

 ## Programming as a Career

There is a shortage of qualified personnel in the computer field but, paradoxically, there are many people at the front end trying to get entry-level programming jobs. Before you join their ranks, consider the advantages of the computer field and what it takes to succeed in it.

The Joys of the Field

Although many people make career changes into the computer field, few choose to leave it. In fact, surveys of computer professionals, especially programmers, consistently report a high level of job satisfaction. There are several reasons for this contentment. One is the challenge—most jobs in the computer industry are not routine. Another is security, since established computer professionals can usually find work. And that work pays well—you will probably not be rich, but you should be comfortable. The

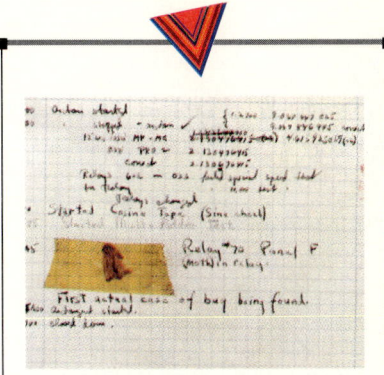

The First "Bug" Was Real

Computer literacy books are bursting with bits and bytes and disks and chips and lessons on writing programs in BASIC. All this is to provide quick enlightenment for the computer illiterate. But the average newly computer literate person has not been told about the bugs.

It is a bit of a surprise, then, to find that the software you are using does not always work quite right. Or, perhaps the programmer who is doing some work for you cannot seem to get the program to work correctly. Both problems are "bugs," errors that were introduced unintentionally into a program when it was written.

The term *bug* comes from an experience in the early days of computing. One summer day in 1945, according to computer pioneer Grace M. Hopper, the Mark I computer came to a halt. Working to find the problem, computer personnel actually found a moth inside the machine (see photo above). They removed the offending bug, and the computer was fine. From that day forward, any mysterious problem or glitch was said to be a bug.

computer industry has historically been a rewarding place for women and minorities. And, finally, the industry holds endless fascination since it is always changing.

What It Takes

You need, of course, some credentials, most often a two- or four-year degree in computer information systems or computer science. The requirements and salaries vary by the organization and the region, so we will not dwell on these here. Beyond that, the person most likely to land a job and move up the career ladder is the one with excellent communication skills, both oral and written. These are also the qualities that can be observed by potential employers in an interview. Promotions are sometimes tied to advanced degrees (an M.B.A. or an M.S. in computer science).

Open Doors

The overall outlook for the computer field is promising. The Bureau of Labor Statistics shows, through the 1990s, a 72 percent increase in programmers and a 69 percent increase in systems analysts. Further, these two professions are predicted to be the number two and number three high-growth jobs. (In case you are curious, the number one high-growth job area is predicted to be the paralegal profession.) The reasons for continued job increase in the computer field are more computers, more applications of computers, and more computer users.

Traditional career progression in the computer field was a path from programmer to systems analyst to project manager. This is still a popular direction, but it is complicated by the large number of options open to computer professionals. Computer professionals sometimes specialize in some aspect of the industry, such as data communications, database management, personal computers, graphics, or equipment. Others may specialize in the computer-related aspects of a particular industry, such as banking or insurance. Still others strike out on their own, becoming consultants or entrepreneurs.

 ## Programming Languages

At present, there are over 200 programming languages—and these are the ones that are still being used. We are not counting the hundreds of languages that for one reason or another have fallen by the wayside over the years. Where did all these languages come from? Do we really need to complicate the world further by adding programming languages to the Tower of Babel of human languages?

Initially, programming languages were created by people in universities or in government and were devised for special functions. Some languages endured because they served special purposes in science, engineering, and the like. However, it soon became clear that some standardization was needed. It made sense for those working on similar tasks to use the same language.

There are several languages in common use today, and we will discuss the most popular ones later in the chapter. Before we turn to specific languages, however, we need to discuss levels of language.

Keeping Up

If you want to be a computer professional, your formal education is merely the beginning. In the ever-changing computer field, you must take responsibility for your ongoing education. There are a variety of formal and informal ways of keeping up: college or on-the-job classes, workshops, seminars, conventions, exhibitions, trade magazines, books, and professional organizations.

Organizations are particularly important; by attending a monthly meeting you can exchange ideas with other professionals, make new contacts, and hear a speaker address some current topic. Here are some of the principal professional societies:

- **ACM.** The Association for Computing Machinery is a worldwide society devoted to developing information processing as a discipline.
- **ASM.** The Association for Systems Management keeps members current on developments in systems management and information processing.
- **AWC.** The Association of Women in Computing is open to professionals interested in promoting the advancement of women in the computer industry.
- **DPMA.** The Data Processing Management Association, one of the largest of the professional societies in the computer field, is open to all levels of information management personnel. The group seeks to encourage high standards and a professional attitude toward data processing.

Levels of Language

Programming languages are said to be "lower" or "higher," depending on how close they are to the language the computer itself uses (0s and 1s—low) or to the language people use (more English-like—high). We will consider five levels of language. They are numbered 1 through 5 to correspond to levels, or generations. In terms of ease of use and capabilities, each generation is an improvement over its predecessors. The five generations of languages are

1. Machine language
2. Assembly languages
3. High-level languages
4. Very high-level languages
5. Natural languages

Note the time line for the language generations in Figure 6-3. Let us look at each of these categories.

Machine Language

Humans do not like to deal in numbers alone—they prefer letters and words. But, strictly speaking, numbers are what machine language is. This lowest level of language, **machine language,** represents data and program instructions as 1s and 0s—binary digits corresponding to the on and off electrical states in the computer.

An example of machine language is shown in Figure 6-4. This is a language taken from a mainframe computer. Each type of computer has its own machine language. In the early days of computing, programmers had rudimentary systems for combining numbers to represent instructions such as add and compare. Primitive by today's standards, the programs were not convenient for people to read and use. The computer industry quickly moved to develop assembly languages.

Assembly Languages

Today, **assembly languages** are considered very low level—that is, they are not as convenient for people to use as more recent languages. At the time they were developed, however, they were considered a great leap forward. To replace the 1s and 0s used in machine language, assembly languages use mnemonic codes, abbreviations that are easy to remember: A for Add, C for Compare, MP for Multiply, STO for storing information in memory, and so on. Although these codes are not English words, they are still—from the standpoint of human convenience—preferable to numbers (0s and 1s) alone. Furthermore, assembly languages permit the use of names—perhaps RATE or TOTAL—for memory locations instead of actual address numbers. Just like machine language, each type of computer has its own assembly language.

The programmer who uses an assembly language requires a translator to convert the assembly language program into machine language. A translator is needed because machine language is the only language the computer can actually execute. The translator is an **assembler program,** also referred to as an assembler. It takes the programs written in assembly lan-

Supergeeks

It is not about the first prize silver bowl or the $6000 scholarship. It is not even about the corporate gurus who may hire them. It's about bragging rights.

At the annual International Collegiate Programming Contest, three-member programming teams from 35 schools, from Bulgaria and Taiwan to Slovakia and New York, flex their brains to solve programming problems. Each team, permitted the use of a single personal computer, is given a sealed envelope containing eight programs to write. Past examples were a crossword puzzle, a program to sort library books, and a navigation problem to guide ships through ports.

Each time a team successfully completes a program, a colored balloon is raised over its area. At the end of the allotted five hours, the team with the most balloons wins.

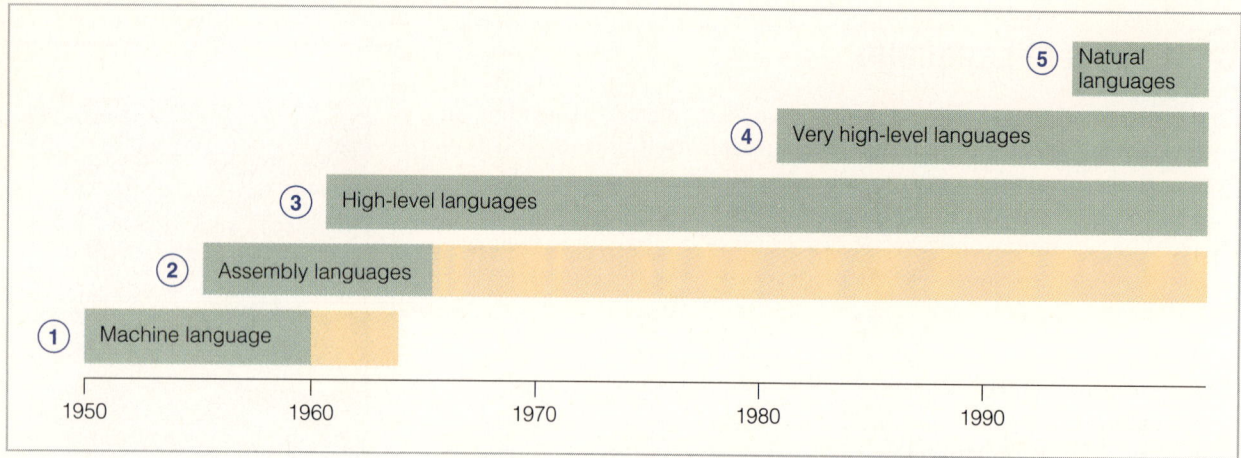

Figure 6-3 Language generations on a time line. The darker shading indicates the period of greater use by applications programmers; the lighter shading indicates the time during which a generation faded from primary use.

```
FD   71   431F   4153
F3   63   4267   4321
96   F0   426D
F9   10   41F3   438A
47   40   40DA
47   F0   4050
```

Figure 6-4 Machine language. True machine language is all binary—only 0s and 1s—but since an example would take too much space here, we are showing an example of machine language in the hexadecimal (base 16) numbering system. (The letters A through F in hexadecimal represent the numbers 10 through 15 in the decimal system.) The computer commands shown, taken from machine language for an IBM mainframe computer, are operation codes instructing the computer to divide two numbers, compare the quotient, move the result into the output area of the system, and set up the result so it can be printed.

guage and turns them into machine language. Programmers need not worry about the translating aspect; they need only write programs in assembly language. The translation is taken care of by the assembler.

Although assembly languages represent a step forward, they still have many disadvantages. A key disadvantage is that assembly language is detailed in the extreme, making assembly programming repetitive, tedious, and error prone. This drawback is apparent in the program in Figure 6-5. Assembly language may be easier to read than machine language, but it is still tedious.

High-Level Languages

The first widespread use of **high-level languages** in the early 1960s transformed programming into something quite different from what it had been. Programs were written in an English-like manner, thus making them more convenient to use. As a result, a programmer could accomplish more with less effort, and programs could now direct much more complex tasks.

These so-called third-generation languages spurred the great increase in data processing that characterized the 1960s and 1970s. During that time the number of mainframes in use increased from hundreds to tens of thousands. The impact of third-generation languages on our society has been enormous.

Of course, a translator is needed to translate the symbolic statements of a high-level language into computer-executable machine language; this translator is usually a compiler. There are many compilers for each language and one for each type of computer. Since the machine language generated by one computer's COBOL compiler, for instance, is not the machine language of some other computer, it is necessary to have a COBOL compiler for each type of computer on which COBOL programs are to be run. Keep in mind, however, that even though a given program would be compiled to different machine language versions on different machines, the source program itself—the COBOL version—can be essentially identical on each machine.

```
                PRINT   NOGEN
PROG8           START   0
CARDFIL         DTFCD   DEVADDR=SYSRDR,RECFORM=FIXUNB,IOAREA1=CARDREC,C
                        TYPEFLE=INPUT,BLKSIZE=80,EOFADDR=FINISH
REPTFIL         DTFPR   DEVADDR=SYSLST,IOAREA1=PRNTREC,BLKSIZE=132
BEGIN           BALR    3,0                     REGISTER 3 IS BASE REGISTER
                USING   *,3
                OPEN    CARDFIL,REPTFIL         OPEN FILES
                MVC     PRNTREC,SPACES          MOVE SPACES TO OUTPUT RECORD
READLOOP        GET     CARDFIL                 READ A RECORD
                MVC     OFIRST,IFIRST           MOVE ALL INPUT FIELDS
                MVC     OLAST,ILAST             TO OUTPUT RECORD FIELDS
                MVC     OADDR,IADDR
                MVC     OCITY,ICITY
                MVC     OSTATE,ISTATE
                MVC     OZIP,IZIP
                PUT     REPTFIL                 WRITE THE RECORD
                B       READLOOP                BRANCH TO READ AGAIN
FINISH          CLOSE   CARDFIL,REPTFIL         CLOSE FILES
                EOJ                             END OF JOB
CARDREC         DS      0CL80                   DESCRIPTION OF INPUT RECORD
IFIRST          DS      CL10
ILAST           DS      CL10
IADDR           DS      CL30
ICITY           DS      CL20
ISTATE          DS      CL2
IZIP            DS      CL5
                DS      CL3
PRNTREC         DS      0CL132                  DESCRIPTION OF OUTPUT RECORD
                DS      CL10
OLAST           DS      CL10
                DS      CL5
OFIRST          DS      CL10
                DS      CL15
OADDR           DS      CL30
                DS      CL15
OCITY           DS      CL20
                DS      CL5
OSTATE          DS      CL2
                DS      CL5
OZIP            DS      CL5
SPACES          DC      CL132' '
                END     BEGIN
```

Figure 6-5 Assembly language. This example shows the IBM assembly language BAL used in a program for reading a record and writing it out again. The left column contains symbolic addresses of various instructions or data. The second column contains the actual operation codes to describe the kind of activity needed; for instance, MVC stands for move characters. The third column describes the data on which the instructions are to act. The far right column contains English-like comments related to the line or lines opposite. This entire page of instructions could be compressed to a few lines in a high-level language.

Some languages are created to serve a specific purpose, such as controlling industrial robots or creating graphics. Many languages, however, are extraordinarily flexible and are considered to be general-purpose. In the past the majority of programming applications were written in BASIC, FORTRAN, or COBOL—all general-purpose languages. In addition to these three, another popular high-level language is C, which we will discuss later.

Very High-Level Languages

Languages called **very high-level languages** are often known by their generation number, that is, they are called **fourth-generation languages** or, more simply, **4GLs.**

Definition

Will the real fourth-generation languages please stand up? There is no consensus about what constitutes a fourth-generation language. The 4GLs are essentially shorthand programming languages. An operation that requires hundreds of lines in a third-generation language such as COBOL typically requires only five to ten lines in a 4GL. However, beyond the basic criterion of conciseness, 4GLs are difficult to describe.

Characteristics

Fourth-generation languages share some characteristics. The first is that they make a true break with the prior generation—they are basically nonprocedural. A **procedural language** tells the computer how a task is done: Add this, compare that, do this if something is true, and so forth—a very specific step-by-step process. The first three generations of languages are all procedural. In a **nonprocedural language**, the concept changes. Here, users define only what they want the computer to do; the user does not provide the details of just how it is to be done. Obviously, it is a lot easier and faster just to say what you want rather than how to get it. This leads us to the issue of productivity, a key characteristic of fourth-generation languages.

Productivity

Folklore has it that fourth-generation languages can improve productivity by a factor of 5 to 50. The folklore is true. Most experts say the average improvement factor is about 10—that is, you can be ten times more productive in a fourth-generation language than in a third-generation language. Consider this request: Produce a report showing the total units sold for each product, by customer, in each month and year, and with a subtotal for each customer. In addition, each new customer must start on a new page. A 4GL request looks something like this:

```
TABLE FILE SALES
SUM UNITS BY MONTH BY CUSTOMER BY PRODUCT
ON CUSTOMER SUBTOTAL PAGE BREAK
END
```

Even though some training is required to do even this much, you can see that it is pretty simple. The third-generation language COBOL, however, typically requires over 500 statements to fulfill the same request. If we define productivity as producing equivalent results in less time, then fourth-generation languages clearly increase productivity.

Downside

Fourth-generation languages are not all peaches and cream and productivity. The 4GLs are still evolving, and that which is still evolving cannot be fully defined or standardized. What is more, since many 4GLs are easy to use, they attract a large number of new users, who may then overcrowd the computer system. One of the main criticisms is that the new languages lack the necessary control and flexibility when it comes to planning how you want the output to look. A common perception of 4GLs is that they do not make efficient use of machine resources; however, the benefits of getting a program finished more quickly can far outweigh the extra costs of running it.

Benefits
Fourth-generation languages are beneficial because

- They are results-oriented; they emphasize what instead of how.
- They improve productivity because programs are easy to write and change.
- They can be used with a minimum of training by both programmers and nonprogrammers.
- They shield users from needing an awareness of hardware and program structure.

It was not long ago that few people believed that 4GLs would ever be able to replace third-generation languages. These 4GL languages are being used, but in a very limited way. Figure 6-6 illustrates a 4GL called Focus.

Query Languages
A variation on fourth-generation languages are **query languages,** which can be used to retrieve information from databases. Data is usually added to databases according to a plan, and planned reports may also be produced. But what about a user who needs an unscheduled report or a report that differs somehow from the standard reports? A user can learn a query language fairly easily and then be able to input a request and receive the resulting report right on his or her own terminal or personal computer. A standardized query language, which can be used with several different commercial database programs, is Structured Query Language, popularly known as SQL. Other popular query languages are Query-by-Example, known as QBE, and Intellect.

Natural Languages

The word *natural* has become almost as popular in computing circles as it has in the supermarket. Fifth-generation languages are, as you may guess, even more ill-defined than fourth-generation languages. They are most often called **natural languages** because of their resemblance to the "natural" spoken English language. And, to the manager new to computers for whom these languages are now aimed, natural means human-like. Instead of being forced to key correct commands and data names in correct order, a manager tells the computer what to do by keying in his or her own words.

    A manager can say the same thing any number of ways. For example, "Get me tennis racket sales for January" works just as well as "I want January tennis racket revenues." Such a request may contain misspelled words, lack articles and verbs, and even use slang. The natural language translates human instructions—bad grammar, slang, and all—into code the computer understands. If it is not sure what the user has in mind, it politely asks for further explanation.

    Natural languages are sometimes referred to as knowledge-based languages, because natural languages are used to interact with a base of

```
TABLE FILE ENROLL
HEADING CENTER
"FOSTER UNIVERSITY"
"<SEMESTER> <YR>              <CNAME>      <CNUM>"
"</2"
PRINT LNAME AND FNAME
BY CNUM NOPRINT PAGE-BREAK
BY SSN
IF SEMESTER EQ 'FALL'
IF YR EQ 96
END
```

(a)

```
                    FOSTER UNIVERSITY

        FALL 96        DIFFERENTIAL EQUATIONS        MAT482

        SSN             LNAME           FNAME
        ___             _____           _____

        101-30-6873     PICKERING       LISA
        124-98-9472     BURTON          MICHAEL
        164-46-3825     ELSLIP          JON
        182-34-0826     OKANO           NOBUKO
        190-89-7463     OFFICER         KEITH
```

(b)

Figure 6-6 An example of Focus. (a) The code here produces (b) the Foster University roster for a differential equations class. To print field names in the heading, you must use angle brackets (<>) as delimiters. The sequence </2 tells Focus to skip two lines after the heading. The program begins a new page every time the course number (CNUM) changes. The IF clauses print records only if they pertain to the fall semester of 1996.

knowledge on some subject. The use of a natural language to access a knowledge base is called a **knowledge-based system.**

Consider this request that could be given in the 4GL Focus: "SUM ORDERS BY DATE BY REGION." If we alter the request and, still in Focus, say something like "Give me the dates and the regions after you've added up the orders," the computer will spit back the user-friendly version of "You've got to be kidding" and give up. But some natural languages can

```
Hello
How may I help you?
Who are my customers in Chicago?
Just a sec.  I'll see.
The customers in that city are:
 I.D.                Name

 Ballard             Ballard and Sons, Inc.
 Fremont             Henry Fremont Associates
 Greenlake           Greenlake Consortium
 Wallingford         Wallingford, Inc.
What can I do for you now?
What is Fremont's balance?
Hang on.  I'll see.
Accounts Receivable   563.47
Unapplied Credit       79.16
         Balance      484.31
What else can I do for you?
Give me Fremont's phone number!
Please wait while I check the files.
   (312) 789-5562
What can I do for you now?
```

Figure 6-7   A natural language. This package, called Cash Management System, uses a language that is so "natural" that some might think it a little too cute, as in "Just a sec." You can follow the dialogue more easily by noting that, in this demonstration, the command from the user is in color and the response from the computer is not.

handle such a request. Users can relax the structure of their requests and increase the freedom of their interaction with the data.

Here is a typical natural language request:

```
REPORT THE BASE SALARY, COMMISSIONS AND YEARS OF
SERVICE BROKEN DOWN BY STATE AND CITY FOR SALESCLERKS
IN NEW JERSEY AND MASSACHUSETTS.
```

You can hardly get closer to conversational English than that.

An example of a natural language is shown in Figure 6-7. Natural languages excel at easy data access. Indeed, the most common application for natural languages is interacting with databases.

Choosing a Language

How do you choose the language with which to write your program? There are several possibilities:

- In a work environment, your manager may decree that everyone on your project will use a certain language.
- You may use a certain language, particularly in a business environment, based on the need to interface with other programs; if two programs are to work together, it is easiest if they are written in the same language.
- You may choose a language based on its suitability for the task. For example, a business program that handles large files may be best written in the business language COBOL.
- If a program is to be run on different computers, it must be written in a language that is portable—suitable on each type of computer—so that the program need be written only once.

- You may be limited by the availability of the language. Not all languages are available in all installations or on all computers.
- The language may be limited to the expertise of the programmer; that is, the program may have to be written in a language the available programmer knows.
- Perhaps the simplest reason, one that applies to many amateur programmers, is that they know the language called BASIC because it came with—or was inexpensively purchased with—their personal computers.

 ## Major Programming Languages

The following sections on individual languages will give you an overview of the third-generation languages in common use today: FORTRAN, COBOL, BASIC, Pascal, Ada, and C. Table 6-1 summarizes these languages and their applications.

This chapter will present programs written in these six languages. You will also see output produced by each program. Each program is designed to find the average of three numbers; the resulting average is shown in the sample output matching each program. Since all six programs perform the same task, you will see some of the differences and similarities among the languages. We do not expect you to understand these programs; they are here merely to let you glimpse each language. Figure 6-8 presents the flowchart and pseudocode for the task of averaging numbers. As we discuss each language, we will provide a program for averaging numbers that follows the logic shown in this figure.

FORTRAN: The First High-Level Language

Developed by IBM and introduced in 1954, **FORTRAN**—for FORmula TRANslator—was the first high-level language. FORTRAN is a scientifically oriented language—in the early days use of the computer was primarily associated with engineering, mathematical, and scientific research tasks.

FORTRAN is noted for its brevity, and this characteristic is part of the reason why it remains popular. This language is very good at serving its primary purpose, which is execution of complex formulas such as those used in economic analysis and engineering. Although in the past it was considered limited in regard to file processing or data processing, its capabilities have been greatly improved.

Table 6-1 Applications of some important programming languages.	
Language	Application
FORTRAN—FORmula TRANslator (1954)	Scientific
COBOL—COmmon Business-Oriented Language (1959)	Business
BASIC—Beginner's All-purpose Symbolic Instruction Code (1965)	Education, business
Pascal—named after French inventor Blaise Pascal (1971)	Education, systems, scientific
Ada—named after Ada, the Countess of Lovelace (1980)	Military, general
C—evolved from the language B, from Bell Labs (1972)	Systems, general

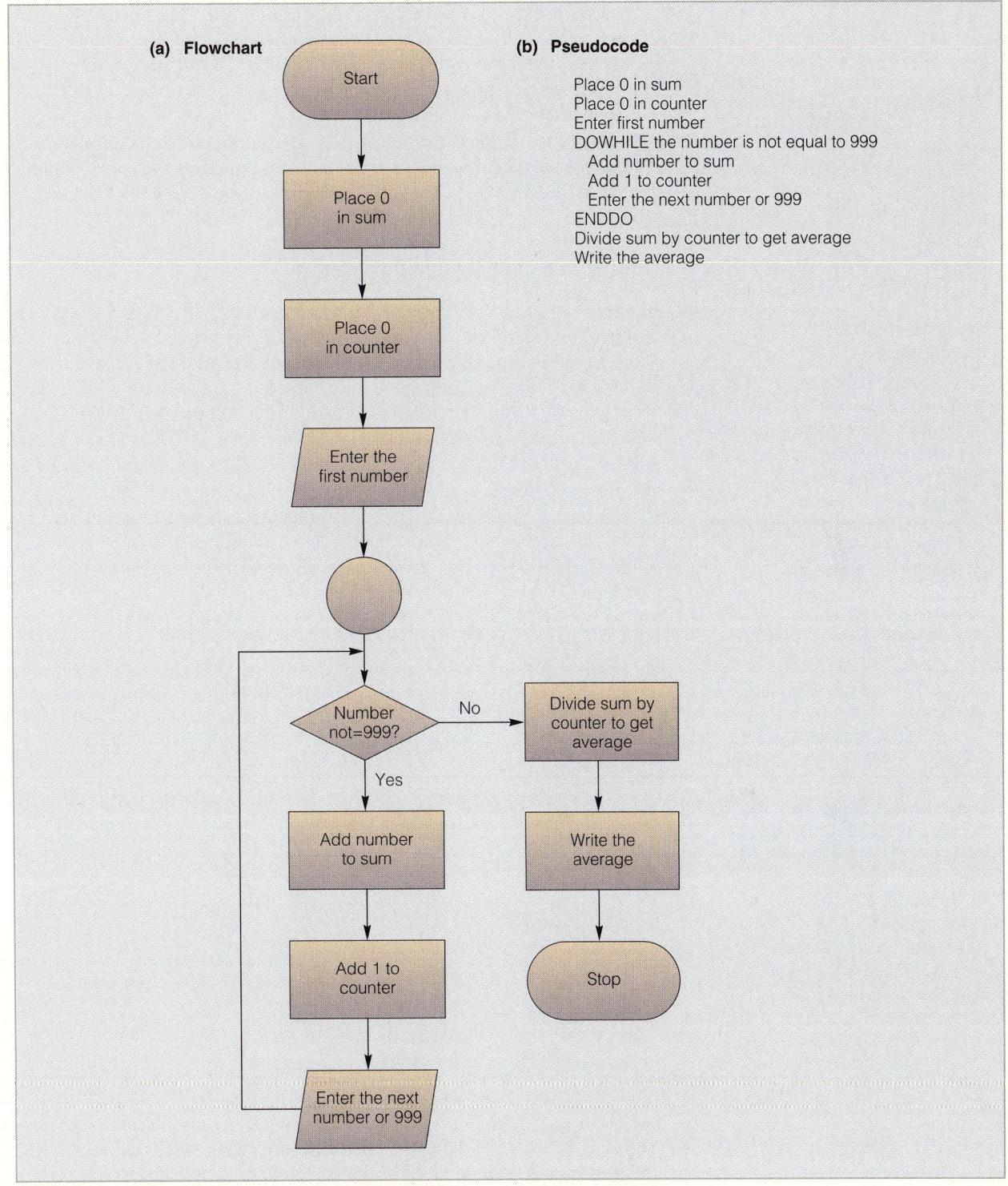

Figure 6-8 Flowchart and pseudocode for averaging numbers. (a) This flowchart, along with (b) matching pseudocode, shows the logic for a program to let a user enter numbers through the keyboard; the program then averages the numbers. The user can make any number of entries, one at a time. To show when he or she is finished making entries, the user enters 999. The logic to enter the numbers forms a loop: entering the number, adding it to the sum, and adding 1 to the counter. When 999 is keyed, the loop is exited. Then the machine computes the average and displays it on the screen. This logic is used for the programs, in various languages, shown in Figures 6-9 through 6-14.

A Programming Pioneer: Grace M. Hopper

When Grace M. Hopper died in 1992 at the age of 85, she left behind the legacy of a true programming pioneer—involvement with the first computers and the first business programming language.

As a Phi Beta Kappa graduate of Vassar College with an M.A. and Ph.D. from Yale University, Hopper joined the U.S. Naval Reserve in 1943. She was assigned to the Bureau of Ordinance Computation Project at Harvard, where she learned to program the first large-scale digital computer, the Mark I.

In 1948 Hopper became senior mathematician at the Eckert-Mauchly Computer Corporation. Later she became senior programmer on the team that created the first commercial large-scale electronic computer, UNIVAC I. In the 1950s she coauthored the COBOL compiler and programming language.

Among all her achievements, the one she liked best was her promotion to the rank of Rear Admiral in the U.S. Naval Reserve.

Not all programs are organized in the same way. Organization varies according to the language used. In many languages (such as COBOL), programs are divided into a series of parts. FORTRAN programs are not composed of different parts (although it is possible to link FORTRAN programs together); a FORTRAN program consists of statements one after the other. Different types of data are identified as the data is used. Descriptions for data records appear in format statements that accompany the READ and WRITE statements. Figure 6-9 shows a FORTRAN program and a sample output from the program.

COBOL: The Language of Business

In the 1950s FORTRAN had been developed, but there was still no accepted high-level programming language appropriate for business. The U.S. Department of Defense in particular was interested in creating such a standardized language, and so it called together representatives from government and various industries, including the computer industry. These representatives formed **CODASYL**—**CO**nference of **DA**ta **SY**stem **L**anguages. In 1959 CODASYL introduced **COBOL**—for **CO**mmon **B**usiness-**O**riented **L**anguage.

The U.S. government offered encouragement by insisting that anyone attempting to win government contracts for computer-related projects had to use COBOL. The American National Standards Institute first standardized COBOL in 1968 and, in 1974, issued standards for another version known as **ANSI-COBOL**. After more than seven controversial years of industry debate, the standard known as **COBOL 85** was approved, making COBOL a more usable modern-day software tool. The principal benefit of standardization is that COBOL is relatively machine independent—that is, a program written for one type of computer can be run with only slight modifications on another type for which a COBOL compiler has been developed.

COBOL is very good for processing large files and performing relatively simple business calculations, such as payroll or interest. A noteworthy feature of COBOL is that it is English-like—far more so than FORTRAN or BASIC. The variable names are set up in such a way that, even if you know nothing about programming, you can still understand what the program does. For example:

```
IF SALES-AMOUNT IS GREATER THAN SALES-QUOTA
    COMPUTE COMMISSION = MAX-RATE * SALES-AMOUNT
ELSE
    COMPUTE COMMISSION = MIN-RATE * SALES-AMOUNT.
```

Once you understand programming principles, it is not too difficult to add COBOL to your repertoire. COBOL can be used for just about any task related to business programming; indeed, it is especially suited to processing alphanumeric data such as street addresses, purchased items, and dollar amounts—the data of business. However, the feature that makes COBOL so useful—its English-like appearance and easy readability—is also a weakness because a COBOL program can be incredibly verbose. A programmer seldom knocks out a quick COBOL program. In fact, there is hardly such a thing as a quick COBOL program; there are just too many program lines to write, even to accomplish a simple task. For speed and simplicity, BASIC, FORTRAN, and Pascal are probably better bets.

```
C       FORTRAN PROGRAM
C       AVERAGING INTEGERS ENTERED THROUGH THE KEYBOARD
        WRITE (6,10)
        SUM = 0
        COUNTER = 0
        WRITE (6,60)
        READ (5,40) NUMBER
  1     IF (NUMBER .EQ. 999) GOTO 2
        SUM = SUM + NUMBER
        COUNTER = COUNTER + 1
        WRITE (6,70)
        READ (5,40) NUMBER
        GO TO 1
  2     AVERAGE = SUM / COUNTER
        WRITE (6,80) AVERAGE
  10    FORMAT (1X, 'THIS PROGRAM WILL FIND THE AVERAGE OF
       *'INTEGERS YOU ENTER ',/1X, 'THROUGH THE ',
       *'KEYBOARD. TYPE 999 TO INDICATE END OF DATA.',/)
  40    FORMAT (I3)
  60    FORMAT (1X, 'PLEASE ENTER A NUMBER ')
  70    FORMAT (1X, 'PLEASE ENTER THE NEXT NUMBER ')
  80    FORMAT (1X, 'THE AVERAGE OF THE NUMBERS IS ',F6.2)
        STOP
        END
```
(a)

```
THIS PROGRAM WILL FIND THE AVERAGE OF INTEGERS YOU ENTER
THROUGH THE KEYBOARD.  TYPE 999 TO INDICATE END OF DATA.
PLEASE ENTER A NUMBER      6
PLEASE ENTER THE NEXT NUMBER      4
PLEASE ENTER THE NEXT NUMBER     11
PLEASE ENTER THE NEXT NUMBER    999
THE AVERAGE OF THE NUMBERS IS    7.00
```
(b)

Figure 6-9 FORTRAN program and sample output. This program is interactive, prompting the user to supply data. (a) The first two lines are comments, as they are in the rest of the programs in this chapter. The WRITE statements send output to the screen in the format called for by the second numeral in the parentheses, which represents the line number containing the format. The READ statements accept data from the user and place it in location NUMBER, where it can be added to the accumulated total, SUM. The IF statement checks for 999 and, when 999 is received, diverts the program logic to statement 2, where the average is computed. The average is then displayed. (b) This screen display shows the interaction between program and user.

As you can see in Figure 6-10, a COBOL program is divided into four parts called divisions. The *identification division* identifies the program by name and often contains helpful comments as well. The *environment division* describes the computer on which the program will be compiled and executed. It also relates each file of the program to the specific physical device, such as the tape drive or printer, that will read or write the file. The *data division* contains details about the data processed by the program, such as type of characters (whether numeric or alphanumeric), number of characters, and placement of decimal points. The *procedure division* contains the statements that give the computer specific instructions to carry out the logic of the program.

It has been fashionable for some time to criticize COBOL: It is old-fashioned, cumbersome, and inelegant. In fact, some companies, devoted to fast, nimble program development, are converting to the more trendy language C. But COBOL, with more than 30 years of staying power, is still famous for its clear code, which is easy to read and debug.

BASIC: For Beginners and Others

BASIC—Beginners' All-purpose Symbolic Instruction Code—is a common language that is easy to learn. Developed at Dartmouth College, BASIC was introduced by John Kemeny and Thomas Kurtz in 1965 and was originally intended for use by students in an academic environment. In the late 1960s it became widely used in interactive time-sharing environments in universities and colleges. The use of BASIC has extended to business and personal computer systems.

The primary feature of BASIC is one that may be of interest to many readers of this book: BASIC is easy to learn, even for a person who has never programmed before. Thus, the language is used often to train students in the classroom. BASIC is also used by nonprogramming people, such as engineers, who find it useful in problem solving. For many years, BASIC was looked down on by "real programmers," who complained that it had too many limitations and was not suitable for complex tasks. Newer versions, such as Microsoft's QuickBASIC, include substantial improvements. An example of a BASIC program and its output are shown in Figure 6-11.

Pascal: The Language of Simplicity

Named for Blaise Pascal, the seventeenth-century French mathematician, **Pascal** was developed as a teaching language by a Swiss computer scientist, Niklaus Wirth, and first became available in 1971. Since that time it has become quite popular, first in Europe and now in the United States, particularly in universities and colleges offering computer science programs.

The foremost feature of Pascal is that it is simpler than other languages —it has fewer features and is less wordy than most. In addition to the popularity of Pascal in college computer science departments, the language has also made large inroads in the personal computer market as a simple yet sophisticated alternative to BASIC. Over the years new versions have improved on the original capabilities of Pascal. Today, Borland's Turbo Pascal leads the Pascal world because its designers eliminated most of the drawbacks of the original Pascal. Turbo Pascal is used by the business community and is often the choice of nonprofessional programmers who

```
******************************************************
 IDENTIFICATION DIVISION.
******************************************************
 PROGRAM-ID.  AVERAGE.
* COBOL PROGRAM
* AVERAGING INTEGERS ENTERED THROUGH THE KEYBOARD.
******************************************************
 ENVIRONMENT DIVISION.
******************************************************
 CONFIGURATION SECTION.
 SOURCE-COMPUTER.          H-P 9000.
 OBJECT-COMPUTER.          H-P 9000.
******************************************************
 DATA DIVISION.
******************************************************
 FILE SECTION.
 WORKING-STORAGE SECTION.
 01 AVERAGE         PIC ---9.99.
 01 COUNTER         PIC 9(02)       VALUE ZERO.
 01 NUMBER-ITEM     PIC S9(03).
 01 SUM-ITEM        PIC S9(06)      VALUE ZERO.
 01 BLANK-LINE      PIC X(80)       VALUE SPACES.
******************************************************
 PROCEDURE DIVISION.
******************************************************
 100-CONTROL-ROUTINE.
     PERFORM 200-DISPLAY-INSTRUCTIONS.
     PERFORM 300-INITIALIZATION-ROUTINE.
     PERFORM 400-ENTER-AND-ADD
             UNTIL NUMBER-ITEM = 999.
     PERFORM 500-CALCULATE-AVERAGE.
     PERFORM 600-DISPLAY-RESULTS.
     STOP RUN.
 200-DISPLAY-INSTRUCTIONS.
     DISPLAY
        "THIS PROGRAM WILL FIND THE AVERAGE OF INTEGERS YOU ENTER".
     DISPLAY
        "THROUGH THE KEYBOARD. TYPE 999 TO INDICATE END OF DATA.".
     DISPLAY BLANK-LINE.
 300-INITIALIZATION-ROUTINE.
     DISPLAY "PLEASE ENTER A NUMBER".
     ACCEPT NUMBER-ITEM.
 400-ENTER-AND-ADD.
     ADD NUMBER-ITEM TO SUM-ITEM.
     ADD 1 TO COUNTER.
     DISPLAY "PLEASE ENTER THE NEXT NUMBER".
     ACCEPT NUMBER-ITEM.
 500-CALCULATE-AVERAGE.
     DIVIDE SUM-ITEM BY COUNTER GIVING AVERAGE.
 600-DISPLAY-RESULTS.
     DISPLAY "THE AVERAGE OF THE NUMBERS IS ",AVERAGE.
```

(a)

```
            THIS PROGRAM WILL FIND THE AVERAGE OF
            INTEGERS YOU ENTER THROUGH THE KEYBOARD.
            TYPE 999 TO INDICATE END OF DATA.

            PLEASE ENTER A NUMBER
            6
            PLEASE ENTER THE NEXT NUMBER
            4
            PLEASE ENTER THE NEXT NUMBER
            11
            PLEASE ENTER THE NEXT NUMBER
            999
            THE AVERAGE OF THE NUMBERS IS    7.00
```

(b)

Figure 6-10 COBOL program and sample output. The purpose of the program and its results are the same as those of the FORTRAN program, but (a) the look of the COBOL program is very different. Note the four divisions. In particular, note that the logic in the procedure division uses a series of PERFORM statements, which divert action to other places in the program. After a prescribed action has been performed, the computer returns to the procedure division, to the statement after the one that was just completed. DISPLAY writes to the screen, and ACCEPT takes user input. (b) This screen display shows the interaction between program and user.

```
'BASIC PROGRAM
'AVERAGING INTEGERS ENTERED THROUGH THE KEYBOARD
PRINT "THIS PROGRAM WILL FIND THE AVERAGE OF INTEGERS YOU ENTER"
PRINT "THROUGH THE KEYBOARD. TYPE 999 TO INDICATE END OF DATA."
PRINT
SUM=0
COUNTER=0
CLS
PRINT "PLEASE ENTER A NUMBER"
INPUT NUMBER
DO WHILE NUMBER <> 999
     SUM=SUM+NUMBER
     COUNTER=COUNTER+1
     PRINT "PLEASE ENTER THE NEXT NUMBER"
     INPUT NUMBER
LOOP
AVERAGE=SUM/COUNTER
PRINT "THE AVERAGE OF THE NUMBERS IS"; AVERAGE
END
```
(a)

```
THIS PROGRAM WILL FIND THE AVERAGE OF INTEGERS YOU ENTER
THROUGH THE KEYBOARD. TYPE 999 TO INDICATE END OF DATA.

PLEASE ENTER A NUMBER
?6
PLEASE ENTER THE NEXT NUMBER
?4
PLEASE ENTER THE NEXT NUMBER
?11
PLEASE ENTER THE NEXT NUMBER
?999
THE AVERAGE OF THE NUMBERS IS   7
```
(b)

Figure 6-11 BASIC program and sample output. (a) PRINT displays data right in the statement on the screen. INPUT accepts data from the user. (b) This screen display shows the interaction between program and user.

need to write their own programs. An example of a Turbo Pascal program and its output are shown in Figure 6-12.

Ada: Named for the Countess

Is any software worth over $25 billion? Not any more, according to Defense Department experts. In 1974 the U.S. Department of Defense had spent that amount on all kinds of software for a hodgepodge of languages for its needs. The answer to this problem turned out to be a new language

```pascal
PROGRAM AverageofNumbers;
(*Pascal Program
  averaging integers entered through the keyboard*)

USES
   crt;

VAR
   counter, number, sum : integer;
   average : real ;

BEGIN (*main*)
    WRITELN ('THIS PROGRAM WILL FIND THE AVERAGE OF INTEGERS YOU ENTER');
    WRITELN ('THROUGH THE KEYBOARD. TYPE 999 TO INDICATE END OF DATA.');
    WRITELN;
    sum :=0;
    counter :=0;
    WRITELN ('PLEASE ENTER A NUMBER');
    READLN (number);
    WHILE number <> 999 DO
        Begin   (*while loop*)
            sum := sum + number;
            counter := counter + 1;
            WRITELN ('PLEASE ENTER THE NEXT NUMBER');
            READ (number);
        END; (*while loop*)
    average := sum / counter;
    WRITELN ('THE AVERAGE OF THE NUMBERS IS ', average:6:2);
END.  (*main*)
```
(a)

```
THIS PROGRAM WILL FIND THE AVERAGE OF INTEGERS YOU ENTER
THROUGH THE KEYBOARD. TYPE 999 TO INDICATE END OF DATA.

PLEASE ENTER A NUMBER
6
PLEASE ENTER THE NEXT NUMBER
4
PLEASE ENTER THE NEXT NUMBER
11
PLEASE ENTER THE NEXT NUMBER
999
THE AVERAGE OF THE NUMBERS IS   7.00
```
(b)

Figure 6-12 Pascal program and sample output. (a) Comments are from (* to *). Each variable name must be declared. The symbol := assigns a value to the variable to its left; the symbol < > means not equal to. WRITELN by itself puts a blank line on the screen. (b) This screen display shows the interaction between program and user. The program was written in Turbo Pascal.

called **Ada**—named for Countess Ada Lovelace, "the first programmer" (see Appendix B). Sponsored by the Pentagon, Ada was originally intended to be a standard language for weapons systems, but it has also been used successfully for commercial applications. Introduced in 1980, Ada has the support not only of the defense establishment but also of such industry heavyweights as IBM and Intel, and Ada is even available for some personal computers. Although some experts have said Ada is too complex, others say that it is easy to learn and that it will increase productivity. Indeed, some experts believe that it is by far a superior commercial language to such standbys as COBOL and FORTRAN.

Widespread use of Ada is considered unlikely by many experts. Although there are many reasons for this (the military services, for instance, have different levels of enthusiasm for it), probably its size—which may hinder its use on personal computers—and complexity are the greatest barriers. Although the Department of Defense is a market in itself, Ada has not caught on to the extent that Pascal and C have, especially in the business community. An example of an Ada program and its output are shown in Figure 6-13.

C: A Portable Language

A language invented by Dennis Ritchie at Bell Labs in 1972, **C** produces code that approaches assembly language in efficiency while still offering high-level language features. C was originally designed to write systems software but is now considered a general-purpose language. C contains some of the best features from other languages, including Pascal. C compilers are simple and compact. A key attraction is that it is independent of the architecture of any particular machine, a fact that contributes to the portability of C programs. That is, a C program can be run on more than one type of computer after it has been compiled for that machine.

Although C is simple and elegant, it is not simple to learn. It was developed for gifted programmers, and the learning curve may be steep. Straightforward tasks may be solved easily in C, but complex problems require mastery of the language.

An interesting sidenote is that the availability of C on personal computers has greatly enhanced the value of personal computers for budding software entrepreneurs. A cottage software industry can use the same basic tool—the language C—used by established software companies such as Microsoft and Borland. Today C is fast being replaced by its enhanced cousin, C++, a language that we will discuss shortly. An example of a C++ program and its output are shown in Figure 6-14.

The languages just described are the major ones used. Many of them occupy their privileged positions for no reason other than they got there first or they were backed by powerful organizations. Even so, as noted in Table 6-2, other languages have flourished in certain niches. Notice that many of them are special-purpose languages, a fact that helps to account for their more limited use.

 ## Object-Oriented Programming

The approach called **object-oriented programming (OOP)** is relatively new and distinctly different. Since this topic is an important emerging trend, it deserves its own section. We will introduce the concepts and terminology of object technology. There is no expectation, however, that you will understand exactly how object-oriented programming works; even professional programmers can take months to gain that knowledge.

```
-- ADA PROGRAM
-- AVERAGING INTEGERS ENTERED THROUGH THE KEYBOARD.
with TEXT_IO; use TEXT_IO;
procedure AVERAGE is
    package INT_IO is new INTEGER_IO (INTEGER);
    AVERAGE:              FLOAT                    ;
    COUNTER:              INTEGER       :=       0;
    NUMBER:               INTEGER                  ;
    SUM:                  INTEGER       :=       0;
begin
    PUT_LINE("THIS PROGRAM WILL FIND THE AVERAGE OF INTEGERS YOU ENTER");
    PUT_LINE("THROUGH THE KEYBOARD. TYPE 999 TO INDICATE END OF DATA.");
    NEW_LINE;
    PUT("PLEASE ENTER A NUMBER");
    INT_IO.GET(NUMBER);
    while NUMBER /= 999 loop
        SUM := SUM + NUMBER;
        COUNTER := COUNTER + 1;
        PUT("PLEASE ENTER THE NEXT NUMBER");
        INT_IO.GET(NUMBER);
    end loop;
    AVERAGE := SUM/COUNTER;
    PUT("THE AVERAGE OF THE NUMBERS IS");
    FLO_IO.PUT(AVERAGE);
end AVERAGE;
```
(a)

```
THIS PROGRAM WILL FIND THE AVERAGE OF INTEGERS YOU ENTER
THROUGH THE KEYBOARD.  TYPE 999 TO INDICATE END OF DATA.
PLEASE ENTER A NUMBER      6
PLEASE ENTER THE NEXT NUMBER      4
PLEASE ENTER THE NEXT NUMBER     11
PLEASE ENTER THE NEXT NUMBER    999
THE AVERAGE OF THE NUMBERS IS    7.00
```
(b)

Figure 6-13 Ada program and sample output. (a) Comments begin with a double hyphen. Ada requires that each variable be declared before the logic begins. NEW_LINE displays a blank line, and PUT_LINE displays data on the screen. The symbol /= means is not equal to. (b) This screen display shows the interaction between program and user.

```
// C++ PROGRAM
// AVERAGING INTEGERS ENTERED THROUGH THE KEYBOARD

#include <iostream.h>
main ()
{
   float average;
   int number, counter = 0; int sum = 0;
   cout << "THIS PROGRAM WILL FIND THE AVERAGE OF INTEGERS YOU ENTER \ n";
   cout << "THROUGH THE KEYBOARD.  TYPE 999 TO INDICATE END OF DATA. \ n \ n";
   cout << "PLEASE ENTER A NUMBER";
   cin >> number;
   while (number !=999)
      {
         sum := sum + number;
         counter ++;
         cout << "\nPLEASE ENTER THE NEXT NUMBER";
         cin >> number;
      }
   average = sum / counter;
   cout << "\nTHE AVERAGE OF THE NUMBERS IS " << average
}
```
(a)

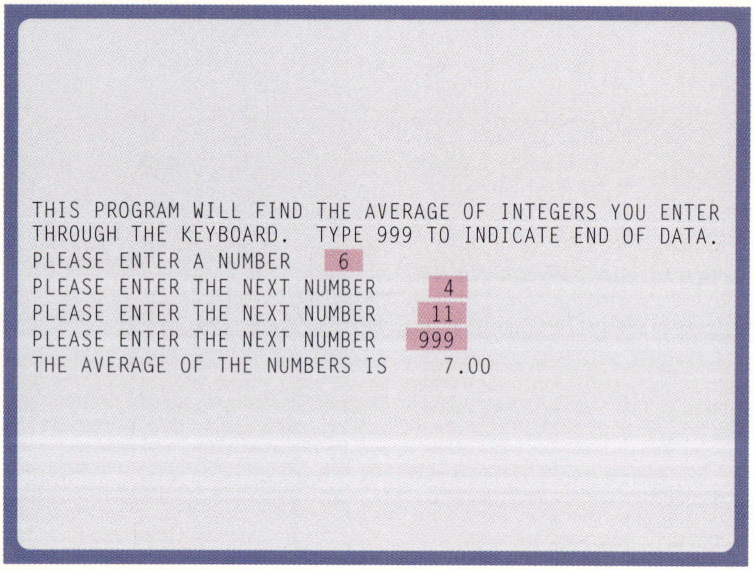

```
THIS PROGRAM WILL FIND THE AVERAGE OF INTEGERS YOU ENTER
THROUGH THE KEYBOARD.  TYPE 999 TO INDICATE END OF DATA.
PLEASE ENTER A NUMBER      6
PLEASE ENTER THE NEXT NUMBER    4
PLEASE ENTER THE NEXT NUMBER    11
PLEASE ENTER THE NEXT NUMBER    999
THE AVERAGE OF THE NUMBERS IS   7.00
```
(b)

Figure 6-14 C++ program and sample output. (a) The symbol // marks comment lines. All variable names, such as number, must be declared. The command cout sends output to the screen, and cin takes data from the user. (b) This screen display shows the interaction between program and user. The program was written in Turbo C++.

What Is an Object?

Consider items that, in everyday parlance, we might call objects, for instance, a tire or a cat. Now affix known facts to those everyday objects. Without trying to be exhaustive, we can say that a tire may be round and black and a cat has four feet and fur. Taking this further, each object also has functions: a tire can roll or stop or go flat, and a cat can eat or purr or howl. In the world of object-orientation, an object includes the item itself and also related facts and functions. More formally, in a programming environment, an **object** is a self-contained unit that contains both data and

Table 6-2

ALGOL	Standing for ALGOrithmic Language, ALGOL was developed primarily for scientific programming and is considered the forerunner of PL/I and Pascal.
APL	Introduced by IBM in 1968, APL—short for A Programming Language is a powerful, interactive language particularly suited to table handling—that is, to processing groups of related numbers in a table.
FORTH	Designed for real-time control tasks (such as guiding astronomical telescopes) as well as assorted business and graphics program, FORTH is available on almost every kind of computer, from micros to mainframes.
LISP	Short for list processing, LISP is designed to process nonnumeric data—that is, symbols, such as characters or words. It is a popular language for writing programs dealing with artificial intelligence.
LOGO	Used mostly in schools to teach problem-solving skills.
Modula-3	Almost identical to Pascal, specifically designed to write systems software.
PILOT	Most often used to write computer-aided instruction and is especially suited for such instructional tasks as drills and tests.
PL/I	An abbreviation for Programming Language One, PL/I was designed as a compromise for both scientific and business use.
PROLOG	Probably the most popular of a short list of artificial intelligence programming languages, PROLOG (PROgramming LOGic) has received attention as a tool for natural-language programming.
RPG	The problem-oriented language RPG—for Report Program Generator—was designed to use special input forms to produce business reports.

related facts and functions—the instructions to act on that data. This is in direct contrast to traditional programming, in which procedures are defined in the program separate from the data.

The word that is used to describe an object's self-containment is *encapsulation*: An object **encapsulates** both data and its related instructions. In an object, related facts are called **attributes,** and the instructions that tell the data what to do are called **methods** or **operations.** A specific occurrence of an object is called an **instance;** your pet kitty Tschugar is an instance of the object cat.

Beginnings: Boats as Objects

Object orientation was first conceived in 1969 by Dr. Kristin Nygaard, who was trying to develop a computer model of boats passing through Norwegian fjords. As Dr. Nygaard wrestled with the complex components of waves, tides, an irregular coastline, and moving boats, he hit upon the idea of isolating each component into autonomous elements—objects—and then modeling the relationships among the elements. Consider the object boats, shown in Figure 6-15. The object called boat consists of the boat itself, its attributes, and its methods—descriptions of the things it does, such as float or sink. We should note, however, that in practice few objects have an inner life and can invoke their own methods spontaneously. Thus, methods in most cases are actions from the outside that change the state of the object.

Using object-oriented programming, programmers define classes of objects. Each **class** contains the characteristics that are unique to objects of that class. In Figure 6-15, for example, a boat object is an instance of the boat class. In addition to classes, objects may be formed from subclasses. Objects are arranged hierarchically in classes and subclasses by their dominant characteristics. In Figure 6-15, some kinds of boats—sailboats, powerboats, and canoes—are subclasses of the object boat.

An object in a subclass automatically possesses all the characteristics of the class from which it is derived; this property is called **inheritance.** The subclass object canoe, for example, contains not only its own characteristics, such as a need to be paddled, but also characteristics such as the ability to float or sink inherited from the higher object class called boats. The characteristics from the class in which a subclass is included need not be repeated in each subclass. This means that, in a programming environment, a programmer would not have to repeat the instructions for characteristics that are inherited, saving both time and money. Even more savings are accrued by the ability to reuse objects.

Reuse

As object technology is used by an organization, the organization gradually builds a library of classes. Once a class has been created, tested, and found useful, it can be used again. In fact, classes may be used and reused in future program applications. Since each class is self-contained, it need not be altered for use in future applications. This reduces errors significantly, since new programs can be constructed largely of pretested error-free classes. Of course, organizations will not reap the benefits of reuse until they are a few projects down the line.

Activating the Object

Since an object is self-contained, how do you get it to do something? A command, called a **message,** is sent to the object from outside it. The message tells the object what needs to be done, and just how it is done may be contained in the object's methods. For example, the message "move" could be sent to any of the boat subclass objects—sailboat, powerboat, or canoe. This brings up a fancy word that goes a long way toward revealing the value of object technology: *polymorphism*. When a message is sent, the property of **polymorphism** allows an individual object receiving the message to know how, using its own methods, to process the message in the appropriate way for that particular object. For example, when the message "move" is received, the object sailboat knows it is supposed to move under power of sail, the object powerboat knows that it moves by means of a motor, and the object canoe knows that it moves by being paddled. In each case the object merely had to be told to move and it did move—using its own built-in methods.

Object-Oriented Languages

The object-oriented language that currently dominates the market is **C++,** which is the object-oriented version of the programming language C (recall

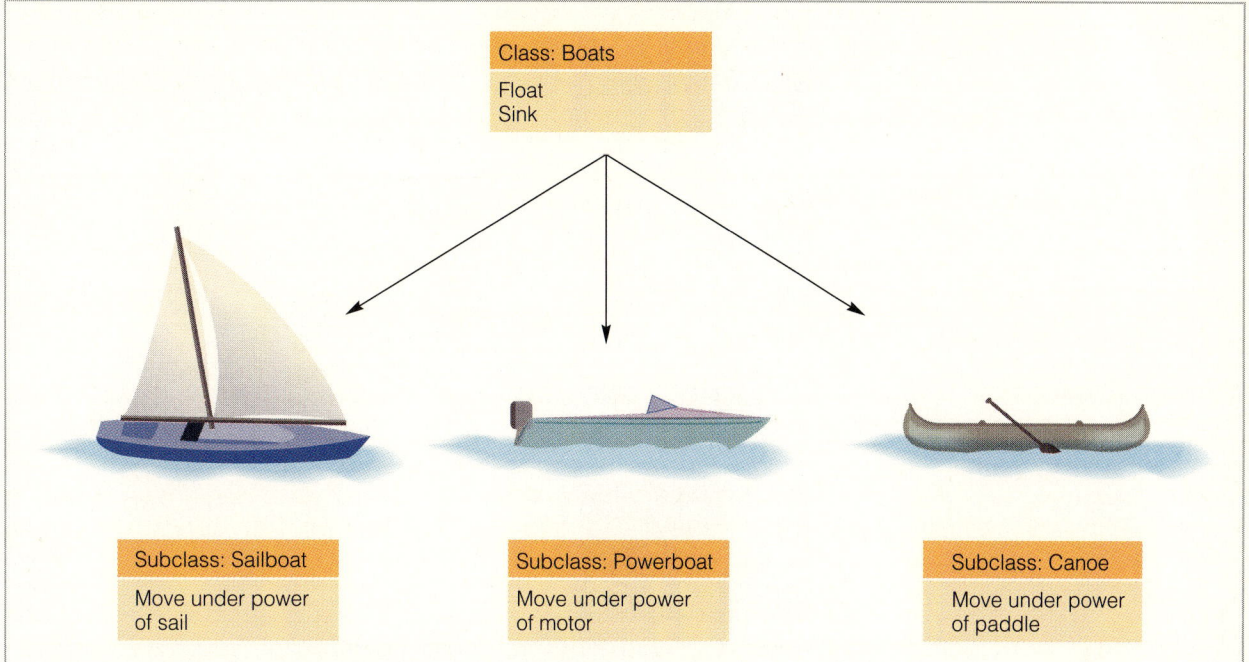

Figure 6-15 Object classes and subclasses. The subclasses sailboat, powerboat, and canoe inherit the characteristics float and sink from the higher class object boat. Furthermore, each subclass, under the property of polymorphism, can respond to the message "move" by using its own methods.

Figure 6-14). Versions of C++ are available for large systems and personal computers.

The language called **Smalltalk,** which was one of the first in the market, is making inroads. Smalltalk signaled a dramatic departure from traditional computer languages because it supports an especially visual system. Smalltalk works by using a keyboard to enter text, but all other interaction takes place through a mouse and icons.

There are other languages that have object-oriented versions, notably Pascal and Microsoft's Visual BASIC (Figure 6-16). A key question is where this leaves all the COBOL programmers. Versions of object-oriented COBOL are now emerging. Many COBOL programmers would like to approach object-oriented programming using the language they know best. However, some experts feel that programmers entrenched in traditional methods should make a complete break by moving to a language such as Smalltalk or C++. A caveat: Although C++ supports object-oriented programming, it is also a procedural language. Some people mistakenly think that if they use C++ they are doing object-oriented programming; this is not necessarily the case.

Object Technology in Business

When businesses approach object technology, they are more likely to be interested in invoices and payroll checks than cats or boats. Business items have their own attributes and methods, which can be coded into objects. Once the objects exist, they can inherit characteristics from objects in higher classes. For example, subclass objects relating to a customer account could inherit the address of the customer, which need not then be

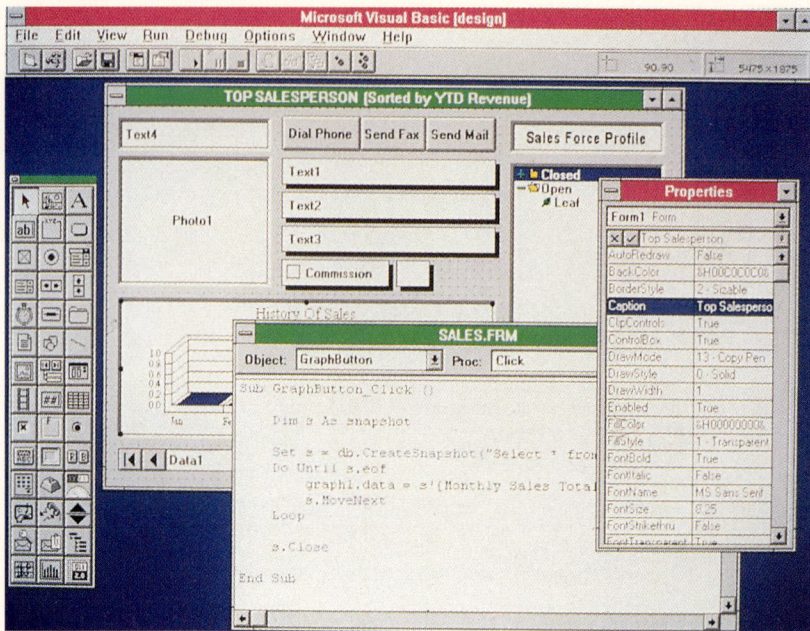

Figure 6-16 Visual BASIC. On this screen you can see part of the program, as well as the visual result being achieved.

repeated in each subclass object. In the fashion of object technology, business objects can also respond to messages, and, of course, be reused.

Although conversion to object-oriented technology is well underway, the change has been slow in coming. There are several reasons for this, including the reluctance of managers to commit money and resources to a technology they find difficult to understand. In fact, the concepts of object-oriented programming are not easy to grasp, at least at first, even by experienced programmers; object orientation is counter-intuitive to the traditional line-by-line methods that programmers have used for years. Furthermore, as we have noted, several languages do not yet offer object-oriented capabilities. Even when all these elements are in place—supportive management, eager programmers, and appropriate languages—there is still the matter of training, which can take months.

Nevertheless, current research shows a clear picture. A properly planned object-oriented system shows superior results in terms of flexibility and costs and, most important, user satisfaction.

Some Advice

If you plan to become a professional programmer, there is a good chance that you will take your first steps in BASIC or Pascal or even C, and then receive more formal training in a language used in business organizations—possibly COBOL or some other language. By all means, if you can, include an object-oriented language such as C++. Many new programmers go into the job world with these language tools. But notice something: All these are third-generation languages.

What about fourth-generation languages? Partly because no single 4GL has dominated the market, 4GLs are offered at only a few schools.

MAKING THE RIGHT CONNECTIONS

Networked Children

Fifth graders in Wauwatosa, Wisconsin, are corresponding via e-mail with pen-pals in all 50 states. They ask probing questions, from "What is your state's most important problem?" to "How much does a pizza cost?" This activity has paid several dividends, from increased student interest in geography to a greater understanding of how people live in large cities.

Early educational software was numbingly dull, used mostly for rote arithmetic or grammar lessons. Today's software is sophisticated and enticing, but the real breakthrough is the possibility of connecting to the world outside the classroom walls. Schools all over the land are gearing up to take advantage of Internet access, where they can plug into the Library of Congress, write to an astronaut, and even send questions to Thomas Jefferson (tj@uva386.schools.virginia.edu). Students find the Internet more fun than picking up an encyclopedia.

Although your training in third-generation languages is valuable because it teaches you to think analytically, you must be alert for opportunities on the job to learn one or more fourth-generation languages. Most people who program in 4GLs learned to do so from employer-sponsored classes. Be ready to take the opportunity when it comes your way.

The essential power of any computer is managed by its operating system. In the next chapter we describe operating systems, the background software that allows computers to manage their own resources.

CHAPTER REVIEW

Summary and Key Terms

- A **programming language** is a set of rules for instructing the computer what operations to perform.
- A programmer converts solutions to the user's problems into instructions for the computer. These instructions are called a **program.** Writing a program involves defining the problem, planning the solution, coding the program, testing the program, and documenting the program.
- Defining the problem means discussing it with the users or a systems analyst to determine the necessary input, processing, and output.
- Planning can be done by using a **flowchart,** which is a pictorial representation of the step-by-step solution, and by using **pseudocode,** which is an English-like outline of the solution. Pseudocode is not executable on the computer.
- Coding the program means expressing the solution in a programming language. Programmers usually use a **text editor,** which is somewhat like a word processing program, to create a file that contains the program.
- Testing the program consists of desk-checking, translating, and debugging. The rules of a programming language are referred to as its **syntax. Desk-checking** is a mental checking or proofreading of the program before it is run. A **walkthrough** is a process in which a group of programmers—your peers—review your program and offer suggestions in a collegial way. In translating, a **translator** program converts the program into a form the computer can understand and in the process detects programming language errors, which are called **syntax errors.** A common translator is a **compiler,** which translates the entire program at one time and gives error messages called **diagnostics.** The original program, called a **source module,** is translated to an **object module,** to which prewritten programs may be added during the **link/load phase** to create an executable **load module. Debugging** is running the program to detect, locate, and correct mistakes—**logic errors.**
- Typical **documentation** contains a detailed written description of the programming cycle and the program along with the test results and a printout of the program.
- Programming languages are described as being lower or higher, depending on how close they are to the language the computer itself uses (0s and 1s—low) or to the language people use (more English-like—high). There are five main levels, or generations, of languages: (1) machine languages, (2) assembly languages, (3) high-level languages, (4) very high-level languages, and (5) natural languages.
- **Machine language,** the lowest level, represents data as 1s and 0s—binary digits corresponding to the on and off electrical states in the computer.
- **Assembly languages** use letters as mnemonic codes to replace the 0s and 1s of machine language. An **assembler program** is used to translate the assembly language into machine language.
- **High-level languages** are written in an English-like manner. Each high-level language requires a different compiler, or translator program, for each type of computer on which it is run.
- **Very high-level languages,** also called **fourth-generation languages (4GLs),** are basically nonprocedural. A **nonprocedural language** defines only what the computer should do, without detailing the procedure. A **procedural language** tells the computer specifically how to do the task.
- A variation on 4GLs are **query languages,** which can be used to retrieve data from databases.

- Fifth-generation languages are often called **natural languages** because they resemble "natural" human language. A system that uses a natural language to access a knowledge base is called a **knowledge-based system.**
- A user may choose a language based on a number of factors, including management decision, interfacing, suitability, portability, availability, and programmer expertise.
- The first high-level language, **FORTRAN** (FORmula TRANslator), is a scientifically oriented language that was introduced by IBM in 1954. Its brevity makes it suitable for executing complex formulas.
- **COBOL** (COmmon Business-Oriented Language) was introduced in 1959 by **CODASYL** (COnference of DAta SYstem Languages) as a standard programming language for business. The American National Standards Institute (ANSI) standardized COBOL in 1968, again in 1974 (in a version called **ANSI-COBOL**), and more recently in a version known as **COBOL 85.** A COBOL program has four divisions: identification, environment, data, and procedure.
- When **BASIC** (Beginners' All-purpose Symbolic Instruction Code) was developed at Dartmouth and introduced in 1965, it was intended for instruction. Now its uses include business and personal computer applications.
- **Pascal,** named for the French mathematician Blaise Pascal, first became available in 1971. It is popular in college computer courses.
- **Ada,** named for Countess Ada Lovelace, was introduced in 1980 as a standard language for weapons systems. Although it also has commercial uses, experts disagree regarding how easy it is to learn.
- Invented at Bell Labs, **C** offers high-level language features such as structured programming. C code is almost as efficient as assembly language, and it is suitable for writing portable programs that can run on more than one type of computer.
- Other important languages include **ALGOL,** for ALGOrithmic Language, for scientific programming; **APL,** for A Programming Language, used for table handling; **FORTH,** designed for real-time tasks; **LISP,** short for list processing, used to process nonnumeric data; **LOGO,** designed for use in schools to teach problem-solving skills; **Modula-3,** a Pascal look-alike designed to write systems software; **PILOT,** used for computer-aided instruction; **PL/I,** short for Programming Language One, designed for scientific and business uses; **PROLOG,** for PROgramming LOGic, popular for natural language programming; and **RPG,** for Report Program Generator, a language for producing business reports.
- The approach called **object-oriented programming (OOP)** uses **objects,** self-contained units that contain both data and related facts and functions—the instructions to act on that data. An object **encapsulates** both data and its related instructions. In an object, related facts are called **attributes,** and the instructions that tell the data what to do are called **methods** or **operations.** A specific occurrence of an object is called an **instance.**
- An object **class** contains the characteristics that are unique to that class. Objects are arranged hierarchically in classes and subclasses by their dominant characteristics. An object in a subclass automatically possesses all the characteristics of the class to which it belongs; this property is called **inheritance.**
- Once an object has been created, tested, and found useful, it can be used and reused in future program applications.
- A command called a **message,** telling what—not how—something is to be done, activates the object. **Polymorphism** means that an individual object receiving a message knows how, using its own methods, to process the message in the appropriate way for that particular object.
- The object-oriented language that currently dominates the market is **C++,** which is the object-oriented version of the programming language C. Versions of C++ are available for large systems and personal computers. The language called **Smalltalk,** which supports an especially visual system, is making inroads.

Discussion Questions

1. It has been noted that, among other qualities, good programmers are detail-oriented. Why might attention to detail be important in the programming process?
2. In addition to insisting on proper documentation, managers encourage programmers to write straightforward programs that another programmer could easily follow. Discuss occasions in which a programmer may have to work with a program written by another programmer. Under what circumstances might a programmer completely take over the care of a program written by another? If you inherited someone else's program, about which you know nothing, would you be dismayed to discover minimal documentation?
3. Should students taking a computer literacy course be required to learn some programming?

Student Study Guide

Multiple Choice

1. The presence of both data and its related instructions in an object is
   a. C++
   b. orientation
   c. encapsulation
   d. inheritance
2. In preparing a program, one should first
   a. plan the solution
   b. document the program
   c. code the program
   d. define the problem
3. During the development of a program, drawing a flowchart is a means to
   a. plan the solution
   b. define the problem
   c. code the program
   d. analyze the problem
4. An English-like language that one can use as a program design tool is
   a. BASIC
   b. PL/I
   c. pseudocode
   d. Pascal
5. In preparing a program, desk-checking and translating are examples of
   a. coding
   b. testing
   c. planning
   d. documenting
6. The process of detecting, locating, and correcting logic errors is called
   a. desk-checking
   b. debugging
   c. translating
   d. documenting
7. Comments in the program itself are part of
   a. compiling
   b. linking
   c. translating
   d. documenting
8. A COBOL program has how many divisions
   a. four
   b. five
   c. two
   d. seven
9. The first high-level language to be introduced was
   a. COBOL
   b. Pascal
   c. FORTRAN
   d. Ada
10. The ability of an object to interpret a message using its own methods is called
    a. polymorphism
    b. inheritance
    c. encapsulation
    d. messaging
11. The language named for a French mathematician is
    a. C
    b. Pascal
    c. Ada
    d. Modula-3
12. Specifying the kind of input, processing, and output required for a program occurs when
    a. planning the solution
    b. coding the program
    c. flowcharting the problem
    d. defining the problem
13. Error messages provided by a compiler are called
    a. bugs
    b. translations
    c. diagnostics
    d. mistakes
14. After stating the solution to a problem in pseudocode, the next step would be
    a. testing the program
    b. documenting the program
    c. coding the program
    d. translating the program
15. The highest-level languages are called
    a. 4GLs
    b. assembly
    c. high-level
    d. natural
16. To activate an object, send
    a. a message
    b. a method
    c. an instance
    d. an attribute
17. Popular object-oriented languages:
    a. Pascal, Modula-3
    b. LOGO, PROLOG
    c. C++, Smalltalk
    d. COBOL, BASIC
18. Software that translates assembly language into machine language is
    a. a binary translator
    b. an assembler
    c. a compiler
    d. a link-loader
19. A standardized business language is
    a. CODASYL
    b. COBOL
    c. BASIC
    d. Ada
20. In developing a program, documentation should be done
    a. as the last step
    b. only to explain errors
    c. throughout the process
    d. only during the design phase
21. A fourth-generation language used for database retrieval:
    a. high-level language
    b. query language
    c. assembly language
    d. procedural language

22. A language designed to generate routine business reports is
   a. COBOL   c. LISP
   b. RPG   d. ALGOL
23. The lowest level of programming language is
   a. nonprocedural language   c. assembly language
   b. BASIC   d. machine language
24. An assembly language uses
   a. English words   c. mnemonic codes
   b. 0s and 1s   d. binary digits
25. The language Smalltalk is
   a. procedural oriented   c. document oriented
   b. problem oriented   d. object oriented

True/False

T F 1. The usual reason for choosing a programming language is simply the one the programmer likes best.
T F 2. Developing a program requires just two steps, coding and testing.
T F 3. A flowchart is an example of pseudocode.
T F 4. Desk-checking is the first phase of testing a program.
T F 5. A translator is a form of hardware that translates a program into language the computer can understand.
T F 6. Modula-3 is a Pascal look-alike for writing systems software.
T F 7. Debugging is the process of locating program logic errors.
T F 8. The highest level of language is natural language.
T F 9. Pseudocode can be used to plan and execute a program.
T F 10. 4GLs increase clarity but reduce productivity.
T F 11. PL/I was designed as a compromise for scientific and business uses.
T F 12. Pascal is particularly easy to use because it has fewer features than most languages.
T F 13. COBOL is divided into four parts called areas.
T F 14. BASIC is especially suited for large and complex programs.
T F 15. FORTRAN stands for FORms TRANsfer.
T F 16. Expressing a problem solution in Pascal is an example of coding a program.
T F 17. Diagnostic messages are concerned with improper use of the programming language.
T F 18. An assembler program translates high-level language into assembly language.
T F 19. An object subclass inherits characteristics from higher object classes.
T F 20. A specific occurrence of an object is called an instance.
T F 21. Polymorphism means that an object knows how, using its own methods, to act on an incoming message.
T F 22. Another name for a high-level language is 4GL.
T F 23. A query language is a type of assembly language.
T F 24. FORTRAN is primarily used in scientific environments.
T F 25. Low-level languages are tied more closely to the computer than are high-level languages.

Fill-In

1. What type of language is used to access databases? _____

2. What type of language replaced machine language by using mnemonic codes? _____

3. What object-orientation property permits a subclass to retain the characteristics of a higher class? _____

4. What level of language is Focus? _____

5. What is the name for a translator that translates high-level languages into machine language? _____

6. The language used in schools to teach problem-solving skills: _____

7. The rules of a programming language are called its _____

8. How many levels of language were described in the chapter? _____

9. A source module is translated into a(n) _____

10. Which language is specifically designed to write systems software? _____

11. List the divisions of a COBOL program.
   a. _____
   b. _____
   c. _____
   d. _____

12. The object-orientation property that permits an object to use its own methods to act on a message is _____

13. Two commonly used OOP languages:
    a. _____
    b. _____

14. Which language was sponsored by the Pentagon? _____

15. Name the language that is popular in computer science programs. _____

16. An object module is link-loaded into a _____

17. What kind of language tells the computer *what* needs to be done, as opposed to *how*? _____

18. What are languages that resemble spoken languages called? _____

19. Which high-level language is scientifically oriented? _____

20. The command that activates an object is called _____

21. Which of the five steps of the programming process is best done throughout the process? _____

22. What are two common methods of planning the solution to a problem?
    a. _____
    b. _____

23. List the three phases of testing a program.
    a. _____
    b. _____
    c. _____

24. What is the next step after a programmer has planned the solution? _____

25. What is the term for the error messages that a translator provides? _____

Answers

Multiple Choice

1. c	6. b	11. b	16. a	21. b
2. d	7. d	12. d	17. c	22. b
3. a	8. a	13. c	18. b	23. d
4. c	9. c	14. c	19. b	24. c
5. b	10. a	15. d	20. c	25. d

True/False

1. F	6. T	11. T	16. T	21. T
2. F	7. T	12. T	17. T	22. F
3. F	8. T	13. F	18. F	23. F
4. T	9. F	14. F	19. T	24. T
5. F	10. F	15. F	20. T	25. T

Fill-In

1. query language
2. assembly language
3. inheritance
4. very high-level language
5. compiler
6. LOGO
7. syntax
8. five
9. object module
10. Modula-3
11. a. identification division
    b. environment division
    c. data division
    d. procedure division
12. polymorphism
13. C++, Smalltalk
14. Ada
15. Pascal
16. load module
17. nonprocedural
18. natural language
19. FORTRAN
20. message
21. documentation
22. a. flowcharting
    b. writing pseudocode
23. a. desk-checking
    b. translating
    c. debugging
24. code the program
25. diagnostics

PLANET INTERNET

Career in Gear

Many people, whether college students or experienced employees, dread the prospect of facing the world and begging for a job. Resources for this onerous task are often limited to the classified ads and perhaps a placement center. You have no such resource limitation if you have the Internet. A number of services, both commercial and nonprofit, specialize in matching employers with job seekers. However, although assistance is available for first-time job seekers, it would be fair to say that the jobs posted on the Internet lean toward experienced people in the computer field.

More schooling. Perhaps your career move requires further schooling before you are ready to enter the job market. You may seek help in taking the tests required as part of the application process. One informative site on the Internet is the Kaplan Education Center.

Online help for students. The Online Career Center is a nonprofit site that offers career support services, such as which keywords to use to search for jobs from thousands of companies. For a modest fee you can place your resume online for 90 days. Career Mosaic, whose on-screen logo is shown here, lists high-powered employers and offers a variety of support for job seekers. The Interactive Employment Network also features a searchable list of jobs and, in addition, lets job seekers key in a location and get the advertisements for that geographic area. If you are seeking a job in the Midwest, an attractive option is the Jobweb, which permits a job seeker to fill out a form online to create a listing on its pages. Most interesting of all, at least from a graphics point of view, is the Job Monster Board, whose logo is also shown here. Guided by amusing monster graphics characters, you can examine employer profiles, check out career events and information, or even submit your resume. The Monster Board lets you search its employer database by location, industry, company, discipline, and job title.

Your home page résumé. Some job seekers have taken advantage of Web exposure by developing their own home page résumés. This goes so far beyond the traditional résumé that a new name should be invented for it. To see what we mean, check the résumé home pages listed at the B/C site. You will find, typically, that the candidate includes a nice photo and then offers perhaps a 10- to 15-line résumé. Why so short? Each résumé line has links! For example, one line may refer to classes taken, with classes being a link. A potential employer merely clicks on the word *classes* to pop up a list of classes the job seeker has taken. Similarly, links can be made to intern work, laboratory assignments, work experience, extracurricular activities, and so forth. A person developing such a home page from scratch can make the résumé as variable as desired.

Internet Exercises

1. **Structured exercise.** Hooking up to the B/C site with URL http://www.aw.com/bc/planet/, take a look at the Monster Board. Check out the Cyberzone link especially designed to provide college students with opportunities and discussion groups.
2. **Freeform exercise.** To expand your set of resources, begin with the Whole Internet Catalog. Find and click on a careers link.

When Linda Ronquillo was taking a night class in applications software at a community college four years ago, she did not have to worry much about the operating system, the necessary software in the background. The college personal computers were on a network that managed all the computers. As Linda sat down to begin work, the computer screen showed a menu of numbered choices reflecting the software packages available: 1. WordPerfect, 2. Microsoft Word, 3. Microsoft Excel, and so forth. At the bottom of the screen, Linda was instructed to type the number of her chosen selection; if she typed 1, for example, the system put her into WordPerfect. Linda did have to learn operating system commands to prepare her own diskettes and save data on them so she could take her work with her, but she had little other contact with the operating system.

On the job as a supervisor in airport freight, Linda needed to use word processing, spreadsheets, and database software packages on her IBM personal computer. But no one had set up a menu shortcut here, so, with a little advice from colleagues, she learned what she needed to know about the operating system called MS-DOS. She learned, among other things, to execute the software she needed to use and to take care of her data files—copy files from one disk to another and sometimes rename or delete them. She eventually felt fairly comfortable with her operating system knowledge.

Eighteen months later, Linda was informed by the company personal computer manager that all personal computers were going to be switched to Microsoft Windows 95, an operating system that included a user-friendly interface. Despite assurances that the new system would be colorful and easy to use, Linda was less than thrilled to be making another change. But she did not say so. She knew that being a computer user meant being willing to adjust to change. So Linda learned to use a mouse and mastered icons, overlapping windows, pull-down menus, and other mysteries.

Approximately six months later, Linda took a job at another airline freight company. Part of the reason she was hired was her response to the revelation that the new company used Macintosh computers, which, she knew, used another operating system altogether. Linda said, "Oh, I have learned several systems. It shouldn't be any problem learning another." She was right, of course. In fact, she thought the Macintosh operating system was the easiest of them all.

Chapter 7

Operating Systems
Software in the Background

LEARNING OBJECTIVES

- Know the functions of an operating system
- Understand the basics of a personal computer operating system
- Appreciate the advantages of an operating environment, especially Microsoft Windows
- Understand the need for network operating systems
- Understand the need for resource allocation on large computers
- Understand multiprocessing, multiprogramming, and time-sharing
- Understand the principles of memory management
- Appreciate the need for service programs
- Distinguish generic operating systems from proprietary operating systems

OPERATING SYSTEMS: HIDDEN SOFTWARE

OPERATING SYSTEMS FOR PERSONAL COMPUTERS: AN OVERVIEW

A LOOK AT MS-DOS
  A Brief Disk Discussion
  Types of Files

OPERATING ENVIRONMENTS
  Microsoft Windows: An Overview
  Multiprogramming/Multitasking: Comparing Mainframe Operating Systems to Windows
  Software Applications with Windows

WINDOWS 95

NETWORK OPERATING SYSTEMS

OPERATING SYSTEMS FOR LARGE COMPUTERS: AN OVERVIEW

RESOURCE ALLOCATION
  Sharing the Central Processing Unit
  Sharing Memory
  Sharing Storage Resources
  Sharing Printing Resources

SERVICE PROGRAMS

GENERIC OPERATING SYSTEMS
  The UNIX Graduates
  Is UNIX a Standard?

DO I REALLY NEED TO KNOW ALL THIS?

Operating Systems: Hidden Software

When a brand new computer comes off the factory assembly line, it can do nothing. The hardware needs software to make it work. Are we talking about applications software such as word processing or spreadsheet software? Partly. But an applications software package does not communicate directly with the hardware. Between the applications software and the hardware is a software interface—an operating system. An **operating system** is a set of programs that lies between applications software and the computer hardware. Figure 7-1 gives a conceptual picture of operating system software as an intermediary between the hardware and the applications software. Incidentally, the term **system software** is sometimes used interchangeably with *operating system*, but system software means all programs related to coordinating computer operations. System software does include the operating system, but it also includes programming language translators, which we mentioned in Chapter 6, and service programs, which we will discuss briefly in this chapter.

Note that we said that an operating system is a *set* of programs. The most important program in the operating system, the program that manages the operating system, is the **supervisor program**, most of which remains in memory and is thus referred to as *resident*. The supervisor controls the entire operating system and loads into memory other operating system programs (called *nonresident*) from disk storage only as needed (Figure 7-2).

An operating system has three main functions: (1) manage the computer's resources, such as the central processing unit, memory, disk drives, and

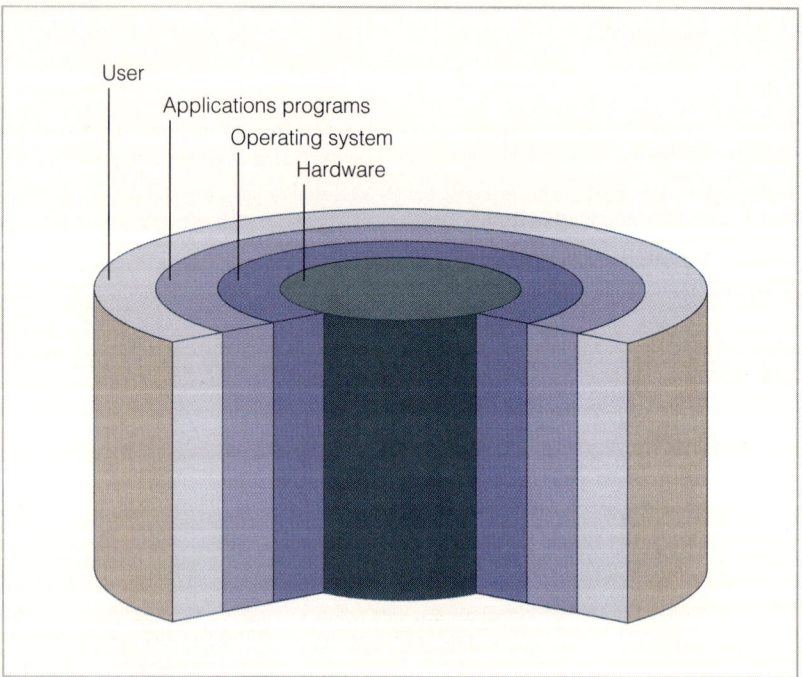

Figure 7-1 A conceptual diagram of an operating system. Closest to the user are applications programs—software that helps a user compute a payroll or play a game or calculate the trajectory of a rocket. The operating system is the set of programs between the applications programs and the hardware.

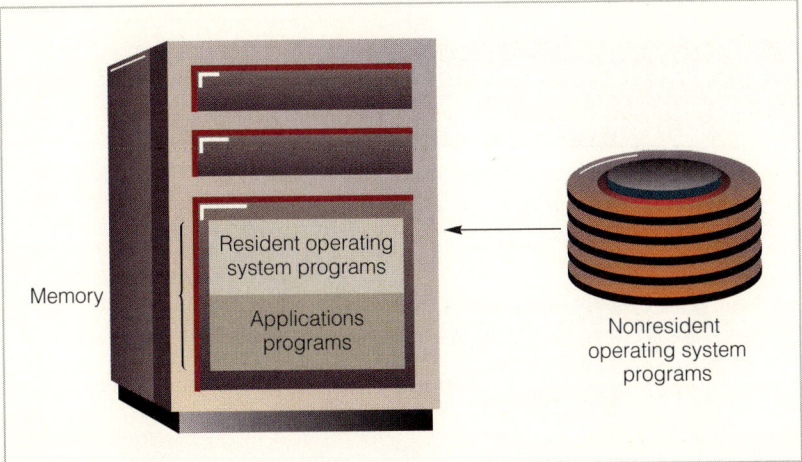

Figure 7-2 Retrieving operating system programs from disk. The supervisor program of the operating system is resident in memory and calls in nonresident operating system programs from the disk as needed.

printers, (2) establish a user interface, and (3) execute and provide services for applications software. Keep in mind, however, that much of the work of an operating system is hidden from the user; many necessary tasks are performed behind the scenes. In particular, the first listed function, managing the computer's resources, is taken care of without the user being aware of the details. Furthermore, all input and output operations, although invoked by an applications program, are actually carried out by the operating system. Although much of the operating system functions are hidden from view, you will know when you are using an applications software package, and this requires that you invoke—call into action—the operating system. Thus you both establish a user interface and execute software.

Operating systems for mainframe and other large computers are complex indeed, since they must keep track of several programs from several users all running in the same time frame. Although some personal computer operating systems—most often found in business or learning environments—can support multiple programs and users, most are concerned only with a single user. We begin by focusing on the interaction between a single user and a personal computer operating system.

 ## Operating Systems for Personal Computers: An Overview

If you peruse software offerings at a retail store, you will generally find the software grouped according to the computer, probably IBM (that is, IBM compatible) or Macintosh, on which the software can be used. But the distinction is actually finer than the differences among computers: Applications software—word processing, spreadsheets, games, whatever—are really distinguished by the operating system on which the software can run.

Generally, an application program can run on just one operating system. Just as you cannot place a Nissan engine in a Ford truck, you cannot take a version of WordPerfect designed to run on an IBM machine and run

Tips for the Macintosh® Computer

The Macintosh Applications Interface

Computers based on the DOS operating system are fundamentally different from Macintosh computers. The main difference has to do with the computer's hardware, especially the design of the processor. But the difference that most users notice first is the Macintosh's graphical user interface (GUI). While DOS-based programs have traditionally employed an interface that is textual (the user has to type in commands), the Mac's interface works by creating a graphical "world," displaying pictures of disks, folders, and even a trash can. By clicking a mouse, you can pick up items in this world, move them around, drop or discard them, using the same intuitive skills of grabbing and handling objects that you learned as a child.

A typical GUI makes use of four basic elements: icons (pictures that symbolize items like files or disks); a pointer (a cursor that you move around the screen by using a mouse); windows (rectangular areas on the screen that you can use to display information, move easily, or resize by using the pointer); and pull-down menus (each of which offers a list of commands that you can choose by using the pointer).

These four elements together provide an interface that is surprisingly easy to learn. For instance, to copy a file from one disk to another, you can simply "grab" the file's icon with the pointer and "drag" it onto the destination disk's icon. The Macintosh was the first example of this interface to succeed in the computer market,

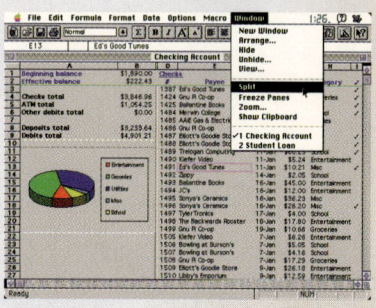

(a) When you switch on Balloon Help you get an instant description about any object in the Mac environment by positioning the pointer over the object. (b) This spreadsheet document was created to balance a checkbook but has a look and feel similar to that of other applications.

although its design had been created years earlier by researchers at XEROX's Palo Alto Research Center. Microsoft provides a similar GUI for DOS-based machines, called Microsoft Windows.

Another advantage of using the Mac is consistency. Before these interfaces became popular, every program had its own way of interacting with the user. Some provided menus, but you had to use special keystrokes to activate them. If you wanted to learn a new software package, you could count on having to learn a whole new set of keystrokes and commands. Apple was aware of this problem when they developed the first Macintosh and went about solving it in two ways.

First, they provided a set of programs, called the Macintosh Toolbox, which is dedicated to providing standard windows, icons, menus, and pointers. The Toolbox is available for any application to use, so when a programmer creates a new application, they do not have to trouble with the design of windows and menus because the details will be taken care of by the Toolbox. The payoff is that virtually every application will have windows and menus that look and behave in exactly the same way.

In addition to the Toolbox, Apple created a set of standards to which applications should adhere. For instance, most Macintosh applications provide File and Edit menus, each of which contains standard commands for working with documents. A newer standard is the availability of Balloon Help. When you activate this feature you can get information on different components of an application simply by positioning the pointer over it; a cartoon-style balloon then appears with a description of that element.

This consistency between applications gives the impression, quite falsely, that all Macintosh software is produced by a single manufacturer. In fact, hundreds of companies as well as individuals are developing software for the Mac, and virtually all of them make a special effort to adhere to the Macintosh standard interface.

it on an Apple Macintosh. The reason is that IBM personal computers and others like them usually use Microsoft's operating system, called MS-DOS (for Microsoft disk operating system) or perhaps the more friendly version called Microsoft Windows 95. Macintoshes use an entirely different operating system, called the Macintosh operating system and produced by Apple. Most personal computers are limited to one of these two. The operating systems are different because the central processing units are different. Software makers must decide for which operating system to write an applications software package, although some make versions of their software for each operating system.

Users do not set out to buy operating systems; they want computers and the applications software to make them useful. However, since the operating system determines what software is available for a given computer, many users observe the high volume of software available for MS-DOS machines and make their computer purchases accordingly. Others prefer the user-friendly style of the Macintosh operating system and choose Macs for that reason.

Although operating systems differ, many of their basic functions are similar. We will show some of the basic functions of operating systems by examining MS-DOS.

A Look at MS-DOS

Most users today have a computer with a hard disk drive. When the computer is turned on, the operating system will be loaded from the hard drive into the computer's memory, thus making it available for use. The process of loading the operating system into memory is called bootstrapping, or **booting** the system. The word *booting* is used because, figuratively speaking, the operating system pulls itself up by its own bootstraps. When the computer is switched on, a small program (in ROM—read-only memory) automatically pulls up the basic components of the operating system from the hard disk. From now on, we will refer to MS-DOS by its commonly used abbreviated name, DOS, pronounced to rhyme with *boss*.

The net observable result of booting DOS is that the characters C> (or possibly C:\>) appear on the screen. The C refers to the disk drive; the > is a **prompt**, a signal that the system is *prompting* you to do something. At this point you must give some instruction to the computer. Perhaps all you need to do is key certain letters to make the application software take the lead. But it could be more complicated than that because C> is actually a signal for direct communication between the user and the operating system.

Although the prompt is the only visible result of booting the system, DOS also provides the basic software that coordinates the computer's hardware components and a set of programs that lets you perform the many computer system tasks you need to do. To execute a given DOS program, a user must issue a **command**, a name that invokes a specific DOS program. Whole books have been written about DOS commands, but we will consider just a few that people use for ordinary activities. Some typical tasks you can do with DOS commands are prepare (format) new diskettes for use, list the files on a disk, copy files from one disk to another, and erase files from a disk.

A Brief Disk Discussion

Since many DOS commands involve files on disk, we are particularly concerned about disk drives in this chapter.

Disk Drive Configuration
Two kinds of disk drives are usually associated with a personal computer: diskette drives and hard disk drives, or hard drives. A third possibility is a CD-ROM drive. A common configuration is a diskette drive as drive A and a hard drive as drive C. If you have a second diskette drive (perhaps one for 3½-inch diskettes and one for 5¼-inch diskettes), the second

Introducing Mr. Gates

Anyone who has not figured out what the trend is for operating systems has missed the heavy press coverage of billionaire Bill Gates, the still-young founder of the Microsoft Corporation, the largest software company in the world. Mr. Gates has graced the covers of *Time, Fortune,* and any other news magazine you could name. Each time he is interviewed, Mr. Gates makes a pitch for his vision of the future, which features, among other things, Microsoft Windows 95. (Incidentally, Microsoft was denied a patent on the word "windows"; the word was considered simply too common.)

The MS-DOS Commands You Will Use Most

These instructions assume you are using a computer with diskette drive A and hard drive C. Instructions for you to type are in this `typeface`.

FORMAT (Prepare an unformatted disk for use).

A new diskette may need to be formatted before it can be used. Caution: *Never* format hard disk drive C; formatting destroys all data on a disk.

1. Insert the blank diskette in drive A.
2. `C:\>CD\DOS`
   Change to the DOS directory.
3. `C:\DOS>FORMAT A:`
   Type the command to format.
4. When asked, press Enter to confirm that a diskette is present and again to skip volume label.

DIR (Directory).

In no time at all, most computer users have lots of files on lots of disks; you can easily forget where files are. DIR produces an on-screen list of file names. /P and /W add further options.

`C:\>DIR`   Lists one line per file, with name, size in bytes, and date/time created

`C:\>DIR/W`   Lists file names only, in five columns across the screen.

`C:\>DIR/P`   Lists one line per file, a page at a time; press any key to continue.

`C:\>DIR/W/P`   Lists file names only in five columns, a page at a time.

COPY (Make a copy of a file).

One important reason to copy a file is to produce a backup copy. Another is to copy a data file generated on a community (school or office) hard disk to your own diskette. If we assume the file to be copied is on the current drive, in this case drive C, C need not be mentioned in the command. And, if you want the new file to have the same name in its new location, which is usually the case, you need not key it again on the new drive, in this case A.

`C:\>COPY MRKTDATA.SUM A:`   Copies file MRKTDATA.SUM on drive C to drive A.

`C:\>COPY *.* A:`   Copies all files in the directory to drive A.

DEL or ERASE (Delete a file).

When your disk gets cluttered with files you no longer want, it is time to clean house. Use DEL followed by the name of the file you want to delete.

RENAME (Give a file a new name).

If you decide to change a file name, use the RENAME command, followed by the old name and then the new name. Assume that a file named MRKTDATA.SUM is on a diskette in drive A.

`A:\>RENAME MRKTDATA.SUM SSDATA.CHT`
New name is SSDATA.CHT.

Other simple commands.

The following three commands can be invoked by simply keying the commands themselves, without the need for any additional information: ***CHKDSK***, meaning check disk, causes a screen display of information about the status of the disk, including number of files, number of bytes used in files, and number of bytes available for use. ***CLS*** clears the screen. When you key ***TIME*** the proper time appears on the screen. If you wish, you can key in a new time; this is convenient for switching back and forth between daylight savings time and standard time. ***VER*** will provide the current version number of the operating system.

---

diskette drive will be drive B. If you have a CD-ROM drive, that will probably be drive D. Configurations vary, but the four most common are shown in Figure 7-3.

The Default Drive

Consider the MS-DOS command DIR, which displays a list of files. How does DOS know which drive to look at when you type DIR? Just which files do you want to list? If you do not specify a particular drive, DOS will look at the default drive.

The **default drive**, also called the **current drive**, is the drive that the computer is currently using. Only one disk drive at a time can be the default drive. If you have a hard disk drive, then that drive—drive C—starts out as the current drive and usually stays as the current drive. DOS uses the

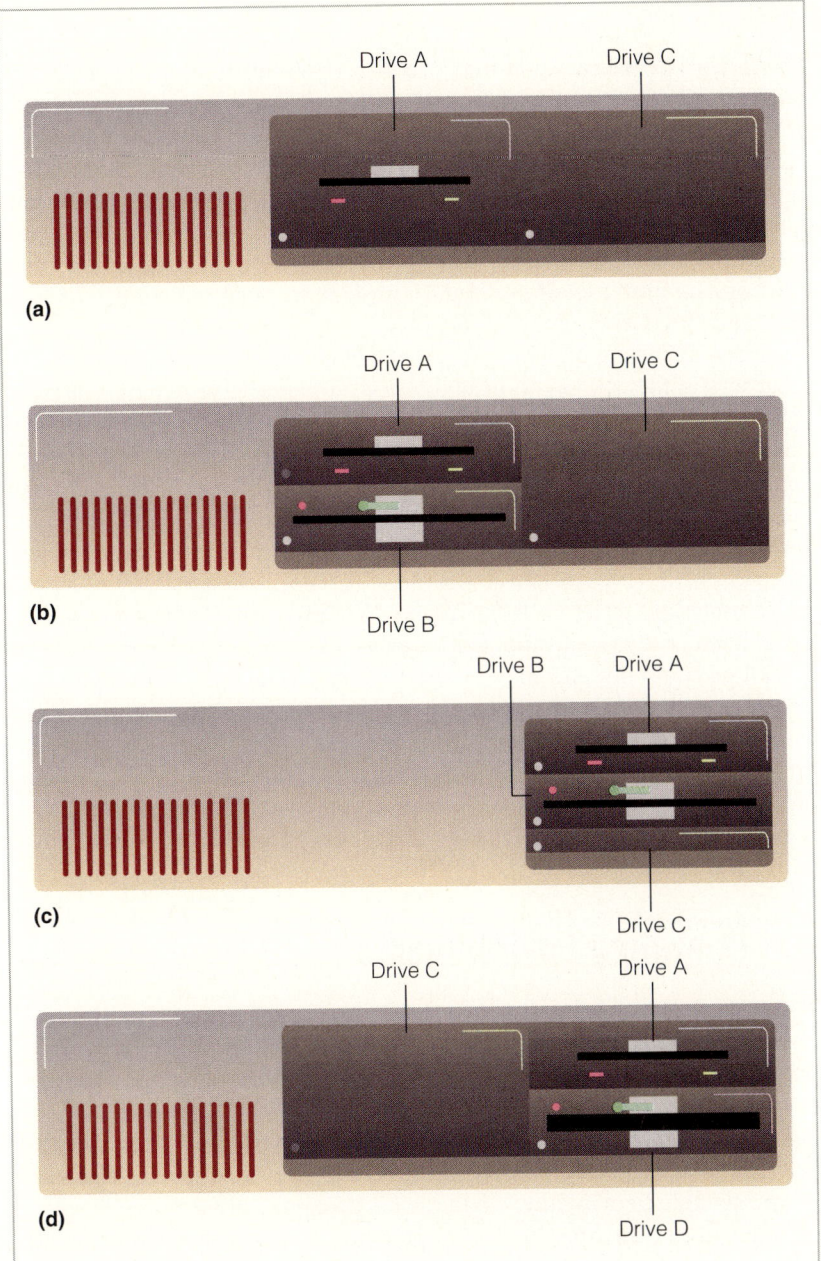

Figure 7-3 Disk drive configurations. As you use different computers, you may see several different types of disk drive combinations. The following are common: (a) Drive A for 3½-inch diskette on the left, hard drive C on the right. (b) Drive A for 3½-inch diskette on the top left; drive B for 5¼-inch diskette stacked on lower left; hard drive C on the right. (c) Drive A for 3½-inch diskette on the top right; drive B for 5¼-inch diskette stacked on lower right; hard drive C may be on bottom right or, in some cases, further back under the housing (in any case, the indication that a drive is being used—read or written—is an indicator light on the front panel, no matter where the actual drive is located). (d) Any of the first three may be augmented by a CD-ROM drive, which is usually drive D. Here we show hard drive C on the left, diskette drive A top right, and CD-ROM drive D bottom right.

prompt to remind you which drive is the current drive. If you see C> on the screen, then the current drive is C.

You can change the default to another drive if you wish. After the prompt, type the letter of the desired drive, followed by a colon, and then press Enter. Suppose, for example, that the default drive is currently drive C (as you can see from C> on the screen), but you want to access files on a diskette in drive A. To change the default drive to A, type A: (the letter A followed by a colon) and then press Enter. (You can, by the way, type either an upper- or lowercase A—DOS recognizes both.) Now the screen should show A>, and you can access files on the diskette in the A drive.

Types of Files

The three types of files you may use are (1) system files, (2) applications software files, and (3) data files. System files include the operating system programs. Applications software files are the software needed for an application, such as word processing. Data files hold data that is related to applications software, such as a memo you typed using word processing software.

When are these files used? The system files are used to start the computer system and, as you proceed, to provide services and control of software and files. Generally speaking, unless you invoke some command for a specific need such as copying a file, you will not deal directly with the operating system files. If you do need to use the operating system files directly, you do so by issuing a command that invokes the program name. Applications software files are invoked by you whenever you use that specific application but, again, you will probably have little direct interaction with the applications files themselves. Data files are used with applications software, either to supply input data or, more likely, to store the files you create.

Data files are different from system and applications software files. To begin with, the system and applications software may belong to your school or company and may be used by several people. Although some people use input data files created by their school or company, users most often deal with output data files. Output data files usually contain data created by you and may be used only by you, especially in an academic environment. Once you place your personal data files on diskettes, then the files are in your exclusive control, to use, to delete, or to take home with you.

Operating Environments

Today there is another—some say better—way to interact with the computer's operating system. Figure 7-4 tells the story: Another layer, called the **operating environment,** has been added to separate the operating system and the user. This layer is often called a **shell** because it forms a "coating" over the operating system. The operating environment creates a new way of doing business and even presents a new screen appearance—one more palatable to many users than the C> prompt.

When using an operating environment, you see pictures or simply worded choices or both instead of C> or some other prompt. Instead of having to *know* some command to type, you have only to make a selection from the choices available on the screen. Apple's Macintosh paved the way for simple interfaces between users and the operating system, and Microsoft has defined the operating environment standard for DOS-based computers with Microsoft Windows.

Microsoft Windows: An Overview

Microsoft Windows—Windows, for short—started out as a shell. Windows uses a colorful graphics interface that, among other things, eases access to the operating system. The feature that makes Windows so easy to use is a **graphical user interface** (GUI—pronounced "*goo*-ee"), in which users work with on-screen pictures called **icons** and with **menus** rather than with keyed-in commands (Figure 7-5). The menus in Figure 7-5 are

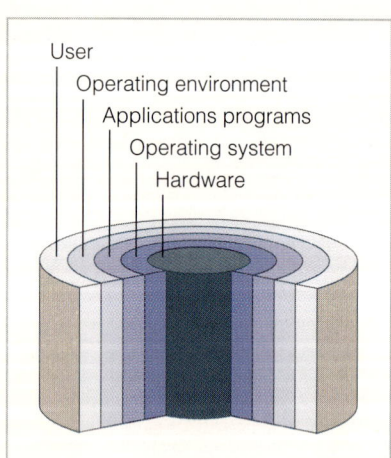

Figure 7-4 Operating environments. This illustration is identical to Figure 7-1, except that an environment layer has been added to shield the user from having to know commands of the operating system.

called **pull-down menus** because they appear to pull down like a window shade from the original selection. Some menus, in contrast, called **pop-up menus** originate from a selection on the bottom of the screen. Furthermore, icons and menus encourage pointing and clicking with a mouse, an approach that can make computer use both fast and easy.

To enhance ease of use, Windows is usually set up so that the colorful Windows display is the first thing a user sees when the computer is turned on. DOS is still there, under Windows, but a user need never see C> during routine activities. The user points and clicks among a series of narrowing choices until arriving at the desired software.

Although the screen presentation and user interaction are the most visible evidence of change, Windows offers changes that are even more fundamental. To understand these changes more fully, it is helpful at this point to make a comparison between traditional operating systems for large computers and Windows.

Multiprogramming/Multitasking: Comparing Mainframe Operating Systems to Windows

Is the operating system action the same for big and small computers? The answer is yes—and no. Yes, if you are considering computer action from the perspective of the central processing unit (CPU). No, if you are taking the viewpoint of the user—who accesses the CPU and other computer resources via the operating system.

Large Computers

Considering the CPU for just a moment, recall how a computer executes instructions. The computer usually has a single processor, which can do only one thing at a time, that is, handle only one instruction at a time. Since, invariably, other tasks are associated with running a program, such as reading from disk or printing, it would be wasteful if the CPU sat idle while these tasks were being accomplished. To maximize CPU use, the

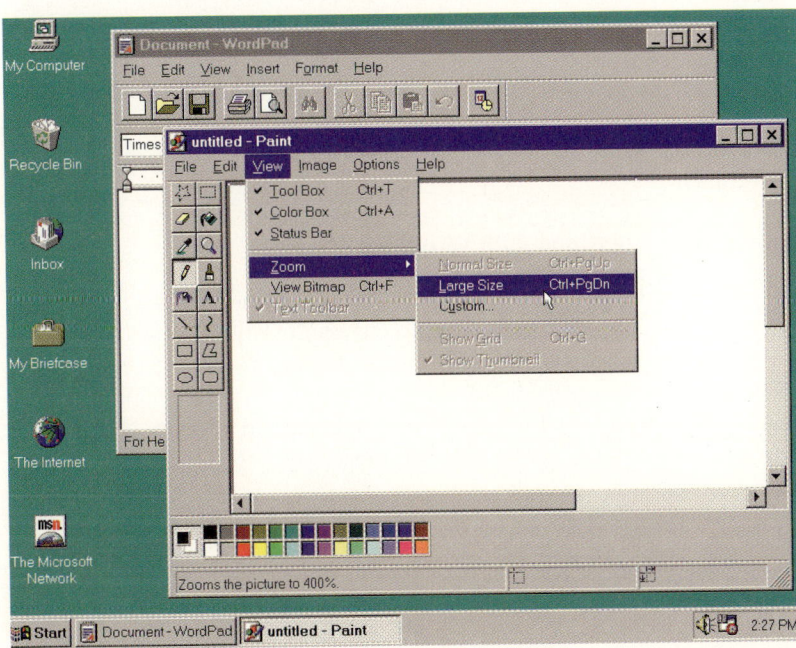

Figure 7-5 Icons and menus. Windows 95 starting page icons are shown on the left. The window in the foreground showing the Paint program demonstrates a series of pull-down menus.

large computer's operating system includes a feature called **multiprogramming,** which permits the running of several programs in the same time frame, or **concurrently.** That is, multiprogramming permits several unrelated programs, probably from many different users, to compete for the processor. Remember that we are talking about large computers now—supercomputers, mainframes, and minicomputers.

Although programs are said to run concurrently, this does not mean that they run simultaneously. In fact, the programs take turns using the CPU. For example, one program could be using the CPU while another program prints a record. Amazingly, the operating system keeps track of everything and makes sure that the programs do not get entangled. From the point of view of the user, his or her program was executed by the computer just as if the computer and all its resources belonged exclusively to that user. (In reality, sometimes a large computer is so overloaded that time delays make the shared nature of the machine more obvious.)

Personal Computers

Personal computers also have a CPU that handles just one instruction at a time. Computers using the MS-DOS operating system without a shell are limited not only to just one user at a time but also to just one program at a time. If, for example, a user were using a word processing program to write a financial report and wanted to access some spreadsheet figures, he or she would have to perform a series of arcane steps: exit the word processing program, enter and use and then exit the spreadsheet program, and then re-enter the word processing program to complete the report. This is wasteful in two ways: (1) the CPU is often idle because only one program is executing at a time, and (2) the user is required to move inconveniently from program to program.

The solution to this problem is a direct descendant of multiprogramming, an operating system approach called **multitasking.** The idea is the same as multiprogramming: Let several programs compete for the use of the CPU. In industry jargon, and for all practical purposes, we consider that these programs are running at the same time.

A key feature of Windows is its multitasking capability. From the user's perspective, several programs can be running at the same time. Using the financial report example described above, the user could access the spreadsheet program without closing down or leaving the word processing program. It is possible to run many programs at once in a multitasking environment.

Software Applications with Windows

Just what will Windows do for your favorite software application? Although you can tell Windows to access your existing software, you will not get the full benefits of Windows unless you use a software version especially designed for use with Windows. Dozens of software manufacturers have written programs for Windows. Look around your local computer store and you will see an entire section reserved just for software written for Windows.

Windows 95

As millions of users can attest, Windows is an unqualified success. But even a popular software product can be improved, hence the version called

All About Bob

Bob is yet another software overlay, dubbed a social interface, to be superimposed on Windows. The screen shows home-like rooms. A study, for example, shows a desk with pen and paper which, when clicked, starts the user on the task of writing a letter. Microsoft offers Bob to demystify computers for the burgeoning mass of at-home computer novices.

One noteworthy feature of Bob is what it lacks: an instruction manual. Instead, Bob offers on-screen cartoon characters, each with a personality, that volunteer advice and steer users through the programs. Bob is actually a set of programs, including a letter-writing aid, a home organizer, a calendar, an address book, electronic mail, and even a checkbook with an option for electronic bill-paying.

And the name Bob? The name was suggested by an ad agency. Surveys show it is "down home" and friendly.

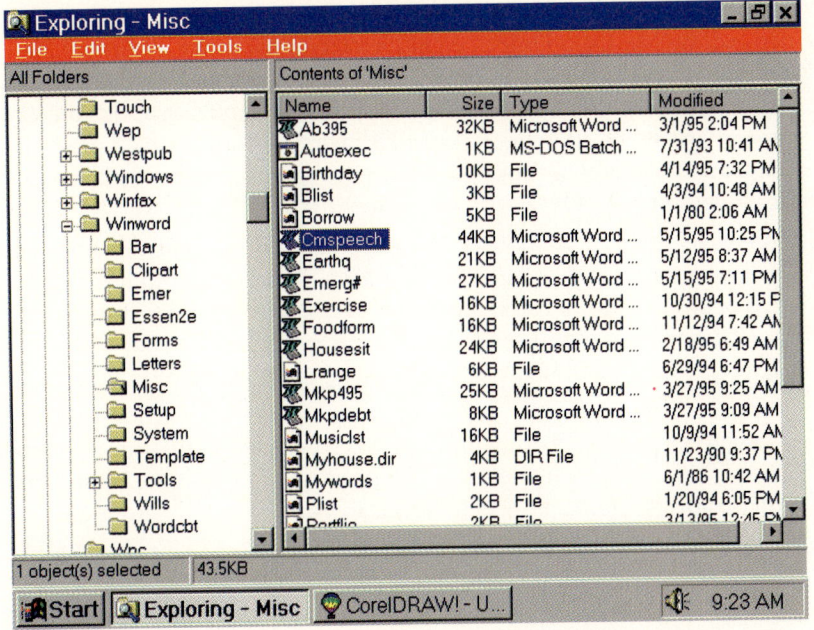

(a)

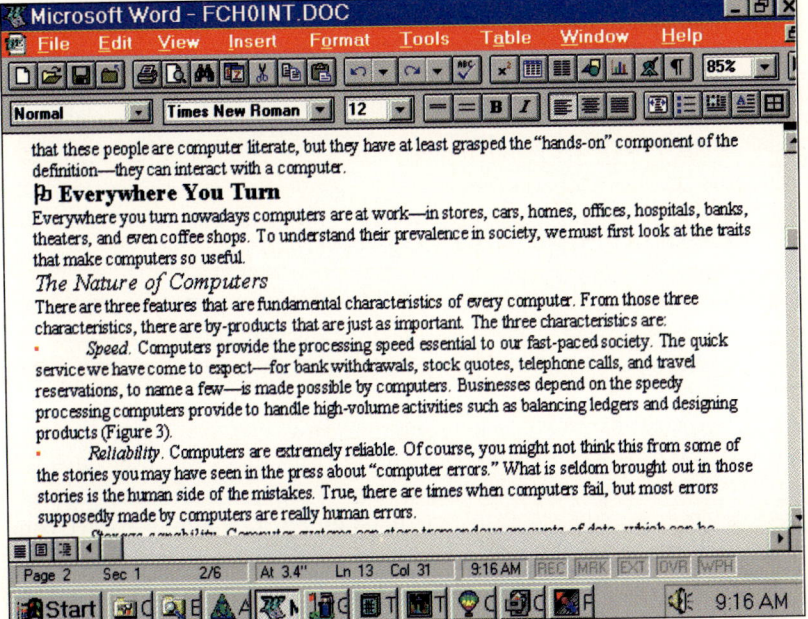

(b)

Figure 7-6 Windows 95. A key feature of Windows 95 is its task bar, which runs along the bottom of the screen. The task bar has a button for each program that has been invoked. By using a mouse to click the appropriate button, a user can quickly and easily switch to that program. As more programs are used, the buttons become smaller to accommodate the greater number. (a) Only two programs are in use here, as indicated by the buttons shown—the Windows Explorer program and the CorelDRAW program. The Explorer program, which can be used to organize and list files, is the program currently showing on the screen. (b) Look at the task bar at the bottom of the screen. Since many programs are in use, the buttons representing the programs are smaller in size. Microsoft Word is the program shown at the moment, as you can see from the top of the screen and also from the icon on the depressed button on the task bar.

Windows 95. Note, however, that Windows 95 is *not* a shell; it is a self-contained operating system, and thus requires no pre-installed DOS. Here are some highlights of Windows 95:

User convenience. You can't miss it: The Start button is in the lower-left corner just waiting to be clicked (Figure 7-6). From this launch you can find a program or a file. Want to revise a memo? Just click its file name to retrieve it—the word processing program is automatically invoked too. As another example of convenience, long file names, up to 255 characters, are permitted; this will be welcomed by users who have found the MS-DOS limit of eight characters per name a burden. Furthermore, the Windows 95 Explorer program can find a file by name or type or even text content. Perhaps the greatest convenience, as you can see in Figure 7-6, is the task bar, an array of buttons for each program in use. You can click from program to program as easily as changing channels on your TV.

A new look. Once Windows 95 is installed, all current software applications get a new look, complete with fancy borders, but run just as they did before Windows 95 was installed. Furthermore, even text fonts will look different—smoother and more streamlined.

Information center. Acknowledging the increasing role of data communications, Windows 95 puts all communications activities—e-mail, downloads, and so forth—in a single screen icon. Furthermore, Windows 95 includes software that makes it easier to configure computers for networks and, in particular, for the Internet.

Plug and play. Anyone who has added a new component—perhaps a modem or a sound card—to an existing computer knows that it may not work correctly right away. The reason is that the new component must be configured to the system, a process that may involve some tricky software and even hardware manipulations. Windows 95 supports **plug and play,** a concept that lets the computer configure itself when a new component is added. However, for plug and play to become a reality, hardware components must also feature the plug and play standard. Once a peripheral is built to the plug and play standard, it can be installed by simply plugging it in and turning on the computer.

▶ Network Operating Systems

An extension of operating systems for personal computers is a **network operating system (NOS),** which is designed to let computers on a network share resources such as hard disks and printers. A network operating system is similar to a standard operating system such as MS-DOS but includes special features for handling network functions. In addition to resource sharing, a NOS supports data security (does this user have the right to that data?), troubleshooting (oops—computer XYZ on the network failed to receive a message intended for it), and administrative control (track the online hours and number of messages to and from each computer).

In a client/server relationship, parts of the NOS (mostly file access and management programs) run on the server computer, while other NOS components, such as software that permits requests to the server and messages to other computers, run on the client computers.

One of the network operating system's main tasks is to make the resources appear as if they are running from the client computer. Whether issuing commands, running applications software, or sending jobs to a printer, the role of the NOS is to make the desired services appear to be local to that client computer. The whole point of a client/server system is to provide expanded services to individual users at their own networked computers; the network operating system is the software that makes it possible.

Operating Systems for Large Computers: An Overview

Large computers—mainframes—were around for over two decades before anyone thought to make a personal computer. Those big computers usually were owned by businesses and universities, which made them available to many users. So, rather than the scenario with which you may be familiar—one person per personal computer at a time—a large computer is used by many people. This presents special problems, which must be addressed by the operating system.

Since a large computer can handle many programs from many users, questions arise from computer users when they first realize that their program is "in there" with all those other programs.

Question: If my program and another program both want to use the *central processing unit* at the same time, what decides which program gets it first?

Answer: The operating system.

Question: If several other programs are in *memory* at the same time as my program, what keeps the programs from getting mixed up with one another?

Answer: The operating system.

Question: I know that for *storage* big computers use big disk packs that can hold files for several users—what keeps the files in some kind of accessible order?

Answer: The operating system.

Question: Well, the *printer* must be a problem. If we all need it at the same time, what prevents everyone's hard copy output from coming out in one big jumble?

Answer: The operating system!

This litany may sound repetitive, but it does make a point: The operating system anticipates these problems—and many others—so that you, as a user, can share the computer's resources with others with minimum concern about the details of how it is done. Notice that the questions above address sharing problems regarding the central processing unit, memory, storage, and the printer. The problems related to shared resources must be handled by the operating system.

Resource Allocation

On a large computer with many users, shared resources are said to be allocated. **Resource allocation** is the process of assigning computer resources to certain programs for their use. Those same resources are deallocated—removed—when the program using them is finished, and then they are re-allocated elsewhere. We begin with how the central processing unit is allocated to programs.

Sharing the Central Processing Unit

Since most computers have a single central processing unit, all programs running on the computer must share it. The sharing process is controlled by the operating system. Two approaches to sharing are multiprogramming and time-sharing.

Multiprogramming Revisited

We examined multiprogramming briefly when comparing it with the multitasking features of Microsoft Windows. Now we can consider it in greater detail. But first, let us distinguish multiprogramming from multiprocessing. **Multiprocessing** refers to the use of a powerful computer with more than one central processing unit, so that multiple programs can run simultaneously, each using its own processor.

However, if there is only one central processing unit (the usual case), it is not physically possible for more than one program to use it at the same time. Remember that multiprogramming means that two or more programs are being executed concurrently on a computer. What this really means is that the programs are taking turns; one program runs for a while and then another one. The key word here is *concurrently* as opposed to *simultaneously*. One program could be using the CPU while another does something else, such as sending output to the printer. Concurrent processing means that two or more programs are using the central processing unit in the same time frame—during the same hour, for instance—but not at the exact same time. In other words, concurrent processing allows one program to use one resource while another program uses another resource; this gives the illusion of *simultaneous* processing. As a result, there is less idle time for the computer system's resources. Concurrent processing is effective because CPU speeds are so much faster than input/output speeds. During the time it takes to execute a read instruction for one program, for example, the central processing unit can execute many—hundreds or even thousands—of calculation instructions for another program.

Multiprogramming is **event-driven.** This means that programs share resources based on events that take place in the programs. Normally, a program is allowed to complete a certain activity (event), such as a calculation, before relinquishing the resource (the central processing unit, in this example) to another program that is waiting for it.

The operating system implements multiprogramming through a system of interrupts. An **interrupt** is a condition that causes normal program processing to be suspended temporarily. Suppose, for example, that several programs are running on a large computer, two of them a payroll program and an inventory management program. When the payroll program needs to read the next employee record, that program is interrupted—or, in a sense, has interrupted itself—while the operating system takes over to do the actual reading by communicating with the disk drive. Meanwhile, since the payroll program is not using the CPU, the operating system may allocate the CPU to the inventory program to do some calculations. Once the record has been read for the payroll program, the interrupt is over and the payroll program may resume executing, perhaps calculating the employee's overtime pay, subject to the availability of the CPU. The CPU may be available right away but, since programs in a multiprogramming environment may be assigned priorities, the CPU may be allocated instead to a different program of higher-priority status.

However, the point of this discussion is not to clarify what program does what when. Rather, it is to show that sharing activities are being managed by the operating system in the background. Although it may appear

ONE JUMP AHEAD

Invisible Computers

Computer professionals who specialize in product research sometimes use the phrase *ubiquitous computing*, meaning that computing is everywhere. But they are not talking about the computer-on-a-desk type of everywhere. Ubiquitous computing means that computers are a part of the physical fabric of our daily spaces but not visible to the casual observer. Computers will be embedded in the environment around us, in classroom blackboards, kitchen countertops, floors, office walls, and more. These feats will be physically possible with sophisticated wireless networks.

Walls and floors, for example, will know which people are present and configure the environment to their needs. From an environmental point of view, when people walk into a room, computers in the walls sense this and balance the heat or air conditioning system accordingly. The same ideas can be expanded to meet convenience in the home and working needs on the job.

What will people think about all this? They probably will not think

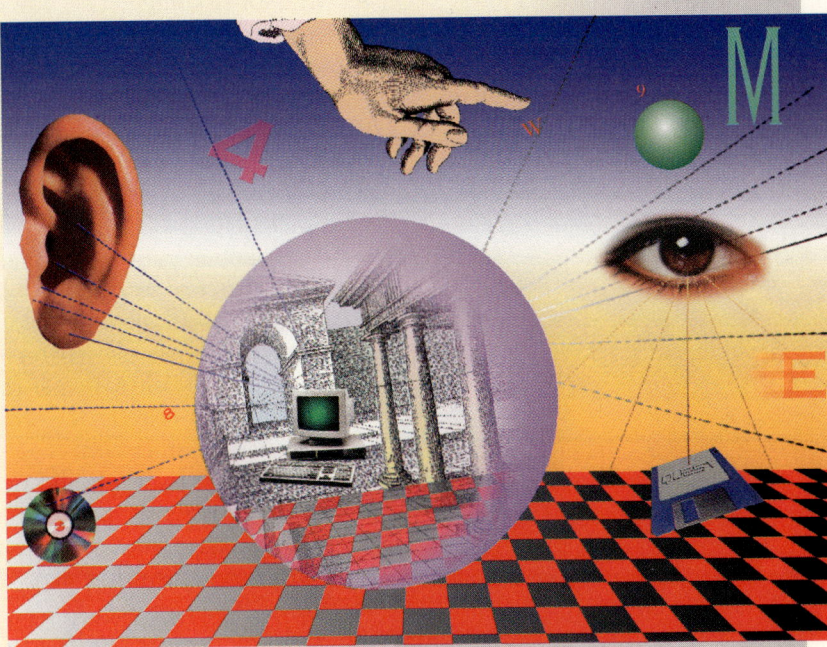

about it at all. People will not see or think about tomorrow's computers any more than they see or think about computers in their cars today. Furthermore, people do not ask what operating system is used to run computers in their cars. Look at it another way. Computer technology is finally getting so powerful that it can become invisible.

---

to the user that a program is being run continuously from start to finish, in fact it is being interrupted continuously.

In large computer systems, programs that run in an event-driven multiprogramming environment are usually batch programs. Typical examples are programs for payroll, accounts receivable, sales and marketing analysis, financial planning, quality control, and stock reporting.

Time-sharing

Time-sharing, a special case of multiprogramming, is usually **time-driven** rather than event-driven. A common approach is to give each user a **time slice**—a fraction of a second—during which the computer works on a single user's tasks. However, the operating system does not wait for completion of the event; at the end of the time slice—that is, when time is up—the resources are taken away from that user and given to someone else. This is hardly noticeable to the user: When you are sitting before a terminal in a time-sharing system, the computer's response time will be quite short—fractions of a second—and it may seem as if you have the computer to yourself.

Response time is the time between your typed computer request and the computer's reply. Even if you are working on a calculation and the operating system interrupts it, sending you to the end of the line until other users have had their turns, you may not notice that you have been deprived of service. Not all computer systems give ideal service all the time, however; if a computer system is trying to serve too many users at the same time, response time may slow down noticeably.

You should realize that, generally speaking, you as the user do not have control over the computer system. In a time-sharing environment the operating system has actual control because it controls the users by allocating time slices. Giving the users the processor in turns is called **round-robin scheduling.** However, sometimes a particular user will, for some reason, be entitled to a higher priority than other users. Higher priority translates to faster and better service. A common method of acknowledging higher priority is for the operating system to give that user more turns. Suppose, for example, that there are five users who would normally be given time slices in order: A-B-C-D-E-A-B-C-D-E, and so forth. However, if user B is assigned a higher priority, the order could be changed to A-B-C-B-D-B-E-B, giving B every other turn.

Typical time-sharing applications are those with many users, each of whom has a series of randomly occurring actions, each of them brief: credit checking, point-of-sale systems, and airline reservation systems. Each of these systems has many users, perhaps hundreds, who need to share the system resources.

Sharing Memory

What if you have a very large program for which it might be difficult to find space in memory? Or what if several programs are competing for space in memory? These questions are related to memory management. **Memory management** is the process of allocating memory to programs and of keeping the programs in memory separate from each other.

There are many methods of memory management. Some systems simply divide memory into separate areas, each of which can hold a program. The problem is how to know how big the areas, sometimes called **partitions** or **regions,** should be; at least one of them should be large enough to hold the largest anticipated program. Some systems use memory areas that are not of a fixed size—that is, the sizes can change to meet the needs of the current assortment of programs. In either case—whether the areas are of a fixed or variable size—there is a problem with unused memory between programs. When these memory spaces are too small to be used, space is wasted.

Foreground and Background

Large all-purpose computers often divide their memory into foreground and background areas. The **foreground** is generally for programs that have higher priority and therefore receive more CPU time. A typical foreground program is in a time-sharing environment, with the user at a terminal awaiting response. That is, a foreground program is interactive, with the CPU often unused while the user is entering the next request. Thus, there is CPU time available for the waiting background programs. The **back-**

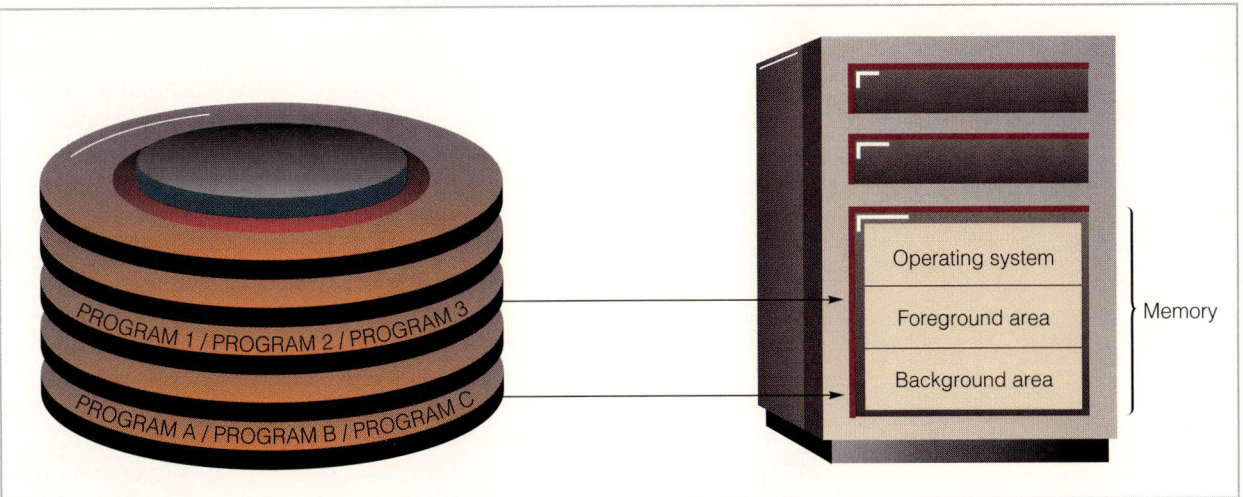

Figure 7-7 Programs waiting in queues. These programs are waiting on disk in queues organized by program class. That is, time-sharing programs (1 through 3) wait in their own queue for a foreground area to open up, and batch programs (A through C) wait in a queue for a background area to be free.

ground, as the name implies, is for programs with less pressing schedules and, thus, lower priorities and less CPU time. Typical background programs are batch programs in a multiprogramming environment. Foreground programs are given privileged status—more turns for the central processing unit and other resources—and background programs take whatever they need that is not currently in use by another program. Programs waiting to run are kept on the disk in **queues** suitable to their job class (Figure 7-7).

Virtual Storage

Many computer systems manage memory by using a technique called **virtual storage** (also called **virtual memory**). The virtual storage concept means that part of the program is stored on disk and is brought into memory for execution only as needed (since only one part of a program can be executing at any given time, the parts not presently needed are the parts left on the disk). The user appears to be using more memory space than is actually the case. Since only part of the program is in memory at any given time, the amount of memory needed for a program is minimized. Memory, in this case, is considered **real storage,** while the secondary storage holding the rest of the program (hard disk, most likely) is considered virtual storage.

Virtual storage can be implemented in a variety of ways. Consider the paging method, for example. Suppose you have a very large program, which means there will be difficulty finding space for it in the computer's shared memory. If your program is divided into small pieces, it will be easier to find places to put those pieces. This is essentially what paging does. **Paging** is the process of dividing a program into equal-size pieces called **pages** and storing them in equal-size memory spaces called **page frames.** All pages and page frames are the same fixed size—typically, 2K or 4K bytes. The pages are stored in memory in *noncontiguous* locations—that is, locations not necessarily next to each other.

Even though the pages are not right next to each other in memory, the operating system is able to keep track of them. It does this by using a **page table,** which, index-like, lists each page that is part of the program and the corresponding beginning memory address where it has been placed.

Memory Protection

In a multiprogramming environment it is theoretically possible for the computer, while executing one program, to destroy or modify another program by transferring to the wrong memory locations. That is, without protection, one program might accidentally hop into the middle of another, causing destruction of data and general chaos. This, of course, is not permitted. To avoid this problem, the operating system confines each program to certain defined limits in memory. If a program inadvertently attempts to enter some memory area outside its limits, the operating system terminates the execution of that program. This process of keeping one program from straying into another is called **memory protection.**

Sharing Storage Resources

If you have used a personal computer, much of your interaction with the operating system probably related to storage: listing, copying, or deleting files. Your hard drive, or even one you shared with other users, held multiple files. The operating system associated with the personal computer could keep track of the files and respond to your commands to handle these files.

The situation is similar for disk files associated with programs run on a large computer. It is the operating system that keeps track of which file is where and also responds to commands to manipulate files. But the situation is complicated by the possibility that more than one user may want to read or write a record from the same disk pack at the same time. Again, it is the operating system that keeps track of the input and output requests and processes them, as appropriate, usually in the order received. Any pro-

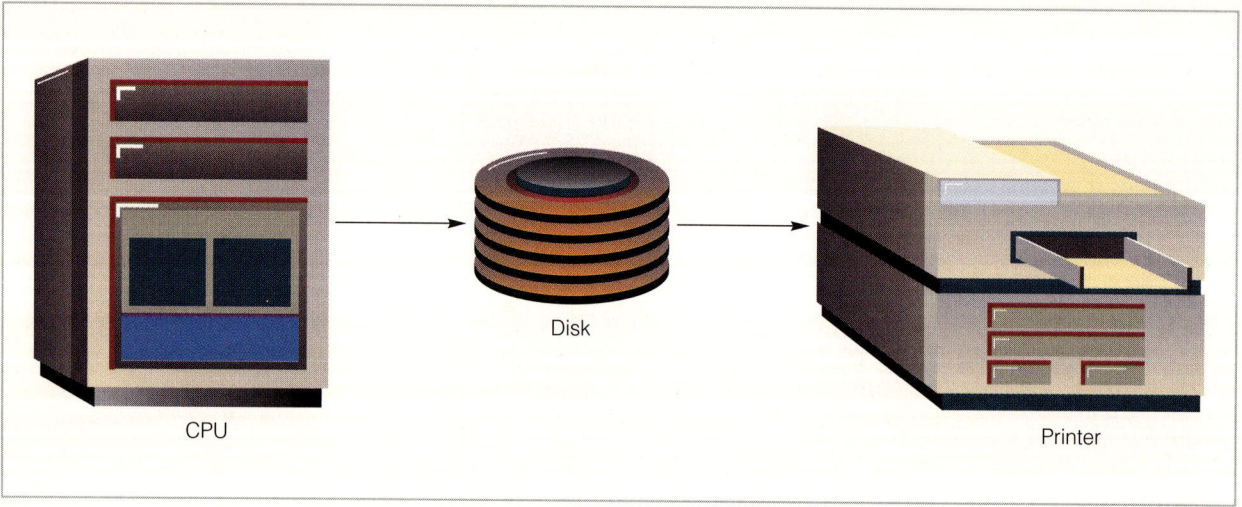

Figure 7-8 Spooling. Program output that is destined for the printer is written first to a disk—spooled—and later transferred to a printer.

MAKING THE RIGHT CONNECTIONS

Nature Conservancy Online

The Nature Conservancy is a nonprofit organization that buys land, particularly shore and woodlands, to preserve it forever for the public. As all nonprofit organizations must, the conservancy solicits donations to support its cause. The conservancy has found a method to receive donations that is reliable and thrifty: contributions online.

A conservancy donor who agrees to be a sustaining member completes a form agreeing to a stipulated monthly dollar amount to be transferred electronically from the donor's bank account to the conservancy's bank account. For this service, the conservancy pays 40¢ to a service bureau and 10¢ to the bank, for a total of 50¢ per transaction, as compared to 92¢ per transaction for processing checks that come through the mail.

The conservancy has enjoyed other benefits from its online approach, including increased contributions, improved membership retention, and lowered administrative costs. A benefit that is particularly satisfying to the conservancy is that online contributions eliminate paper, thus saving trees.

gram instruction to read or write a record is routed to the operating system, which processes the request and then returns control to the program.

Sharing Printing Resources

Suppose a half dozen programs are active, but the computer has only one printer. If all programs took turns printing out their output a line or two at a time, interspersed with the output of other programs, the resulting printed report would be worthless. To get around this problem, a process called **spooling** is used: Each program writes onto a disk each file that is to be printed. Or, to be more accurate, the program thinks it is writing to the printer, but the operating system intercepts that output and sends it instead to the disk. When the entire file is on the disk, spooling is complete, and the disk files are printed intact (Figure 7-8).

Spooling also addresses the problem of relatively slow printer speeds. Writing a record on disk is much faster than writing that same record on a printer. A program, therefore, completes execution more quickly if records to be printed are written temporarily on disk instead. Once the file to be printed is transferred to disk, the program can continue executing. The actual printing can be done at some later time, when printing will not slow program execution. Some installations use a separate (usually smaller) computer dedicated exclusively to the printing of spooled files; some print off-hours or overnight so that smaller, more immediate jobs can use the printer during the day.

Service Programs

Most of the resource allocation tasks just described are done by the operating system without involvement by a user. For example, activities such as paging and spooling go on without explicit commands from users. But the operating system can also perform explicit services at the request of the user.

Why reinvent the wheel? Duplication of effort is what **service programs,** also known as **utilities,** are supposed to avoid. Such prewritten programs perform many standard chores, such as copying a file, sorting a file, or merging two files into one file. As we have noted, strictly speaking, these utilities are not part of the operating system.

Generic Operating Systems

Early operating systems were developed by the manufacturers who made the computers. The operating system was considered a **proprietary operating system** because it was used exclusively with that type of hardware. Even today, in the mainframe world, operating systems are usually defined by the hardware vendor. But some changes have occurred in recent years. The trend is toward what is called a **generic operating system,** sometimes called a **portable operating system,** that works with more than one manufacturer's computer system. Several generic operating systems are available today. Here we will discuss a generic operating system that is particularly influential: UNIX.

The UNIX Graduates

UNIX, a multiuser time-sharing operating system, was developed in 1971 by Ken Thompson and Dennis Ritchie at AT&T's Bell Laboratories for use on its DEC minicomputers. In the late 1970s Bell gave away UNIX to many colleges and universities, and students became accustomed to using it. Consequently, when many of these schools' graduates entered the work force, they began agitating for the acceptance of UNIX in industry, thus producing what is known as the "UNIX graduate" phenomenon.

Though not everyone agrees, many consider UNIX to be the most sophisticated operating system available.

Is UNIX a Standard?

When something like an operating system is accepted by a majority of users in the computer world, it is said to be a **standard.** Many believe that UNIX is a standard among large-system users. Key UNIX supporters—the scientific community, the federal government, the aerospace industry—often named UNIX in their bid specifications to computer manufacturers. In effect, they said, "If you want our business, you better offer a system that includes UNIX." Vendors who could not offer UNIX-supported hardware were effectively cut out of the bidding process. This was a powerful incentive to offer UNIX with a hardware system. Today UNIX runs on everything from the Cray-2 supercomputer to personal computers.

Do I Really Need to Know All This?

The answer to that question depends on how you expect to use a computer. If your primary use of a computer is as a tool to enhance your other work, then you may have minimum interaction with an operating system. In that case, whether you are using a personal computer or a mainframe, you will learn to access the application software of choice very quickly.

But there are other options. In fact, there are far more options than we are able to present in this introductory chapter. As a sophisticated user, you can learn your way around the operating system of any computer you might be using. If you plan to be a programmer, then there is no question about whether you need to know everything in this chapter; you will need to know this and much more.

Now that we have examined two major topics related to programming—the programming process and operating systems—we can put these tasks in perspective. Our next topic is systems analysis and design, which shows you the big picture.

CHAPTER REVIEW

Summary and Key Terms

- An **operating system** is a set of programs that lies between applications software and the computer hardware. **System software** means all programs related to coordinating computer operations, including the operating system, programming language translators, and service programs.
- The **supervisor program,** most of which remains in memory, is called *resident*. The supervisor controls the entire operating system and loads into memory *nonresident* operating system programs from disk storage as needed.
- An operating system has three main functions: (1) manage the computer's resources, such as the central processing unit, memory, disk drives, and printers, (2) establish a user interface, and (3) execute and provide services for applications software.
- In general, a personal computer applications software package can run on just one operating system. However, a software maker may make different versions for different operating systems.
- Loading the operating system into memory is called **booting** the system.
- In the on-screen displays A> and C>, the A and C refer to disk drives. The > is a **prompt,** a signal that the system is waiting for you to give an instruction to the computer.
- To execute a given DOS program, a user must issue a **command,** a name that invokes a specific DOS program.
- The **default drive,** also called the **current drive,** is the drive that the computer is currently using.
- Disks may hold system files, applications software files, or data files.
- Some operating systems provide pictures or simply worded choices, or both, instead of giving a prompt. In effect, these pictures and choices form a user-friendly "coating," or **shell.** They create a comfortable **operating environment** for the user, who does not have to remember or look up the appropriate commands.
- A key product is Microsoft Windows, software with a colorful **graphical user interface (GUI).** Windows offers on-screen pictures called **icons** and both **pull-down** and **pop-up menus,** both of which encourage pointing and clicking with a mouse, an approach that can make computer use faster and easier.
- On mainframe operating systems, **multiprogramming** permits the running of several programs in the same time frame, or **concurrently**. That is, multiprogramming permits several unrelated programs, probably from many different users, to compete for the processor. Using Windows, **multitasking** lets several programs compete for the use of the CPU; for all practical purposes, we consider that these programs are running at the same time.
- Among the features of Microsoft Windows 95 is **plug and play,** a concept that lets the computer configure itself when a new component is added.
- A **network operating system (NOS)** is designed to let computers on a network share resources such as hard disks and printers. A NOS supports resource sharing, data security, troubleshooting, and administrative control. Parts of the NOS run on the server computer, while other NOS components run on the client computers.
- **Resource allocation** is the process of assigning computer resources to certain programs for their use.
- **Multiprocessing** means that a computer with more than one central processing unit can run multiple programs simultaneously, each using its own processor.
- Multiprogramming is running two or more programs concurrently on the same computer. Multiprogramming is **event-driven,** meaning that one program is allowed to use a particular resource (such as the central processing unit) to complete a certain activity (event) before relinquishing the resource to another program. In multiprogramming, the operating system

- uses **interrupts**, which are conditions that temporarily suspend the execution of individual programs.
- **Time-sharing** is a special case of multiprogramming in which several people use one computer at the same time. Time-sharing is **time-driven**—each user is given a **time slice** in which the computer works on that user's tasks before moving on to another user's tasks. **Response time** is the time between the user's typed computer request and the computer's reply. The system of having users take turns is called **round-robin scheduling.**
- **Memory management** is the process of allocating memory to programs and of keeping the programs in memory separate from each other. Some systems simply divide memory into separate areas, sometimes called **partitions** or **regions**, each of which can hold a program. Large all-purpose computers often divide memory into a **foreground** area for programs with higher priority and a **background** area for programs with lower priority. Programs waiting to be run are kept on the disk in **queues.**
- In the **virtual storage** (or **virtual memory**) technique of memory management, part of the application program is stored on disk and is brought into memory only when needed for execution. Memory is considered **real storage**; the secondary storage holding the rest of the program is considered virtual storage.
- Virtual storage can be implemented in several ways. **Paging** divides a program into equal-size pieces (**pages**) that fit exactly into corresponding noncontiguous memory spaces (**page frames**). The operating system keeps track of page locations using an index-like **page table.**
- In multiprogramming, **memory protection** is an operating system process that defines the limits of each program in memory, thus preventing programs from accidentally destroying or modifying one another.
- In multiprogramming, all file operations are handled via the operating system.
- **Spooling** writes each file to be printed temporarily onto a disk instead of being printed immediately. When this spooling process is complete, all the appropriate files from a particular program can be printed intact.
- **Service programs,** also called **utilities,** are prewritten standard programs that perform many file-handling tasks such as copying or sorting files.
- A **proprietary operating system** is used exclusively with the computer hardware for which it was written. A **generic operating system,** also called a **portable operating system,** is one that works with more than one manufacturer's computer system.
- **UNIX,** a multiuser, time-sharing operating system, has been described as a generic operating system. Many consider UNIX a **standard,** the operating system accepted by a majority of users of large computer systems.

Discussion Questions

1. How would your access to computers be affected if there were no operating systems?
2. How would you explain the rapid acceptance of Microsoft Windows?
3. What kinds of operating systems might you expect to use in your career? Personal computer operating system? Large computer operating system? Network operating system? All of these? Will it depend on the type of job you have?

Student Study Guide

Multiple Choice

1. An operating system is a
   a. set of users
   b. set of programs
   c. form of time-sharing
   d. supervisor program
2. In multiprogramming, two or more programs can be executed
   a. by optimizing compilers
   b. simultaneously
   c. with two computers
   d. concurrently
3. Time-sharing of resources by users is usually
   a. based on time slices
   b. event-driven
   c. based on input
   d. operated by spooling
4. Management of an operating system is handled by
   a. an interpreter
   b. utility programs
   c. the supervisor program
   d. the CPU
5. The process of allocating main memory to programs and keeping the programs in memory separate from each other is called
   a. memory protection
   b. virtual storage
   c. memory management
   d. real storage
6. UNIX is an example of a(n)
   a. memory management
   b. NOS
   c. generic operating system
   d. utility program
7. The technique in shared systems that avoids interspersed printout from several programs is
   a. paging
   b. slicing
   c. queuing
   d. spooling
8. The technique whereby part of the program is stored on disk and is brought into memory for execution as needed is called
   a. memory allocation
   b. virtual storage
   c. interrupts
   d. prioritized memory
9. An operating system used exclusively with the manufacturer's computer:
   a. DOS
   b. UNIX
   c. proprietary
   d. NOS
10. A portable operating system:
    a. generic
    b. allocated
    c. backup
    d. utility
11. Another name for an operating environment is
    a. page
    b. shell
    c. layer
    d. supervisor
12. Loading the operating system into a personal computer is called
    a. booting
    b. interrupting
    c. prompting
    d. paging
13. Which one of the following is a graphical shell?
    a. UNIX
    b. utility program
    c. page
    d. GUI
14. In multiprogramming, the process of confining each program to certain defined limits in memory is called
    a. spooling
    b. program scheduling
    c. time-sharing
    d. memory protection
15. The corresponding memory spaces for pages are called
    a. page utilities
    b. page blocks
    c. page frames
    d. page modules
16. The time between the user's request and the computer's reply
    a. concurrent time
    b. allocation time
    c. response time
    d. event time
17. An on-screen picture:
    a. page
    b. icon
    c. NOS
    d. spool
18. Take-a-turn time-sharing:
    a. spooling
    b. round-robin scheduling
    c. interfacing
    d. prompting
19. Page frames are typically
    a. 1K or 2K bytes
    b. 2K or 3K bytes
    c. 3K or 4K bytes
    d. 2K or 4K bytes
20. The memory area for programs with highest priority:
    a. frame
    b. page table
    c. foreground
    d. default drive
21. Programs waiting to be run are kept on disk in
    a. page frames
    b. shells
    c. the background
    d. queues
22. Prewritten standard file-handling programs are called
    a. pull-down menus
    b. supervisors
    c. pages
    d. utilities
23. The signal that the computer is awaiting a command from the user:
    a. prompt
    b. event
    c. time slice
    d. interrupt
24. Another name for virtual memory is
    a. virtual storage
    b. background
    c. foreground
    d. utility
25. NOS refers to
    a. the default drive
    b. operating system for a network
    c. booting
    d. round-robin scheduling

True/False

T F 1. A DOS program is invoked by issuing a command.
T F 2. The most important program in an operating system is the supervisor program.
T F 3. Multiprogramming means that two or more programs can run simultaneously.
T F 4. Time-sharing is effective because input/output speeds are so much faster than CPU speeds.
T F 5. A proprietary operating system will run on any personal computer.
T F 6. Background programs are usually batch programs.
T F 7. Virtual storage is a technique of memory management that appears to provide users with more memory space than is actually the case.
T F 8. With the virtual storage technique, secondary storage is considered real storage.
T F 9. The default drive is also called the current drive.
T F 10. In a network operating system, some functions are performed by the server and others by the client computers.

T F 11. Spooling is a process that results in interspersed printout from several programs.
T F 12. Shell is another name for page.
T F 13. An operating system includes system software, programming language translators, and service programs.
T F 14. Utility programs avoid duplication of effort.
T F 15. In a given memory system, all page frames are the same size.
T F 16. Time-sharing is both event-driven and time-driven.
T F 17. Resource allocation means a NOS distributes most of its functions to the client computers.
T F 18. UNIX is a generic operating system that can be run on both large and small computers.
T F 19. Loading the operating system into memory is called booting.
T F 20. Windows is a popular graphical user interface.
T F 21. A portable operating system is usually a proprietary operating system.
T F 22. Paging divides a program into pieces of various sizes to fit in the available memory spaces.
T F 23. Response time is the time it takes a program to run.
T F 24. Round-robin scheduling gives each user the processor in turn.
T F 25. Multiprocessing is simultaneous processing.

Fill-In

1. NOS stands for _____

2. What operating system program remains resident in memory? _____

3. What term is used for the time between a user's request at the terminal and the computer's reply? _____

4. What type of system lets two or more programs execute concurrently? _____

5. What are the program pieces called in the virtual storage technique of paging? What are the corresponding memory spaces called?

   a. _____

   b. _____

6. Simultaneous processing of more than one program using more than one processor is called _____

7. Multiprogramming is considered to be _____ – driven

8. Having users take turns in a time-sharing environment is what type of scheduling? _____

9. In paging, what is needed to keep track of the various pages? _____

10. The generic operating system some consider a standard: _____

11. Which process of an operating system avoids interspersing the printout from several programs? _____

12. A shell program that overlays the operating system to provide a more friendly environment: _____

13. In time-sharing, each user is given a unit of time called a _____

14. The computer that pioneered simple operating system interfaces: _____

15. Another name for service programs: _____

16. Loading the operating system into memory is called _____

17. An operating system written for and used exclusively with a vendor's computer is called _____

18. How are programs kept on disk while waiting to be run? _____

19. If memory is divided into foreground and background areas, time-sharing applications are likely to be where: _____

20. What does GUI stand for? _____

21. In multiprogramming, a condition that temporarily suspends program execution: _____

22. Keeping programs in memory separate is called _____

23. Microsoft's operating environment is called _____

24. MS-DOS uses the > symbol as a _____

25. Another name for the default drive is _____

Answers

Multiple Choice
1. b	6. c	11. b	16. c	21. d
2. d	7. d	12. a	17. b	22. d
3. a	8. b	13. d	18. b	23. a
4. c	9. c	14. d	19. d	24. a
5. c	10. a	15. c	20. c	25. b

True/False
1. T	6. T	11. F	16. F	21. F
2. T	7. T	12. F	17. F	22. F
3. F	8. F	13. F	18. T	23. F
4. F	9. T	14. T	19. T	24. T
5. F	10. T	15. T	20. T	25. T

Fill-In
1. network operating system
2. the supervisor program
3. response time
4. multiprogramming
5. a. pages
   b. page frames
6. multiprocessing
7. event
8. round-robin
9. page table
10. UNIX
11. spooling
12. operating environment
13. time slice
14. Macintosh
15. utilities
16. booting
17. proprietary
18. in queues
19. foreground
20. graphical user interface
21. interrupt
22. memory protection
23. Windows
24. prompt
25. current drive

PLANET INTERNET

Images, Icons, and Flags

Since many people are interested in colorful graphics, it is not surprising that the subject is well covered on the Internet. A good place to begin is the site called Images, Icons, and Flags. This site has dozens of links to nature and travel scenes, medical images, and museum archives. Shown here is a fractal artwork (art generated by variations on an original pattern—the fractal) done by pioneer Benoit Mandelbrot. You can find this by linking to a site called the Fractal Art Gallery.

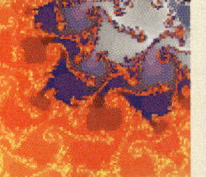
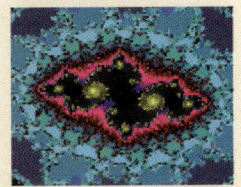

Images. Graphics images may be art or photos or some combination of the two. Art may be produced entirely on the computer. A photo or artwork devised externally may be shown on a computer screen once it has been input to the computer in some form.

Icons. An icon is a small symbolic picture, usually artwork but possibly a photo. Icons are everywhere, especially in advertising—think of the Pillsbury Doughboy or the swoosh that identifies Nike running shoes. On the Net, people use icons on their home pages. Many users like to use icons in computer-produced newsletters, advertising, and even correspondence. A large collection of icons is on the Net, and most of them are freely available for your use.

Flags. The Net supplies broad flag coverage, including colorful depictions of the flags themselves. The original site we referenced (Images, Icons, and Flags) has links to international flags, maritime flags, and semaphore flags.

Keeping it for yourself. Many files, including graphics images, can be downloaded—moved—from the source computer to your computer. Also, we'll mention briefly that you may have available, or can purchase, screen capture software. Using this software, you can save anything that you can see on your screen to a disk file.

Permission. Many artists and photographers put their works on the Internet to be shared freely by all. Others, usually professionals, state that their works are copyrighted and may be used only with permission. An example is The Stock Solution, whose logo is shown here. The Stock Solution is an agency that represents professional photographers whose works can be seen right on your screen. But you must agree to a lease fee before you can make any further use of the photos.

Waiting and waiting. Anyone who wants to explore images on the Net needs to understand that it is a time-consuming process. This could be a problem if you use shared computer resources. Unlike text, which is mostly white space with occasional black markings, color images are made up of tiny tightly-packed dots, sometimes hundreds of thousands of them for a single image. Moving an image from the source computer to your screen takes time. Your best defense is a speedy modem. Also, consider getting on the Net at odd hours when there are fewer users.

Internet Exercises

1. **Structured exercise.** Begin with the B/C URL http://www.aw.com/bc/planet/ and click on semaphores; look up the semaphore alphabet.
2. **Freeform exercise.** From the starting point of Images from Various Sources, choose three links and bring images to your screen.

Jean Winston, a second-year college student, worked part time as an assistant librarian at the Montlake Branch of the Carthage Public Library, which has a central library and four branches. The various libraries cooperated with each other, particularly in the exchange of books needed by library patrons. Unfortunately, the exchange system was cumbersome and unreliable. From her brief study of computers, Jean suspected that a computer system would probably improve service, but she also knew that money was scarce and funds were unlikely to be available for innovations.

The next year, however, the city passed a comprehensive bond issue that included money for the library. The library manager at the central branch immediately engaged an outside company to study how computers could be used in a library system. Jean was delighted but discovered, to her surprise, that the full-time librarians were not. Although the employees were actually in favor of the idea of computers, they were apprehensive in two ways: They worried about whether the computer would prove so efficient that it would eliminate some of their jobs, and they fretted about their own ability to use computers.

This reluctance was soon obvious to the systems analyst who visited the libraries to study the way the current system worked and to interview the library employees. Over time and many discussions, however, the systems analyst was able to reassure the librarians. A new system, whose components included self-service terminals from which customers could order a book from any branch, eventually was put in place.

Chapter 8

Systems Analysis and Design
The Big Picture

LEARNING OBJECTIVES

- Understand the terms *system, analysis,* and *design*
- Know the principal functions of the systems analyst
- Know the phases of the systems development life cycle
- Become acquainted with data-gathering and analysis tools
- Become acquainted with systems design tools
- Understand the concept of prototyping
- Understand the value of CASE tools
- Understand the predominance of maintenance

THE SYSTEMS ANALYST
   The Analyst and the System
   The Systems Analyst as Change Agent
   What It Takes to Be a Systems Analyst

HOW A SYSTEMS ANALYST WORKS: OVERVIEW OF THE SYSTEMS DEVELOPMENT LIFE CYCLE

PHASE 1: PRELIMINARY INVESTIGATION
   Problem Definition
   Wrapping Up

PHASE 2: SYSTEMS ANALYSIS
   Data Gathering, Data Analysis
   System Requirements
   Report to Management

PHASE 3: SYSTEMS DESIGN
   Preliminary Design
   Prototyping
   CASE Tools
   Detail Design

PHASE 4: SYSTEMS DEVELOPMENT
   Scheduling
   Programming
   Testing

PHASE 5: IMPLEMENTATION
   Training
   Equipment Conversion
   File Conversion
   System Conversion
   Auditing
   Evaluation, Maintenance

PUTTING IT ALL TOGETHER: IS THERE A FORMULA?

The Systems Analyst

As the opening tale depicts, people are often nervous when they are about to be visited by a systems analyst. A systems analyst with any experience, however, knows that people are uneasy about having a stranger pry into their job situations and that they may be nervous about computers. Before we discuss how the systems analyst helps people address change, let us begin with a few basic definitions.

The Analyst and the System

Although we will describe a systems project more formally later in the chapter, let us start by defining what we mean by the words *system, analysis,* and *design*. A **system** is an organized set of related components established to accomplish a certain task. There are natural systems, such as the cardiovascular system, but many systems have been planned and deliberately put into place by people. For example, a fast food franchise has a system for serving a customer, including taking an order, assembling the food, and collecting the charge. A **computer system** is a system that has a computer as one of its components.

Systems analysis is the process of studying an existing system to determine how it works and how it meets user needs. Systems analysis lays the groundwork for improvements to the system. The analysis involves an investigation, which in turn usually involves establishing a relationship with the client for whom the analysis is being done and with the users of the system. The **client** is the person or organization contracting to have the work done. The **users** are people who will have contact with the system, usually employees and customers. For instance, in a fast food system, the client is probably the franchise owner or manager, and the users are both the franchise employees and the customers.

Systems design is the process of developing a plan for an improved system, based on the results of the systems analysis. For instance, an analysis of a fast food franchise may reveal that customers stand in unacceptably long lines waiting to order. A new system design might involve plans to have employees press buttons that match ordered items, causing a display on an overhead screen, that can be seen by other employees who can quickly assemble the order.

The **systems analyst** normally performs both analysis and design. (The term *systems designer* is not common, although it is used in some places.) In some computer installations a person who is mostly a programmer may also do some systems analysis and thus have the title **programmer/analyst**. Traditionally, most people who have become systems analysts have done so by way of programming. Starting out as a programmer helps the analyst appreciate computer-related problems that arise in analysis and design work. As you will see, programmers often depend on systems analysts for specifications from which to design programs.

A systems analysis and design project does not spring out of thin air. There must be an *impetus*—motivation—for change and related *authority* for the change. The impetus for change may be the result of an internal force, such as the organization's management deciding a computer could be useful in warehousing and inventory, or an external force, such as government reporting requirements or customer complaints about billing (Figure 8-1). Authority for the change, of course, comes from higher management.

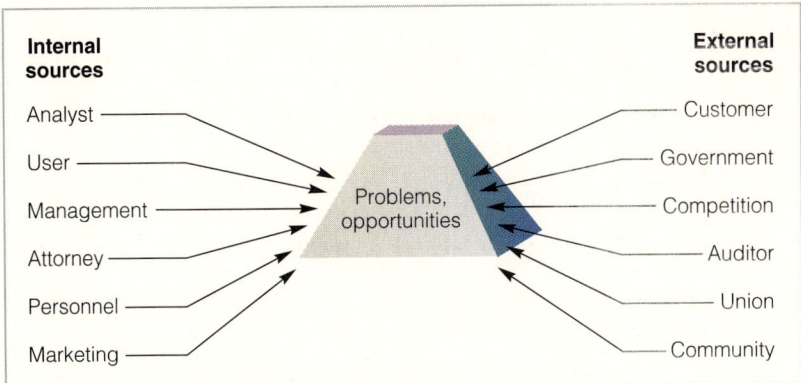

Figure 8-1 Impetus for change. Internal or external sources can initiate a system change.

The Systems Analyst as Change Agent

The systems analyst fills the role of **change agent.** That is, the analyst must be the catalyst or persuader who overcomes the natural inertia and reluctance to change within an organization. The key to success is to involve the people of the client organization in the development of the new system. The common industry phrase is **user involvement,** and nothing can be more important to the success of the system. Some analysts like to think in terms of who "owns" the system. If efforts toward user involvement are successful the user begins to think of the system as *my* system, rather than *their* system. Once that happens, the analyst's job becomes much easier.

The finest system in the world will not suffice if users do not perceive it as useful. Users must be involved in the process from beginning to end. The systems analyst must monitor the user pulse regularly to make sure that the system being planned is one that will meet user needs.

What It Takes to Be a Systems Analyst

Not every computer professional aspires to the job of systems analyst. Before we can understand what kind of person might make a good systems analyst, we need to look at the kinds of things an analyst does. The systems analyst has three principal functions:

- **Coordination.** An analyst must coordinate schedules and system-related tasks with a number of people: the analyst's own manager; the programmers working with the system; the system's users, from clerks to top management; the vendors selling the computer equipment; and a host of others, such as mail-room employees handling mailings and carpenters doing installation.
- **Communication, both oral and written.** The analyst may be called upon to make oral presentations to clients, users, and others involved with the system. The analyst provides written reports—documentation—on the results of the analysis and the goals and means of the design. These documents may range from a few pages long to a few inches thick.
- **Planning and design.** The systems analyst, with the participation of members of the client organization, plans and designs the new system. This function involves all the activities from the beginning of the project until the final implementation of the system.

Checking the Classified Ads

Compare these two classified advertisements:

> **Position Wanted as Systems Analyst.** Expertise in systems design. Good technical skills—programming, database design, data communications. Experienced on variety of hardware. References available. Jim. 937-4783.
>
> **Wanted: Systems Analyst.** Strong user orientation. Ability to assess system impact on user departments. Implementation skills, especially motivation and training of users. Send resume to Athens Chemical, P. O. Box 5, Eugene, OR 97405.

Could Jim be the person that Athens is looking for? Possibly, but you would not know it from their classified ads, which appear to have no common ground. If Jim were asked if he was user-oriented, he would—if he is smart—say yes, and Athens, of course, would be seeking an employee with technical skills as well as user orientation.

Why the contrast in their ads? Systems analysts are often most proud of their technical achievements and think that these skills will be appreciated. But Athens, like other companies, has been around the block a few times and knows that technical skills mean little if the total effort does not serve the user. Managers, in survey after survey, consistently list user needs as their top priority.

ONE JUMP AHEAD

Community Computing

For some time, computers have assisted both businesses and individuals in the community. The great untapped group is the community itself. These two examples suggest what is to come.

The neighborhood takes shape. Urban planners have long used computers to lay out new designs for streets, stores, business areas, and parks. Now citizens are getting into the act, participating via computer in designing changes to their own neighborhoods. Based on scanned photos, an interactive computer simulation system (a Los Angeles neighborhood is shown here) lets people travel through three-dimensional, photo-realistic computer images of their community, suggest changes, and see those changes reflected on the screen. The expectation is that this approach will revolutionize and democratize how communities are planned.

Online with city hall. Burned out street light? Missed garbage pickup? Need a business permit? No need to hang on the phone and be switched from department to department. In a growing movement citizens can use their computers to send messages directly to city hall. In fact, many online hookups are interactive, so that citizens can engage in spirited exchanges with city officials. Some sophisticated systems provide on-screen icons that need only be clicked to engage the proper department. Alas, this service is not free to individual citizens; subscribers usually pay a modest annual fee. However, public access computers are often installed in libraries and schools.

When we look at these principal functions, the kind of personal qualities that are desirable in a systems analyst becomes apparent: An *analytical mind* and *good communication skills*. Perhaps not so obvious, however, are qualities such as *self-discipline* and *self-direction*—a systems analyst often works without close supervision. An analyst must have good *organizational skills* to be able to keep track of all the facts about the system. An analyst also needs *creativity* to envision the new system. Finally, an analyst needs the *ability to work without tangible results*. There can be long dry spells when the analyst moves numbly from meeting to meeting, when it can seem that little is being accomplished.

Let us suppose that you are blessed with these admirable qualities and that you have become a systems analyst. You are given a job to do. How will you go about it?

How a Systems Analyst Works: Overview of the Systems Development Life Cycle

Whether you are investigating how to improve a bank's customer relations, or how to track inventory for a jeans warehouse, or how to manage egg production on a chicken ranch—or any other task—you will proceed by using the **systems development life cycle (SDLC)**. The systems development life cycle can be described in five phases:

1. Preliminary investigation—determining the problem
2. Analysis—understanding the existing system

3. Design—planning the new system
4. Development—doing the work to bring the new system into being
5. Implementation—converting to the new system

These simple explanations for each phase will be expanded to full-blown discussions in subsequent sections; each phase is summarized in Table 8-1. As you read about the phases of a systems project, follow the Swift Sport Shoes inventory case study, which is presented in accompanying boxes. Although space limitations prohibit us from presenting a complete analysis

Table 8-1 Systems development life cycle

Phase	Focus
Phase 1: Preliminary investigation	True nature of problem Problem scope Objectives

Phase 2: Systems analysis	Data gathering
	    Written documents
	    Interviews
	    Questionnaires
	    Observation
	    Sampling
	Data analysis
	    Charts
	    Tables
	System requirements

Phase 3: Systems design	Alternative candidates Output Input Files Processing Controls Backup
Phase 4: Systems development	Programming Testing
Phase 5: Implementation	Training Equipment conversion File conversion System conversion Auditing Evaluation Maintenance

and design project, this case study gives the flavor of the real thing. Let us begin at the beginning.

 ## Phase 1: Preliminary Investigation

The **preliminary investigation**—often called the **feasibility study** or **system survey**—is the initial investigation, a brief study of the problem. It consists of the groundwork necessary to determine if the systems project should be pursued. You, as the systems analyst, need to determine what the problem is and what to do about it. The net result will be a rough plan for how—and if—to proceed with the project.

Before you can decide whether to proceed, you must be able to describe the problem. To do this, you will work with the users. One of your tools will be an **organization chart,** which is a hierarchical drawing showing the organization's management by name and title. Figure 8-2 shows an example of an organization chart. Many organizations already have such a

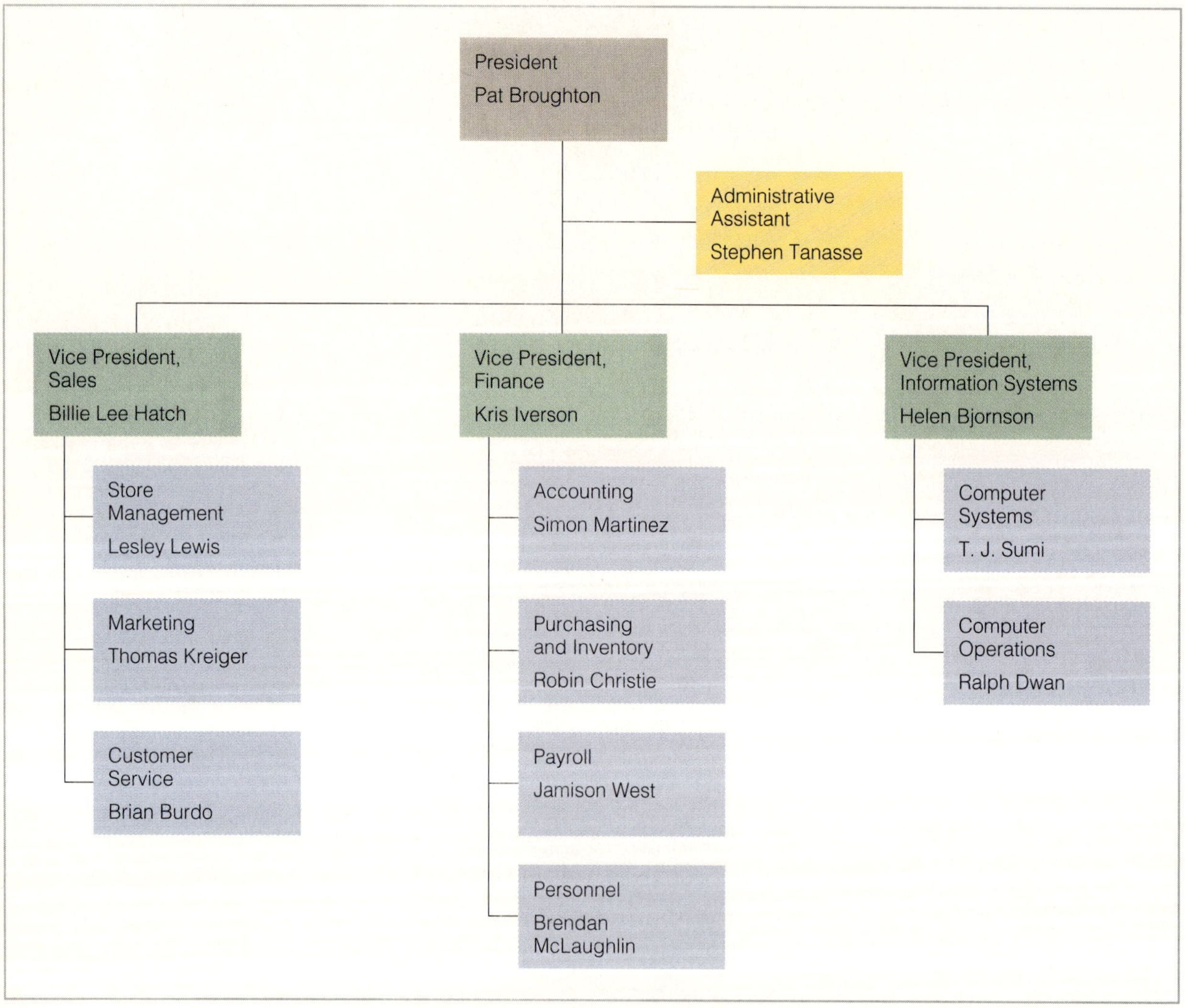

Figure 8-2 An organization chart. The chart shows the lines of authority and formal communication channels. This example shows the organizational setup for Swift Sport Shoes, a chain of stores.

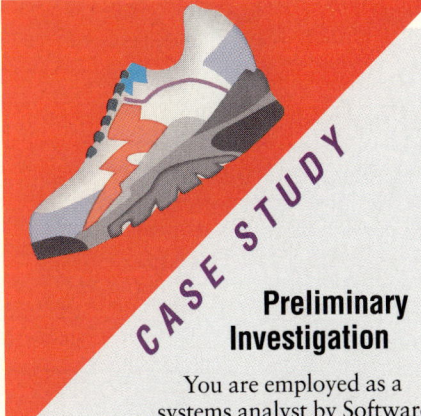

CASE STUDY

Preliminary Investigation

You are employed as a systems analyst by Software Systems, Inc., a company offering packaged and custom software as well as consulting services. Software Systems has received a request for a systems analyst; the client is Swift Sport Shoes, a chain of stores carrying a huge selection of footwear for every kind of sport. Your boss hands you this assignment, telling you to contact company officer Kris Iverson.

In your initial meeting with Mr. Iverson, who is vice president of finance, you learn that the first Swift store opened in San Francisco in 1974. The store has been profitable since the second year. Nine new stores have been added in the city and nearby shopping malls. These stores also show a net profit; Swift has been riding the crest of the fitness boom. But even though sales have been gratifying, Mr. Iverson is convinced that costs are higher than they should be.

In particular, Mr. Iverson is disturbed about inventory problems, which are causing frequent stock shortages and increasing customer dissatisfaction. The company has a superminicomputer at headquarters, where management offices are. Although there is a small information systems staff, their experience is mainly in batch processing for financial systems. Mr. Iverson envisions more sophisticated technology for an inventory system and figures that outside expertise is needed to design it. He introduces you to Robin Christie, who is in charge of purchasing and inventory. Mr. Iverson also tells you that he has sent a memo to all company officers and store managers, indicating the purpose of your presence and his support of a study of the current system. Before the end of your visit with Mr. Iverson, the two of you construct the organization chart shown in Figure 8-2.

In subsequent interviews with Ms. Christie and other Swift personnel, you find that deteriorating customer service seems to be due to lack of information about inventory supplies. Together, you and Ms. Christie determine the problem definition, as shown in Figure 8-4. Mr. Iverson accepts your report, in which you outline the problem definition and suggest a full analysis.

chart and can give you a copy. If the chart does not exist, you must ask some questions and then make it yourself. Constructing such a chart is not an idle task. If you are to work effectively within the organization, you need to understand what are the lines of authority through the formal communication channels.

Problem Definition: Nature, Scope, Objectives

Your initial aim is to define the problem. You and the users must come to an agreement on these points: You must agree on the nature of the problem and then designate a limited scope. In the process you will also determine what the objectives of the project are. Figure 8-3 shows an overview of the problem definition process, and Figure 8-4 gives an example related to the Swift Sport Shoes project.

Nature of the Problem

Begin by determining the true nature of the problem. Sometimes what appears to be the problem turns out to be, on a closer look, only a symp-

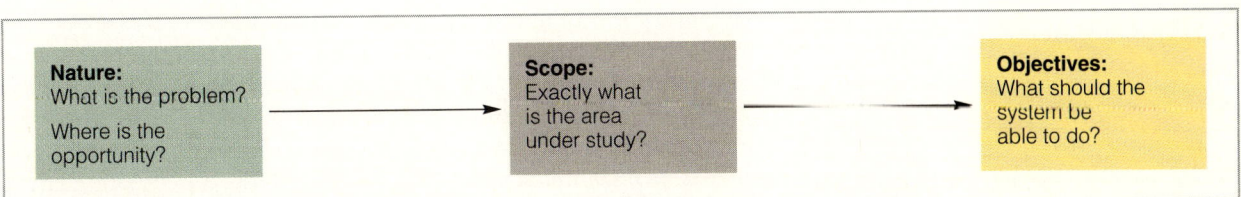

Figure 8-3 Problem definition overview.

> **SWIFT SPORT SHOES: PROBLEM DEFINITION**
>
> True Nature of the Problem
>
> The nature of the problem is the existing manual inventory system. In particular:
>
> – Products are frequently out of stock
>
> – There is little interstore communication about stock items
>
> – Store managers have no information about stock levels on a day-to-day basis
>
> – Ordering is done haphazardly
>
> Scope
>
> The scope of the project will be limited to the development of an inventory system using appropriate computer technology.
>
> Objectives
>
> The new automated inventory system should provide the following:
>
> – Adequate stock maintained in stores
>
> – Automatic stock reordering
>
> – Stock distribution among stores
>
> – Management access to current inventory information
>
> – Ease of use
>
> – Reduced operating costs of the inventory function

Figure 8-4 Problem definition. The nature and scope of the problem along with system objectives are shown for the Swift Sport Shoes system.

tom. For example, suppose you are examining customer complaints of late deliveries. Your brief study may reveal that the problem is not in the shipping department, as you first thought, but in the original ordering process.

Scope

Establishing the scope of the problem is critical because problems tend to expand if no firm boundaries are established. Limitations are also necessary to stay within the eventual budget and schedule. So in the beginning the analyst and user must agree on the scope of the project: what the new or revised system is supposed to do—and not do. If the scope is too broad, the project will never be finished, but if the scope is too narrow, it may not meet user needs.

Objectives

You will soon come to understand what the user needs—that is, what the user thinks the system should be able to do. You will want to express these needs as objectives. Examine the objectives for the Swift inventory process. The people who run the existing inventory system already know what such a system must do. It remains for you and them to work out how this can be achieved on a computer system. In the next phase, the systems analysis phase, you will produce a more specific list of system requirements based on these objectives.

Wrapping Up the Preliminary Investigation

The preliminary investigation, which is necessarily brief, should result in some sort of report, perhaps only a few pages long, telling management what you found and listing your recommendations. Furthermore, money is always a factor in all go/no-go decisions: Is the project financially feasible? At this point management has three choices: They can (1) drop the matter; (2) fix the problem immediately, if it is simple; or (3) authorize you to go on to the next phase for a closer look.

Phase 2: Systems Analysis

Let us suppose management has decided to continue. Remember that the purpose of systems analysis is to understand the existing system. A related goal is to establish the system requirements. The best way to understand a system is to gather all the data you can about it; this data must then be organized and analyzed. During the systems analysis phase, then, you will be concerned with (1) data gathering and (2) data analysis. Keep in mind that the system being analyzed may or may not already be a computerized system.

Data Gathering

Data gathering is expensive and requires a lot of legwork and time. There is no standard procedure for gathering data because each system is unique. But there are certain sources that are commonly used:

- Written documents
- Interviews
- Questionnaires
- Observation
- Sampling

Sometimes you will use all these sources, but in most cases it will be appropriate to use some and not others. All references to data-gathering techniques assume that you have the proper authority and the cooperation of the client organization before proceeding.

Written Documents
These include procedures manuals, reports, forms, and any other kind of material bearing on the problem that you find in the organization. You may find very few documents and no trail to follow. Sometimes the opposite is true: You find so many documents that it is difficult to know how to sift through them. Thus, judgment is required, or you will spend hours reading outdated reports or manuals that no one follows. However, take time to get a copy of each form an organization uses.

Interviews
This method of data gathering has advantages and disadvantages. A key advantage is that interviews are flexible; as the interviewer, you can change the direction of your questions if you discover a productive area of investigation. Another bonus is that you can probe with open-ended questions that people would balk at answering on paper. You will find that some respondents yield more information in an interview than they would if they had to commit themselves in writing. You can also observe the

Some Tips for Successful Interviewing

- Plan questions in advance—even if you vary from them during the interview.
- Dress and behave in a business-like manner.
- Avoid technical jargon.
- Respect the respondent's schedule. Make an appointment; do not just drop in.
- Listen carefully to the answers and observe the respondent's voice inflection and body movements for clues to evaluate responses.
- Avoid office gossip and discussion of the respondent's personal problems.

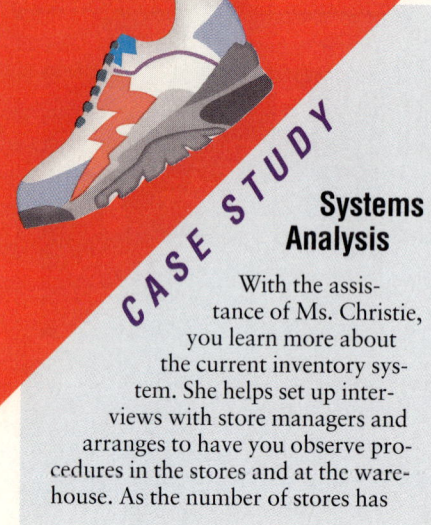

CASE STUDY

Systems Analysis

With the assistance of Ms. Christie, you learn more about the current inventory system. She helps set up interviews with store managers and arranges to have you observe procedures in the stores and at the warehouse. As the number of stores has increased, significant expansion has taken place in all inventory-related areas: sales, scope of merchandise, and number of vendors.

Out-of-stock situations are common. The stock shortages are not uniform across all ten stores, however; frequently, one store will be out of an item that the central warehouse or another store has on hand. The present system is not able to recognize this situation and transfer merchandise. There is a tendency for stock to be reordered only when the shelf is empty or nearly so. Inventory-related costs are significant, especially those for special orders of some stock items. Reports to management are minimal and often too late to be useful. Finally, there is no way to correlate order quantities with past sales records, future projections, or inventory situations.

During this period you also analyze the data as it is gathered. You prepare data flow diagrams of the various activities relating to inventory. Figure 8-6 shows the general flow of data to handle purchasing in the existing system. You prepare various decision tables, such as the one shown in Figure 8-7b.

Your written report to Mr. Iverson includes the list of system requirements in Figure 8-8.

---

respondent's voice inflection and body motions, which may tell you more than words alone. Finally, of course, there is the bonus of getting to know clients better and establishing a rapport with them—an important factor in promoting user involvement in the system from the beginning. Interviews have certain drawbacks. They are unquestionably time-consuming and therefore expensive. You will not have the time or the money to interview large numbers of people. If you need to find out about procedures from 40 mail clerks, for example, you are better off using a questionnaire.

There are two types of interviews, structured and unstructured. A **structured interview** includes only questions that have been planned and written out in advance. The interviewer sticks to those questions and asks no others. A structured interview is useful when it is desirable—or required by law—to ask identical questions of several people. However, the **unstructured interview** is often more productive. An unstructured interview includes prepared questions, but the interviewer is willing to vary from the line of questioning and pursue other subjects if they seem appropriate.

Questionnaires

Unlike interviews, questionnaires can be used to get information from large groups. They allow people to respond anonymously—the respondents just complete forms and turn them in—and presumably, they respond more truthfully. Questionnaires may be used to verify information gathered from interviewing just a few of the users. However, due to the large number of respondents, sometimes a trend or problem pattern emerges that would not be evident from a small number of interviews. Questionnaires do have disadvantages, however. Some people will not return questionnaires because they are wary of putting anything on paper, even anonymously. And the questionnaires you do get back may contain biased answers.

There are many types of questionnaires; the ballot-box type (in which the respondent simply checks off "yes" or "no") and the qualified response (in which one rates agreement or disagreement with the question on a scale from, say, 1 to 5) are two common examples. In general, people prefer a questionnaire that is quick and simple. Analysts also prefer simple questionnaires because their results are easier to tabulate. If you have long, open-ended questions, such as "Please describe your job functions," you should probably save them for an interview.

Observation
As an analyst and observer, you go into the organization and watch who interrelates with whom. In particular, you observe how data flows: from desk to desk, fax to fax, or computer to computer. Note how data comes into and leaves the organization. Initially, you make arrangements with a group supervisor and make everyone aware of the purpose of your visit. Be sure to return on more than one occasion so that the people under observation become used to your presence. One form of observation is **participant observation;** in this form the analyst temporarily joins the activities of the group. This practice is not especially popular but can be useful in studying a complicated organization.

Sampling
You may need to collect data about quantities, costs, time periods, and other factors relevant to the system. How many phone orders can be taken by an order entry clerk in an hour? If you are dealing with a major mail-order organization, such as L. L. Bean in Maine, this type of question may be best answered through a procedure called sampling: Instead of observing all 125 clerks filling orders for an hour, you pick a sample of 3 or 4 clerks. Or, in a case involving a high volume of paper output, such as customer bills, you could collect a random sample of a few dozen bills. Although the actual methods are beyond the scope of this book, we need to mention that there are statistical techniques that can determine exactly what sample size will yield accurate results.

Data Analysis

Your data-gathering processes will probably produce an alarming amount of paper and a strong need to get organized. It is now time to turn your attention to the second activity of this phase, data analysis. What, indeed, are you going to do with all the data you have gathered? There are a variety of tools—charts and diagrams—used to analyze data, not all of them appropriate for every system. You should become familiar with the techniques, then use the tools that suit you at the time. We will consider two typical tools: data flow diagrams and decision tables.

The reasons for data analysis are related to the basic functions of the systems analysis phase: to show how the current system works and to determine the system requirements. In addition, data analysis materials will serve as the basis for documentation of the system.

Data Flow Diagrams
A **data flow diagram (DFD)** is a sort of road map that graphically shows the flow of data through a system. It is a valuable tool for depicting present procedures and data flow. Although data flow diagrams can be used in the design process, they are particularly useful for facilitating communication between you and the users during the analysis phase. Suppose, for example, you spend a couple of hours with a McDonald's franchise manager,

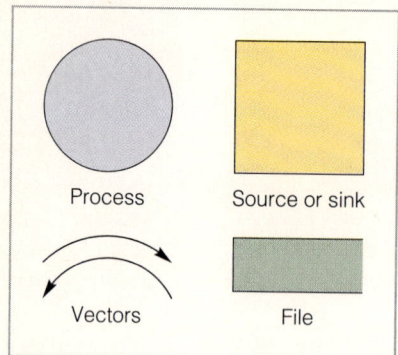

Figure 8-5 Data flow diagram symbols.

talking about the paperwork that keeps the burgers and the customers flowing. You would probably make copious notes about what goes on where. But that is only the data-gathering function—now you must somehow analyze your findings. You could come back on another day with pages of narrative for the manager to review or, instead, show an easy-to-follow picture. Most users would prefer the picture.

There are a variety of notations for data flow diagrams. The notation used here has been chosen because it is informal and easy to draw and read. The elements of a data flow diagram are processes, files, sources and sinks, and vectors, as shown in Figure 8-5. Note also the DFD for Swift Sport Shoes (Figure 8-6) as you follow this discussion.

Processes, represented by circles, are the actions taken on the data—comparing, checking, stamping, authorizing, filing, and so forth. A **file** is a repository of data—a disk file, a set of papers in a file cabinet, or even mail in an in-basket or blank envelopes in a supply bin. In a DFD a file is represented by an open-ended box.

A **source** is a data origin outside the system under study. An example is a payment sent to a department store by a charge customer; the customer is a source of data. A **sink** is a destination for data going outside the system; an example is the bank that receives money deposits from the accounts receivable organization. A source or a sink is represented by a square. **Vectors** are simply arrows, lines with directional notations showing the flow of data. A vector must come from or go to a process circle, or bubble.

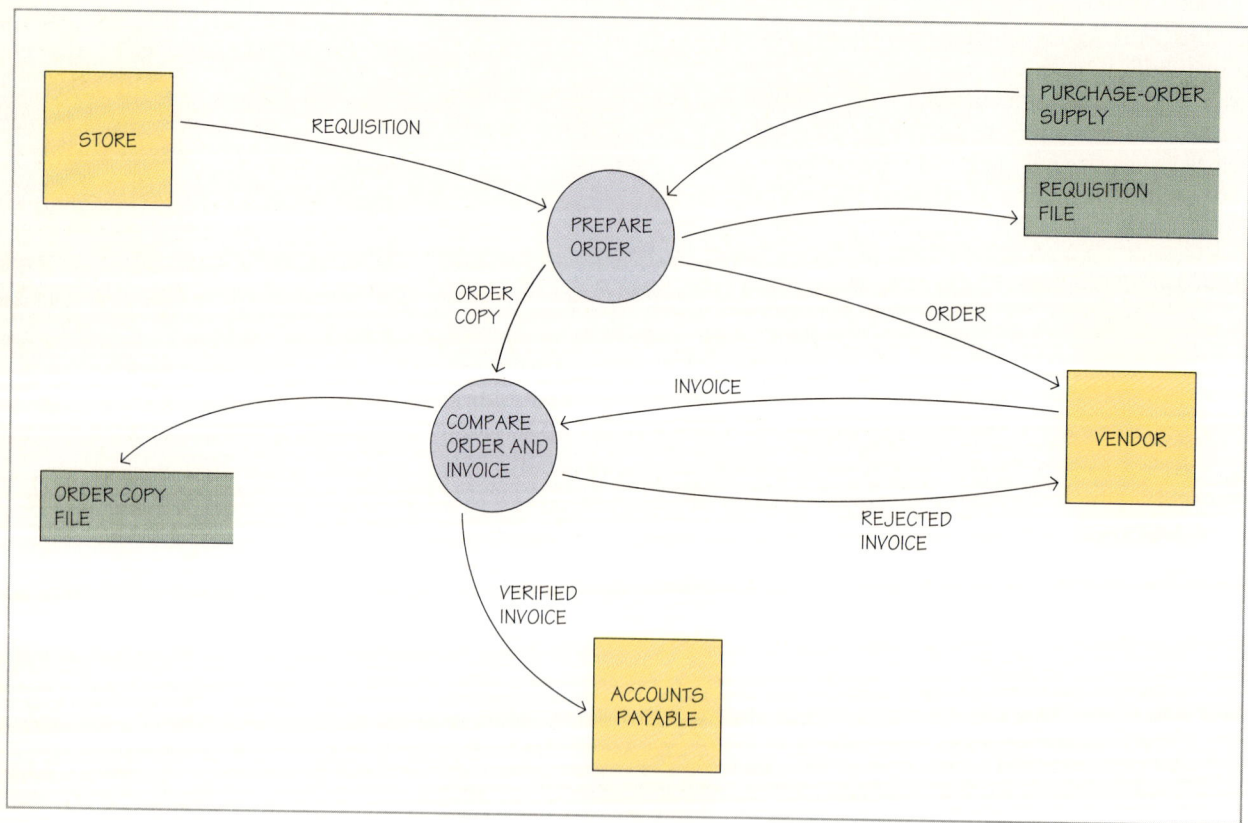

Figure 8-6 A data flow diagram. This "map" shows the current flow of data in the purchasing department at Swift Sport Shoes. The diagram (greatly simplified) includes authorization for purchases of goods, purchase-order preparation, and verification of the vendor's invoice against the purchase order. Note that the stores, vendors, and accounts payable are in square boxes because they are outside the purchasing department.

Decision Tables

A **decision table,** also called a **decision logic table,** is a standard table of the logical decisions that must be made regarding potential conditions in a given system. Decision tables are useful in cases that involve a series of interrelated decisions; their use helps to ensure that no alternatives are overlooked. Programmers can code portions of a program right from a decision table. Figure 8-7a shows the format of a decision table; Figure 8-7b gives an example of a decision table that applies to the Swift Sport Shoes system.

We present data flow diagrams and decision tables here because they are easy to understand. However, they are by no means the only data analysis tools, so you may encounter others.

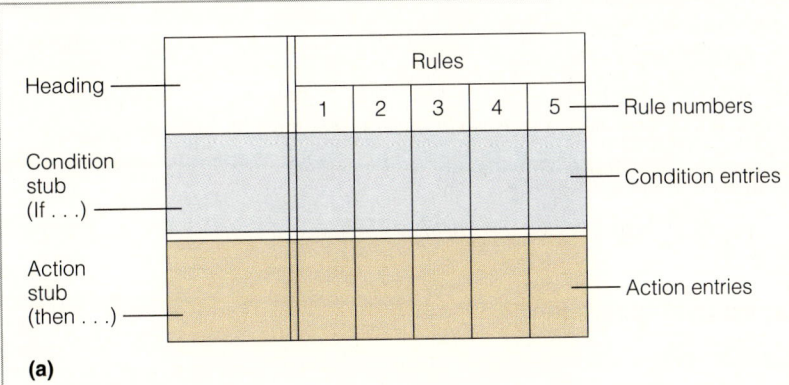

Order procedure	Rules					
	1	2	3	4	5	6
Valid requisition	Y	Y	Y	Y	Y	N
Available warehouse	Y	N	N	N	N	—
Available another store	—	Y	N	N	N	—
Required order volume	—	—	Y	N	N	—
Special customer order	—	—	—	Y	N	—
Transfer goods from warehouse	X					
Transfer goods from store		X				
Determine vendor			X			
Send purchase order			X			
Hold requisition				X	X	
Send back-order notice				X		
Reject requisition						X

(b)

Figure 8-7 Decision tables. (a) The format of a decision table. The table is organized according to the logic that "If this condition exists or is met, then do this." (b) A decision table example. This decision table, which describes the current ordering procedure at Swift Sport Shoes, takes into consideration whether a requisition for goods from a store is valid, the availability of the wanted goods in the warehouse or some other Swift store, whether the quantity ordered warrants an inventory order, and if the order is a special order for a customer. Examine rule 4. The requisition is valid, so we proceed. The desired goods are not available in either the warehouse or in another store, so they must be ordered. However, there is not the required volume of customer demand to place a standard inventory order now, so the requisition is put on hold until there is. (In other words, this order will be joined with others.) And, finally, since this is a special customer order and the order is on hold, a back-order notice is sent.

System Requirements

As we mentioned, the purpose of gathering and analyzing data is twofold: to understand the system and, as a by-product of that understanding, to establish the **system requirements**, a detailed list of the things the system must be able to do. The description of the system was quite broad in the preliminary investigation phase, but now you are ready to list precise system requirements. You need to determine and document specific user needs. A system that a bank teller uses, for example, needs to be able to retrieve a customer record and display it on a screen within five seconds.

If appropriate, you may be able to open the users' eyes to system capabilities they may not have considered. Users who are used to leafing through computer-produced printed reports, for example, might find screen output easier and more timely. On the other hand, as an analyst, you may have to reign in unrealistic user expectations. A user may have seen or heard about computer systems that are fancier or more sophisticated than what circumstances or budget allow.

The importance of accurate requirements cannot be overemphasized, because the design of the new system will be based on the system requirements. Furthermore, the analyst and management must come to clear agreement on the system requirements, since a misunderstanding can result in a poor evaluation of the new system and even cause a delay in project completion. Note the requirements for the Swift system shown in Figure 8-8.

Report to Management

When you have finished the systems analysis phase, you present a report to management. This comprehensive report, part of the continuing process of documentation, summarizes the problems you found in the current system, describes the requirements for the new system, includes a cost analysis, and makes recommendations on what course to take next. If the project is significant, you may also make a formal presentation, including visual displays. If implementing the new system presents significant problems, this might be a good point for management to stop the project, because the investment is still small relative to the amount that will be invested when programming begins. If management decides to pursue the project, you move on to phase 3.

SWIFT SPORT SHOES: REQUIREMENTS

The requirements for the Swift Sport Shoes inventory system are as follows:

- Capture inventory data from sales transactions
- Implement automatic inventory reordering
- Implement a standardized interstore transfer system
- Provide both on-demand and scheduled management reports
- Provide security and accounting controls throughout the system
- Provide a user-oriented system whose online usage can be learned by a new user in one training class
- Reduce operating costs of the inventory function by 20%

Figure 8-8 System requirements. These are the requirements for an inventory system for Swift Sport Shoes.

Phase 3: Systems Design

The systems design phase is the phase in which you actually plan the new system. This phase is divided into two subphases: **preliminary design**, in which the analyst establishes the new system concept, followed by **detail design**, in which the analyst determines exact design specifications. The reason this phase is divided into two parts is that an analyst wants to make sure management approves the overall plan before spending time and money on the details of the new system.

Preliminary Design

The first task of preliminary design is to review the system requirements and then consider some of the major aspects of a system. Should the system be centralized or distributed? Should the system be online? Can the system be run on the users' personal computers? How will input data be captured? What kind of reports will be needed? The questions can go on and on.

A key question that should be answered early on is whether packaged software should be purchased, as opposed to having programmers write custom software. That is, instead of designing, developing, and implementing a new system from scratch, you may be able to obtain an existing system—**acquisition by purchase**—that meets your client's requirements. This may be tricky because clients often think that their problems are unique. However, if the new system falls into one of several major categories, such as accounting or inventory control, then many software vendors offer packaged solutions. A packaged solution should meet at least 75 percent of client requirements. For the remaining 25 percent, the client can adjust ways of doing business to match the package software or, more expensively, **customize**, or alter, the packaged software to meet the client's special needs. Finally, some systems analysts, especially those who work independently as consultants, specialize in certain types of problems—accounting is a good example—and become expert in acquiring and implementing certain commercial software packages to solve those problems.

A related possibility is **outsourcing**, which means turning the system over to an outside agency to develop. Large organizations that employ their own computer professionals may outsource certain projects, especially if the subject matter is one in which a reputable outsourcing firm specializes. The outsourcing company then turns the completed system over to the client. Some organizations outsource most or all of their computer projects, preferring to avoid bearing the costs of keeping their own staff. (In fact, organizations who do not retain their own computer professionals usually outsource the entire project from its inception; this is the case, for example, in the accompanying case study, in which Swift Sport Shoes engages Systems Software, Inc.)

If you proceed with an in-house design, then, together with key personnel from the user organization, you determine an overall plan. In fact, it is common to offer alternative plans, called **candidates**. Each candidate meets the user's requirements but with variations in features and costs. The chosen candidate is usually the one that best meets the user's current needs and is flexible enough to meet future needs. The selected plan is expanded and described so that it can be understood by both the user and the analyst.

At this stage it is wise to make a formal presentation of the selected plan, or possibly all the alternatives. The point is that you do not want to commit time and energy to—nor does the user want to pay for—a detailed design until you and the user agree on the basic design. Such presentations often include a drawing of the system from a user's perspective, such as the one shown in Figure 8-9 for the Swift Sport Shoes system. This is the time to emphasize system benefits—see the list in Figure 8-10.

Prototyping

The idea of building a prototype—a sort of guinea-pig model of the system—has taken a sharp upward turn in popularity recently. Considered from a systems viewpoint, a **prototype** is a limited working system—or subset of a system—that is developed quickly, sometimes in just a few days. A prototype is a working model, one that can be tinkered with and fine-tuned. The idea is that users can get an idea of what the system might be like before it is fully developed. If they are not satisfied, they can revise their requirements before a lot has been invested in developing the new system.

Could you adopt this approach to systems development? It seems at odds with this chapter's systems development life cycle, which promotes doing steps in the proper order. And yet, some analysts in the computer industry are making good use of prototypes. We need to ask how and why. The "how" begins with prototyping tools.

Prototyping Tools

The prototype approach exploits advances in computer technology and uses powerful high-level software tools. These software packages allow analysts to build systems quickly in response to user needs. In particular, recall the fourth-generation languages we discussed in Chapter 6. One of their key advantages is that they can be used to produce something quickly. The systems produced can then be refined and modified as they are used, in a continual process, until the fit between user and system is acceptable.

Why Prototyping?

Many organizations use prototyping on a limited basis. For example, an organization may make a prototype to demonstrate a particular screen output or an especially complex or questionable part of a design. That is, prototyping does not necessarily have the scope of the final system. Some organizations develop throwaway prototypes that they use only to get a grip on the requirements; then they begin again and go through the systems development life cycle formally. Other organizations start with a prototype and keep massaging it until it becomes the final and accepted version. In either case a prototype forces users to get actively involved.

Prototyping is a possibility if you work in an organization that has quick-build software and management support for this departure from traditional systems procedures.

Presentations

Presentations often come at the completion of a phase, especially the analysis and design phases. They give you an opportunity to formalize the project in a public way and to look good in front of the brass. The full range of presentation techniques—using visuals, planning logistics, keeping the audience focused, communicating effectively, and minimizing stage fright—must be topics for another book, but we can consider presentation content here.

- **State the problem.** Although you do not want to belabor the problem statement, you do want to show you understand it.
- **State the benefits.** These are a new system's whole reason for being, so your argument here should be carefully planned. Will the system improve accuracy, speed turnaround, save processing time, save money? The more specific you can be, the better. Use terminology appropriate to your audience; do not lapse into technical jargon.
- **Explain the analysis/design.** Here you should give a general presentation and then be prepared to take questions about details. Remember that higher management will not be interested in hearing all the details.
- **Present a schedule.** How long is it going to take to carry out the plan? Give your audience the time frame.
- **Estimate the costs.** The costs include development costs (those required to construct the system) and operating costs (those ongoing costs of running the system). You will also need to tell your audience how long it is going to be before they get a return on their original investment.
- **Answer questions.** A good rule of thumb is to save half the allotted time for questions.

Chapter Eight ▼ Systems Analysis and Design

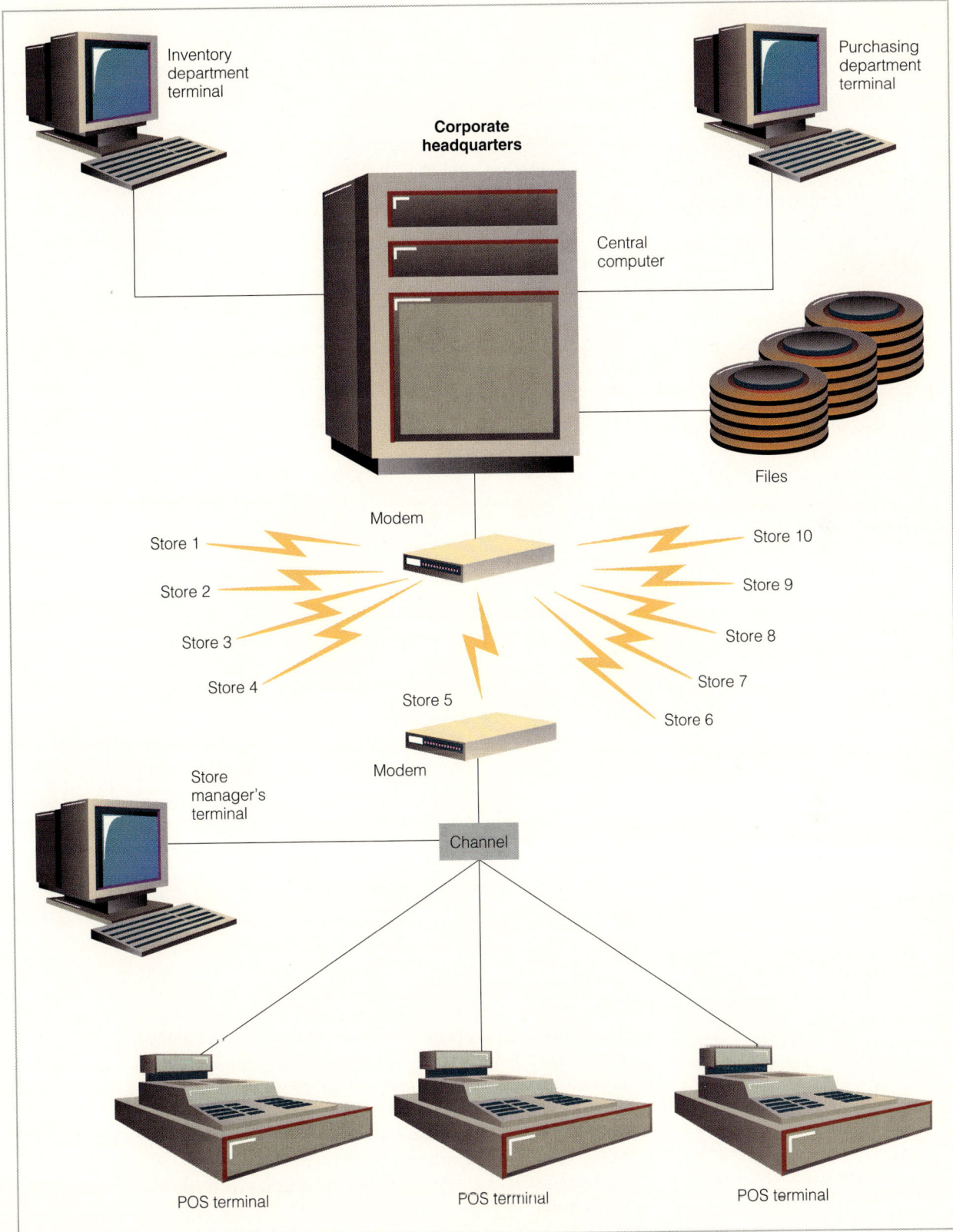

Figure 8-9 Overview of the system. This overview shows the Swift Sport Shoes inventory system from a user's point of view. Input data is from point-of-sale (POS) terminals. Except for local data validation, processing takes place in the central computer. All storage files are located at the central site. Output is in the form of screen displays and printed reports.

CASE STUDY

Systems Design

The store managers, who were uneasy at the beginning of the study, are by now enthusiastic participants in the design of the new system they are counting on for better control of their inventory. As part of the preliminary design phase, you offer three alternative system candidates for consideration.

The first is a centralized system, with all processing done at the headquarters computer and batch reports generated on a daily basis. This system provided little control in the stores and thus was not considered seriously, but it was offered because of its relatively low cost. The third takes the opposite approach, placing all processing in the stores on their own computers. This approach was attractive to the store managers but did not give the headquarters staff as much control or vision as they needed. The second candidate, the one eventually selected, is a networked system that includes processing at the central headquarters site; however, data will be edited locally, at the individual stores, before transmission to the central site. This fairly simple system is appropriate for the size of the organization, with only 10 stores, but will continue to be workable for growth to 20 stores.

The second alternative makes use of point-of-sale (POS) terminals at the store checkout counters, where inventory data is captured as a by-product of the sale. There will be continuous two-way data transmission between the nine stores and the central site. All files will be maintained at the central site. Output will be in two forms: printed reports and on-demand status reports on terminal screens available to store managers locally and to department managers in the headquarters office. Figure 8-9 shows the overall design from a user's viewpoint. The key ingredient of the proposed solution is an automatic reorder procedure: The computer generates orders for any product shown to be below the preset reorder mark.

You make a formal presentation to Mr. Iverson and other members of company management. Slides you prepared on a personal computer (with special presentation software) accent your points visually. After a brief statement of the problem, you list anticipated benefits to the company; these are listed in Figure 8-10. You explain the design in general terms and describe the expected costs and schedules. With the money saved from the reduced inventory expenses, you project that the system development costs will be repaid in three years. Swift Sport Shoes management accepts your recommendations, and you proceed with the detail design phase.

You design printed reports and screen displays for managers; samples are shown in Figures 8-12 and 8-13. There are many other exacting and time-consuming activities associated with detail design. Although space prohibits discussing them, we list some of these tasks here to give you the flavor of the complexity: You must plan the use of wand readers to read stock codes from merchandise tags, plan to download (send) the price file daily to be stored in the POS terminals, plan all files on disk with regular backups on tape, design the records in each file and the methods to access the files, design the data communications system, draw diagrams to show the flow of the data in the system, and prepare structure charts of program modules. Figure 8-15 shows a skeleton version of a systems flowchart that represents part of the inventory processing. Some of these activities, such as data communications, require certain expertise, so you may be coordinating with specialists. Several systems controls are planned, among them a unique numbering system for stock items and validation of all data input at the terminal.

You make another presentation to managers and more technical people, including representatives from information systems. You are given the go-ahead.

Prototype Results

What is the net result of making a prototype of a system? What will it produce for users? A prototype of a whole system will initially include minimum input data, no validation checks, incomplete files, limited security checks, sketchy reports, and minimum documentation. But actual software uses real data to produce real output. Remember that prototyping is an iterative process; the system is changed again and again based on the lessons learned by creating the prototype.

The computer industry is looking beyond prototyping to a future using CASE tools.

CASE Tools

CASE tools turn traditional systems approaches upside down. The set of software known as **CASE**—for **computer-aided software engineering**—

SWIFT SPORT SHOES: ANTICIPATED BENEFITS
–Better inventory control
–Improved customer service
–Improved management information
–Reduced inventory costs
–Improved employee morale

Figure 8-10 Benefits. Benefits are usually closely tied to the system objectives. These are the anticipated benefits of the new Swift Sport Shoes inventory system.

tools goes beyond the concept of prototyping and has become a significant factor in the development of systems. CASE tools provide an automated means of designing and changing systems (Figure 8-11). In fact, integrated CASE tools can automate most of the systems development life cycle.

CASE software is available for personal computers. The systems analyst can produce designs right on the computer screen. Thus, a key ingredient of a package of CASE tools is a graphics interface. What is more, that screen is usually part of a personal computer. Other important CASE ingredients are a data store—often called a data dictionary or even an encyclopedia—and the ability to generate a program automatically, right from the design.

CASE tools have several advantages. Foremost is the ability to note inconsistencies in the system design. Such tools can also make global changes related to a single change; for example, if you change a name in one place, the CASE tool automatically changes it throughout the design specifications. CASE tools provide consistency, speed, increased productivity, and cost savings.

CASE tools, however, are not cure-alls. To begin with, they really only have value for new systems; an estimated 80 percent of computer organization time is devoted to the maintenance of existing systems. Also, CASE standards have not been established, and the result is a hodgepodge of methodologies from a variety of vendors. Keep in mind, also, that no CASE tool, or any other tool, will help if you do not know what you are doing. Good tools do not necessarily create good systems.

Detail Design

Let us say that the users have accepted your design proposal—you are on your way. You must now develop detailed design specifications, or a detail design. This is a time-consuming part of the project, but it is relatively straightforward.

In this phase every facet of the system is considered in detail. Here is a list of some detail design activities: designing output forms and screens, planning input data forms and procedures, drawing system flowcharts, planning file access methods and record formats, planning database interfaces, planning data communications interfaces, designing system security controls, and considering human factors. This list is not comprehensive, nor will all activities listed be used for all systems. Some analysts choose to plan the overall logic at this stage, preparing program structure charts, pseudocode, and the like.

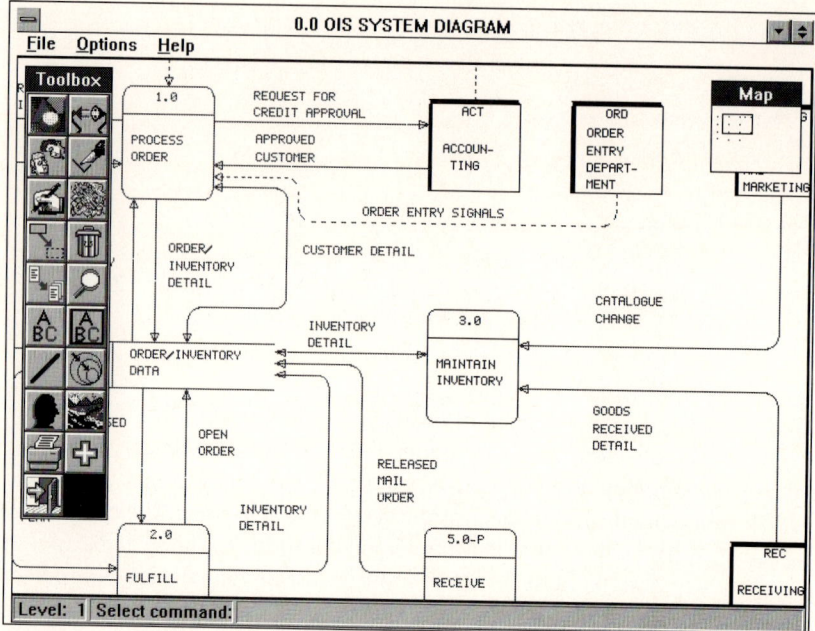

Figure 8-11 CASE tools. Here is a data flow diagram produced by Excelerator, a commercially available CASE package. Notice that Excelerator uses rounded squares for data flow diagram processes.

Normally, in the detail design phase, parts of the system are considered in this order:

- Output requirements
- Input requirements
- Files and databases
- Systems processing
- Systems controls and backup

Output Requirements

Before you can do anything, you must know exactly what the client wants the system to produce—the output. As an analyst, you must also consider the *medium* of the output—paper, computer screen, microfilm, and so on. In addition, you must determine the *type* of reports needed (summary, exception, and so on) and the *contents* of the output—what data is needed for the reports. What *forms* the output will be printed on is also a consideration; they may need to be custom-printed if they go outside the organization to customers or stockholders. You may wish to determine the report format by using a **printer spacing chart**, which shows the position of headings, the spacing between columns, and the location of date and page numbers (Figure 8-12). You may also use screen reports, mock-ups on paper of how the screen will respond to user queries. A sample screen report is shown in Figure 8-13.

Input Requirements

Once your desired output is determined, you must consider what kind of input is required to produce it. First you must consider the input *medium:* Will you try to capture data at the source via point-of-sale (POS) terminals? Must the input be keyed from a source document? Next you must consider *content* again—what fields are needed, the order in which they

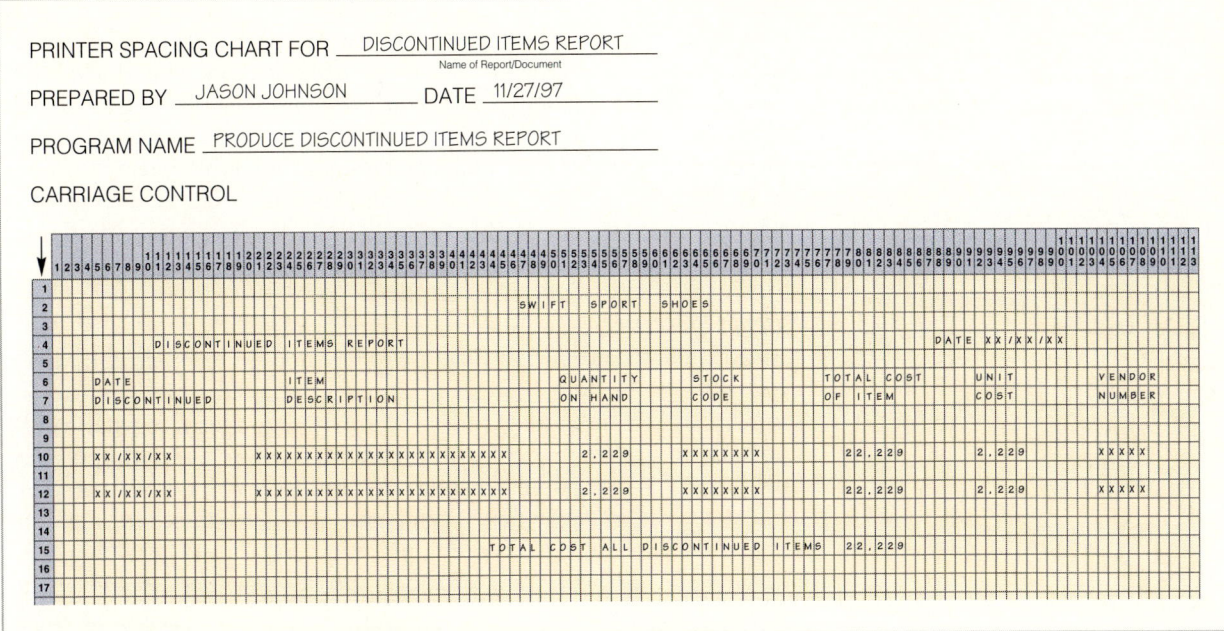

Figure 8-12 Example of a printer spacing chart. This chart shows how a systems analyst wishes the report format to look—headings, columns, and so on—when displayed on a printer. This example shows discontinued items, a report that is part of the new Swift Sport Shoes system. Xs represent alphabetic data, and 9s represent numeric data.

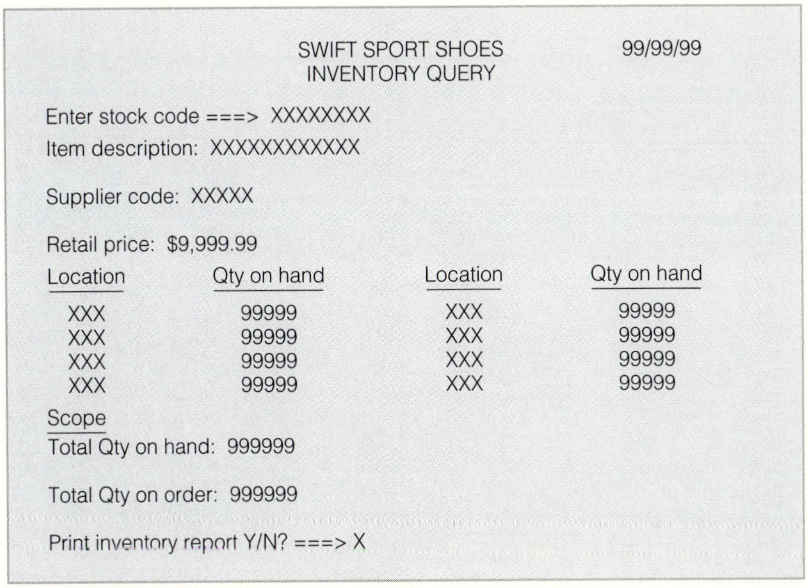

Figure 8-13 Example of a screen report. This screen report layout has been designed as part of the Swift Sport Shoes system. The purpose of the screen is to give information about how much of a given stock item is in each store. The report shows an approximation of what the user will see on the screen after entering a stock code.

come, and the like. This in turn may involve designing *forms* that will organize data before it is entered. You need to plan some kind of input *validation* process, a check that data is reasonable as well as accurate—you would not expect a six-figure salary, for example, for someone who works in the mail room. Finally, you need to consider input *volume*, particularly the volume at peak periods. Can the system handle it? A mail-order house, for instance, may have to be ready for higher sales of expensive toys in the December holiday season than at other times of the year.

Files and Databases

You need to consider how the files in your computer system will be organized: sequentially, directly, or by some other method. You also need to decide how the files should be accessed. They might be organized as indexed files but be accessed directly or sequentially, for example. You need to determine the format of records making up the data files. If the system has one or more databases, or accesses databases used in other systems, then you will have to coordinate your design efforts with the database administrator, the person responsible for controlling and updating databases.

Systems Processing

Just as you drew a data flow diagram to describe the old system, now you need to show the flow of data in the new system. One method is to use standard ANSI flowchart symbols (Figure 8-14) to illustrate what will be done and what files will be used. Figure 8-15 shows a resulting **systems flowchart.** Another popular way to describe processing is the structure chart described in Appendix A. Note that a systems flowchart is not the same as the logic flowchart used in programming. The systems flowchart describes only the big picture; a logic flowchart represents the flow of logic within a single program.

Systems Controls and Backup

To make sure data is input, processed, and output correctly and to prevent fraud and tampering with the computer system, you will need to institute appropriate controls. In a batch system, in which data for the system is

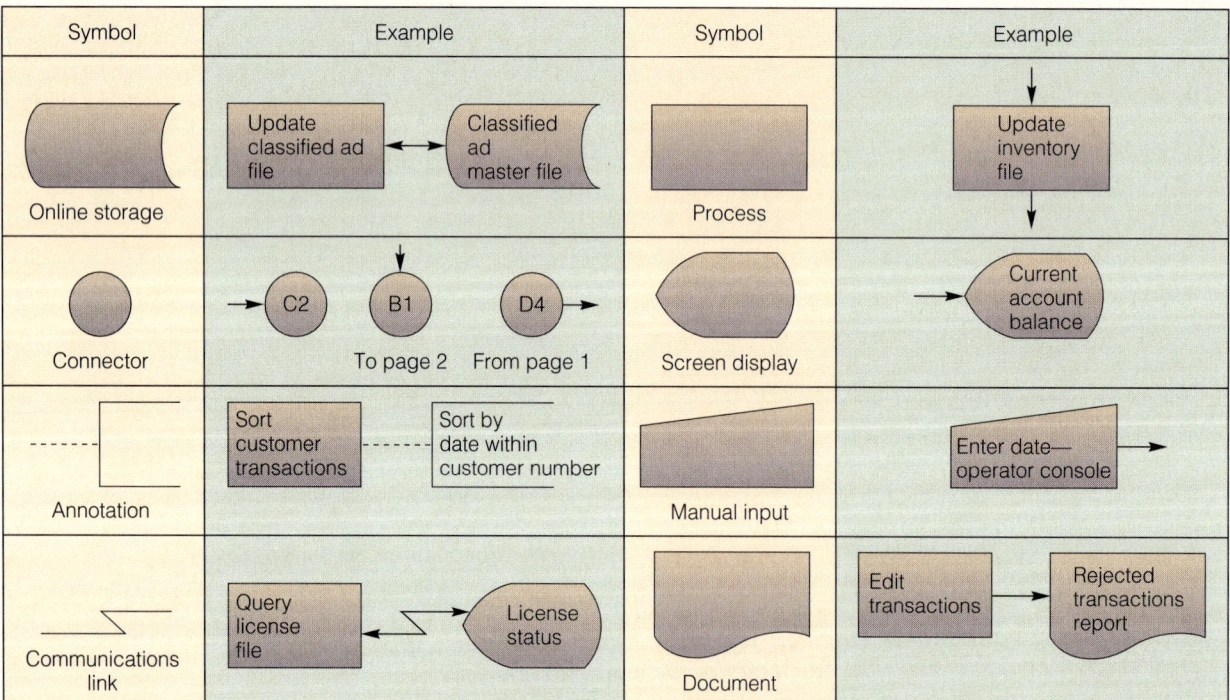

Figure 8-14 ANSI systems flowchart symbols. These are some of the symbols recommended by the American National Standards Institute for systems flowcharts, which show the movement of data through a system.

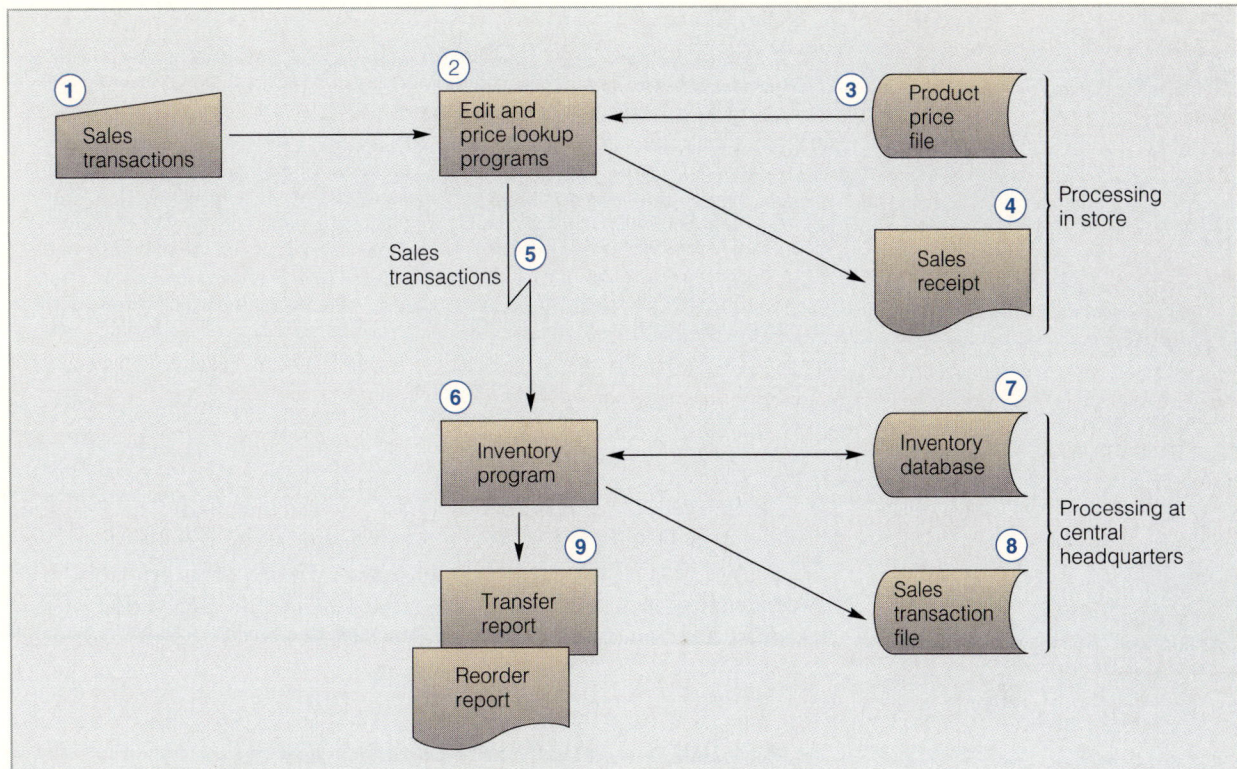

Figure 8-15 Systems flowchart. This very simplified systems flowchart shows part of the processing for the new Swift Sport Shoes inventory system. Note that the top half of the drawing shows processing that occurs in the store. The processing takes place in a POS terminal while the customer waits. The bottom part of the drawing shows processing that is done on the computer at the central headquarters site. The clerk ① inputs sales transaction data, which ② is edited by the POS terminal processor. The POS terminal also looks up the item price from the ③ files downloaded earlier in the day from the central site, then ④ prints a sales receipt. That takes care of the customer. Meanwhile, ⑤ the sales transaction data is sent over data communications lines to the central computer, which ⑥ processes it for inventory purposes by updating the ⑦ inventory database, placing the ⑧ sales transaction in its own file for later auditing and for producing ⑨ transfer and reorder reports as needed.

processed in groups, begin with the source documents, such as time cards or sales orders. Each document should be serially numbered so the system can keep track of it. Documents are time-stamped when received and then grouped in batches. Each batch is labeled with the number of documents per batch; these counts are balanced against totals of the processed data. The input is controlled to make sure data is accurately converted from source documents to machine-processable form. Data input to online systems is backed up by **system journals,** files that record every transaction processed at each terminal, such as an account withdrawal through a bank teller. Processing controls include the data validation procedures we mentioned in the section on input requirements.

It is also important to plan for the backup of system files; copies of transaction and master files should be made on a regular basis. These file copies are stored temporarily in case the originals are inadvertently lost or damaged. Often the backup copies are stored off site for added security. We will focus on security in Chapter 11.

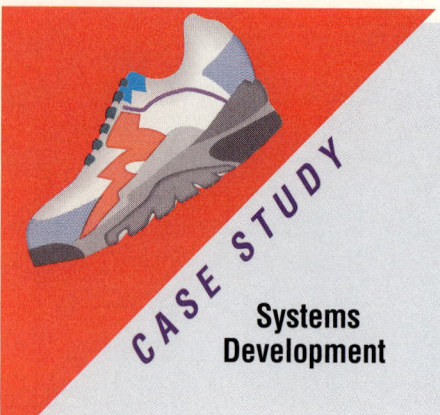

CASE STUDY
Systems Development

Working with Dennis Harrington of the information systems department, you prepare a Gantt chart, as shown in Figure 8-16. This chart shows the schedule for the inventory project. Program design specifications are prepared using pseudocode, the design tool Mr. Harrington thinks will be most useful to programmers. The programs will be written in C++, since that is the primary language of the installation and it is suitable for this application. Three programmers are assigned to the project.

You work with the programmers to develop a test plan. Some inventory data, both typical and atypical, is prepared to test the new system. You and the programmers continue to build on the documentation base by implementing the pseudocode and by preparing detailed data descriptions, logic narratives, program listings, test data results, and related material.

As before, the results of this phase are documented. The resulting report, usually referred to as the detail design specifications, is an outgrowth of the preliminary design document. The report is probably large and detailed. A presentation often accompanies the completion of this stage. Unless something unexpected has happened, it is normal to proceed now with the development of the system.

 Phase 4: Systems Development

Finally, the system is actually going to be developed. As a systems analyst you prepare a schedule to monitor the principal activities in **systems development**—programming and testing.

Scheduling

Figure 8-16 shows what is known as a **Gantt chart**, a bar chart commonly used to depict schedule deadlines and milestones. In our example the chart shows the work to be accomplished over a given period. It does not, however, show the number of work hours required. If you were the supervisor,

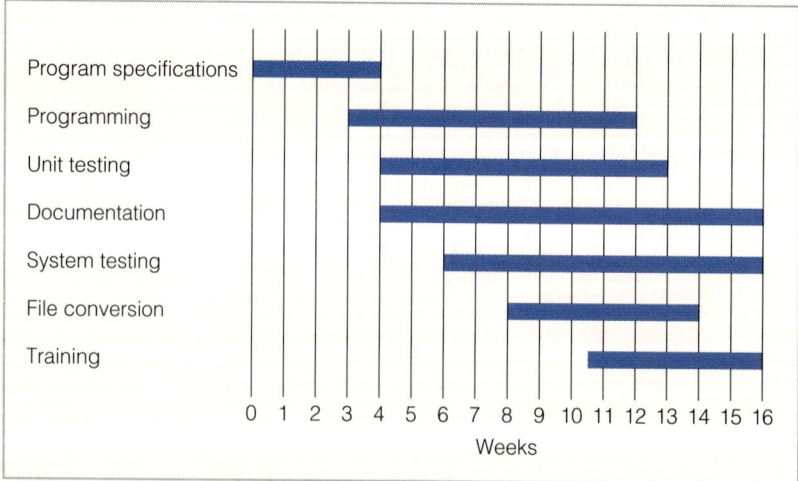

Figure 8-16   Gantt chart. This bar chart shows the scheduled tasks and milestones of the Swift Sport Shoes project. Notice that some phases overlap.

it would be common practice for you to ask others on the development team to produce individual Gantt charts of their own activities.

A Gantt chart is a simple device, easy to understand and implement. However, organizations that want more control use **project management software,** which offers additional features, such as allocating people and resources to each task, monitoring schedules, and producing status reports (Figure 8-17). In particular, project management software tracks the impact of schedule slides on future dependent activities. For example, if a program will depend on the design of certain files, then a delay in file

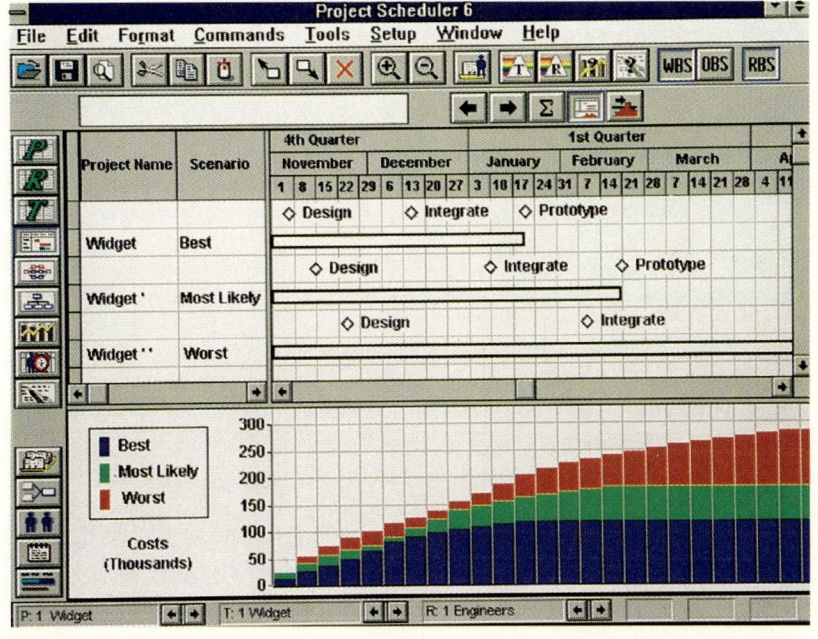

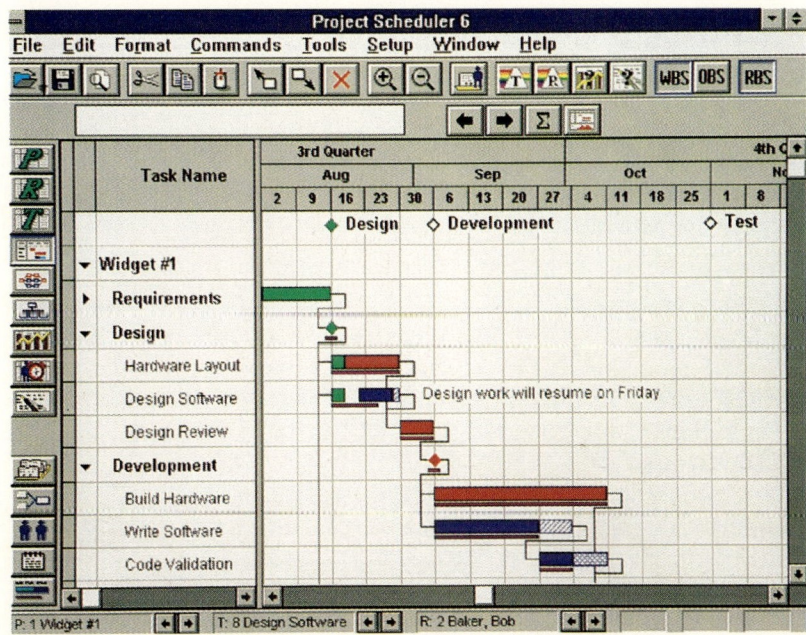

Figure 8-17 Project management software. (a) This chart shows both the planned schedule and anticipated costs. (b) This chart shows phases of a project in various stages of completion.

Programming

Until this point there has been no programming, that is, unless some prototyping was done. So, usually, before programming begins, you need to prepare detail design specifications. Program development tools must be considered. Some of this work may already have been done as part of the design phase, but usually programmers participate in refining the design at this point. Design specifications can be developed through detailed logic flowcharts and pseudocode, among other tools.

Testing

Would you write a program and then simply turn it over to the client without checking it over first? Of course not. Thus, programmers perform **unit testing,** by which they individually test their own program pieces (units), using test data. Programmers try even bad data so that they can be confident that their program can handle it appropriately. This is followed by **system testing,** which determines whether all the program units work together satisfactorily. During this process the development team uses test data to test every part of the programs. Finally, **volume testing** uses real data in large amounts. Volume testing sometimes reveals errors that do not show up with test data, especially errors in storage or memory usage. In particular, volume testing of online systems will reveal problems that are likely to occur only under heavy usage.

As in every phase of the project, documentation is required. Indeed, documentation is an ongoing activity (as the Gantt chart in Figure 8-16 shows). In this phase documentation describes the program logic and detailed data formats.

Phase 5: Implementation

You may think that implementation means stopping the old system and starting the new system. You are not alone. Many companies believe that also, but they find out that there is much more to it. Even though **implementation** is the final phase, a good deal of effort is still required, including the following activities:

- Training
- Equipment conversion
- File conversion
- System conversion
- Auditing
- Evaluation
- Maintenance

Training

Often systems analysts do not give training the attention it deserves, because they are so concerned about the computer system itself. But a system can be no better than the people using it. A good time to start training

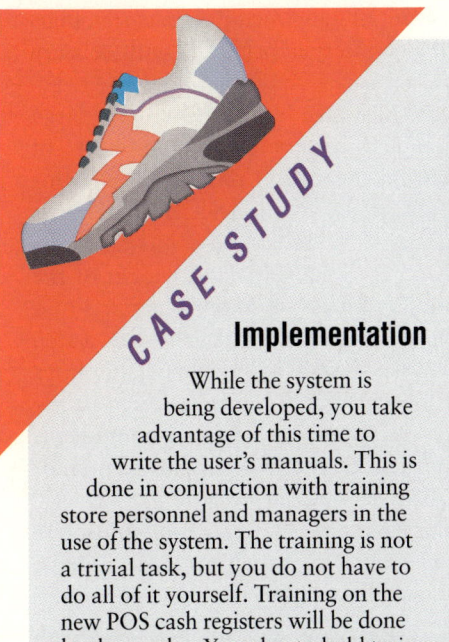

CASE STUDY

Implementation

While the system is being developed, you take advantage of this time to write the user's manuals. This is done in conjunction with training store personnel and managers in the use of the system. The training is not a trivial task, but you do not have to do all of it yourself. Training on the new POS cash registers will be done by the vendor. You plan to hold training classes for the people who will use the local micros to run programs and send data to the computer at headquarters. You will have separate classes to teach managers to retrieve data from the system via terminal commands. In both cases training will be hands-on. Company personnel should find the training enjoyable because the on-screen dialogue is user friendly—the user is instructed clearly every step of the way.

File conversion is painful. One evening after closing time, the staff works into the evening to take inventory in the stores. Temporary personnel are hired to key an inventory master file from this data. Transactions for the master file are accumulated as more purchases are made, up until the time the system is ready for use; then the master will be updated from the transactions generated by the POS terminals. After discussing the relative merits of the various system conversion methods, you and Ms. Christie agree that a pilot conversion would be ideal. Together you decide to bring up the original store first, then add other stores to the system one or two at a time.

To evaluate the new system, Mr. Iverson puts together a local team consisting of Ms. Christie, a programmer, and an accountant. Since your documentation is comprehensive, it is relatively easy for the team to check the system completely to see if it is functioning according to specifications. The evaluation report notes several positive results: out-of-stock conditions have almost disappeared (only two instances in one store in one month), inventory transfer among stores is a smooth operation, and store managers feel an increased sense of control. Negative outcomes are relatively minor and can be fixed in a system maintenance operation.

—for at least a few of the users—is at some point during the testing, so that people can begin to learn how to use the system even as the development team is checking it out. Do not be concerned that these users will see a not-yet perfect system; users actually gain confidence in a budding system as errors get fixed and the system improves every day.

An important training tool is the user's manual, a document prepared to aid users not familiar with the computer system. The user's manual can be an outgrowth of the other documentation. Some organizations employ technical writers to create the user's manual while the system is being developed. But documentation for the user is just the beginning. Any teacher knows that students learn best by doing. Besides, users are as likely to read a thick manual as they are to read a dictionary. The message is clear: Users must receive hands-on training to learn to use the system. The trainer must prepare exercises that simulate the tasks users will be required to do. For example, a hotel clerk learning a new online reservation system is given typical requests to fulfill—a family of four for three nights—and uses a terminal to practice. The user's manual is used as a reference guide. Setting all this up is not a trivial task. The trainer must consider class space, equipment, data, and the users' schedules.

Equipment Conversion

Equipment considerations vary from almost none to installing a main frame computer and all its peripheral equipment. If you are implementing a small- or medium-size system on established equipment in a major information systems department, then perhaps your equipment considerations will involve no more than negotiating scheduled run time and disk space.

If you are purchasing a moderate amount of equipment, such as terminals or personal computers, then you will be concerned primarily with delivery schedules and compatibility.

A major equipment purchase, on the other hand, demands a large amount of time and attention. The planning for such a purchase, of course, must begin long before the implementation phase. For a major equipment purchase you will need site preparation advice from vendors and other equipment experts. You may be considering having walls moved! You will need to know the exact dimensions and weight of the new equipment to fit it through doors and locate it in various parts of a building. You may need to protect the systems from damage by dirt, water, and fire. You may need to consider electrical capacity and wiring hookups as well as new flooring—probably flooring that is raised to hide cabling and ease access for repairs to large computers and related equipment. Finally, most medium to large machines need air conditioning and humidity control.

Personal computer systems are less demanding, but they too require site planning in terms of the availability of space, accessibility, and cleanliness. And, as the analyst, you may be the one who does the actual installation.

File Conversion

This activity may be very tricky if the existing files are handled manually. The data must be prepared in such a way that it is accessible to computer systems. All the contents of the file drawers in the personnel department, for instance, must now be keyed, or possibly scanned, to be stored on disk. Some scheme must be used to input the data files and keep them updated. You may need to employ temporary help. The big headache during this process is keeping all file records up-to-date when some are still manual and some have been keyed in preparation for the new system.

If you are modifying an existing computer system and thus have files already in computer-accessible form, you may need to have a program written to convert the old files to the format needed for the new system. This is a much speedier process than having to key in data from scratch. Nevertheless, it is not unusual for file conversion to take a long time.

System Conversion

This is the stage in which you actually "pull the plug" on the old system and begin using the new one. There are four ways of handling the conversion.

Direct conversion means the user simply stops using the old system and starts using the new one—a somewhat risky method, since there is no other system to fall back on if anything goes wrong. This procedure is best followed only if the old system is in unusable condition. A **phased conversion** is one in which the organization eases into the new system one step at a time so that all the users are using some of the system. In contrast, in a **pilot conversion** the entire system is used by some of the users and is extended to all users once it has proved successful. This is most useful when a company has several branch offices or separate divisions. In **parallel conversion**—the most prolonged and expensive method—the old and new systems are operated simultaneously for some time, until users are satisfied that the new system performs to their standards.

System conversion is often a time of stress and confusion for all concerned. As the analyst, your credibility is on the line. During this time users

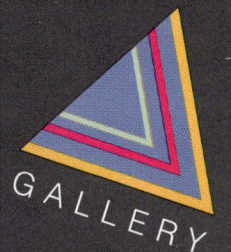

COMPUTER GRAPHICS

GALLERY

The fanciful computer artwork that opens this gallery, called Bladud, the Vain Prince, was inspired by illuminated Celtic scriptures of Ireland. The computer artist is Ron Richey. The computer artworks on this page were created by various artists.

GALLERY

4

5

GALLERY

Women graphic artists used to be rare. Today computer art by women can be found in every aspect of the graphics universe, from magazine covers to posters to works that adorn corporate walls. On these two pages we will recognize the artists by name.

7. Helen Golden calls this intriguing computer-produced artwork The White Door.
8. Karin Schminke named this computer artwork Nemus.
9. Dorothy Krause, a professor of computer graphics at the Massachusetts College of Art, calls this work White Dove.
10. Pamela Hobbs made this computer-produced poster promoting a Tori Amos concert.
11. Diane Fenster produced this work called Falling in Love Online.
12. Bonny Lhotka, whose works hang in corporate halls, begins a computer work by assembling miscellaneous materials from her art lab and scanning them into the computer where she completes the work. The final result may later be blown up to a large size. This work, called XI, is 100x52 inches.
13. Wendy Grossman features pyramids, pharoahs, and camels in this fascinating computer artwork.

7

8

9

10

11

GALLERY

Whether the colorful work is in a magazine, on a poster, or on television, chances are that it was created on a computer by a computer graphics artist. Here are some samples of their works.

14. A dramatic Academy Awards screen is prepared for the annual presentation.
15. DELL Computer features a different artist each time they put together a new advertising layout. This ad was done by artist Wendy Grossman.
16-17-18. These logos were all computer-created.
19. Even toothbrush commercials get the computer graphics treatment.
20. This scene was created on the computer for use on a packaging box for a toy cycle.
21. This computer work was created for the cover of *Apple DEVELOP* magazine.
22. This colorful corporate identification was created for a retail leisure clothing company.

14

15

16

17

18

19

20

21

22

GALLERY

23

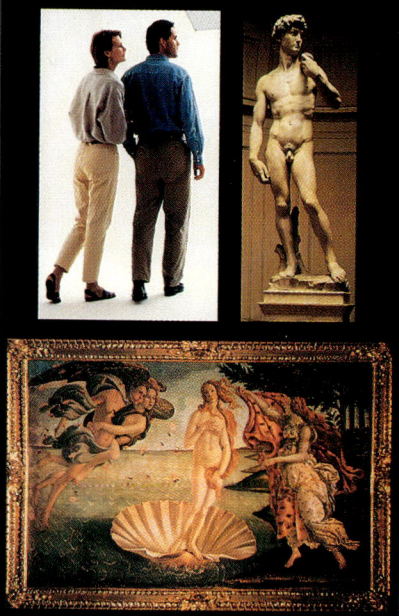
24

25

The works on this page show how photos can be manipulated by the computer. Begin with 23, a photo of a building interior. Then consider 24, photos of strolling tourists, a statue, and a painting. These four photos have been scanned into the computer and manipulated to become 25, a museum with artworks and tourists to view them. Note, in particular, the adjusted shape of the painting and the computer artist's addition of clouds in the skylight.

26. Here is the intriguing result of computer imaging of four photos. The original photos were of trees, a sunset, a swan, and a red world logo.

27. The artist has produced various computer-manipulated versions of an original photo of a child.

26

27

MAKING THE RIGHT CONNECTIONS

Unknown Consequences

Historically speaking, a technology breakthrough that holds the promise of wonderful benefits for society carries with it unimaginable and sometimes undesirable changes as well. Television, for example, disseminates entertainment and information but also brings violence into the home and encourages mindless couch potatoing. Similarly, the telephone gives us easy communication but also brings phone solicitations during dinner and obscene callers in the night.

We already know some unpleasant consequences of networking, including junk e-mail, anonymous users representing themselves falsely, and even digitized pornography. We can only guess at yet unknown consequences. Perhaps, at some future time when we are all online, we will become so busy pounding the keyboard that no one will be listening. Some social commentators worry that, ironically, connected users may become *dis*connected from society as the need for face-to-face contact diminishes. Perhaps anonymous superficial relationships will, for some people, replace the real thing.

are often doing double duty, trying to perform their regular jobs and simultaneously cope with a new computer system. Problems seem to appear in all areas, from input to output. Clearly, this is a period when your patience is needed.

Auditing

Security violations, whether deliberate or unintentional, can be difficult to detect. Data begins from some source, perhaps a written source document or a transaction, for which there must be a record log. Eventually, the data —no matter how it was originally keyed—is part of the system on some media, probably disk. Once the data is on disk, it is possible for an unauthorized person to alter it in some illicit way. How would anyone know that the disk files had been changed and, in fact, no longer match the original source documents from which the data came? To guard against this situation, the systems analyst designs an **audit trail** to trace output back to the source data. In real-time systems security violations can be particularly elusive unless all transactions are recorded on disk for later references by auditors. Modern auditors no longer shuffle mountains of paper; instead, they have computer programs of their own to monitor applications programs and data.

Evaluation

Is the system working? How well is it meeting the original requirements, budgets, schedules, and so forth? Out of such evaluation will come adjustments that will improve the system. Approaches to evaluation vary. Sometimes the systems analyst and someone from the client organization evaluate the system against preset criteria directly related to the requirements that were determined during the systems analysis phase. Some organizations prefer to bring in an independent evaluating team on the assumption that independent members will be free from bias and expectations.

Maintenance

Many consider maintenance to be a separate phase, one that begins only when the initial system effort is implemented and complete. In any case **maintenance** is an ongoing activity, one that lasts the lifetime of the system. Monitoring and necessary adjustments continue so that the computer produces the expected results. Maintenance tasks also include making revisions and additions to the computer system.

Students who hope to become programmers may envision themselves writing programs for new systems, but they are more likely, at least at first, to be maintaining programs in existing systems. This type of work has advantages and disadvantages. It can be an advantage to learn from programs written by others but, unless strict programming procedures have been in place since the program's inception, a distinct disadvantage to unravel layered changes added by many programmers. Some maintenance programmers find satisfaction in having a deep understanding of a complicated system, so that everyone knows they can fix any problem. Contrast this with participating in the development of a new system, which has unknown perils at every turn and is scrutinized regularly for schedule and budget overruns.

As more computer systems are implemented, organizations obviously have an increased number of systems to maintain. As we have noted, an estimated 80 percent of a computer organization's time is devoted to the maintenance of existing systems; thus, in most computer installations, a high percentage of personnel are dedicated to maintenance. This necessarily limits the number of personnel available for developing new systems. The net result is often a backlog of development projects.

Putting It All Together: Is There a Formula?

The preceding discussion may leave the impression that, by simply following a recipe, a system can be developed. In fact, novice analysts sometimes have the impression that there is a formula for developing systems. Each system is unique, however, so no one formula can fit every project. It would be more correct to say that there are guidelines.

Historically, even analysts who followed the guidelines were not always successful in developing systems. Systems analysts have been embarrassed to find that they were not always good at estimating time, so schedules constantly slipped. (Budget overruns are one of the obvious results of sliding schedules.) Some observers, in fact, think that systems analysis is so ambiguous that analysts do not even know when they are finished. Sometimes it seems that the definition of project completion is the point at which analysts have run out of time on the schedule.

Another frequent problem has been imperfect communication between analysts and users. Poor communication results in poorly defined specifications, which, in turn, results in a supposedly complete system that does not do what the user expects. In addition, by-guess-and-by-gosh methods of analysis and design have often been used instead of formal tools.

Out of these experiences, however, have come some solutions. Managers have become more sophisticated—and more realistic—in planning schedules and budgets. Analysts have learned to communicate with users and to recognize the cyclic nature of the systems analysis process. In addition to the analysis and design approach described here (which is considered the traditional way of creating a system), there are other, newer approaches, which are beyond the scope of this book. If you pursue a career in systems analysis, you will no doubt encounter these approaches and find them useful.

Being a systems analyst can be important work. An analyst is in a position to help institute fundamental changes that alter business operations, work habits, and use of time. As we suggested at the beginning of this chapter, however, a systems analyst must be sensitive to the possible effects of his or her work on people's lives. The real danger, it has been remarked, is not that computers will begin to think like people, but that people will begin to think like computers.

▼ ▲ ▼

This chapter has addressed a broad spectrum of systems change, taking into account its effects on the entire organization. In the next chapters we will consider the entire business picture and the place of the computer in that picture.

CHAPTER REVIEW

Summary and Key Terms

- A **system** is an organized set of related components established to accomplish a certain task. A **computer system** has a computer as one of its components. A **client** requests a **systems analysis**, a study of an existing system, to determine both how it works and how well it meets the needs of its **users**, who are usually employees and customers. Systems analysis can lead to **systems design**, the development of a plan for an improved system. A **systems analyst** normally does both the analysis and design. Some people do both programming and analysis and have the title **programmer/analyst**. The success of the project requires both impetus and authority within the client organization to change the current system.
- The systems analyst must be a **change agent** who encourages **user involvement** in the development of a new system.
- The systems analyst has three main functions: (1) **coordinating** schedules and task assignments, (2) **communicating** analysis and design information to those involved with the system, and (3) **planning and designing** the system, with the help of the client organization. A systems analyst should have an analytical mind, good communication skills, self-discipline and self-direction, good organizational skills, creativity, and the ability to work without tangible results.
- The **systems development life cycle (SDLC)** can be described in five phases: (1) preliminary investigation, (2) analysis, (3) design, (4) development, and (5) implementation.
- Phase 1, **preliminary investigation**, also known as the **feasibility study** or **system survey**, is the preliminary investigation of the problem to determine how—and if—an analysis and design project should proceed. Aware of the importance of establishing a smooth working relationship, the analyst refers to an **organization chart** showing the lines of authority within the client organization. After determining the **nature and scope of the problem**, the analyst expresses the users' needs as **objectives**.
- In phase 2, systems analysis, the analyst gathers and analyzes data from common sources such as written documents, interviews, questionnaires, observation, and sampling.
- The analyst must evaluate the relevance of **written documents** such as procedure manuals and reports. **Interview** options include the **structured interview**, in which all questions are planned and written in advance, and the **unstructured interview**, in which the questions can vary from the plan. **Questionnaires** can save time and expense and allow anonymous answers, but response rates are often low. Another method is simply **observing** how the organization functions, sometimes through **participant observation**, which is temporary participation in the organization's activities. Statistical **sampling** is also useful, especially when there is a large volume of data.
- The systems analyst may use a variety of charts and diagrams to analyze the data. A **data flow diagram (DFD)** provides an easy-to-follow picture of the flow of data through the system. The elements of a DFD are processes, files, sources and sinks, and labeled vectors. **Processes** are the actions taken on the data. A **file** is a repository of data. A **source** is a data origin outside the system; a **sink** is a destination for data going outside the system. **Vectors** are arrows indicating the direction in which the data travels. Another common tool for data analysis is the **decision table**, or **decision logic table**, a standard table indicating alternative actions under particular conditions.
- The analysis phase also includes preparation of **system requirements**, a detailed list of the things the system must be able to do.
- Upon completion of the systems analysis phase, the analyst submits to the client a report that includes the current system's problems and requirements, cost analysis, and recommendations about what course to take next.
- In phase 3, systems design, the analyst submits a general preliminary design for the client's approval before proceeding to the specific detail design.

- **Preliminary design** begins with reviewing the system requirements, followed by considering **acquisition by purchase** (perhaps to be **customized** for the client) or **outsourcing** to an outside firm, or in-house development with, perhaps, alternative **candidates**. The analyst presents the plan in a form the users can understand.
- The analyst may also develop a **prototype,** a limited working system or part of a system that gives users a preview of how the new system will work. Software for **CASE—computer-aided software engineering**—provides an automated means of designing systems. CASE tools provide a graphics interface to produce both analysis and design on the screen, some kind of data store, program code generators, and the possibility of global changes.
- **Detail design** normally involves considering the parts of the system in the following order: output requirements, input requirements, files and databases, systems processing, and systems controls and backup. **Output requirements** include the medium of the output, the type of reports needed, the contents of the output, and the forms on which the output will be printed. The analyst might determine the report format by using a **printer spacing chart,** which shows the position of headings, columns, dates, and page numbers. **Input requirements** include the input medium, the content of the input, and the design of data entry forms. The analyst also plans an input validation process for checking whether the data is reasonable, and the analyst makes sure that the system can handle variations in input volume. The organization of **files and databases** must be specified. **Systems processing** must also be described, perhaps by using a **systems flowchart** that uses ANSI flowchart symbols to illustrate the flow of data or by using the hierarchical organization of a structure chart. The analyst must also spell out **systems controls and backup.** Data input to online systems must be backed up by **system journals,** files that record transactions made at the terminal. Processing controls involve data validation procedures. Finally, copies of transaction and master files should be made regularly.
- Phase 4, **systems development,** consists of scheduling, programming, and testing. Schedule deadlines and milestones are often shown on a **Gantt chart. Project management software** allocates people and resources, monitors schedules, and produces status reports. The programming effort involves selecting the program language and developing the design specifications. Programmers then do **unit testing** (individual testing of their own programs), which is followed by **system testing** (the assessment of how the programs work together). **Volume testing** tests the entire system with real data. Documentation of phase 4 describes the program logic and the detailed data formats.
- Phase 5, **implementation,** includes **training,** to prepare users of the new system; **equipment conversion,** which involves ensuring compatibility and providing enough space and electrical capacity; **file conversion,** making old files accessible to the new system; system conversion; **auditing,** the design of an **audit trail** to trace data from output back to the source documents; **evaluation,** the assessment of system performance; and **maintenance,** the monitoring and adjustment of the system.
- **System conversion** may be done in one of four ways: **direct conversion,** immediately replacing the old system with the new system; **phased conversion,** easing in the new system a step at a time; **pilot conversion,** testing the entire system with a few users and extending it to the rest when proved successful; and **parallel conversion,** operating the old and new systems concurrently until the new system is proved successful.

Discussion Questions

1. Which qualities of a systems analyst do you consider to be the most important?
2. Would the following most likely be good projects for acquisition by purchase, outsourcing, or in-house development?
    a. An inventory control system for a pizza franchise
    b. A payroll system for a small retailer

c. A system to network and provide basic software offerings for 13 office personal computers
d. A system to draw airplane galley installation assembly diagrams for an airline manufacturer
e. A system to process market research data gathered for new toys to be produced by the country's largest toy manufacturer
f. A system to permit networked artists to collaborate by computer on artistic ventures
g. A system to manage patient appointments, dental records, and billing for a clinic with four dentists
h. A system to track traffic tickets issued by the state patrol and convey this information to the state drivers' licensing agency
i. A system to perform automated check-writing and expense tracking for a funeral home
j. A system to install a terminal in the field office of each franchisee, to be connected to the central headquarters of a truck rental company for the purpose of tracking truck locations

3. Should system evaluation be done by the analyst and the client organization or by an independent evaluating team?

Student Study Guide

Multiple Choice

1. Testing of each individual program or module is called
   a. program testing
   b. system testing
   c. volume testing
   d. unit testing
2. The preliminary investigation of a systems project is also called a(n)
   a. analysis survey
   b. feasibility study
   c. systems design
   d. evaluation
3. The people who will have contact with the system, such as employees and customers, are
   a. programmers
   b. users
   c. systems analysts
   d. clients
4. The SDLC is defined as a project involving
   a. two phases
   b. three phases
   c. four phases
   d. five phases
5. Phase one of a systems project involves
   a. a system survey
   b. a systems analysis
   c. data gathering
   d. questionnaires
6. The person who fills the role of change agent is the
   a. systems user
   b. administrator
   c. systems analyst
   d. client
7. The scope and true nature of the problem is determined during
   a. the second phase
   b. the systems development
   c. the preliminary investigation
   d. the systems analysis
8. A chart of positions and departments within an organization is
   a. a data flow diagram
   b. an organization chart
   c. a project management report
   d. a Gantt chart
9. In a data flow diagram, a destination for data outside the system is called a
   a. vector
   b. source
   c. file
   d. sink
10. In the course of a systems project, systems design
    a. follows systems analysis
    b. precedes systems analysis
    c. follows development
    d. is the fourth phase
11. Positioning of headings and columns for the report format, considered during systems design, might use
    a. a record layout
    b. a decision table
    c. an organization chart
    d. a printer spacing chart
12. Programming and testing are elements of
    a. systems analysis
    b. implementation
    c. systems development
    d. systems design
13. Data gathering and data analysis take place
    a. after the system survey
    b. during systems design
    c. after systems analysis
    d. during evaluation
14. The kind of interview where all questions are planned in advance is called
    a. preplanned
    b. observation
    c. structured
    d. unstructured
15. The entire new system is used by a portion of the users:
    a. direct conversion
    b. file conversion
    c. pilot conversion
    d. parallel conversion
16. A systems analyst would observe the flow of data and interrelations of people within an organization during
    a. detail design
    b. systems analysis
    c. preliminary design
    d. a system survey
17. Used to ensure that no alternative is overlooked during data analysis:
    a. data flow diagram
    b. Gantt chart
    c. organization chart
    d. decision table
18. The phase following detail design is
    a. preliminary investigation
    b. systems development
    c. implementation
    d. system conversion
19. Scheduling deadlines and milestones can be shown on a
    a. system survey
    b. prototype
    c. decision table
    d. Gantt chart

20. Turning an entire project over to an outside firm for development is called
    a. auditing
    b. preliminary investigation
    c. outsourcing
    d. prototyping
21. The person who requests study or work on a system is the
    a. client
    b. change agent
    c. analyst
    d. user
22. The data gathering vehicle that permits high-volume anonymous answers:
    a. observation
    b. questionnaire
    c. unstructured interview
    d. structured interview
23. In the preliminary design phase, the analyst may prepare alternative
    a. candidates
    b. questionnaires
    c. organization charts
    d. decision tables
24. A set of software tools for automated design is called
    a. prototypes
    b. CASE
    c. data flow diagrams
    d. Gantt charts
25. A plan to trace data to its source is called
    a. an audit trail
    b. sampling
    c. a vector
    d. volume testing

True/False

T F 1. Systems analysis is the process of developing a plan for an approved system.
T F 2. Users are people who will have contact with the system.
T F 3. A systems analyst normally performs both analysis and design.
T F 4. Documentation is the least important aspect of a systems project.
T F 5. A feasibility study needs to be conducted following data gathering.
T F 6. Questionnaires are usually a more expensive form of data gathering than are interviews.
T F 7. An organization chart shows the flow of data through an organization.
T F 8. A decision table can help ensure that no alternative is overlooked.
T F 9. In some cases it is possible to acquire a new system by purchasing it.
T F 10. Input requirements should be considered prior to considering output requirements.
T F 11. Problem definition includes the nature of the problem, its scope, and the objectives of the system.
T F 12. Prototyping tools include powerful high-level software.
T F 13. Project management software can be used to monitor the allocation of both people and resources.
T F 14. A systems flowchart is the same as a logic flowchart.
T F 15. A Gantt chart is a bar chart that depicts deadlines and milestones.
T F 16. A prototype is a complete nonworking model of the computer system.
T F 17. File conversion is one form of system conversion.
T F 18. Maintenance of the system should take only a short time if the previous work was done with care.
T F 19. Poor communication between analysts and users can result in a system that does not do what the user expected.
T F 20. CASE tools permit both automated design and automated revision of a system.
T F 21. Even if unit testing is done thoroughly, system testing is necessary.
T F 22. Most people are uncomfortable with change.
T F 23. Input to online systems should be backed up with system journals.
T F 24. A phased conversion is riskier than a direct conversion.
T F 25. A source is a data origin outside the system.

Fill-In

1. What is the process called that evaluates a presently existing system to determine how it works and how it meets user needs? _____
2. List the three principal functions of a systems analyst.
   a. _____
   b. _____
   c. _____
3. What data analysis tool is used to illustrate information flow within a system? _____
4. As related to data, what are the two major steps of the systems analysis phase?
   a. _____
   b. _____
5. What is the total procedure to develop a new project through the five phases called? _____
6. Who is the person or the organization that contracts to have a systems analysis done? _____
7. What are files whose records represent transactions processed by online systems? _____

8. Which type of interview permits variation from planned questions? _____

9. Since a systems analyst brings change to an organization, what is the analyst often referred to as? _____

10. What is the by-product of understanding the system in the systems analysis phase? _____

11. In addition to questionnaires and interviews, name three sources of information for data gathering.

    a. _____

    b. _____

    c. _____

12. In data flow diagrams, what is the name for a repository of data? _____

13. Auditing and evaluation are part of which phase? _____

14. Which method of conversion is the most prolonged and expensive? _____

15. Name two other terms for the preliminary investigation.

    a. _____

    b. _____

16. What does CASE stand for? _____

17. Gathering a representative data subset is called _____

18. What are alternative plans for a new system design called? _____

19. During which phase should programming begin? _____

20. What word describes an organized set of related components that accomplish a certain task? _____

Answers

Multiple Choice

1. d	6. c	11. d	16. b	21. a
2. b	7. c	12. c	17. d	22. b
3. b	8. b	13. a	18. b	23. a
4. d	9. d	14. c	19. d	24. b
5. a	10. a	15. c	20. c	25. a

True/False

1. F	6. F	11. T	16. F	21. T
2. T	7. F	12. T	17. F	22. T
3. T	8. T	13. T	18. F	23. T
4. F	9. T	14. F	19. T	24. F
5. F	10. F	15. T	20. F	25. T

Fill-In

1. systems analysis
2. a. coordination
   b. communication
   c. planning and design
3. data flow diagram
4. a. data gathering
   b. data analysis
5. systems development life cycle
6. the client
7. system journals
8. unstructured
9. change agent
10. system requirements
11. a. written documents
    b. observation
    c. sampling
12. file
13. implementation
14. parallel conversion
15. a. feasibility study
    b. system survey
16. computer-aided software engineering
17. sampling
18. candidates
19. systems development
20. system

PLANET INTERNET

Not Quite Perfect Yet

The Internet has been heaped with well-deserved praise. But still there are concerns. To begin with, no one really knows exactly who is out there on the Net and what they are doing online. No one can truly compile statistics, much less establish content standards or control behavior. It's a little worrisome. On the other hand, many users find the freewheeling, no-controls aspect of the Internet appealing. Some fear that eventually the government, or perhaps a liaison of governments, will try to tame the Internet.

There really are some behavior problems. True. But there are behavior problems in any aspect of society, from the playground to the boardroom. Those who abuse the Net are, relatively speaking, small in number. Besides, the community of users monitors behavior on the Net. *Netiquette* refers to appropriate behavior in network communications, such not typing in caps (IT'S LIKE SHOUTING) and sticking to the subject at hand in a discussion group. Netiquette rules are published in various places on the Net and in every book about the Net. Users who commit serious sins may be subjected to flaming. *Flaming* refers to angry e-mail, sometimes by the hundreds or even thousands, directed to someone on the Net who has done something egregious, such as mass advertising. Meanwhile, some user/philosophers have occasionally produced statements of principle about the Internet. The Cyber Rights Home Page, sponsored by an organization called Computer Professionals for Social Responsibility, focuses on a variety of Net-related social issues, including online privacy and free speech.

Uselessness. Some people consider some home pages useless. In fact, there is a site called Useless Pages that maintains a listing of pages the site manager deems useless. However, many people are willing to pay for the connection to a Web server in order to promote a home page they fancy, such as Pete's Pond, shown here. Others put out birth or wedding announcements, complete with photos. One useless page does nothing except count the number of times the page is accessed. If you are not interested, simply skip such pages. No harm done.

Overburdened. It is just a matter of time before exponential user growth overwhelms the most popular sites. What's New and Cool Site of the Day are two of the most heavily accessed Web sites; they may soon be almost inaccessible. In addition to access problems, some individuals on the Net become overburdened by input from other Net users. You may see poignant messages on sites pleading for no more e-mail or, at least, explaining that they get so much e-mail that they cannot respond to it.

No guarantees. The Internet is full of misinformation. Just because something is on the Internet does not mean it is true. If someone steps up to announce that the government uses black helicopters to spy on us or that tapes sound better if you soak them in water first, you need not accept such information as fact. It's not that people intend to be wrong, it's just that they sometimes are. If you are doing serious research on the Internet, be sure to back it up from other sources, especially non-Internet sources.

Internet Exercises

1. **Structured exercise.** Begin with the B/C URL http://www.aw.com/bc/planet/ and read the Cyber Rights manifesto.
2. **Freeform exercise.** Beginning with the Useless Pages site, spend a few minutes checking out pages you will never have to visit again.

Workplace Tools

When Jeff Santerre was studying for his degree in finance, he took on computers, too. He knew he had to master spreadsheets and more sophisticated financial software. Furthermore, he was convinced that word processing would be useful for memos and reports. When he was hired by an international bank, he anticipated that he would use a personal computer on a fairly regular basis. What he did not anticipate was that the computer would be the major tool on his desk and that he would use it constantly for everything he did.

On a typical work morning, Jeff turns on his desktop computer, which is part of a network of office computers, as soon as he walks into his office. While he removes his coat and unpacks his briefcase, the computer is displaying a list of options. He selects Today's Calendar, and the screen displays "10:30 AM—Carston meeting re new trust" and "1:00 PM—lunch with T. Morales, Lakeside Cafe." The meeting notice reminds him to ask trust manager Kendall Barnes to bring some supplemental reports to the Carston meeting. He uses e-mail to compose a memo quickly and then sends it to Kendall via the office computer network. Next, he selects the option Read Mail from the screen and checks the list of incoming messages. Jeff sees that his computer has received a memo from his boss. He decides that it needs immediate attention and displays it on the screen. He sees that the boss is calling for an emergency meeting in the conference room at 9:00. Jeff stores the other messages so he can read them later, and he heads for his first meeting of the day.

As the day moves on, Jeff uses the computer to fetch client data from his database, to send information about a loan applicant to a business bureau via his fax modem, posts messages to clients in Switzerland and Sweden via the Internet, and plots strategies for a retiree using spreadsheet software. He prints reports on the laser printer. And, finally, he takes home diskettes of office data for evening work on his home computer. Jeff can use all this technology as casually as he uses the telephone or the copy machine. Jeff is a prime example of a personal computer user on the job. His employer is committed to a full range of computers—from supercomputers through mainframes and minicomputers to, literally, thousands of personal computers networked around the world. There are few limits to the uses of computers in a complex business enterprise.

Chapter 9

Computers in the Workplace
Large, Small, and Networked

LEARNING OBJECTIVES

- Understand the variety of ways major organizations use computers
- Appreciate the evolution of the use of both large and small computers
- Appreciate the impact of computers on the way workers do their jobs
- Understand the significance of groupware
- Become familiar with the types of software used in the workplace
- Learn what software might be most useful to a small business

BIG GUYS: USERS OF LARGE COMPUTER SYSTEMS

PERSONAL COMPUTERS IN THE WORKPLACE
- Evolution of Personal Computers in the Workplace
- The Impact of Personal Computers
- Where Personal Computers Are Almost a Job Requirement
- The Name of the Game Is Speed

HOW COMPUTERS CHANGE THE WAY WE WORK
- Groupware
- Computerized Meetings

SOFTWARE AT WORK
- Communications Software
- Graphics at Work
- Decision-Making Software
- Storage and Retrieval Software
- Vertical Market Software
- What Every Small Business Should Know

WHAT COULD POSSIBLY BE NEXT?

The Big Payoff

For years computers did not seem to be living up to their much-ballyhooed potential. This economic mystery was examined in articles that appeared in business journals, often with titles such as "Computer Rip-off: No Improved Productivity." Part of the problem seemed to be the difficulty of quantifying the benefits of computers. Since computer services produce unmeasurable output, they are harder to tally than, say, cars or dishwashers. In addition, such articles often hinted at mismanagement. Now, however, the sages are spinning another tale.

Recent studies of Fortune 500 companies show an eye-popping 67 percent productivity increase from investments in computer technology. So, what was the problem before? Old studies, it turns out, relied on obsolete data from the 1970s and early 1980s —too early to get the good news. The latest studies are based on data from the 1990s.

Big Guys: Users of Large Computer Systems

The emphasis in this book—indeed, the emphasis most often used in the computer industry—has been on personal computers. We will examine personal computers in the workplace in some detail, but first we must acknowledge the role of large computers.

The companies that use large computer systems today are mostly the same companies that used them 30 or 40 years ago: corporations that have big budgets. These organizations—typically banks, insurance firms, and aerospace companies—were the pioneers. The computers they used were potent by then-current standards, even though today's personal computers are more powerful than early mainframes were. In the 1950s and early 1960s, many companies obtained big computers because they believed that computers would provide a competitive edge. This turned out to be true, but it was a long time before managers really understood the true promise of computers.

Early users were somewhat uncertain about how to use the new tool. In fact, to them the computer was useful for no more than clerical tasks. Pioneering applications for many companies were payroll and accounting systems. The idea was to save labor costs by having the computer do some of the work. Many organizations, including the government, used computers as "number crunchers," machines that ground away at formulas. When computers began to be used interactively, business people saw that the computer could be used as a service tool, giving information such as instant reservations or bank balances.

Today, large computer systems are used in every conceivable way, from research to manufacturing. Midrange computers make number crunching and computerized services available to medium-size companies. Another important role for large and midrange computers is as a host for data needed by networked personal computers. It is, in fact, the highly affordable personal computer that has really opened up computing for the worker.

Personal Computers in the Workplace

Personal computers are not limited to the government and large corporations but are everywhere in the workplace, no matter the industry: retail, finance, insurance, real estate, health care, education, government, legal services, sports, politics, publishing, transportation, manufacturing, agriculture, construction, and on and on (Figure 9-1). In fact, it would be difficult to find an industry where computers are excluded. No company interested in increasing productivity would exclude them and still expect to compete in today's marketplace.

Evolution of Personal Computers in the Workplace

Personal computer use seems to have evolved in three phases. Personal computers were first used in business by individual users to transform work tasks. Constantly retyped documents, for example, gave way to quickly modified word-processed documents. The much-erased manual spreadsheet became the automatically recalculated electronic spreadsheet.

Figure 9-1 Personal computer users. Whether absorbed in work in a private office or working as part of a team, these workers are all making use of a personal computer.

And overflowing file drawers were transformed into automated databases. These changes gave a significant boost to individual productivity and can be considered the first phase of the evolution of personal computer use. Some organizations are still in this phase.

Many more organizations have entered the second phase. That is, they have gone beyond personal computer use by individuals. The second phase involves transforming a work group or department. This department-oriented phase probably embraces a network and may also include personal computer access to large computer systems. This phase requires planning and structure. Most personal computers in the workplace today are networked.

The third phase of personal computer evolution is the most dramatic, calling for the transformation of the entire business. Practically speaking, however, the third phase is really just an extension of the earlier phases: Each individual and each department uses computers to enhance the company as a whole. Few companies have come close to this idyllic state. This three-stage transformation—individual, department, and business—broadly describes a company's progress at blending its computer and business goals.

The Impact of Personal Computers

In the decades to come, personal computers will continue to alter the business world radically, much as the automobile did. For more than 50 years, the automobile fueled the economy, spawning dozens of industries, from oil companies to supermarkets. Other businesses, like real estate and restaurants, were transformed by the mobility provided by the car. Personal computers are having a similar effect for two reasons: (1) They have brought the cost of computing down to the level of a mass-produced consumer product, and (2) they have worked their way into most business organizations.

Now that computers are in businesses, let us consider who in business really needs to use them.

Where Personal Computers Are Almost a Job Requirement

Who must know how to use a personal computer to perform some part of a job? As we have already implied, the answer may soon be everyone. But we are not close to that landmark yet. Even if you can see that your intended job is in the must-know category, it is likely that you can receive some on-the-job training. Let us look at some of the jobs that might require personal computer knowledge.

- Real estate broker, attorney, doctor, auto mechanic, or anyone who needs to search for information in a variety of ways
- Accountant, tax planner, medical researcher, farmer, psychologist, budget manager, financial planner, stockbroker, or anyone who needs to analyze data
- Advertising copywriter, secretary, author, teacher, student, legislator, reporter, or anyone who needs to write and change documents
- Designer, editor, nurse, members of the military, engineer, retail sales manager, or anyone who needs to share data with other workers
- Project leader, construction manager, reservations agent, trucker, factory supervisor, or anyone who must keep track of schedules
- Insurance salesperson, fitness consultant, political candidate, sports manager, or anyone who needs to give a compare-the-results sales pitch

This list is a bit deceptive: Notice that many of the jobs mentioned would probably fit in more than one category.

The Name of the Game Is Speed

Not the computer game—the business game. From California to Maine, a principal topic among management consultants and business school professors is speed. Why? Who cares? And, if there really is a good reason for speed, how is it gained?

The *why* question has the most straightforward answer: Speed kills the competition. If your product gets to market first, you get the sale. Lag behind and get left out.

Who cares? Managers, of course, and—perhaps surprisingly—employees. Employee satisfaction improves when employees are working for a responsive, successful company. Also, using computers to speed up operations gives employees more responsibility and flexibility.

So, how is speed achieved? Do we just put out an order to step on the gas? Probably not—that would just speed up the mess and burn out

ONE JUMP AHEAD

Kicking the Telephone Habit

Suppose you have been on a business trip for four days. When you return to your office, your computer indicates that you have 27 e-mail messages. By coincidence, your telephone indicates that you have exactly 27 voice mail messages. Now, which set of messages would you rather tackle first? Which will give you the most immediate information? If you instinctively go for the "real person" voice mail, prepare to spend some time listening, first to figure out who the callers are and what they want and, perhaps, wishing the speakers could be a little more concise.

The e-mail messages, on the other hand, are much more under your control. Your computer screen shows a line of information for each message: date sent, e-mail address of sender, and a brief note about content. You will doubtless be more interested in a

message from your boss about "Year-end bonus" than in a message from a colleague regarding "Baby shower for Sue." You can read the important messages first, save some messages for a more convenient time, and possibly ignore some messages altogether.

In the not-so-distant future, traditional telephone functions will be absorbed by the computer. The computer will become a sort of universal mailbox, offering access to voice mail, e-mail, and faxes at the click of a mouse button. There will be a line of information about each message: its type (voice, e-mail, or fax), its length in minutes or pages, the phone number or e-mail address of the sender, and possibly even the sender's name if you have it in your computer files. To hear a spoken message, click the phone icon on your screen, and the phone will ring and deliver the message. You can interrupt the message and later resume where you left off. You can pause, back up, or fast forward any message, using on-screen icons that look like the controls of a tape recorder.

And the phone? It is still around, but mostly for outgoing calls at your convenience.

---

machines and workers. There are many ways to speed up operations, including providing worker incentives, reducing the number of approvals needed for action, and—of course—putting the computer to work whenever possible.

By providing timely access to data, computers let us spend less time checking and rechecking data—and more time getting things done. This inspires informed decision making and improves overall productivity. Today's computer systems speed memos, documents, and graphs to workers throughout the organization. This kind of direct people-to-people communication enriches every aspect of a business. There is no question that computers are changing the way we work.

How Computers Change the Way We Work

Computers are changing the way both individuals and organizations work. In addition to increasing overall productivity, computers have had a fundamental impact on the organization itself.

Executives were among the first to notice the change. An oil company executive observed that her secretary is no longer the keeper of all knowledge because workers now prepare their own documents without the secretary even seeing them. But perhaps the most significant way that computers have impacted how people work is the software for networked computers called groupware.

Groupware

Groupware, also called **collaborative software,** can be defined generally as any kind of software that lets a group of people share things or track things together. Using that general definition, some people might say that electronic mail is a form of groupware. But simply sending data back and forth by e-mail has inherent limitations for collaboration, the most obvious being confusion if there are more than two group members. To work together effectively on a project, the data being used must be in a central place that can be accessed and changed by anyone working on the project. That central place is a database, or databases, on disk.

A popular groupware package called Notes combines electronic mail, networking, and database technology (Figure 9-2). Using such groupware, business people can work with one another and share knowledge or expertise unbounded by factors such as distance or time zone differences. Notes can be installed on all computers on the network, including Windows-based personal computers, Apple Macintoshes, and UNIX-based machines.

A broad picture of groupware goes beyond the Notes model of e-mail, networking, and a database. A variety of add-on software, software written to work in conjunction with Notes, is available. For example, scheduling software can, via the network, examine the personal calendar file of each potential attendee to find a meeting time that all have open (Figure 9-3).

Collaboration

Groupware distributes knowledge from those who have it to those who need it. This is different from e-mail, where a knowledge seeker would have to express a particular need before someone would be alerted to pass along information or expertise. With groupware, each person contributes to the databases, and the knowledge seeker queries the database for what he or she needs. A user need not, of course, wade through all available information; he or she can browse through just the desired databases. For example, a user could ask to see only entries pertaining to environmental issues relating to salmon spawning or to liability coverage for company engineers.

Companies must define what they want to do with groupware and how they want to use it. Left uninstructed, users may set up groupware databases just because the technology exists. That is backwards; the problem

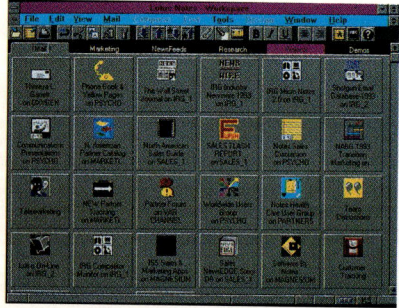
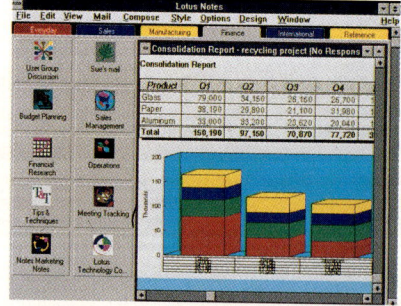
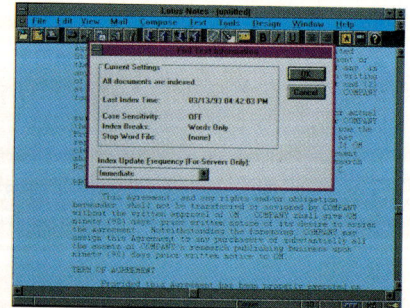

Figure 9-2 Notes. These three screens are produced by a groupware package called Notes. Groupware lets workers use the computer to collaborate on a project.

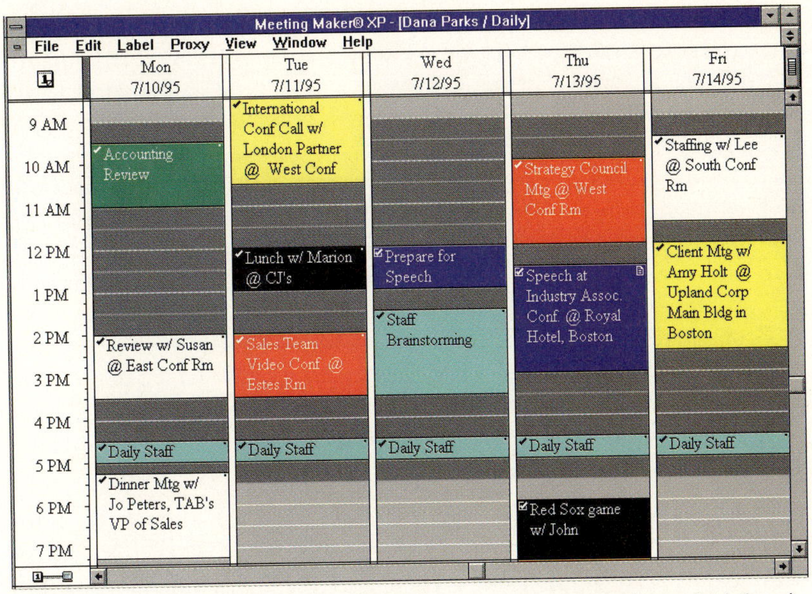

Figure 9-3 Scheduling software. Scheduling software, used on a network, can check the calendars of potential meeting attendees, searching for a common unscheduled time for which to plan a meeting.

to solve comes first and then the technology. Groupware is most often used by a team for a specific project. A classic example is a bid prepared by Price Waterhouse, an accounting firm, for a consulting contract. They had just a few days to put together a complex proposal, and the four people needed to write it were in three different states. They were able to work together using their computers and Notes, which permitted a four-way dialogue on-screen. They also extracted key components of the proposal from existing company databases, including the résumés of company experts and borrowed passages from similar successful proposals.

A key to Notes' effectiveness in a group setting is its ability to **replicate** data to all users, copying data from computer to computer throughout the network. The effect is that all users have the most recent version of the database. Thus, each user knows what the others are thinking. Notes even permits more than one group member to work on the same database at the same time.

To appreciate the complexity of replication, imagine that you are on the road and have just convinced a partner to join in a project as a subcontractor. Using your laptop computer, you call up the office computer and download whatever project databases you need. You tap away, adding comments and numbers. Meanwhile, your project colleagues in other locations may be making their own changes to the very same databases. What happens when you call back to the office computer and send your changes to the database that has, in the interim, been altered by others? Notes can reconcile the results, no matter how complex. If there are conflicts, Notes notifies the participants and asks them to resolve the differences, either keeping one version or perhaps both.

Crossing Boundaries
Although groupware emphasizes collaboration within the organization, many organizations are expanding that notion beyond the company walls.

If collaborative software works well for us, they seem to be saying, can we not harness that same power to work with our suppliers and even our customers? The answer is emphatically *yes*. Companies see a big opportunity to transform the way they do business across boundaries with their suppliers and customers.

To the immediate question—are company secrets at risk—the answer is no because security is built into the software. It is possible to connect two systems together and still retain appropriate security for each enterprise. In fact, any information that is confidential can be restricted so that only authorized people, even within the enterprise, can call it up.

Groupware began in large corporations because Notes sales representatives initially concentrated on customers who would invest the time and money required to appreciate the potential of the technology. Now businesses large and small are using groupware. They know that if they are not using it, their competitors are.

Computerized Meetings

Although groupware is, in some ways, an ongoing meeting, most organizations find that they still need the in-person kind of meeting, an activity some workers dread. But consider the words of employee Jonas MacDowell: "I never thought I would actually like a meeting." Jonas has just been to a new kind of meeting, one that injects technology into the typical meetings Jonas disliked. Many workers consider meetings boring and unproductive. Furthermore, studies have shown that as much as 30 percent of corporate overhead is burned up in traditional meetings. Read on to see how technology changed Jonas' mind about meetings.

Managers have responded to the excess meeting problem in various ways. One manager removed the chairs from a conference room, forcing attendees to stand up throughout the meeting. This was effective in speeding up the meeting, but the accomplishments were suspect.

Enter technology. A software package called Meeting Meter dramatically refutes the hazy notion that meetings do not really cost the company anything. Using a laptop computer, a manager can feed the program guesstimates of the salaries of each attendee. The meter program then acts as a money clock, displaying the running salary cost of the meeting. For example, for a meeting of two managers (earning approximately $35 per hour each), seven professionals ($30), two assistants ($20), and one secretary ($10), the clock would display $495 after 1½ hours. The clock encourages attendees to focus on the subject at hand and to become more active participants.

Jonas' organization, a product design group, went a step further. Their meetings used to be endless, with talkers filibustering, daydreamers gazing at the ceiling, and Jonas springing tears from suppressed yawns. The group now uses a meeting facility that has keyboards and screens connected to a computer. Although meetings include some oral discussion, most of the action takes place on the computer screens (Figure 9-4). The computer is also used to let everyone at the meeting secretly vote on issues. Users are sometimes surprised, for example, when a vote reveals unanimity on a certain issue; since everyone is in favor, they can dispense with passionate speeches and move to the next issue. The greatest benefit, however, is that people tend to digress less when using a keyboard than they do orally. Finally, all meeting ideas have, at meeting's end, been recorded in the computer for future reference.

Figure 9-4 A new kind of conference room. When meetings are held in computerized conference rooms, each participant can use a keyboard for his or her own input and follow what others are saying on an individual screen.

Software at Work

We have discussed groupware in some detail, but there are many other types of online software and services that workers value. Consider some specific business applications for the software tools in the marketplace.

Communications Software

If, as an individual, you have a computer, a modem, and some communications software, you have the capability to access any other computer system similarly configured. But it is businesses that are the major users of communications software.

Online Reservations
Need a reservation? Do not call your travel agent—reach for your computer. For a fee any business can have direct access to the major airlines reservations systems used by travel agents and airlines around the world. Individuals can have access to these same systems. Those who use airline reservations systems for business or personal travel have immediate access to airfares at any hour of any day. The reservation system provides reports on an airline's on-time performance, uses a personalized profile to sort through flights and seating arrangements, spells out applicable restrictions, and summarizes travel arrangements.

Weather Forecasting
We have long relied on the media, both television and print, to keep us informed about the weather. This service is adequate for most of us. Some

businesses, however, are so dependent on the weather that they need constantly updated information. Online services offer analysis of live weather data, including air pressure, fog, rain, and wind direction and speed. Businesses that depend on the weather include agriculture, amusement parks, ski areas, and transportation companies, all of which make business decisions based on weather forecasts.

The Stock Exchange

Stock portfolios can be managed by software that takes quotations online directly from established market monitors such as Dow Jones. The software keeps records and offers quick and accurate investment advice. And, of course, the stock exchange itself is a veritable beehive of computers.

We communicate by computer, but, increasingly, we also use computer output in the form of pictures.

Graphics at Work

Computer graphics can delight and entertain and inform. Business people can enhance a message by using graphics to express numbers in an easily understood form. But, sometimes, producing graphics is the chief function of a computer system or an integral part of the job of the worker who uses the system (Figure 9-5). Here we list some of the many workers who depend on computer graphics to do their jobs.

Researchers

Some people worry a little about earthquakes; others get paid to worry about them. Researchers in the field of earthquake prediction use graphics in several different ways to help them visualize the forces that cause temblors. The "photo" in Figure 9-5a is a computer-generated graphic that helps scientists understand underground formations whose movements result in quakes. Computers can also digitize and assemble satellite photos, and the graphics that result help researchers spot geological patterns and shifts. These two examples are of government research, but private firms also do an enormous amount of research to strengthen their product lines.

Artists and Designers

As a tool of their craft, artists use sophisticated software to produce stunning computer art and animation. A clear business application of graphics is design. Everyone from architects to engineers to fashion designers can use the computer to design and simulate products.

Musicians

The old movies feature inspired composers hanging over the piano in the middle of the night. First we see a few fingers dabbling at the keys and then a pause, a cocked ear, another bit of key tinkling, and, finally, a pencil writing the notes on paper. Many composers work that way today. But some do not. Composers still play the notes, but a computer equipped with listening software captures them and reproduces them as graphic images on the screen. When the composition is completed, the composer will instruct the computer to print out the sheet music.

(a)

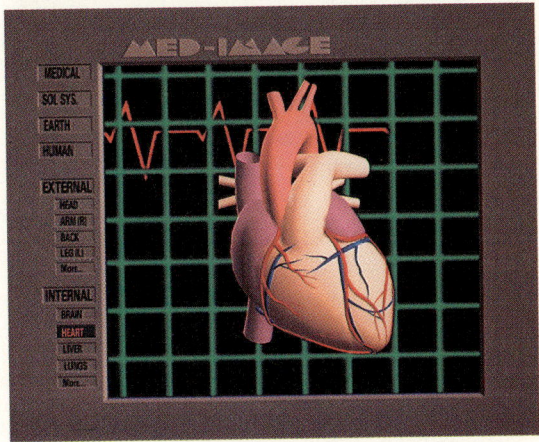
(b)

(c)

Figure 9-5 Graphics software at work. (a) Off the coast of Chile, three rocky plates that make up the ocean floor come together. Computer images can show varying growth rates, which cause one of the plates to slide under another, disappearing the way the steps at the bottom of an escalator seem to vanish. (b) A computer image of the human heart. (c) This map maker merges computer-enhanced satellite images.

Doctors

How do you look inside the body to take its picture? X-rays are one way, of course. But X-ray films can be so fuzzy that interpreting them seems more an art than a science. Modern medical imaging has gotten quite a boost from the computer. Figure 9-5b shows a definitive image of the human heart.

Decision-Making Software

The computerized spreadsheet is a generic type of decision-making software. The ability of spreadsheet software to recalculate automatically lets decision makers explore different possibilities. Beginning with loan amortization, let us consider some of the significant ways that businesses use spreadsheets and other decision-making software.

Loan Amortization

Software for loan amortization determines due dates, payment number, payment amount, principal, interest, accumulated interest, and loan balance. Most loan amortization software also produces yearly and monthly reports.

Break-Even Analysis

Can we afford the new equipment? Should we buy or lease? Should we try to compete in that market? Is the cost worth it? What is the payoff? At what point do we break even? Net present value and break-even software answers these questions and more—analyzing the relationships among variable costs, fixed costs, and income—and produces alternatives to consider. A computer-generated analysis based on actual conditions is a decided improvement over hunches scribbled on the back of an envelope.

Property Management

A type of program sometimes referred to as "the landlord" can be used to manage any income property, whether marina, apartment complex, or shopping mall. The software can record charges and payments for each renter and produce a variety of reports, such as a lease expiration list and tax analysis lists for each property.

No matter the business, getting a job done efficiently often depends on reliable access to stored data—our next topic.

Storage and Retrieval Software

Office workers and salespeople and manufacturers all use computers as tools in their businesses; most of these workers and many others rely on access to stored data.

Crime Detection

A lot of crime detection involves a process of elimination, which can be tedious work. A tedious task, however, is often the kind the computer does best. Once data is entered into a database, then searching by computer is possible. Examples: Which criminals use a particular mode of operation? Which criminals are associates of this suspect? Does license number AXB221 refer to a stolen car? And so on. One specific type of crime-detection computer system is the fingerprint-matching system, which can match crime-scene fingerprints with computer-stored fingerprints.

Sports Statistics

Here is the situation: Tied game, bases loaded, two outs, left-handed batter, bottom of the ninth. If you are coaching the team in the field, do you leave in the right-handed pitcher or pull him for a lefty? The seat-of-the-pants hunch is less common these days. Coaches and managers in professional sports want any help they can get. A wonderful source of help, one that can be carried right to the edge of the field, is the computer. In our current example, the bases-loaded scenario, the manager can check statistics from the batter's past performance against statistics about each available pitcher—in just seconds, of course. All kinds of statistics can be stored in a database and retrieved on the spot.

Performing Arts

People in the performing arts use computers as standard business tools. They find databases particularly useful. Database software, for example,

MAKING THE RIGHT CONNECTIONS

The Virtual Office

The word *virtual* is applied in various computer settings, but it always means the same thing: the appearance of something that really does not exist. The computer, somehow, masks the reality and permits benefits similar to those offered by the real thing. In this discussion of the virtual office, the office as we know it—a physical place with a desk and a chair and office supplies—does not actually exist. But its functions do exist.

Consider the way Nora Mathison runs her sprinkler installation business in Phoenix. She relies on an 800 phone number, voice mail, a cellular phone, and a laptop computer with a fax modem. No building, no office, no desk.

Nora advertises her 800 number in the Yellow Pages; potential customers in the urban/suburban area can use the 800 number to call without charge. When they do, they will be advised to leave a voice mail message. Nora, working on site in some customer's yard, can retrieve her voice mail messages and return the calls on her cell phone. She can use software on her laptop to work up a bid right at a customer site or do the work later and fax the results to the customer. She also uses the laptop for scheduling, work flow, and billing.

In addition to convenience, the virtual office can minimize start-up costs for fledgling entrepreneurs. For business people who spend most of their time out of the office anyway, the virtual office is an ongoing asset.

---

can search for the names of musical pieces—all 20-minute violin pieces by German composers, for example. The American Ballet Theater managers take their computers on tour: One database plots rehearsal schedules; another keeps track of sets, lighting, and costumes. Some sophisticated organizations use databases to coordinate ticket sales with fund-raising; as a patron, this probably means that you will receive your tickets with great efficiency and then be solicited for a donation.

Legal Services
You have seen the formal photograph of the judge, the attorney, the politician—each with a solid wall of law books in the background. Those books are not just decoration, however. Workers in any law office need to be able to research legal precedents and related matters. But why not take the information in those books and just "drop it" into a computer? That is, in essence, what has been done. The books have been converted into computer-accessible databases, and the result is that legal research time has decreased significantly. Two common computerized legal research systems are LEXIS and WESTLAW, available in most law libraries and law firms.

Vertical Market Software

Vertical market software describes software written especially for a particular group of customers, such as accountants or doctors. Some software makers specialize in computer systems for such markets (Figure 9-6). This

Figure 9-6 Vertical market software. Some software is designed specifically for vertical markets—a specific type of worker, such as accountant, engineer, lawyer, or real estate agent. These accountants are using software specifically geared to the needs of accountants.

user-oriented software usually presents options with a series of easy-to-follow menus that minimize training needed. Here are some examples of businesses where you might find vertical market software.

Auto Repair Shops

Designed in conjunction with people who understand the auto repair business, this all-in-one software for an auto shop can prepare work orders, process sales transactions, produce invoices, evaluate sales and profits, track parts inventory automatically, print reorder reports, and update the customer mailing list.

Videotape Rental Stores

An important goal for a videotape rental store is fast service. One concept for fast service is simple enough: Let the computer match customers and tapes they rent. Here is how it works. Each regular customer receives a

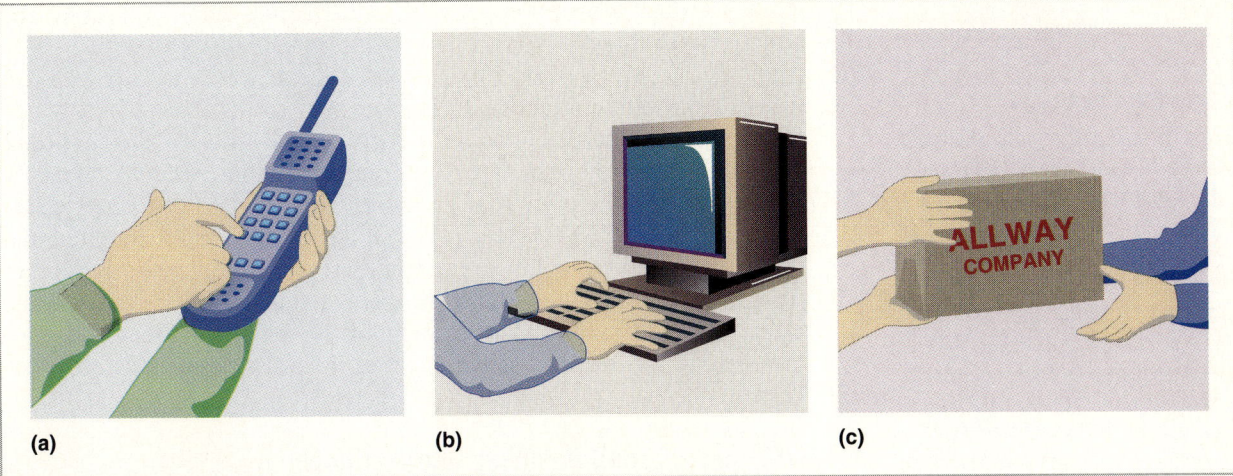

Figure 9-7 Making returns easy. (a) A customer calls a mail-order house to says that she wants to return a purchase. (b) The mail-order house employee inputs purchase information and retrieves customer data, which is sent by computer network to a courier service. (c) The next day a courier picks up the package at the customer's home or office or other prearranged location.

card with an identifying bar code on it; each rental tape is also bar-coded. At rental time the customer's bar code and the tape's bar code are scanned, and that information is used to print an invoice. The entire transaction takes only a few seconds. (The scanning system can be overridden by typing in names and other data for new or cardless customers.) When the tape is returned, a clerk needs only to scan the tape label; the system automatically credits the proper customer with the return. The software can produce a variety of reports, such as reports of overdue rentals and mailing lists.

Mail-Order Houses
Many people like the convenience of shopping by catalog. That convenience diminishes when customers have to return merchandise. This annoyance has been addressed by software that puts a next-day pickup in motion (Figure 9-7). The customer uses an 800 (toll free) number to call the mail-order house. A company employee inputs the customer-supplied merchandise identifier to the computer, which retrieves customer information and sends it via network to United Parcel Service (or a similar carrier). The carrier picks up the merchandise at the customer's home, office, or some other prearranged site the next day.

Beauty Salons
Does your hairdresser really remember exactly how to do your hair and that you like yard work and movies? Maybe. But it is more likely that a card is on file somewhere, listing your preferences. In some shops that "card" is stored in the computer. Before you arrive, this data can be pulled up on a screen. After you leave, the hairdresser immediately updates your customer history. In addition, the computer credits your stylist for providing the service and uses this data to calculate the stylist's commission. The computer can also produce reports, include sales summaries by period,

product inventories, appointment reminder cards, thank-you cards, and promotional letters.

Mailing List Services

Traditionally, managers ignored the cost of mailings as a nickel-and-dime expense. Now, managers view their company mailing list as a target for cost reduction. There are many software packages to generate mailing lists; cost savings can result from trimming and focusing them. "Hunter-killer" software roots out incomplete addresses and duplicates, even if they are not quite identical in appearance. This kind of software can also use addresses "intelligently." For example, if the mailing consists of a lawn care circular, the software can eliminate addresses that include an apartment number.

What Every Small Business Should Know

Suppose that, as a fledgling entrepreneur with some computer savvy, your ambition is to be as competitive as one personal computer will let you be. You know you cannot afford expensive software, but you also know that there is a wealth of moderately priced software that can enhance all aspects of your business, whether it is a product or a service.

This presentation, rather than by software type, is offered according to business functions—things you will want to be able to do. The computer can help.

- **Accounting.** Totaling the bottom line must be the number one priority for any business. If you are truly on your own, or have just one or two employees, you may be able to get by with simple spreadsheet software to work up a ledger and balance sheet, and generate basic invoices and payroll worksheets. Larger operations can consider a complete accounting package, which produces profit-and-loss statements, balance sheets, cash flow reports, and tax summaries. Most packages will also write and print checks, and some have payroll capability.
- **Writing and advertising.** Word processing is an obvious choice because you will need to write memos and the like. Desktop publishing is not so obvious but can be a real boon to a small business, letting you design and produce advertisements, flyers, and even your own letterhead stationery, business forms, and business cards. A big advantage of publishing your own advertisements or flyers is that you can print small quantities and start anew when your business or address or terminology changes; this is in direct contrast to ordering printed materials where 500 is a typical minimum requirement. Finally, you may think it is worthwhile to publish your own newsletter for customers.
- **Customer service.** Customer service is a byword throughout the business world, but the personal touch is especially important in a small business. Database software can be useful here. Suppose, for example, that you run a pet grooming service. You surely want to keep track of each customer by address, and so forth, for billing and advertising purposes. But this is just the beginning. Why not store data about each pet, too? Think how impressed the customer on the phone will be when you recall that Sadie is a standard poodle, seven years old, and that it is time to have her toenails clipped.
- **Keeping up and making contacts.** Even if you have only one computer, you can still be networked to the outside world. Business connections

Computer User Groups

You have started your own business and even have a computer to keep track of things. The knowledge you have from your computer classes sustains you, but you wish you had some ongoing source for help and even enrichment. Unfortunately, the very people who could be helped the most—new users—often do not even know that user groups exist. User groups are run by volunteers with minimum funds and thus have little money to advertise.

There are hundreds of user groups, from the 23,000-member Boston Computer Society, which has its own building, to the tiny Nicolet Computer Club of Townsend, Wisconsin, which meets in the library. Typically, a group meets once a month to share ideas and knowledge, and also publishes a newsletter. In addition, most groups have subgroups called special interest groups (SIGs) that focus on a single topic, such as graphics or a particular software package.

Some people have the false impression that user groups are for the die-hard techie, but anyone can join. Just show up at a meeting, pay your (nominal) dues, and you belong. Interested? You can probably find out about local user groups at your local computer store or library. Or call (914) 876-6678, a service that will let you search for a user group by state, telephone area code, or zip code.

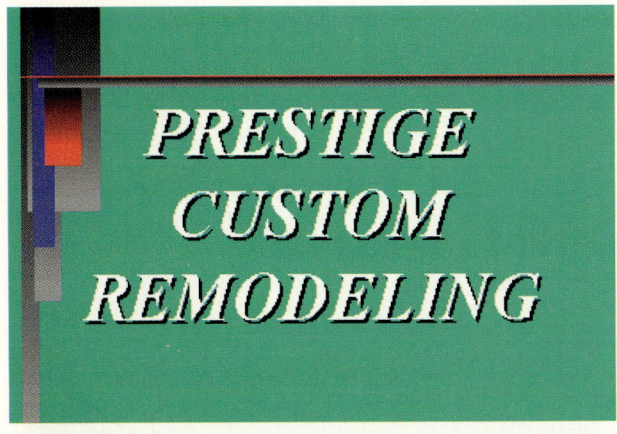

Figure 9-8 Presentation graphics. Even a small business can take advantage of the computer to produce this professional-looking set of computer screens, which can be projected on a wall screen or converted to slides.

are available in many forms from information services and the Internet. For example, CompuServe offers topics called Business Networking, Hot Opportunities, and Business Software.

- **Making sales pitches.** If your business depends on pitching your product or service in some formal way, then presentation software can help you create colorful demonstrations that are the equivalent of an electronic slide show. Presentation software is designed for regular people, not artists, so putting together a slick sequence of text and graphics is remarkably simple (Figure 9-8). Fancy fonts and graphics are provided by the software; you just choose what is appropriate and do the arranging.
- **Project management.** Do you put out a product or plan a service that involves a series of tasks, some of which depend on the completion of prior tasks? For example, Dieter Schrempf is a contractor who specializes in remodeling old houses. Dieter must spend much of his time attracting and then determining specifications with new customers. Meanwhile, Dieter oversees his four employees and various subcontractors as they move the current jobs along. Obviously, plumbing and electrical work must be done before installing sheet rock. Hundreds of other tasks must be performed in order and flow smoothly. Managing

the order of the tasks is one of many advantages of project management software.

Is there more? Always. Individual businesses have unique needs that can probably be filled by some kind if software. However, a business equipped to handle the needs mentioned above will be, overall, well-positioned.

 ## What Could Possibly Be Next?

The future is already in sight. Soon workers will wonder how today's mute, passive boxes were ever called computers. Already some personal computers, or terminals hooked up to bigger machines, can talk, listen, and display live images. No computer is an island: They can call one another and send faxes, mail, and messages. People who cannot leave their computers at home are taking them along—and may even wear them one day. Some computers are disappearing altogether—into the furniture to become part of the desk, the cabinet, the blackboard. It seems to be the fate of the computer to move into the background—and to be everywhere.

Which brings up the subject of our next chapter: Just how do you manage information systems when they are both in and out of the office and even disappearing into the woodwork?

CHAPTER REVIEW

Summary and Key Terms

- The companies that use mainframe computers today are mostly the same companies that were computer pioneers, large corporations that have large budgets. Early users used computers for clerical tasks and number crunching. When computers began to be used interactively, business people saw that the computer could be used as a service tool.
- Midrange computers made computer services available to medium-size companies. But it was the highly affordable personal computer that really opened up computing for the worker.
- A partial list of where computers are used includes retail, finance, insurance, real estate, health care, education, government, legal services, sports, politics, publishing, transportation, manufacturing, agriculture, and construction.
- Personal computer use seems to evolve in three phases: transformation of an individual's productivity, transformation of a department, and transformation of a business.
- Personal computers will radically alter the business world for two reasons: (1) They have brought the cost of computing down to the level of a mass-produced consumer product, and (2) they have worked their way into most business organizations.
- Many workers must know how to use a personal computer to perform some part of a job. If a job requires the use of a personal computer, it is likely that the worker will receive some on-the-job training.
- There are many ways to speed up business operations, including putting the computer to work whenever possible.
- **Groupware**, also called **collaborative software**, can be defined generally as any kind of software that lets a group of people share things or track things together, usually using software that combines electronic mail, networking, and database technology. With groupware, each person contributes to the databases, and the knowledge seeker queries the database for what he or she needs.
- Groupware is most often used by a team for a specific project. A key to the effectiveness of the groupware called Notes is its ability to **replicate** data to all users, copying data from computer to computer throughout the network. The effect is that all users have the most recent version of the database. Although groupware emphasizes collaboration within the organization, many organizations are expanding that notion beyond the company walls to their suppliers and customers.
- Software can display the running salary cost of a meeting, encouraging attendees to focus on the subject at hand and to become more active participants. Some modern meeting facilities have keyboards and screens connected to a computer for use by participants.
- Businesses use software for communications, graphics, decision making, and storage and retrieval of information.
- **Vertical market software** is software for a group of similar customers, such as accountants or doctors.
- Small businesses can get the computer advantage with even one personal computer for such uses as accounting, writing and advertising, customer service, keeping up and making contacts, making sales pitches, and project management.

Discussion Questions

1. Consider these firms. What uses would each have for computers? Mention as many possibilities as you can.
   a. Security Southwestern Bank, a major regional bank with several branches

b. Azure Design, a small graphic-design company that produces posters, covers, and other artwork
c. Checkerboard Taxi Service, whose central office manages a fleet of 160 cabs that operate in an urban area
d. Gillick College, a private college that has automated all student services, such as registration, financial aid, and testing
e. Payton Realty, a realty firm with multiple listings and 27 agents

2. Are you considering a particular career? Discuss how computers are used, or could be used, in that field. How may their use change in the future?

3. Consider these businesses, with a solo owner, perhaps occasional temporary employees, and one personal computer. What business needs might be filled using the computer, and what type of software might fill each need? Incidentally, some of these, such as psychologist, may not seem much like businesses; do they still have business needs?
   a. Plumbing contractor
   b. Writer of children's stories
   c. Photographer
   d. Caterer
   e. Freelance bookkeeper
   f. Wedding planner
   g. Importer
   h. Nail artist
   i. Independent truck driver
   j. Investment planner
   k. Jewelry maker
   l. Roofer
   m. Karate instructor
   n. Psychologist
   o. Window washer

Student Study Guide

True/False

T F 1. The companies that use mainframes today are mostly the same large corporations that were computer pioneers.

T F 2. Early computer users used computers mostly for personal scheduling.

T F 3. Computers are used in formal business offices but not in less formal settings such as sports or agriculture.

T F 4. Many workers must know how to use personal computers to perform their jobs.

T F 5. Vertical market software is used mostly by architects to plan high-rise buildings.

T F 6. The major users of communications software are individuals using bulletin board systems.

T F 7. Personal computer use in business has evolved in five stages, the highest use being an increase in individual worker productivity.

T F 8. A key reason for the extensive use of personal computers in business is their low cost.

T F 9. An advantage of groupware is collaboration.

T F 10. Meetings may be improved with the use of computer technology.

T F 11. Because of their low cost, personal computers were the first computers to be used by corporations.

T F 12. Small businesses rarely can make effective use of computers.

T F 13. Groupware databases can be used within an organization but not by its customers.

T F 14. Notes is used primarily for spreadsheet applications.

T F 15. Computers can be a critical factor in speeding up processes and beating the competition to market.

Answers

True/False
1. T   6. F   11. F
2. F   7. F   12. F
3. F   8. T   13. F
4. T   9. T   14. F
5. F   10. T 15. T

PLANET INTERNET

Free and Not Free

Many people, especially those associated with schools and government organizations, have free access to the Internet. But is the information available on Internet sites also free? Often the answer is yes. This means that the average citizen could go to, say, the local library, get on the Internet for no charge and then also pick up information from the Net free of charge. Free and free. Even users who have to pay for Internet access can get free Net information.

What information is free and what isn't? There are no uniform rules to guide you. Although some information providers make a blanket "help yourself" statement, much information is unaccompanied by a proprietary statement. However, business people and others who do not want their works copied can post clear notices on their pages, using phrases such as *copyrighted, online access only,* and *all rights reserved.*

Freebies. Several categories of information tend to be free. Categories of such information include health care, the environment, science, government agencies, humor, lists of events, family issues, social sciences such as philosophy and religion, the weather, anything to do with space, and most topics found on individual home pages. The heart shown here is the logo for a site that does a serious job of exploring the heart. The planets are from the popular and educational Nine Planets site. The multicolor shot is from a NASA volcano exhibit; this shot was taken from the space shuttle.

Not free. Noticeably missing from the list of freebies are business activities. Although there is always free advice on any topic, businesses by nature want to make a profit, so business products and services are likely to have a fee. What's more, businesses are running, not walking, to the Net, so the proportion of for-a-fee offerings can only increase. Even information on sites related to sports and entertainment may not be freely available. For example, the MTV site cautions, in case anyone wants to borrow their cool logos for an individual home page, that "all logos are trademarks" and "all rights are reserved."

Free forever? Probably not. Generally, when a product or service is offered free, its use may be abused. Users may be charged something, however nominal, just so they will recognize the value of the Internet and use it wisely. The most serious problem, however, is overload. The number of users of the Net is doubling worldwide every few months. In short order, critical mass will demand that some scheme be devised to limit usage in some way, and that approach may include access fees.

Internet Exercises

1. **Structured exercise.** Begin with the B/C URL http://www.aw.com/bc/planet/ and examine the free (save frog lives!) computerized Frog Dissection Kit.
2. **Freeform exercise.** Beginning with the EINet Galaxy directory or your own your favorite online directory, find the maps to track the weather in your home state.

Mick Dalton pursued a business degree with the goal of a career in management. He was uncertain, however, about his career ambitions. He thought that someday he would like to be at the very top of an organization with, perhaps, an office with a stunning view. He thought it was more likely, however, that he would end up somewhere in the middle, reporting to the top bosses but with responsibilities for major activities below him. He assumed that his entry into management would be at the lowest rung on the ladder, in direct contact with the workers, supervising their operations and making sure they had what they needed to do the job.

As it happened, Mick did all these things, but not in the way he expected. While he was in college, he began a computer word processing service, typing up his classmates' term papers and résumés. He used part of his profits to buy a laser printer and desktop publishing software. Thus, he was able to produce professional-looking documents and was able to offer his services to local small businesses. Mick's business-on-the-side grew beyond his expectations; he decided to go into business for himself full time after graduation. Mick's company eventually specialized in the production end of publishing periodicals and paperback books. As the company grew, Mick managed at all levels and, eventually, did indeed have a corner office overlooking the cityscape.

Whether managing your own company or someone else's—whether at the top, middle, or bottom level—the challenge is the same: to use available resources to get the job done on time, within budget, and to the satisfaction of all concerned. Let us begin with a discussion of how managers do this, then see how computer systems can help them.

Chapter 10

Management Information Systems
Classic Models and New Approaches

LEARNING OBJECTIVES

- Know the classic functions of managers—planning, organizing, staffing, controlling, and directing
- Understand the purpose and components of a management information system (MIS)
- Understand how computer networking and related software have flattened the classic management pyramid
- Be aware that many companies use employees in task-oriented teams
- Be aware of sophisticated software for top managers
- Understand the problems and solutions related to managing personal computers

CLASSIC MANAGEMENT FUNCTIONS
MIS FOR MANAGERS
THE NEW MANAGEMENT MODEL
   A Flattened Pyramid
   The Impact of Groupware
   Teamwork
TOP MANAGERS AND COMPUTERS
   Decision Support Systems
   Executive Support Systems
MANAGING PERSONAL COMPUTERS
   The Personal Computer Manager
   Personal Computer Acquisition
   The Information Center
   Dumping Technology on Workers
   Do You Even Know Where Your PCs Are?
LEADING BUSINESS INTO THE FUTURE

Classic Management Functions

Managers historically have had five main functions:

- **Planning,** or devising both short-range and long-range plans for the organization and setting goals to help achieve the plans
- **Organizing,** or deciding how to use resources, such as people and materials
- **Staffing,** or hiring and training workers
- **Directing,** or guiding employees to perform their work in a way that supports the organization's goals
- **Controlling,** or monitoring the organization's progress toward reaching its goals

All managers perform these functions as part of their jobs. The level of responsibility regarding these functions, however, varies with the level of the manager. The levels of managers are traditionally represented as a pyramid, with the fewest managers at the top and the largest numbers at the lowest level (Figure 10-1). Often you will hear the terms *strategic, tactical,* and *operational* associated with high-level managers, middle-level managers, and low-level managers, respectively.

Whether the head of General Electric or of an electrical appliance store, a high-level manager must be concerned with the long-range view—the *strategic* level of management. For this manager, usually called an executive, the main focus is **planning.** Consider a survey showing that Ameri-

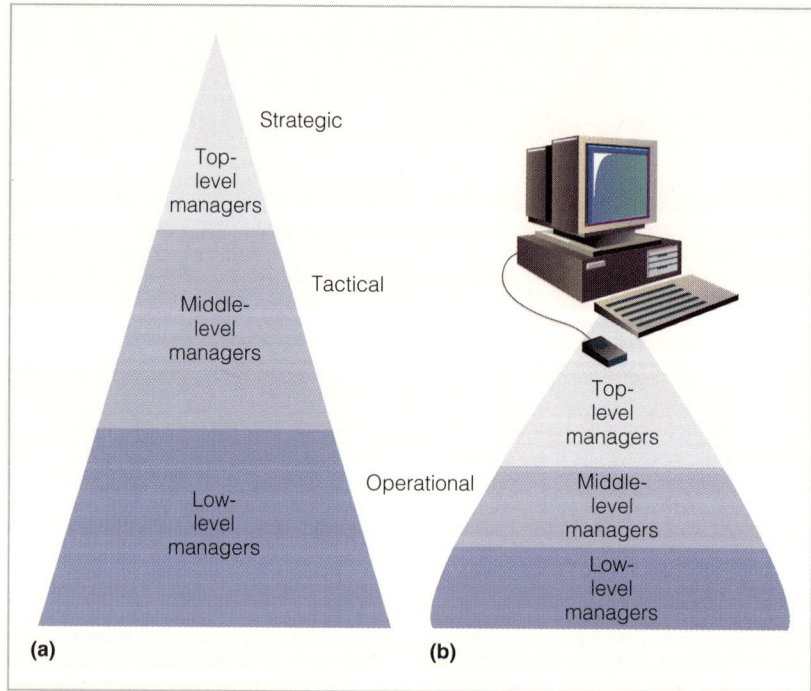

Figure 10-1 The management pyramid. (a) The classic view of management functions involves a pyramid featuring top managers handling strategic long-range planning, middle managers focusing on the tactical issues of organization and personnel, and low-level managers directing and controlling day-to-day operations. (b) The increasing use of networked personal computers in business is squeezing out mid- and low-level managers, thus flattening the pyramid.

cans want family vacations and want the flexibility and economy of a motor vehicle; however, they also want more space than the family car provides. To the president of a major auto company, this information may suggest further opportunities for expansion of the recreational vehicle line.

The middle-level manager of that same company must be able to take a somewhat different view because his or her main concern is the *tactical* level of management. The middle manager will prepare to carry out the visions of the top-level managers, assembling the material and personnel resources to do the job. Note that these tasks focus on **organizing** and **staffing**. Suppose the public is inclined to buy more recreational vehicles. To a production vice president, this may mean organizing production lines using people with the right skills at the right wage and perhaps farming out portions of the assembly that can be done by less expensive, less skilled labor.

The low-level manager, usually known as a supervisor, is primarily concerned with the *operational* level of management. For the supervisor, the focus is on **directing** and **controlling**. Workers must be directed to perform the planned activities, and the supervisor must monitor progress closely. The supervisor—an assembly-line supervisor in our recreational vehicle example—is involved in a number of issues: making sure that workers have the parts they need, checking employee attendance, maintaining quality control, handling complaints, keeping a close watch on the schedule, tracking costs, and much more.

To make decisions about planning, organizing, staffing, directing, and controlling, managers need data that is organized in a way that is useful for them. An effective management information system can provide it.

MIS for Managers

A **management information system (MIS)** may be defined as a set of formal business systems designed to provide information for an organization. (Incidentally, you may hear the term *MIS system*, even though the *S* in the abbreviation stands for *System*; this is an accepted redundancy.) Whether or not such a system is called an MIS, every company has one. Even managers who make hunch-based decisions are operating with some sort of information system—one based on their experience. The kind of MIS we are concerned with here includes one or more computers as components. Information serves no purpose until it gets to its users. Timeliness is important, and the computer can act quickly to produce information.

The extent of a computerized MIS varies from company to company, but the most effective kinds are those that are integrated. An integrated MIS incorporates all five managerial functions—planning, organizing, staffing, directing, and controlling—throughout the company, from typing to top-executive forecasting. An integrated management computer system uses the computer to solve problems for an entire organization, instead of attacking them piecemeal. Although in many companies the complete integrated system is still only an idea, the functional aspects of MISs are expanding rapidly in many organizations.

The **MIS manager** runs the MIS department. This person's position has been called information resource manager, director of information services, chief information officer, and a variety of other titles. In any case whoever serves in this capacity should be comfortable with both computer technology and the organization's business.

The New Management Model

The traditional management pyramid that we discussed earlier means a very specific kind of communication. An executive has time to communicate with perhaps a handful of people. Each of these people can convey information to another five or six people below them. Information trickles down, layer by layer, either in meetings or more informally.

A Flattened Pyramid

Enter the computer network. Networks connect people to people, and people to data. Using e-mail, or perhaps groupware, information can be disseminated companywide as fast as fingers can fly over a computer keyboard. So much for passing along information through traditional hierarchical channels. The dispersion of information via the network has caused the traditional management pyramid to become flatter in structure and more physically distributed.

What are managers, so long the keepers of information, supposed to do now that, via the network, information is so freely available to so many workers? A good part of a manager's job, communicating above and below, has been replaced by the flow of information through the network. Many industries are finding that, to some extent, they can do without middle managers and have eliminated certain positions. Managers on all levels still have plenty to do, but they are doing it a bit differently from the ways of the past. Networks irrevocably alter the nature of managerial authority and work.

The Impact of Groupware

Consider the impact of groupware on worker interaction. As we discussed in Chapter 9, groupware permits information to be assembled in central databases. People working on a project contribute information to a database and can see and use information contributed by others.

The introduction of groupware can be a searing experience for some managers. Two reasons for this are changes in the way information is shared and changes in managerial authority. People acquire power in an organization by knowing things that others do not. Managers may feel threatened by groupware because they are not accustomed to unstructured information sharing. Studies have shown that groupware works best in organizations where there is already a fairly flexible corporate culture—that is, where an attitude of sharing, and even egalitarianism, already exists. Perhaps even more painful, managers may not enjoy being in the electronic spotlight when decisions that were once theirs alone are now fair game for comment and change by everyone involved. Furthermore, in contrast to organizations whose focal point is the manager, groupware supports organizations that are team-based and information-driven.

Consider another change from the old ways. Say a particular aspect of a project requires collaboration between two people, Jack and Jim, from different operational units under different management. The traditional way of doing business is for Jack to go to his manager, and then Jack's manager talks to Jim's manager, who talks to Jim. Then the information flows in reverse, back to Jack. Up, across, down, and then back again. Now it is possible to accomplish the same communication using groupware. Infor-

MAKING THE RIGHT CONNECTIONS

Remote Users

Many companies want their sales representatives out of the office, both to reduce office costs and to make the sales effort more effective. But these representatives must have adequate access to computer data. Offering remote users access to data residing in computers in the office frees them from having to carry around large amounts of data. Sales people using laptop computers can connect to the home office to download pricing information from the mainframe to the laptop computer being used in the field and to upload order entries from the field to the mainframe computer. The connection also manages electronic mail.

MIS managers have a variety of concerns about remote users and their access to information from the company's mainframe. The first concern is security. Remote users, at the least, should use a password when making connections. Training is a consistent problem because road warriors seldom come into the office for extended periods. Finally, they worry about lost or stolen equipment.

mation moves laterally, from worker to worker, saving a roundabout trip through the management maze.

As many managers have discovered, networks make leadership much harder. No longer able to look over their employees' shoulders, managers are learning to rely on other management techniques. They must first give careful attention to the selection and training of employees. Secondly, managers must set clear expectations for their employees. But most importantly, managers must use customer satisfaction as a measuring stick of employee performance in a networked environment.

Teamwork

The availability of networks and groupware coincides nicely with the concept of organizing employees into task-focused teams. Just as the manager is no longer the sole dispenser of wisdom and decisions, so the employee is no longer merely an individual in a static organization. Many companies are organizing their employees in teams. But a team has no permanence; work and people are organized around tasks. When a task is complete the team is disbursed. When a new task is being tackled, a new team is assembled. Each team is composed of people whose skills are needed for the task at hand. In this kind of work environment, reorganization is a way of life (Figure 10-2).

Experts consider eight people an ideal team size. If a team gets much bigger than that, team members spend too much time communicating what is already inside their heads instead of applying that knowledge to their parts of the task. But what if the team is behind schedule? Imagine a status meeting in which it is revealed that a critical activity is behind schedule. The activity under scrutiny is to finalize product specifications for an electronic hoop, a toy that can be manipulated remotely and is expected to be a big hit for the upcoming holiday gift-buying season. This activity is critical because other activities down the line, including manufacturing

(a)

(b)

Figure 10-2 **Business teams.** Whether they are (a) planning a new stadium, or (b) planning financial strategies, many companies put employees in teams.

and promotion, cannot begin until specifications are complete. The most common response to tardy projects is to add more people to the project—exactly the wrong thing to do. If outsiders are belatedly added to an existing team, the project quickly comes to a halt in order to bring the new people on board. And, of course, from that moment forward, there are more people with whom one must communicate.

What is the proper solution? There is no ideal answer, other than to plan better in the first place, but most organizations find that the better part of wisdom is to rely on the commitment of the original team members. The good news, however, is that a properly composed team of an ideal size is less likely to get in trouble. Communication remains easy, and each member retains a strong sense of responsibility and participation while benefiting greatly from the contributions of teammates.

Top Managers and Computers

Since the early days of computing, managers at all levels have had computer support in the form of printed reports. In the last decade most managers, even the most resistant executives, have succumbed to the personal computer. Managers have found personal computer software useful for every aspect of their jobs, from something as simple as sending an e-mail message to complex chores such as designing a compensation package for a thousand employees. For top managers, executives who must have the vision to guide the entire company, sophisticated software is needed.

Decision Support Systems

Imagine yourself as an executive trying to deal with a constantly changing environment, having to consider changes in competition, in technology, in consumer habits, in government regulations, in union demands, and so on.

ONE JUMP AHEAD

Your Electronic Agent

You have an important job and an activity-packed life. You long ago stopped traipsing from store to store shopping for particular goods or services; instead you use catalogs or the yellow pages and the telephone. A new camera? Flowers for a birthday? A Mexican restaurant? A vacation? The next logical step is to dispatch your own electronic agent to track down the best deal and, if you wish, purchase it for you.

The electronic agent, sometimes called a software agent, takes an order and goes shopping for you throughout the network. Suppose you have read some consumer articles and have decided that you want a camera weighing less than a pound with an automatic zoom lens for under $300. You type in these instructions to your computer and let your software agent do the work.

While you go on with other tasks, the agent goes to a directory, finds camera stores, and then sifts through their camera inventories, looking for the required features and price. Eventually, the agent accumulates a list of acceptable choices, which is presented on your personal computer screen at your convenience. You can then choose which one, if any, to purchase. A variation on this option is available: You can give your agent advance authority to purchase a suitable camera with the best price, probably charging it to a credit card.

Someone eavesdropping on a discussion of agents might think that a real person was being described. The agent, of course, is just sophisticated software. Agent software is available today. What is slow in coming, however, is signing up businesses to pay to make their list of goods and services available to the network. It is a chicken-and-egg problem: Users will not flock to an online agency until many merchants and services are online, but the merchants and service providers will not sign up until there are many users.

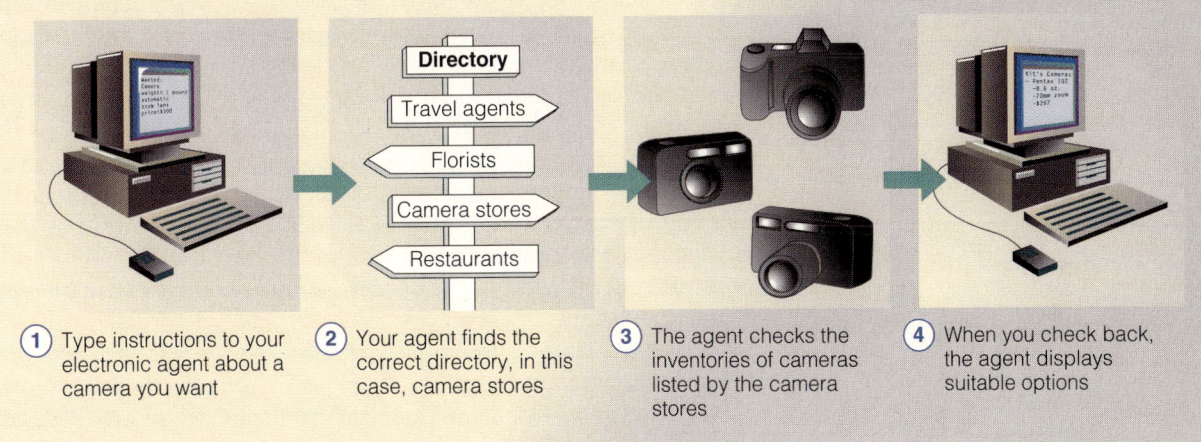

① Type instructions to your electronic agent about a camera you want

② Your agent finds the correct directory, in this case, camera stores

③ The agent checks the inventories of cameras listed by the camera stores

④ When you check back, the agent displays suitable options

How are you going to make decisions about those matters for which there are no precedents? In fact, making one-of-a-kind decisions—decisions that no one has had to make before—is the real test of a manager's mettle. In such a situation you would probably wish you could turn to someone and ask a few "what-if" questions (Figure 10-3).

"What if . . . ?" That is the question business people want answered, especially when considering new situations. A **decision support system (DSS)** is a computer system that supports managers in nonroutine decision-making tasks. The key ingredient of a decision support system is a modeling process. A **model** is a mathematical representation of a real-life system. A mathematical model can be computerized. Like any computer program, the model can use inputs to produce outputs. The inputs to a model are called **independent variables** because they can change; the outputs are called **dependent variables** because they depend on the inputs.

Figure 10-3 Making decisions with the help of a computer. Business people use computers to try out different scenarios, without investing a great deal of time and money.

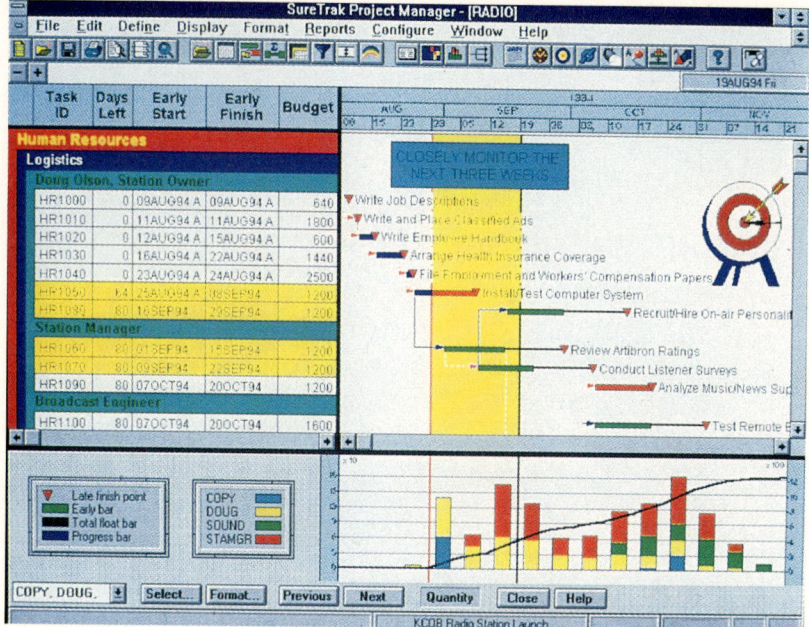

Consider this example. Suppose, as a manager, you have the task of deciding which property to purchase for one of your manufacturing plants. You have many factors to consider: the appraised value, asking price, interest rate, down payment required, and so on. These are all independent variables—the data that will be fed into the computer model of the purchase. The dependent variables, computed on the basis of the inputs, are the effect on your cash resources, long-term debt, and ability to make other investments. To increase complexity, we could add that the availability of workers and nearness to markets are also input factors. Increasing the complexity is appropriate, in fact, because decision support systems often work with problems that are more complex than any one individual can handle.

Using a computer model to reach a decision about a real-life situation is called **simulation.** It is a game of "let's pretend." You plan the independent variables—the inputs—and you examine how the model behaves based on the dependent variables—the outputs—it produces. If you wish, you may change the inputs and continue experimenting. This is a relatively inexpensive way to simulate life situations, and it is considerably faster than the real thing.

The decision-making process must be fast, so the DSS is interactive: The user is in direct communication with the computer system and can affect its activities. In addition, most DSSs cross departmental lines so that information can be pulled from the databases of a variety of sources, such as marketing and sales, accounting and finance, production, and research and development. A manager trying to make a decision about developing a new product, for example, needs information from all these sources.

A decision support system does not replace a MIS; instead, a DSS supplements a MIS. There are distinct differences between them. MIS emphasizes planned reports on a variety of subjects; DSS focuses on decision making. MIS is standard, scheduled, structured, routine; DSS is quite unstructured and available on request. MIS is constrained by the organizational system; DSS is immediate and friendly.

Executive Support Systems

Top-level executives and decision makers face unique decision-making problems and pressures. An **executive support system (ESS)** is a decision support system especially made for senior-level executives. An executive support system is concerned with how decisions affect an entire organization. An ESS must take into consideration

- The overall vision or broad view of company goals
- Strategic long-term planning and objectives
- Organizational structure
- Staffing and labor relations
- Crisis management
- Strategic control and monitoring of overall operations

Executive decision-making also requires access to outside information from competitors, federal authorities, trade groups, consultants, and news-gathering agencies, among others. A high degree of uncertainty and a future orientation are involved in most executive decisions. Successful ESS software must therefore be easy to use, flexible, and customizable.

Several commercial software packages are available for specific modeling purposes. The purpose might be marketing, sales, or advertising. Other packages that are more general provide rudimentary modeling but let you customize the model for different purposes—budgeting, planning, or risk analysis.

 ## Managing Personal Computers

Personal computers burst on the business scene in the early 1980s with little warning and less planning. The experience of the Rayer International Paper Company is typical. One day a personal computer appeared on the desk of engineer Mike Burton—he had brought his in from home. Then accountants Sandy Dean and Mike Molyneaux got a pair of machines—they had squeezed the money for them out of the overhead budget. Nobuko Locke, the personnel manager, got personal computers for herself and her three assistants in the company's far-flung branch offices. And so it went, with personal computers popping up all over the company. Managers realized that the reason for runaway purchases was that personal computers were so affordable: Most departments could pay for them out of existing budgets, so the purchasers did not have to ask anyone's permission.

Managers, at first, were tolerant. There were no provisions for managing the purchase or use of personal computers, and there certainly was no rule against them. And it was soon apparent that these machines were more than toys. Pioneer users had no trouble justifying their purchases—their increased productivity spoke for them. In addition to mastering software for word processing, spreadsheets, and database access, these users declared their independence from the MIS department.

Managers, however, were soon faced with several problems. The first was that no one person was in charge of the headlong plunge into personal computers. The second problem was incompatibility—the new computers came in an assortment of brands and models and did not mesh well. Software that worked on one machine did not necessarily work on another. Third, users were not as independent of the MIS department as they had thought—they needed assistance in a variety of ways. In particular, they

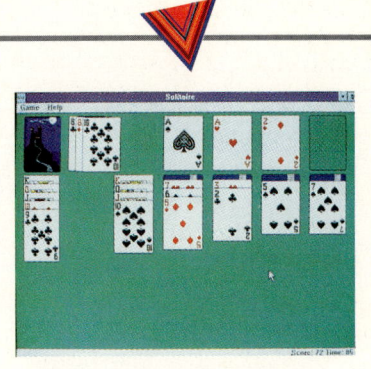

Looking Intently at the Screen

Suppose you saw an office worker with a furrowed brow, obviously involved in his work on the computer. But look more closely—is it a spreadsheet or a database inquiry on the screen? Why no, it is a game of solitaire. This fellow is goofing off at work.

Should he be playing solitaire on his personal computer at the office? Maybe. The game was included as an extra in Microsoft Windows, partly to promote friendliness to home users. Its popularity was rather a surprise, a not especially pleasant one, to the companies whose employees are using Windows.

Some managers have adopted the attitude that a little relaxation with a computer game relieves stress, but many more have reacted negatively, prohibiting employees from playing games at the office. Some have gone so far as to remove all games from the company's personal computers.

Troubleshooting Personal Computers

"The computer is down." The generic phrase, referring to failure of some kind, has entered the public lexicon. If your home computer is down, it is an aggravation. If your personal computer in a corporate setting is down, it is a serious problem. But when the personal computer that runs your small business is down, it is a crisis. Customers know exactly what the phrase means: It means they cannot get what they want when they want it.

Here are some hints about what to do when a standalone—unnetworked—personal computer is down. First, if possible, make a complete backup of all computer files. Keep notes of remedies you try.

Change only one thing at a time; for example, if searching for a bad cable, change only one cable at a time. Try telephone help, if it is available, probably from a software product helpline. If you do talk to a helpline, describe your problem as specifically as possible. For example, rather than saying "I can't get it to print right," say something such as "I am working in Microsoft Word, trying to print a memo on a LaserJet 4P/4MP printer. It does print, but the output is a jumbled mess of unrecognizable letters." Be prepared to give more information, as requested.

---

needed data that was in the hands of the MIS department. In addition, companies were soon past the stage of the initial enthusiasts; they wanted all kinds of workers to have personal computers, and those workers needed training. Furthermore, in just a few years, most companies networked their computers together, bringing a whole new set of responsibilities and problems. Finally, many companies had so many personal computers that they did not know how many, or where they were, or what software was on them. Many organizations solved these management problems in these ways:

- They corrected the management problem by creating a new position called the personal computer manager, which often evolved to the network manager.
- They addressed the compatibility problem by establishing acquisition policies.
- They solved the assistance problem by creating information centers and providing a variety of training opportunities.
- They used software to locate, count, and inventory their personal computers.

Let us examine each of these solutions.

The Personal Computer Manager

The benefits of personal computers for the individual user have been clear almost from the beginning: increased productivity, worker enthusiasm, and easier access to information. But once personal computers move beyond entry status, standard corporate accountability becomes a factor. Large companies are spending millions of dollars on personal computers, and top-level managers want to know where all this money is going. Company auditors begin worrying about data security. The company legal department begins to worry about workers illegally copying software. Before long everyone is involved, and it is clear that someone must be placed in charge of personal computer use. That person is the **personal computer manager.**

There are four key areas that need the attention of this manager:

- **Technology overload.** The personal computer manager must maintain a clear vision of company goals so that users are not overwhelmed by the massive and conflicting claims of aggressive vendors plying their wares. Users engulfed by phrases like *network topologies* and *file gateways* or a jumble of acronyms can turn to the personal computer manager for guidance.
- **Cost control.** Many people who work with personal computers believe the initial costs are paid back rapidly, and they think that should satisfy managers who hound them about expenses. But the real costs entail training, support, hardware and software extras, and communications networks—much more than just the computer itself. The personal computer manager's role includes monitoring all related expenses.
- **Data security and integrity.** Access to corporate data is a touchy issue. Many personal computer users find they want to download (or access) data from the corporate mainframe to their own machines, and this presents an array of problems. Are they entitled to the data? Will they manipulate the data in new ways and then present it as the official version? Will they expect the MIS to take the data back after they have

done who-knows-what with it? The answers to these perplexing questions are not always clear-cut, but at least the personal computer manager will be tuned in to the issues.
- **Computer junkies.** And what about employees who are feverish with the new power and freedom of the computer? When they are in school, these user-abusers are sometimes called hackers; on the job they are often called junkies because their fascination with the computer seems like an addiction. Unable to resist the allure of the machine, they overuse it and neglect their other work. Personal computer managers usually respond to this problem by setting down guidelines for computer use.

The person selected to be the personal computer manager is usually from the MIS area. Ideally, this person has a broad technical background, understands both the potential and limitations of personal computers, and is well known to a diverse group of users.

With the advent of networking, the personal computer manager is often the same person as the **network manager** or, if the network is a local area network, the **LAN manager.** The network manager must keep the network operational. The basic tasks are the ability to let network users share program and data files and resources such as printers. The network manager is responsible for installing all software on the network and making sure that existing software runs smoothly. The network manager also must make sure that backup copies are made of all files at regular intervals. In addition, the network must be kept free from viruses, illegal software intrusions that we will study further in Chapter 11. The greatest challenge may be to make sure that the network has no unauthorized users.

Company managers often underestimate the amount of work it takes to keep even a small network going. In a large company an individual or even an entire team of people may be dedicated to this task. In a small company the network may be managed by someone who already has a full-time job at the company.

Personal Computer Acquisition

As we noted, workers initially purchased personal computers before any companywide or even officewide policies had been set. The resulting compatibility problems meant that they could not easily communicate or share data. Consider this example: A user's budgeting process calls for certain data that resides in the files of another worker's personal computer or perhaps involves figures output by the computer of a third person. If the software and machines these people use do not mesh, compatibility becomes a major problem.

In many companies MIS departments have now taken control of personal computer acquisition. The methods vary, but they often include establishing standards and restricting the number of vendors used. Most companies now have established standards for personal computers, for the software that will run on them, and for data communications. Commonly, users must stay within established standards so they can tie into corporate resources. Some companies limit the number of vendors—sellers of hardware and software—from whom they allow purchases. Managers have discovered they can prevent most user complaints about incompatibility, not to mention getting a volume discount, by allowing products from just a handful of vendors.

The Information Center

The company **information center,** often called by other names such as *support center,* offers help to users in several forms. Although no two are alike, a typical information center gives users support for the users' equipment. The information center is devoted exclusively to giving users service. And, best of all, user assistance is immediate, with little or no red tape.

Information center services often include the following:

- **Software selection.** Information center staff helps users determine which software packages suit their needs.
- **Data access.** If appropriate, the staff helps users get data from the large corporate computer systems for use on the users' computers.
- **Training.** Education is a principal reason for an information center's existence. Classes are usually small, frequent, and on a variety of topics (Figure 10-4). The information center is not the only form of training, however; we will discuss training in more detail shortly.
- **Technical assistance.** Information center staff members stand ready to assist in any way possible, short of actually doing the users' work for them. That help includes advising on company standards for hardware purchases, aiding in the selection and use of software, finding errors, helping submit formal requests to the MIS department, and so forth.

To be successful, the information center must be placed in an easily accessible location. The center should be equipped with personal computers and terminals, a stockpile of software packages, and perhaps a library. It should be staffed with people who have a technical background but whose explanations feature plain English. Their mandate is, the user comes first.

Dumping Technology on Workers

The title is a sham. Any manager knows that simply "dumping" technology—hardware, software, networks, whatever—on workers in the

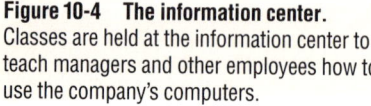

Figure 10-4 The information center. Classes are held at the information center to teach managers and other employees how to use the company's computers.

hope of increased productivity would be a disaster. The first obvious approach is to provide training for the new technology, whatever it is.

Training: Pay Now or Pay Later

Organizations tend to be remiss about training. Years ago, vendors typically included training as part of the hardware or software package. Once training became a separate item with a separate price tag, organizations were more apt to think they could get along without it. Furthermore, although training was once needed for just a few technical workers, now training is needed for entire populations of workers companywide.

Those who do offer training too often rely on the one-shot teacher-in-the-classroom model. This traditional approach, however, does not work well. To begin with, unless the classes are off-site or attendance is rigorously enforced, participation may be sporadic because employees are much more concerned about the real work that has been abandoned on their desks. Furthermore, especially when new software is the topic, training of two days or two weeks, even hands-on in a computer-stocked classroom, yields minimal results.

Workers adopting new technology do need initial training, but they also need follow-up support. One approach that seems to work well is to cultivate home-grown gurus. When confronted with a computer problem, the first instinct of a baffled user is to consult a more knowledgeable friend or colleague. With this in mind, savvy companies, after the first round of training, ask for volunteers who would like to learn more. These users become the office gurus on that technology. Initially, they may not know a lot more than their colleagues, but they are usually a bit ahead and, by sheer numbers of consultations, accumulate more knowledge than other workers.

In-house support, such as the information center described above, can be a big factor in the success of new technology. The best guarantee of workers absorbing training, however, is prior motivation, achieved by getting them involved.

Involving the Workers

The catch-phrase often used is *empowering the workers*. It is a variation on the systems analyst precept of user involvement. Rather than simply installing new technology and training the workers, begin with the workers—the people who will be using the technology. To put it another way, deal with the people at the same time as you deal with the technology.

Paine Webber, a stock brokerage, offers a model approach. Paine Webber wanted to upgrade their brokers' ten-year-old computer system to a network that would offer far more information access and control. The systems analyst began by surveying the attitudes of the 5000-plus brokers. He discovered that approximately one third of the brokers felt the current system met their needs, another third thought they would like some improvements, and the final third thought that the current system was hopelessly outmoded.

The company's response was to build a dazzling new system with the old system built into it. Paine Webber unfolded the new system branch by branch, emphasizing not the wonders of technology but what the system could do for brokers. The instructors were not technical types but specially trained Paine Webber employees who already knew the brokerage business. This worked well for everyone, even those who were initially reluctant.

Workers whose comfort level was the old system could begin with that version, but most of them gradually picked up the features of the new system.

Finally, regarding worker involvement, do not forget the generation gap. Employees who grew up playing Nintendo games have a built-in advantage over their elders, who may show the foot-dragging signs of a precomputer upbringing. In fact, training experts recommend that big-time computerphobes be loosened up by playing computer games. This way they will at least become comfortable with a mouse and with interactions that cause changes on the computer screen.

Do You Even Know Where Your PCs Are?

Many corporate administrators, when put to the test, are embarrassed to admit that they do not know where corporate personal computers are in use in the company; in fact, they do not even know how many there are. One manager, for example, was quite certain that the company had 600 personal computers and an average of 12 users per printer. The reality turned out to be quite different; there were 1100 computers and 1 printer per computer. To make matters even worse, managers have no idea what software is on the computers—Microsoft Excel or WordPerfect, or perhaps the latest incarnation of Doom.

This is a critical problem because administrators have no idea how to budget for their personal computers. Contrary to folklore, personal computers cost much more than their original modest price. The initial purchase price of a corporate personal computer accounts for only 10 percent of its lifetime cost. The rest is spent on troubleshooting, administration, software, and training. If the computers are hidden, then so are the costs of maintaining them. Clearly, administrators must confront the missing-computers problem.

Specialized computer services now offer a sort of lost-and-found for personal computers and related equipment. Corporate personal computers that are networked—and that means most of them—can be counted and interrogated by software set up on the network. The polling software not only counts computers but also determines their components and software.

Once companies get a handle on what they have, they can begin containing costs.

Leading Business into the Future

Who will manage businesses in the future? Someone once remarked, somewhat facetiously, that all top management—presidents, chief executive officers (CEOs), and so forth—should be drawn from the MIS ranks. After all, the argument goes, computers pervade the entire company, and people who work with computer systems can bring broad experience to the job. Today, most presidents and CEOs still come from legal, financial, or marketing backgrounds. But as the computer industry and its professionals mature, that pattern could change.

Another challenge for managers at every level is the security and integrity of corporate data. We investigate security, privacy, and ethical issues in the next chapter.

CHAPTER REVIEW

Summary and Key Terms

- All managers have five main functions: **planning, organizing, staffing, directing,** and **controlling.** A management pyramid shows that top-level managers focus primarily on strategic functions, especially long-range planning; middle-level managers focus on the tactical, especially the organizing and staffing required to implement plans; and low-level managers are concerned mainly with operational functions, controlling schedules, costs, and quality as well as directing personnel.
- A **management information system** (**MIS**) is a set of business systems designed to provide information for decision making. A computerized MIS is most effective if it is integrated.
- The **MIS manager,** a person familiar with both computer technology and the organization's business, runs the MIS department.
- The traditional management pyramid has been flattened by the dissemination and sharing of information over computer networks. The impact of groupware has removed exclusive manager access to information and forced managers to share decision making. Some companies are organizing workers into teams around tasks.
- A **decision support system** (**DSS**) is a computer system that supports managers in nonroutine decision-making tasks. A DSS involves a **model,** a mathematical representation of a real-life situation. A computerized model allows a manager to try various "what-if" options by varying the inputs, or **independent variables,** to see how they affect the outputs, or **dependent variables.** The use of a computer model to reach a decision about a real-life situation is called **simulation.** Since the decision-making process must be fast, the DSS is interactive, allowing the user to communicate directly with the computer system and affect its activities.
- An **executive support system** (**ESS**) is a decision support system for senior-level executives, who make decisions that affect an entire company.
- When personal computers first became popular in the business world, most businesses did not have general policies regarding them, which led to several problems. Many businesses created the position of **personal computer manager** (later called the **network manager** or **LAN manager**) to ensure coordination of personal computers, established acquisition policies to solve the compatibility problem, established **information centers** to provide assistance to users, provided formal and informal training for users, and used software to monitor their existing personal computers.

Discussion Questions

1. Suppose a team of eight people in a construction firm is designing a new hospital. The team members, drawn from several departments, include two engineers, two architects, an electrician, a plumber, a graphics designer, and a planner. How and by whom might the classic management functions be carried out?
2. Describe a problem situation that could be simulated through a decision support system. Specify the input factors and the types of output.
3. What special pressures might be on a network manager?

Student Study Guide

True/False

T F 1. The information center typically offers users training and assistance.
T F 2. The function of the network manager is to help executives with decision support systems.
T F 3. Communication of information is most efficient through the traditional management pyramid.
T F 4. Decision support systems help managers in nonroutine decision-making tasks.
T F 5. A model is a mathematical representation of an artificial situation.
T F 6. Middle-level managers focus on planning.
T F 7. Inputs to a model are called independent variables.
T F 8. Groupware is usually focused on groups of executives.
T F 9. The use of personal computers by managers is declining.
T F 10. Simulation is using a model to answer real-life situations.

Answers

True/False
1. T   6. F
2. F   7. T
3. F   8. F
4. T   9. F
5. F   10. T

PLANET INTERNET

Resources for Living

Need information? Need information fast? Whether commonplace or rare, any information you may need is probably somewhere on the Internet.

Government resources. The government had a head start and has made excellent use of the Internet. We have already mentioned the White House, but you can also use the resources of the Library of Congress, whose opening screen is shown here, or contact the United States House of Representatives or even the CIA. You may peruse recent Supreme Court decisions by topic or by case name. And, although you may have little inclination, you can access the Internal Revenue Service site to get forms or advice. Finally, how would you like a crack at reducing the deficit? The site called Balance the Budget lets you do just that.

News you can use. Consider bits of information you might need in any given week. Weather forecast for your travel destination? Every sort of weather information is available, for regions and individual cities. Buying a new or used car? Pricing information is available just a computer away. As you would expect, consumer information is available on just about any topic. Want to plan ahead for natural calamities? You can get serious advice on food supplies and survival in the event of an emergency from sources on the Net. Would you just like a good book? Project Gutenberg makes books available online.

Finding like-minded folks. *Usenet,* a network intertwined with the Internet, offers *newsgroups,* special groups set up by people who share a common interest. Usenet computers store messages sent by users and periodically forward them to other Usenet computers. Using your browser software, you can access Usenet to read messages contributed by others and perhaps add some of your own. Usenet newsgroups are arranged in topical hierarchies, with focus shifting from broad to narrow. The major categories are quite general, for example, computers or business. Categories are subdivided by topic, which is further subdivided into newsgroups and then subjects. The broadest category by far is the Alternative category, which has hundreds of topics ranging from aromatherapy to Elvis to Rhodesia. Your browser will let you single out newsgroups in which you have a particular interest and then, on command, pick up new messages from just those newsgroups.

Different kinds of people participate in newsgroups. Some are experts who dispense wisdom, and others are neophytes who are there to soak it up. Some users, called *flamers,* like to respond to messages with personal insults. New users are often *lurkers,* people who read messages but want to learn about the group dynamics before contributing anything of their own.

Reaching out. Nonprofit organizations such as Impact Online, whose screen is shown here, use the Net to send their messages. Impact Online helps people get involved with nonprofits nationwide through the use of technology. One advantage of a truly *worldwide* web is the possibility of addressing worldwide issues. The site for Friends of the Earth, an environmental group, is in Britain.

Internet Exercises

1. **Structured exercise.** Begin with the B/C URL http://www.aw.com/bc/planet/ and take your turn at balancing the budget.
2. **Freeform exercise.** Venture into Usenet. Use your browser to access newsgroups and read messages from topics that interest you.

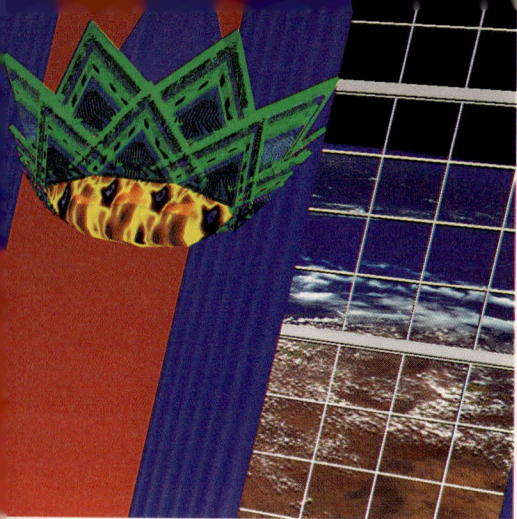

Melvin Pang is the administrative assistant to the head of the Business Division at Ballinger Community College. His responsibilities include setting up meetings, coordinating classes and classrooms, assisting faculty and students, and supervising two secretaries. For these and other tasks, Melvin uses word processing, spreadsheet, and scheduling software on his personal computer, producing dozens of files each week.

Melvin knows that he is responsible for the safety and well-being of his computer files. In fact, he has attended training seminars on this very subject. In particular, he learned that it is prudent to make extra copies of his files, so that his work will not be impaired if the original files on hard disk are accidentally destroyed. As a class assignment, Melvin wrote down all the reasons a person might neglect to back up files properly. His list was as follows: (1) It takes too much time. (2) It is too boring, just nuisance work. (3) I have more pressing tasks. (4) I have used this computer for a year and nothing has gone wrong yet. Melvin vowed to avoid these traps. But, somehow, knowing what he should do was not enough. On a sunny Thursday afternoon, Melvin's hard drive crashed and the files on his hard drive were destroyed. Subsequent inspection revealed that less than 40 percent of his files had been backed up—copied to another place.

Melvin asked himself just one question: Why? Why, indeed! How could he have been so careless, so thoughtless? While Melvin castigates himself, we can reflect on human nature. We tend to think that bad things happen to other people, not us. It is hard to perform chores consistently because something bad *might* happen. Instead, we spend our time on other tasks and put our files at risk.

The good news is that users in a business setting today are likely to have procedures in place for regular file backup. The user most at risk is the individual who uses a computer in another environment, probably home or school. Although this chapter covers a variety of threats to computer systems, the most common by far is the loss of files due to improper backup techniques.

Chapter 11

Computer Issues in the Workplace

Security, Privacy, and Ethics

LEARNING OBJECTIVES

- Become aware of the problem of computer crime, including criminal profiles, types of crimes, and the difficulties of discovery and prosecution
- Become aware of the need for security, including disaster recovery plans, software and data security, and security legislation
- Understand the importance of privacy and how it is affected by the computer age
- Understand the importance of ethics as related to a computer environment

INTANGIBLE ISSUES IN THE WORKPLACE
COMPUTER CRIME
   Who Is the Computer Criminal?
   Types and Methods of Computer Crime
   Discovery and Prosecution
SECURITY: PLAYING IT SAFE
   Identification and Access: Who Goes There?
   When Disaster Strikes: What Do You Have to Lose?
   Disaster Recovery Plan
   Software Security
   Data Security
   Worms and Viruses
   Network Security
   Personal Computer Security
   Prepare for the Worst: Back Up Your Files
PRIVACY: KEEPING PERSONAL INFORMATION PERSONAL
   Passing Your Data Around
   Privacy Legislation
   Privacy in the Workplace
A MATTER OF ETHICS
COPYING SOFTWARE
   OK if I Copy That Software?
   Why Those Extra Copies?
   Site Licensing
   The Battle Continues

Intangible Issues in the Workplace

After some study and some experience, it should seem that a computer is something tangible, something you can handle, even if in limited fashion. The same could probably be said for its trappings—printer, disks, software, and the like. Issues such as security, privacy, and ethics seem murky by comparison. However, each of these three presents a set of dangers that could become real problems for the unwary user.

Here are some simple examples that could apply to you, even if your current computer use is limited to an academic environment. From a security standpoint, your data files (such as term papers or spreadsheets you have created using applications software) could be accidentally destroyed. Privacy could be an issue if someone snoops in your data files and, perhaps, reads a letter you have written. And ethics? Whether you realize it or not, you would have an ethics problem if you let an unauthorized person use software entrusted to you for your own use.

In a workplace environment the problems become more pronounced, due to the larger numbers of people and equipment and the seriousness of the potential problems. Furthermore, as an employee, there is a greater chance that you will bear responsibility for a problem, rather than just being a victim or bystander. There are many, many possibilities, but here are some hypothetical examples:

- What if company data files that are your responsibility are destroyed, but you failed to provide backup copies?
- What if you dump computer printouts containing information about a new product into the trash, where they are later retrieved and passed on to a company competitor?
- What if you send a gossipy e-mail message to a co-worker and then delete it, only to discover that it was read anyway by your manager?
- What if you neglect to secure your personal computer and arrive the next morning to find it stolen?
- What if you fail to guard software disks that are then borrowed and illegally copied?

We will address these issues and many more as we march through security, privacy, and ethics. However, we begin with a fascinating aspect of the security problem: computer crime.

 ## Computer Crime

It was 5 o'clock in the morning, and 14-year-old Randy Miller was startled to see a man climbing in his bedroom window. "FBI," the man announced, "and that computer is mine." So ended the computer caper in San Diego, where 23 teenagers, ages 13 to 17, had used their home computers to invade systems as far away as Massachusetts. The teenagers are **hackers**, people who attempt to gain access to computer systems illegally, usually from a personal computer, via a data communications network.

The term *hacker* used to mean a person with significant computer expertise, but the term has taken on the more sinister meaning with the advent of teenage computer miscreants. In this case the hackers did not use the system to steal money or property. But they did create fictitious accounts and destroyed or changed some data files. The FBI's entry through the window was calculated—they figured that, given even a moment's warning, the teenagers were clever enough to alert each other via computer.

This story—except for the name—is true. Hackers ply their craft for a variety of reasons, especially to show off for their peers and to harass people they do not like. A favorite trick is to turn a rival's telephone into a pay phone, so that when his or her parents try to dial a number an operator would interrupt to say "Please deposit 25¢." A hacker may have more sinister motives, such as getting computer services without paying for them or getting information to sell. However, hackers are only a small fraction of the security problem. The most serious losses are caused by electronic pickpockets who are usually a good deal older and not so harmless. Consider these examples:

- A brokerage clerk sat at his terminal in Denver and, with a few taps of the keys, transformed 1700 shares of his own stock, worth $1.50 per share, to the same number of shares in another company worth ten times that much.
- A Seattle bank employee used her electronic funds transfer code to move certain bank funds to an account held by her boyfriend as a "joke"; both the money and the boyfriend disappeared.
- A keyboard operator in Oakland, California, changed some delivery addresses to divert several thousand dollars' worth of department-store goods into the hands of accomplices.
- A ticket clerk at the Arizona Veteran's Memorial Coliseum issued full-price basketball tickets for admission and then used her computer to record the sales as half-price tickets and pocketed the difference.

These stories point out that computer crime is not always the flashy, front-page news about geniuses getting away with millions of dollars.

Stories about computer crime continue to fascinate the general public. They are "clean" white-collar crimes; no one gets physically hurt. They often feature people beating the system—that is, beating an anonymous, faceless, presumably wealthy organization. Sometimes the perpetrators even fancy themselves as modern-day Robin Hoods, taking from the rich to give to the poor—themselves and their friends. One electronic thief, in fact, described himself as a "one-man welfare agency."

The problems of computer crime have been aggravated in recent years by increased access to computers (Figure 11-1). More employees now have access to computers on their jobs. In fact, computer crime is often just white-collar crime with a new medium: Every time an employee is trained on the computer at work, he or she also gains knowledge that—potentially—could be used to harm the company.

Who Is the Computer Criminal?

Here is what a computer criminal is apt to be like. He (we will use he here, but of course he could be she) is usually someone occupying a position of trust in an organization. Indeed, he is likely to be regarded as the ideal employee. He has had no previous law-breaking experience and, in fact, will not see himself as a thief but as a "borrower." He is apt to be young and to be fascinated with the challenge of beating the system. Contrary to expectations, he is not necessarily a loner; he may operate well in conjunction with other employees to take advantage of the system's weaknesses.

Some "Bad Guy" Tricks

Although the emphasis in this chapter is on preventing rather than committing crime, being familiar with the terms and methods computer criminals use is part of being a prudent computer user. Many of these words or phrases have made their way into the general vocabulary.

Bomb: A virus that sabotages a program to trigger damage based on certain conditions; it is usually set to go off at a later date—perhaps after the perpetrator has left the company.

Data diddling: Changing data before or as it enters the system.

Data leakage: Removing copies of data from the system without a trace.

Piggybacking: Using another person's identification code or using that person's files before he or she has logged off.

Salami technique: Using a large financial system to squirrel away small "slices" of money that may never be missed.

Scavenging: Searching trash cans for printouts and carbons containing not-for-distribution information.

Trapdoor: Leaving, within a completed program, an illicit program that allows unauthorized—and unknown—entry.

Trojan Horse: A virus that covertly places illegal instructions in the middle of a legitimate program. In other words, it appears to do something useful but actually does something destructive in the background.

Zapping: Using an illicitly acquired software package to bypass all security systems.

What motivates the computer criminal? The causes are as varied as the offenders. However, a few frequent motives have been identified. A computer criminal is often a disgruntled employee, possibly a longtime loyal worker out for revenge after being passed over for a raise or promotion. In another scenario an otherwise model employee may commit a crime while suffering from personal or family problems. Not all motives are emotionally based. Some people simply are attracted to the challenge of the crime. In contrast, it is the ease of the crime that tempts others. In many cases the criminal activity is unobtrusive; it fits right in with regular job duties. The risk of detection is often quite low. Computer criminals think they can get away with it. And they do—some of the time.

Types and Methods of Computer Crime

Computer crime falls into three basic categories:

- Theft of computer time for development of software, either for personal use or with the intention of selling it
- Theft, destruction, or manipulation of programs or data
- Alteration of data stored in a computer file

Although it is not our purpose to write a how-to book on computer crime, the margin note called *Some "Bad Guy" Tricks* mentions some criminal methods.

Discovery and Prosecution

Prosecuting the computer criminal is difficult because discovery is often difficult. Many times the crime simply goes undetected. In addition, crimes that are detected are—an estimated 85 percent of the time—never reported to the authorities. By law, banks have to make a report when their computer systems have been compromised, but other businesses do not. Often they choose not to report because they are worried about their reputations and credibility in the community.

Most computer crimes are discovered by accident. For example, a bank employee changed a program to add 10¢ to every customer service charge under $10 and $1 to every charge over $10. He then placed this overage into the last account, a bank account he opened himself in the name of Zzwicke. The system worked fairly well, generating several hundred dollars each month, until the bank initiated a new marketing campaign in which they singled out for special honors the very first depositor—and the very last. In another instance some employees of a city welfare department created a fictitious work force, complete with Social Security numbers, and programmed the computer to issue paychecks, which the employees would then intercept and cash. They were discovered when a police officer investigated an illegally parked overdue rental car—and found the fraudulent checks inside.

Even if a computer crime is detected, prosecution is by no means assured. There are a number of reasons for this. First, some law enforcement agencies do not fully understand the complexities of computer-related fraud. Second, few attorneys are qualified to handle computer crime cases. Third, judges and juries are not always educated about computers and may not understand the value of data to a company.

In short, the chances of committing computer crimes and having them go undetected are, unfortunately, good. And the chances that, if detected,

Disgruntled or militant employee could • Sabotage equipment or programs • Hold data or programs hostage 	Competitor could • Sabotage operations • Engage in espionage • Steal data or programs • Photograph records, documentation, or CRT screen displays 	Data control worker could • Insert data • Delete data • Bypass controls • Sell information 
Clerk/supervisor could • Forge or falsify data • Embezzle funds • Engage in collusion with people inside or outside the company 	System user could • Sell data to competitors • Obtain unauthorized information 	Operator could • Copy files • Destroy files 
User requesting reports could • Sell information to competitors • Receive unauthorized information 	Engineer could • Install "bugs" • Sabotage system • Access security information 	Data conversion worker could • Change codes • Insert data • Delete data 
Programmer could • Steal programs or data • Embezzle via programming • Bypass controls 	Report distribution worker could • Examine confidential reports • Keep duplicates of reports 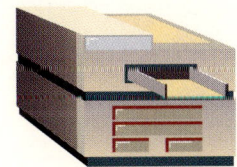	Trash collector could • Sell reports or duplicates to competitors 

Figure 11-1 The perils of increased access. By letting your imagination loose, you can visualize many ways in which people can compromise computer security. Computer-related crime would be far more rampant if all the people in these positions took advantage of their access to computers.

there will be no ramifications are also good: A computer criminal may not go to jail, may not be found guilty if prosecuted, and may not even be prosecuted.

But this situation is changing. Since Congress passed the **Computer Fraud and Abuse Act** in 1986, there has been a growing awareness of computer crime on the national level. This law is supplemented by state statutes; most states have passed some form of computer crime law.

Security: Playing It Safe

As you can see from the previous section, the computer industry has been vulnerable in the matter of security. Computer security once meant only the physical security of the computer itself—guarded and locked doors. But locked doors by no means prevent access, as we have seen. Management interest in security has been heightened, and managers are now rushing to purchase more sophisticated security products.

What is computer security? We may define it as follows: **Security** is a system of safeguards designed to protect a computer system and data from deliberate or accidental damage or access by unauthorized persons. That means safeguarding the system against such threats as natural disasters, fire, accidents, vandalism, theft or destruction of data, industrial espionage, and hackers (Figure 11-2).

Identification and Access: Who Goes There?

How does a computer system detect whether you are the person who should be allowed access to it? Various means have been devised to give access to authorized people—without compromising the system. These means fall into four broad categories: what you have, what you know, what you do, and who you are.

- **What you have.** You may have a key, badge, token, or plastic card to give you physical access to the computer room or a locked-up terminal. A card with a magnetized strip, for example, can give you access to your bank account via a remote cash machine. Taking this a step further, some employees begin each business day by donning an **active badge,** a clip-on identification card with an embedded computer chip. The badge signals its wearer's location—whether legal or otherwise—by sending out infrared signals, which are read by sensors sprinkled around the building. The active badge, which is becoming increasingly common, presents a challenging problem: balancing an employee's privacy against a corporation's desire for efficiency and control.
- **What you know.** Standard what-you-know items are a system password or an identification number for your bank cash machine. Cipher locks on doors require that you know the correct combination of numbers.
- **What you do.** In our daily lives we often sign documents as a way of proving who we are. Though a signature is difficult to copy, forgery is not impossible. For this and other reasons, signatures lend themselves to human interaction better than machine interaction.
- **What you are.** Now it gets interesting. Some security systems use **biometrics,** the science of measuring individual body characteristics. Fingerprinting may seem to be old news, but not when you skip the ink pad and simply insert your finger into an identification machine.

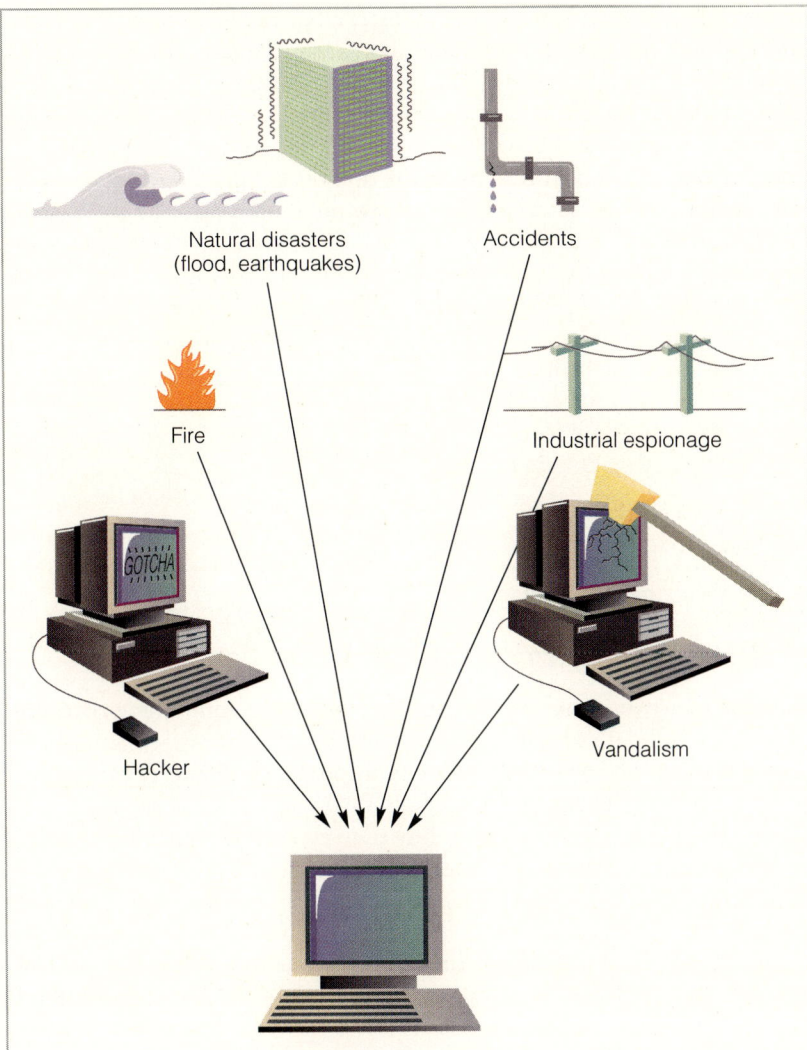

Figure 11-2 Is your computer secure? The computer industry is vulnerable to both natural and man-made disasters.

Another approach is identification by voice recognition. Even newer is the concept of identification by the retina of the eye, which has a pattern that is harder to duplicate than a voice print (Figure 11-3).

Some systems use a combination of the preceding four categories. For example, access to an automated teller machine requires both something you have—a plastic card—and something you know—a personal identification number (PIN).

When Disaster Strikes: What Do You Have to Lose?

In New York a power outage shut down computer operations and effectively halted business, air traffic, and transportation throughout the United States. In Italy armed terrorists singled out corporate and state computer centers as targets for attack and, during a ten-month period, bombed ten such centers throughout the country. In California a poem, a pansy, a bag of cookies, and a message, "Please have a cookie and a nice day," were left at the Vandenberg Air Force Base computer installation—along with five demolished mainframe computers. Computer installations of any kind can

Figure 11-3 Identification. The eye can be a means of personal identification. A user first keys a unique identification code number. The security system then matches the person's unique retina pattern to the individual's computer-stored retina pattern for conclusive identification of authorized users.

be struck by natural or man-made disasters that can lead to security violations. What kinds of problems might this cause an organization?

Your first thoughts might be of the hardware, the computer and its related equipment. But loss of hardware is not a major problem in itself; the loss will be covered by insurance, and hardware can be replaced. The true problem with hardware loss is the diminished processing ability that exists while managers find a substitute facility and return the installation to its former state. The ability to continue processing data is critical. Some information industries, such as banking, could literally go out of business in a matter of days if their computer operations were suspended. Loss of software should not be a problem if the organization has heeded industry warnings—and used common sense—to make backup copies of program files.

A more important problem is the loss of data. Imagine trying to reassemble lost or destroyed master files of customer records, accounts receivable, or design data for a new airplane. The costs would be staggering. We will consider software and data security in more detail later in this chapter. First, however, let us present an overview of disaster recovery, the steps to restoring processing ability.

Disaster Recovery Plan

A **disaster recovery plan** is a method of restoring computer processing operations and data files if operations are halted or files are damaged by major destruction. There are various approaches. Some organizations revert temporarily to manual services, but life without the computer can be difficult indeed. Others arrange to buy time at a service bureau, but this is inconvenient for companies in remote or rural areas. If a single act, such as a fire, destroys your computing facility, it is possible that a mutual aid pact will help you get back on your feet. In such a plan two or more companies agree to lend each other computing power if one of them has a problem. This would be of little help, however, if there were a regional disaster and many companies needed assistance.

Banks and other organizations with survival dependence on computers sometimes form a **consortium,** a joint venture to support a complete computer facility. Such a facility is completely available and routinely tested but used only in the event of a disaster. Among these facilities, a **hot site** is a fully equipped computer center, with hardware, environmental controls, security, and communications facilities. A **cold site** is an environmentally suitable empty shell in which a company can install its own computer system.

The use of such a facility or any type of recovery at all depends on advance planning—specifically, the disaster recovery plan. The idea of such a plan is that everything except the hardware has been stored in a safe place somewhere else. The storage location should be several miles away, so it will not be affected by local physical forces, such as a hurricane. Typical items stored at the backup site are program and data files, program listings, program and operating systems documentation, hardware inventory lists, output forms, and a copy of the disaster plan manual.

The disaster recovery plan should include these items:

- **Priorities.** A list of priorities identifies the programs that must be up and running first. A bank, for example, would give greater weight to account inquiries than to employee vacation planning.

ONE JUMP AHEAD

Just the Standard Insurance, Please

You just got the job you wanted, and now you must wade through the various personnel forms. Hmmm, seems pretty standard. Life insurance, health insurance, dental insurance, computer insurance. Computer insurance? Yes. It may appear under a variety of names, but the policy is basically a form of computer malpractice insurance. If you think this is ludicrous, talk to any doctor who practiced before the advent of medical malpractice insurance.

The idea of malpractice insurance was first suggested for computer professionals. Not that they wanted the

insurance or wanted to pay for it. The need was brought to their attention by the increasing tendency of courts to hold firms, and even individual computer professionals, liable for losses caused by computer system errors. As an alternative to being sued, or perhaps in addition to being sued, computer professionals sometimes lose their jobs. For example, five computer professionals were fired by financial institutions following one New York City power outage because data was irretrievably lost. It is rare today for an employee to be fired for making a computer error. However, these five people did not simply make a "computer error"; they failed to have a plan to back up data files.

Computer law experts have predicted that one day anyone—not just computer professionals—who uses a computer to affect company files will carry computer insurance.

- **Personnel requirements.** The plan should comprise procedures for notifying employees of changes in locations and procedures.
- **Equipment requirements.** A list of needed equipment and where it can be obtained will speed recovery efforts.
- **Facilities.** Most organizations cannot afford consortiums, so the recovery plan should include a list of alternative computing facilities.
- **Capture and distribution.** This part of the plan outlines how input and output data will be handled in a different environment.

Computer installations actually perform emergency drills. At some unexpected moment a notice is given that "disaster has struck," and the computer professionals must run the critical systems at some other site.

Software Security

Software security has been an industry concern for years. Initially, there were many questions: Who owns custom-made software? Is the owner the person who wrote the program or the company for which the author wrote the program? What is to prevent a programmer from taking copies of programs from one job to another? The answer to these questions are well established. If the author of the software—the programmer—is in the employ of the organization, the software belongs to the organization, not the programmer. The programmer may not take the software along to the next job. If the programmer is a consultant, however, the ownership of the software produced should be spelled out specifically in the contract—otherwise, the parties enter extremely murky legal waters.

According to a U.S. Supreme Court decision, software can be copyrighted. Unfortunately, although unauthorized duplication is specifically prohibited by law, software continues to be copied as blatantly as music cassettes. We will examine this issue more closely when we consider ethics later in the chapter.

Data Security

We have discussed the security of hardware and software. Now let us consider the security of data, which is one of an organization's most important assets. Here too there must be planning for security. Usually, this is done by security officers who are part of top management. There are five critical planning areas for data security:

- Determination of appropriate policies and standards. A typical statement of policy might read: "All computer data and related information will be protected against unauthorized disclosure and against alteration or destruction."
- Development and implementation of security safeguards, such as passwords, which we will discuss shortly.
- Inclusion of security precautions at the development stage of new automated systems, rather than after the systems are in use.
- Review of state and federal laws related to security. This is particularly significant in banking.
- Maintenance of historical records associated with computer abuse.

What steps can be taken to prevent theft or alteration of data? There are several data protection techniques; these will not individually (or even collectively) guarantee security, but at least they make a good start.

Secured Waste
Discarded printouts, printer ribbons, and the like can be sources of information to unauthorized persons. This kind of waste can be made secure by the use of shredders or locked trash barrels.

Internal Controls
Internal controls are controls that are planned as part of the computer system. One example is a transaction log. This is a file of all accesses or attempted accesses to certain data.

Auditor Checks
Most companies have auditors go over the financial books. In the course of their duties, auditors frequently review computer programs and data. From a data security standpoint, auditors might also check to see who has accessed data during periods when that data is not usually used. They are also on the lookout for unusual numbers of corrected data entries, usually a trouble sign. What is more, the availability of off-the-shelf audit software—programs that assess the validity and accuracy of the system's operations and output—promotes tighter security because it allows auditors to work independently of the programming staff.

Applicant Screening
The weakest link in any computer security system is the people in it. Unfortunately, employers often spend more time investigating the backgrounds of people who rarely touch the data itself—the bosses who make the strategic decisions—than the people who work with it daily. At the very least, employers should verify the facts that job applicants list on their résumés to help weed out dishonest applicants before they are hired.

Separation of Employee Functions
Should a programmer also be a computer operator? If so, this puts him or her in the position of being able not only to write unauthorized programs

---

Some Gentle Advice on Security

Being a security expert is an unusual job because, once the planning is done, there is not a lot to do except wait for something bad to happen. Security experts are often consultants who move from company to company. They sometimes write books and articles for the trade press in which they usually include long and detailed checklists: Do this, do that, and you will be OK. We cannot attempt such a long set of lists here, but we offer a brief subset that includes some of the most effective approaches.

Beware of disgruntled employees. Ed Street was angry. Seething. How could they pass over him for a promotion again? Well, if they were not going to give him what he deserved, he would take it himself. Ah, the tale is too common. Be forewarned.

Sensitize employees to security issues. Most people are eager to help others. They must be taught that some kinds of help, such as assisting unauthorized users with passwords, are inappropriate. Most security breaches are possible because people are ignorant, careless, or too helpful.

Keep personnel privileges up to date. And, we might add, make sure they are enforced properly. "Hi, Bill, how ya doin'?" "Pretty good, Frank, good to see you." Bill, the guard, has just swept unauthorized Frank into the computer area. Some of the biggest heists have been pulled by people who formerly had legitimate access to secured areas. Often, they can still get in because the guard has known them by sight for years.

but also to run them. By limiting employee duties so that doubling up on job functions is not permitted, a computer organization can restrict the amount of unauthorized access. That is, in an installation where the computers—mainframes or minis—are behind locked doors, only operators have physical access to them. Unfortunately, separation of functions is not practical in a small shop; usually one or more employees perform multiple functions. And, of course, separation of functions does not apply in a personal computer environment.

Consider some simpler examples. Should the person who prepares the checks also be the person who signs them? Clearly not. A more common problem is the employee whose devotion to duty is suspect—the employee is never late, never absent, never sick, and never takes vacations. The employee may not be a criminal, but he or she could be a knowledge hoarder. Many companies require employees to take their vacations, precisely to avoid this sort of situation. It is better to find out that a lone person has a key piece of information while he or she is on a short vacation than to discover that fact after the individual leaves the company.

Passwords

A password is a secret word or number, or a combination of the two, that must be typed on the keyboard to gain access to a computer system. Good data protection systems change passwords often and also compartmentalize information by passwords, so that only authorized persons can have access to certain data. Cracking passwords is the most prevalent method of illicit entry to computer systems.

Built-in Software Protection

Software can be built into operating systems in ways that restrict access to the computer system. One form of software protection system matches a user number against a number assigned to the data being accessed. If a person does not get access, it is recorded that he or she tried to tap into some area to which they were not authorized. Another form of software protection is a user profile: Information is stored about each user, including the files to which the user has legitimate access. The profile also includes each person's job function, budget number, skills, areas of knowledge, access privileges, supervisor, and loss-causing potential. These profiles are available for checking by managers if there is any problem.

Worms and Viruses

These rather unpleasant terms have entered the jargon of the computer industry to describe some of the insidious ways that computer systems and programs can be invaded. A **worm** is a program that transfers itself from computer to computer over a network and plants itself as a separate file on the target computer's disks. One newsworthy worm, originated by Robert Morris when he was a student at Cornell University, traveled the length and breadth of the land through an electronic mail network, shutting down thousands of computers. The worm was injected into the network and multiplied uncontrollably, clogging the memories of infected computers until they could no longer function.

A virus, as its name suggests, is contagious. That is, a **virus**, a set of illicit instructions, passes itself on to other programs with which it comes in contact. A virus may be dealt with by means of a **vaccine**, or **antivirus**, program, a computer program that stops the spread of and often eradi-

---

I've Got a Secret

Employers wish that computer passwords were better-kept secrets. Here are some hints on password use.

- Do not name your password after your child or car or pet, an important date, or your phone number. Passwords that are easy to remember are also easy to crack. Recommended password creation techniques include using at least six characters, embedding at least one nonalphabetic character, and even mixing upper- and lowercase letters. Example: GREEN*frame.

- Change your password often, at least once a month. In some installations, passwords are changed so seldom that they become known to many people, thus defeating the purpose.

- Do not fall for hacker phone scams to obtain your password. Typical ruses are callers posing as a neophyte employee ("Gosh, I'm so confused, could you talk me through it?"), a system expert ("We're checking a problem in the network and need your password."), or even an angry top manager ("This is outrageous! How do I get into these files anyway?").

cates the virus. Viruses seem to show up when least expected. In one instance a call came to a company's Information Center about 5:00 p.m.: The caller's computer was making a strange noise. With the exception of an occasional beep, computers performing routine business chores do not usually make noises. Soon employees were calling from all over the company, all with "noisy" computers. One caller said that it might be a tune coming from the computer's small internal speaker. Finally, one caller recognized a tinny rendition of Yankee Doodle, confirmation that an old virus had struck once again. The Yankee Doodle virus, once attached to a system, is scheduled to go off at 5:00 p.m. every eight days. Viruses, once considered merely a nuisance, are costing American business over $2 billion a year. Unfortunately, viruses are easily transmitted.

Transmitting a Virus

Consider this typical example. A programmer secretly inserts a few viral instructions into a game called Kriss-Kross, which she then offers free to others via a bulletin board. Any takers download the game to their own computers. Now, each time a user runs Kriss-Kross, that is, loads it into memory, the virus is loaded too. The virus stays in memory, infecting any other program loaded. The virus now has spread to other programs, and the process can be repeated again and again. In fact, each newly infected program becomes a virus carrier. Although many viruses are transmitted just this way over bulletin boards, the most common method is by passing diskettes from computer to computer (Figure 11-4).

More insidious viruses attach to the operating system. One virus, called Cascade, causes random text letters to "drop" to a pile at the bottom of the screen (Figure 11-5). Viruses attached to the operating system itself have greater potential for mischief.

Damage from Viruses

The Yankee Doodle virus described earlier is relatively benign, as is the virus that simply displays a peace message. But many viruses do significant damage, often including destruction of files.

Most viruses remain dormant until triggered by some activity. For example, a virus called Jerusalem B activates itself every Friday the 13th and proceeds to erase any file you may try to load from your disk. Another virus includes instructions to add 1 to a counter each time the virus is copied to another disk. When the counter reaches 4, the virus erases all data files. But this is not the end of the destruction, of course; the three copied disks have also been infected.

Prevention

A word about prevention is in order. Although viruses are most commonly passed via diskettes, viruses use many other means to propagate—bulletin boards, local area networks, and electronic mail. If your personal computer has a disk drive, a modem, or a network connector, it is vulnerable. Furthermore, viruses are rampant on some college campuses, a source of considerable annoyance to students. Use these common sense approaches to new files:

- Never install a program unless the diskette comes in a sealed package.
- Be especially wary of software that arrives unexpectedly from companies with whom you have not done business.
- Use virus-scanning software to check any file, no matter what the source, before loading it onto your hard disk.

COMPUTERS AT WORK

GALLERY

In this gallery we will look at some of the ways in which workers put computers to use on the job. The photo that opens this gallery shows a British Airways maintenance worker atop an airplane wing; he will record fuel data on his computer.

When people think of computers in the workplace, they probably envision a traditional office setting. Computers, however, are in all kinds of workplace settings, some of which are shown here.

1. This marine biologist is recording data on water samples on his laptop computer on site.
2. Taking inventory in the supermarket is quick and efficient when using a computer with a pen to identify and record data from the products on the shelf.
3. Whether tracking orders or recipes, food preparers of all sorts use the computer.
4. These workers in the Louvre Museum in Paris are using the computer to catalog statues.
5. The New York Stock Exchange handles millions of transactions each business day with the very necessary help of computers.
6. Blood pressure can be tracked by computer and shown on screen as it is being taken.
7. Indoors, outdoors, on site—any setting is appropriate as long as the computer can come along.
8. The workplace setting for this woman's computer is her home.

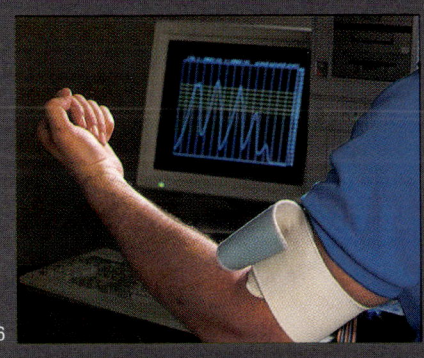

GALLERY

Many workers spend a large part of their time using a computer. Some take their computers with them wherever they go.

9. Financial traders have their own computers, which they use to analyze data and place orders.
10. This emergency medical technician uses maps on an oversized screen to locate emergency callers.
11. These archeologists use their computer to re-create ancient worlds.
12. This worker uses her computer to check a network at Bell Atlantic.
13. These auto racers hold the computer that connects the pit crew with the cars.
14. A doctor studies a computer scan of a chest and abdomen.
15. A United Parcel Service employee asks a customer to acknowledge receipt of a package by using a pen-based computer to accept his signature.
16. This NASA employee uses multiple computer screens to monitor global warming.
17. Workers in oil fields can use their laptop computers to record and analyze production data on-site.

9

10

11

12

13

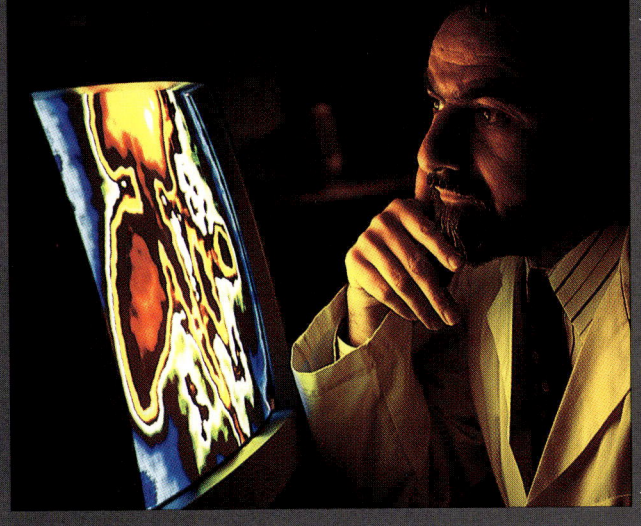

14

15

16

17

5

GALLERY

Computer graphics software lets designers choose from a wide range of colors and styles to create just the image they need, whether for advertising, tourism, or some other useful purpose.

18. This advertisement for Montblanc pens was computer-produced.
19. The use of simulated light falling across the room lends a touch of realism to this computer-produced image, which was used to illustrate a magazine story.
20. This poster was produced using computer graphics, and then the output was enlarged to poster size.
21. This skyline, produced in the image of a rubber stamp, was used as part of a tourist promotion.
22. This appears to be a photo of drawings, but in fact the entire work is a computer-produced work that is part of a tourism promotion.
23. The ducky was added as an attention-getter in this image used to encourage people to keep water pure.
24. A graphics artist used a credit card theme for the cover of a corporate annual report.
25. This exquisite trout drawing was developed to support a fishing contest.

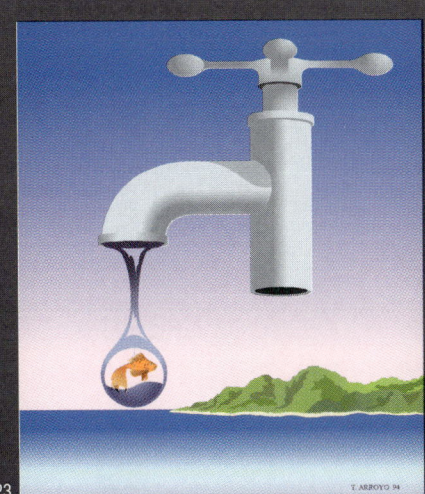

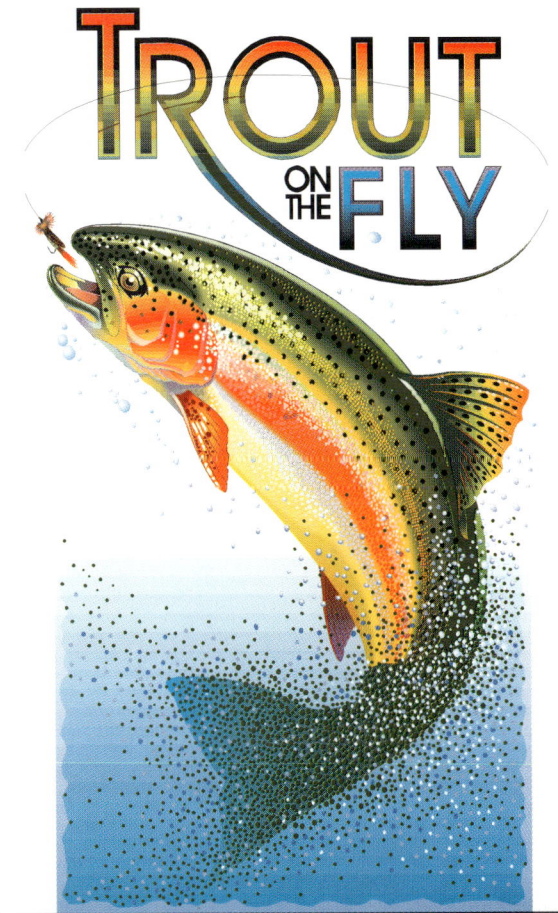

GALLERY

When most people think of robots, they probably have in mind the typical humanoid-shaped automaton of classic science-fiction movies. But robots of all shapes and sizes are performing a variety of serious tasks—without looking much like people.

26. This robot has both the speed and agility to assemble circuit boards.
27. Dances with robots? Not exactly. The human "trainer," wired with dozens of sensors, is using software to choreograph the movements of humans and robots working side by side. Motions of both appear on the screen.
28. In this stark photo the robot is doing serious work—disarming a bomb.
29. Robots are known for their work in automotive factories, but this one is making a washing machine.

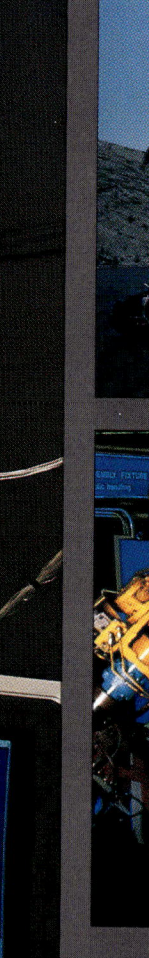

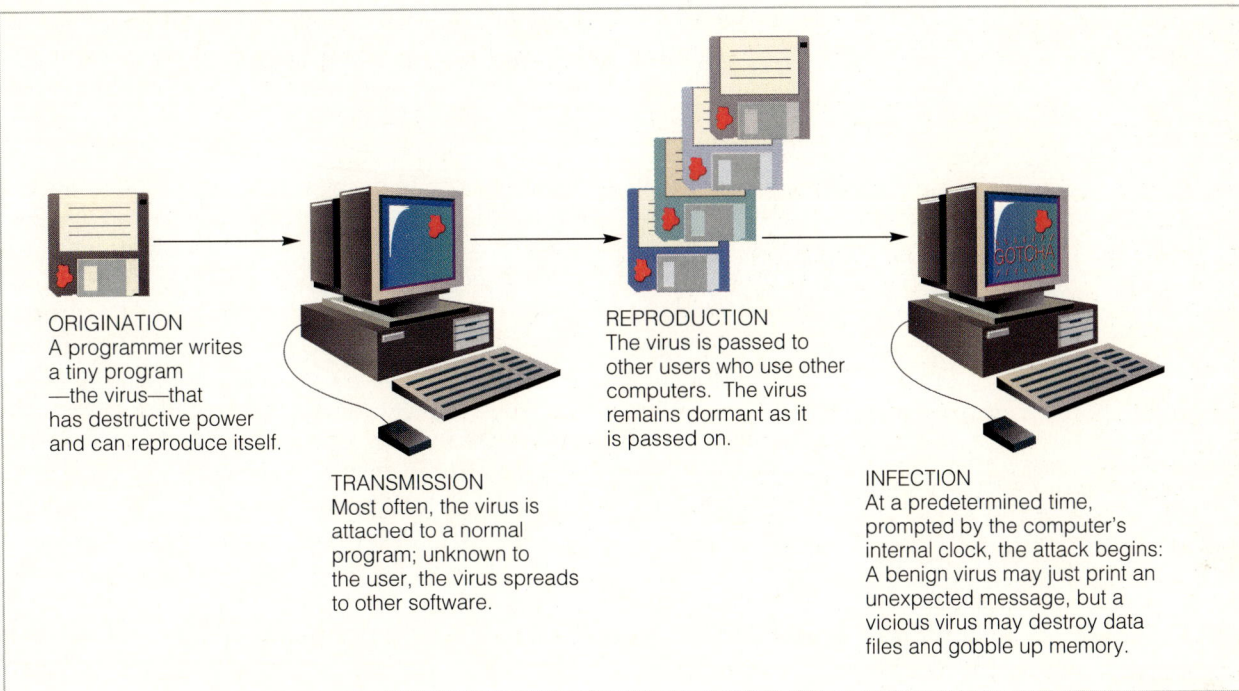

Figure 11-4 An example of a virus invasion.

- If your own diskette was used in another computer, scan it to see if it caught a virus.

Although there have been isolated instances of viruses in commercial software, viruses tend to show up on free software acquired from friends or through electronic bulletin board systems. Antivirus software can be installed to scan your hard disk every time you boot the computer or, if you prefer, at regularly scheduled intervals.

Network Security

Networks pose a unique security problem. Many people have access to the system, often from remote locations. Clearly, the question of security arises: If it is so easy for authorized people to get data, what is to stop unauthorized people from tapping it? To begin with, network operating systems provide basic security features, such as user identification and authentication, probably by password.

Sophisticated network systems can permit network supervisors to assign varying access rights to individual users. All users, for example, could access word processing software, but only certain users could access payroll files. Some network software can limit how many times users can call up a particular file and generate an audit trail of who looked at which files.

One fundamental approach to network security is to dedicate one computer, called a **firewall**, whose sole purpose is to talk to the outside world. A firewall will provide an organization with greatly increased security because only one network computer is accessible to people outside the network, and that one computer accepts only appropriate access. Among its other chores, a firewall computer can call back all remote-access terminals. Don't call us, we'll call you. If it is appropriate to limit access from only certain locations, then the computer can keep a list of valid phone num-

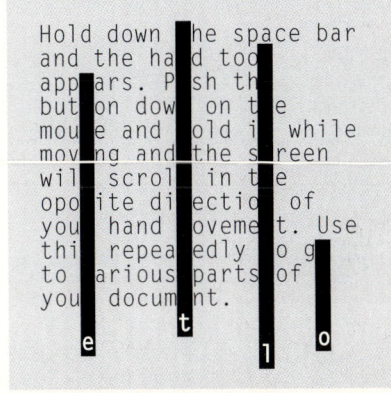

Figure 11-5 **The Cascade virus.** This virus attaches itself to the operating system and causes random letters in text to "drop" to a pile at the bottom of the screen display.

bers. With such a system the computer has to call the caller back for the user to gain remote access, and it will do so only if the user's number is valid. The fact that an unauthorized caller has the computer's phone number is irrelevant. What matters is whether the computer has the caller's number.

In addition to monitoring access to the network, organizations must be concerned about unauthorized people intercepting data in transit, possibly thieves or industrial spies. Data being sent over communications lines may be protected by scrambling the messages—that is, putting them in code that can be broken only by the person receiving the message. The process of scrambling messages is called **encryption.** The American National Standards Institute has endorsed a process called the **Data Encryption Standard (DES),** a standardized public key by which senders and receivers can scramble and unscramble their messages. Although the DES code is well known, companies still use it because the method makes it quite expensive to intercept coded messages. Thus, interlopers are forced to use other methods of gathering data—methods that carry greater risk of detection. Encryption software is available for personal computers. A typical package, for example, offers a variety of security features: file encryption, keyboard lock, and password protection.

Personal Computer Security

One summer evening two men in coveralls with company logos backed a truck up to the building that housed a university computer lab. They showed the lab assistant, a part-time student, an authorization slip to move 23 personal computers to another lab on campus. The assistant was surprised but not shocked, since lab use was light in the summer quarter. The computers were moved, all right, but not to another lab. In another case a ring of thieves mingled with students in computer labs in various west coast universities and stole hundreds of microprocessor chips from the campus computers.

There is an active market for stolen personal computers and their internal components. As these unfortunate tales indicate, personal computer security breaches can be pretty basic. One simple, though not foolproof, remedy is to secure personal computer hardware in place with locks and cables. Also, most personal computers have an individual cover lock that prevents access to internal components.

In addition to theft, personal computer users need to be concerned about the computer's environment. Personal computers in business are not coddled the way bigger computers are. They are designed, in fact, to withstand the wear and tear of the office environment, including temperatures set for the comfort of people. Most manufacturers discourage eating and smoking near computers and recommend some specific cleaning techniques, such as vacuuming the keyboard. The response to these recommendations is directly related to the awareness level of the users.

Most personal computer data is stored on diskettes, which are vulnerable to sunlight, heaters, cigarettes, scratching, magnets, theft, and dirty fingers. The data is vulnerable as well. Hard disks used with personal computers are subject to special problems too. If a computer with a hard disk is used by more than one person, your files on the hard disk may be available for anyone to browse through.

Several precautions can be taken to protect disk data. One is to use a **surge protector,** a device that prevents electrical problems from affecting data files. The computer is plugged into the surge protector, which is plugged into the outlet. Diskettes should be under lock and key. The most critical precaution, however, is to back up your files.

Prepare for the Worst: Back Up Your Files

A computer expert, giving an impassioned speech, said "If you are not backing up your files regularly, you *deserve* to lose them." Strong words. Although organizations recognize the value of data and have procedures in place for backing up data files on a regular basis, personal computer users are not as devoted to this activity. In fact, one wonders why, with continuous admonishments and readily available procedures, some people still leave their precious files unprotected.

What Could Go Wrong

Users could use software incorrectly or simply input data incorrectly; it may be some time before the resulting erroneous data is detected, and you need to go back to the time when the data files were still acceptable. Sometimes the software itself can harm data. A hard disk could physically malfunction, making your files inaccessible. Although none of these are too likely, they certainly do happen. It is even less likely that you lose your hard disk files to fire or flood, but this is also possible. It is most likely that you will accidentally delete some files yourself. One fellow gave a command to delete all files with the file name extension BAK—there were four of them—but accidentally typed BAT instead, inadvertently wiping out 57 files. (Deleted files, we should mention, can probably be recovered using utility software if the action is taken right away, before other data is written over the deleted files.) Finally, there is always the possibility of your files being infected with a virus. Experts estimate that average users experience a significant disk loss every year.

Ways to Back Up Files

Some people simply make another copy of their hard drive files on diskette. This is not too laborious if you do so as you go along. If you are at all vulnerable to viruses, you should back up all your files on a regular basis.

A better way is to back up all your files on a tape. Backing up to a tape drive is safer and faster. You can also use software that will automatically back up all your files at a certain time of day, or on command. Sophisticated users place their files on a mirror hard disk, which simply makes a second copy of everything you put on the original disk; this approach, as you might expect, is expensive.

Keep backed up files in a cool, dry place off site. For those of you with a home computer, this may mean keeping copies of your important files at a friend's house; some people even use a bank safety deposit box for this purpose.

Privacy: Keeping Personal Information Personal

Think about the forms you have willingly filled out: paperwork for loans or charge accounts, orders for merchandise through the mail, magazine subscription orders, applications for schools and jobs and clubs, and on

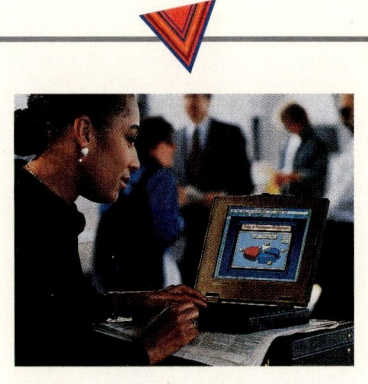

Easy to Steal

Dale McKenna usually does some work on her laptop computer during her frequent business flights. On a brief trip from Washington, D.C., to New York, however, she decided that the flight was too short to bother working, so she stowed her computer with her other carry-on luggage in the back of the plane. After a delay in take-off, a flight attendant announced that passengers would have to move to another plane and that all carry-on luggage would be moved to the other plane by the crew.

Dale made it to the other plane but her computer did not. When Dale discovered the loss she was dismayed not about the lost computer but about the lost files on the hard disk. It took several days to recreate the files that had been produced at the meeting she had just attended. Now, when traveling, Dale keeps copies of files on a diskette in her purse. And she does not let her laptop out of her sight.

and on. There may be some forms you filled out with less delight—for taxes, military draft registration, court petitions, insurance claims, or a stay in the hospital. And remember all the people who got your name and address from your check—fund-raisers, advertisers, and petitioners. We may not have covered all the ways you have supplied information, but we can say with certainty where all this information went: straight to computer files.

Passing Your Data Around

Where is that data now? Is it shared, rented, sold? Who sees it? Will it ever be deleted? Or, to put it more bluntly, is *anything* private anymore? In some cases we can only guess at the answers. It is difficult to say where your data is now, and bureaucracies are not often eager to enlighten you. The data may have been moved to other files without your knowledge. In fact, much of the data is most definitely passed around, as anyone with a mailbox can attest. Even online services sell their subscriber lists, neatly ordered by zip code and computer type.

Own a Car? Your Name and Address Are Probably for Sale

In approximately two thirds of the states it is perfectly legal to obtain the identification of car owners. Anyone can jot down a license plate number and go the motor vehicles department to get the owner's name and address. Some states go even further, selling information—name, address, height, weight, age, vision, and type of car—on computer tapes to all comers, from insurance companies to marketers to detectives. California, spurred by privacy invasions by crazed movie fans, passed a law to keep such information private.

As for who sees your personal data, the answers are not comforting. Government agencies, for example, regularly share data that was originally filed for some other purpose. IRS records, for example, are compared with draft registration records to catch draft dodgers, and also with student loan records to intercept refunds to former students who defaulted on their loans. The IRS created a storm of controversy by announcing a plan to use commercial direct-mail lists to locate tax evaders. Many people are worried about the consequences of this kind of sharing (Figure 11-6). For one thing, few of us can be certain that data about us, good or bad, is deleted when it has served its legitimate purpose.

However, public outrage occasionally derails the mass distribution of private data. The U.S. Postal Service offered to keep customers' mailing lists on disk and generate address labels. As a little side business, however, the service also wanted to sell those same computerized lists to businesses, incidentally creating enormous revenue for the postal service. A private company was also bold, trumpeting the imminent offering of a CD-ROM filled with data about just about everyone in the United States—a CD-ROM intended for sale to marketers. After the public heard of these efforts—and vigorously complained—the projects were canceled.

But the fact remains that, for very little money, anybody can learn anything about anybody—through massive databases. There are matters you want to keep private. You have the right to do so. Although you can do little to stop data about you from circulating through computers, there are some laws that give you access to some of it. Let us see what kind of protection is available to help preserve privacy.

Privacy Legislation

Significant legislation relating to privacy began with the **Fair Credit Reporting Act** in 1970. This law allows you to have access to and gives you the right to challenge your credit records. In fact, this access must be given to you free of charge if you have been denied credit. Businesses usually contribute financial information about their customers to a community credit bureau, which gives them the right to review a person's prior credit record with other companies. Before the Fair Credit Reporting Act, many

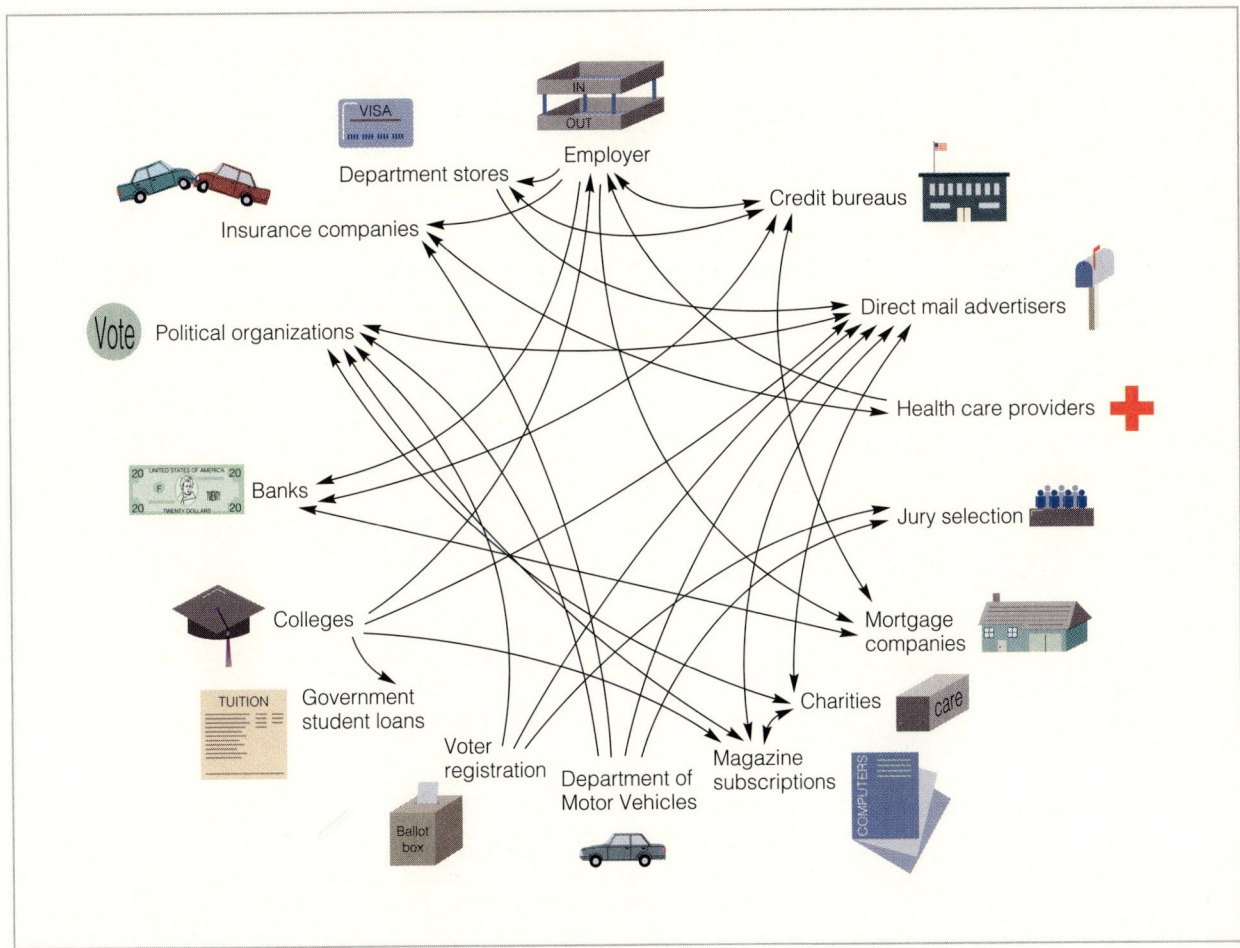

Figure 11-6 Potential paths of data. When an organization acquires data about you, it is often shared with—or sold to—other organizations.

people were—without explanation—turned down for credit because of inaccurate financial records about them. Because of the act, people may now check their records to make sure they are accurate.

The **Freedom of Information Act** was also passed in 1970. This landmark legislation allows ordinary citizens to have access to data about them that was gathered by federal agencies (although sometimes a lawsuit has been necessary to pry data loose).

The most significant legislation protecting the privacy of individuals is the **Federal Privacy Act** of 1974. This act stipulates that there can be no secret personnel files; individuals must be allowed to know what is stored in files about them and how the data is used, and be able to correct it. The law applies not only to government agencies but also to private contractors dealing with government agencies. These organizations cannot obtain data willy-nilly, for no specific purpose; they must justify obtaining it.

A more recent law is the **Video Privacy Protection Act** of 1988, which prevents retailers from disclosing a person's video rental records without a court order; privacy supporters want the same rule for medical and insurance files. Another step in that direction is the **Computer Matching and Privacy Protection Act** of 1988, which prevents the government from comparing certain records in an attempt to find a match. However, most comparisons are still unregulated.

Privacy in the Workplace

Although employees do not have expectations of total privacy at the office, they are often shocked when they discover that the boss has been spying on them. The boss, of course, is not spying at all, merely monitoring. This debate has been heightened recently by the advent of software that lets managers check up on networked employees without their ever knowing that they are under surveillance. With a flick of a mouse button, the boss can silently pull up an employee's current computer screen.

Surveillance software is not limited to checking screens. It can also check on electronic mail, count the number of keystrokes per minute, note the length of a worker's break, and monitor what computer files are used and for how long.

Worker associations complain that workers who are monitored suffer much higher degrees of stress and anxiety than nonmonitored workers. However, vendors defend their products by saying they are not "spy software" but rather products designed for training, monitoring resources, and helping employees.

Privacy groups are lobbying legislators at both the state and federal levels to enact legislation that requires employers to alert employees that they are being monitored.

A Matter of Ethics

The problem of professional computer personnel having access to files has always existed. In theory they could do something as simple as snooping into a friend's salary on a payroll file or as complex as selling military secrets to foreign countries. But the problem has become more tangled as everyday people—not just computer professionals—have daily contact with computers. They have access to files too. Many of those files are on diskettes and may be handled in a careless manner. As we noted earlier, data is the resource most difficult to replace, so increased access is the subject of much concern among security officers.

Where do you come in? As a student you could easily face ethical problems involving access and much more. Consider some of these examples: A nonstudent friend wants to borrow your password to get access to the school computer. Or you know of a student who has bypassed computer security and changed grades for himself and some friends. Perhaps a "computer freak" pal collects software and wants you to copy a software disk used in one of your classes. And so on.

The problems are not so different in the business world. You will recognize that, whether you are a computer professional or a user, you have a clear responsibility to your own organization and its customers to protect the security and privacy of their information. Any compromise of data, in particular, is considered a serious breach of ethics. Many corporations have formal statements saying as much and present them to employees individually for their signatures.

However, the most sizzling ethics issue related to computers in the workplace is the issue of illegal software copies. Lamentations by both

MAKING THE RIGHT CONNECTIONS

You Have No Privacy Whatever

No privacy on the company e-mail, that is. Your employer can snoop into messages you send or receive even if you think you erased them. Companies may fail to convey the message that e-mail, as a company conduit, is not private. Employees are often startled, after the fact, to discover that their messages have been invaded.

Furthermore, some people specialize in extracting deleted messages for use as evidence in court. E-mail can be a dangerous time bomb in every corporation because litigators argue that, more than any other kind of written communication, e-mail

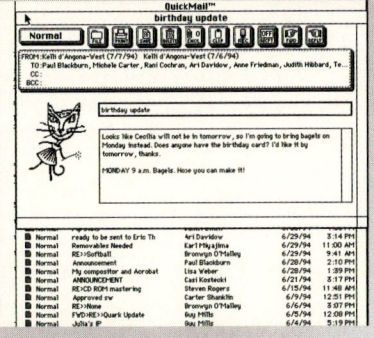
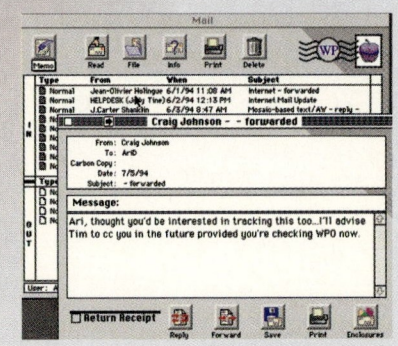

reflects the real, unedited thoughts of the writer.

What to do? It is certainly degrading to have something you thought was private waved in front of you as evidence of malingering. As one computer expert put it, if nothing is private, just say so. Companies have begun doing exactly that. The company policy on e-mail is—or should be—expressed in a clear, written document.

business and the computer industry are so persistent and so loud that we are devoting a separate section to this issue.

 ## Copying Software

Have you ever copied a friend's music tape onto your own blank tape? Many people do so without much thought. It is also possible to photocopy a book. Both acts are clearly illegal, but there is much more fuss over illegal software copying than over copying music or books. Why is this? Well, to begin with, few of us are likely to undertake the laborious task of reproducing *War and Peace* on a copy machine. Another difference is that a copied tape or book is never quite as good as the original; a copied program is identical to the original and works just as well. The other part of the issue is money. A pirated copy of a top-20 tape will set the recording company—and the artist—back about $10. But pirated software may be valued at hundreds of dollars. The problem of stolen software has grown right along with the personal computer industry. Before we discuss industry solutions, we must distinguish among various kinds of software, based on its availability to the public.

OK if I Copy That Software?

Some software is considered to be in the **public domain** because its generous maker, probably an individual at home or an educator, chooses to make it free to all. Software in the public domain is sometimes called **freeware**. Software called **shareware** is also given away free; the maker hopes for voluntary monetary compensation—that is, he or she hopes that you like it well enough to send a contribution. Both public domain software and shareware may be copied freely and given to other people. But the software that people use most often, such as word processing or spreadsheet or database management software, is **copyrighted software,** software that costs money and must not be copied without permission from the manufacturer. Making illegal copies of copyrighted software is called **software piracy.**

Consider this incident. Bill Huston got his computer education at a local community college. One of his courses taught him how to use software on personal computers. He had access to a great variety of copyrighted software in the college computer lab. After graduating, he got a job at a local museum, where he used database software on a personal computer to help them catalog museum wares. He also had his own computer at home.

One day Bill stopped back at the college and ran into a former instructor. After greetings were exchanged, she asked him why he happened to drop by. "Oh," he said, "I just came by to make some copies of software." He wasn't kidding. Neither was the instructor who, after she caught her breath, replied, "You can't do that. It's illegal." Bill was miffed, saying "But I can't afford it!" The instructor immediately alerted the computer lab. As a result of this encounter, the staff strengthened policies on software use and increased the vigilance of lab personnel. In effect, schools must protect themselves from people who lack ethics or are unaware of the law.

There are many people like Bill. He did not think in terms of stealing anything; he just wanted to make copies for himself. But, as the software industry is quick to point out, unauthorized copying *is* stealing because the software makers do not get the revenues to which they are entitled. Furthermore, if software developers are not properly compensated, they may eventually find it not worthwhile to develop new software for our use.

Why Those Extra Copies?

Copying software is not always a dirty trick—there are lots of legitimate reasons for copying. To begin with, after paying several hundred dollars for a piece of software, you will definitely want to make a backup copy in case of disk failure or accident. You might want to copy the program onto a hard disk and use it, more conveniently, from there. Or you might want to have one copy at the office and another to use at home. Many software publishers have no trouble with any of these types of copying. But thousands of computer users copy software for another reason: to get the program without paying for it. And therein lies the problem.

Software publishers first tried to solve the problem by placing **copy protection** on their software—a software or hardware roadblock to make it difficult or impossible to make pirated copies. In effect, copy protection punishes the innocent with the guilty. There was vigorous opposition from software users, who argued that it was unfair to inconvenience paying customers just to outsmart a few thieves. Most software vendors have now dropped copy protection from their software, but they are still vigilant about illegal copies. But the most widespread solution seems to be vendor permission to copy software *legally*, an approach called site licensing.

Site Licensing

Although there is no clear definition industrywide, in general a **site license** permits a customer to make multiple copies of a given piece of software. The customer needing all these copies is usually a corporation, which can

Copyrighted Artwork

Software is not the only product that can be pirated. An interesting copyright issue is computer-generated artwork, called clip art. A library of clip art is often included with products such as word processing and desktop publishing software. But clip art can also be purchased separately, with hundreds or even thousands of images in a package.

Is it all right to use clip art in your neighborhood newsletter? Yes. What about in a magazine ad? Probably OK. Interestingly, licensing agreements accompanying clip art usually stipulate that the art can be used in any published form. The exclusion is using the art in widely distributed electronic form or reselling it. The art above is a clip art image from a package called CorelDRAW!

probably obtain a significant price discount for volume buying. The exact nature of the arrangement between the user and the software maker can vary considerably. Typically, however, a customer obtains the right to make a limited number of copies of a product, agrees to keep track of who uses it, and takes responsibility for copying and distributing manuals to its own personnel. The advantages to the user include a price break, the availability of as many copies as needed, and, of course, freedom from potential lawsuits from the software vendor.

Some software makers, however, oppose site licensing; they do not want to be bogged down in licensing negotiations. Industry leader Microsoft Corporation favors **concurrent licensing,** a system that charges a fee based on the number of users at a given time or perhaps at peak periods. Suppose, for example, that 20 users are on a network but a maximum of 10 would be using Microsoft Excel at a given time. The company could pay for just 10 copies of the software. However, once 10 of the users are using the software at a given moment, an 11th potential user is locked out.

The Battle Continues

The software industry persuaded Congress to amend the **Copyright Act** to raise software piracy from a misdemeanor to a felony. Under the 1992 law a convicted pirate faces the possibility of up to five years' jail time and $250,000 in fines. The industry organization called the **Software Publishers Association (SPA)** pursues software pirates of all kinds. The SPA runs spot checks and audits on corporations large and small, suing for damages when they find, for example, a firm that bought a single copy of a program and then made numerous unlicensed copies for its employees. Although companies want to stay within the law, they may be unaware of illegal software being used within their organizations. The SPA will provide a free program called SPAudit that can be used to create an inventory of installed software, which is the first step to determining if there is any pirated software in the company.

The SPA also has launched a campaign to help educate the public about illegal copying (Figure 11-7). Its most effective tool, however, is its antipiracy hotline ((800) 388-7478). Anyone can call the hotline to report illegal software copying. Of the approximately 30 calls received per day, many are from disgruntled ex-employees of offending companies. The reported culprits, who may have had no criminal intent in their minds, could nevertheless face legal action, stiff fines, and even prison terms.

The issues raised in this chapter are often the ones we think of after the fact —that is, when it is too late. The security and privacy factors are somewhat like insurance that we wish we did not have to buy. But we do buy insurance for our homes and cars and lives because we know we cannot risk being without it. The computer industry also knows that it cannot risk being without safeguards for security and privacy. As a computer user, you will share responsibility for addressing these issues.

In the next chapter we turn to modern trends in the workplace: expert systems, robotics, and virtual reality.

Figure 11-7 Capturing your attention. The Software Publishers Association, an organization that supports software developers, has launched an advertising campaign to warn personal computer users about the possible "rewards" for copying software illegally.

CHAPTER REVIEW

Summary and Key Terms

- The word **hacker** originally referred to an enthusiastic, self-taught computer user, but now the term usually describes a person who gains access to computer systems illegally.
- Computer criminals are likely to be trusted employees with no previous law-breaking experience. Many are motivated by resentment toward an employer, by personal or family problems, by the challenge of beating the system, or by the tempting ease with which the crime can be committed.
- Three basic types of computer crime are (1) theft of computer time for development of software; (2) theft, destruction, or manipulation of programs or data; and (3) alteration of data stored in a computer file.
- Prosecution of computer crime is often difficult because law enforcement officers, attorneys, and judges are unfamiliar with the issues involved. However, in 1986 Congress passed the latest version of the **Computer Fraud and Abuse Act,** and most states have passed some form of computer crime law.
- **Security** is a system of safeguards designed to protect a computer system and data from deliberate or accidental damage or access by unauthorized persons.
- The means of giving access to authorized people are divided into four general categories: (1) **what you have** (a key, badge, or plastic card), (2) **what you know** (a system password or identification number), (3) **what you do** (such as signing your name), and (4) **what you are** (by making use of **biometrics,** the science of measuring individual body characteristics such as fingerprints, voice, and retina). An **active badge,** a clip-on employee identification card with an embedded computer chip, signals its wearer's location—whether legal or otherwise—by sending out infrared signals, which are read by sensors sprinkled around the building.
- Loss of hardware and software is generally less of a problem than loss of data. Loss of hardware should not be a major problem, provided that the equipment is insured and a substitute processing facility is found quickly. Loss of software should not be critical, provided that the owner has taken the practical step of making backup copies. However, replacing lost data can be quite expensive.
- A **disaster recovery plan** is a method of restoring data processing operations if they are halted by major damage or destruction. Common approaches to disaster recovery include relying temporarily on manual services; buying time at a computer service bureau; making mutual assistance agreements with other companies; or forming a **consortium,** a joint venture with other organizations to support a complete computer facility.
- A **hot site** is a fully equipped computer facility with hardware, environmental controls, security, and communications equipment. A **cold site** is an environmentally suitable empty shell in which a company can install its own computer system.
- A disaster recovery plan should include (1) priorities indicating which programs must be running first, (2) personnel requirements specifying where employees should be and what they should do, (3) equipment requirements, (4) information about an alternative computing facility, and (5) specifications for how input and output data will be handled in a different environment.
- If a programmer is employed by an organization, any program written for the organization belongs to the employer. If the programmer is a consultant, however, the contract must clearly state whether it is the organization or the programmer that owns the software. Software can be copyrighted.
- There are five critical planning areas for data security: (1) determination of appropriate policies and standards, (2) development and implementation of security safeguards, (3) inclusion of security precautions during development of new automated systems, (4)

- review of state and federal laws related to security, and (5) maintenance of historical records associated with computer abuse.
- Common means of protecting data are secured waste, internal controls, auditor checks, applicant screening, separation of employee functions, passwords, and built-in software protection.
- A **worm** is a program that transfers itself from computer to computer over a network, planting itself as a separate file on the target computer's disks. A **virus** is a set of illicit instructions that passes itself on to other programs with which it comes in contact. A **vaccine,** or **antivirus,** is a computer program that stops the spread of the virus and eradicates it.
- Network security can be improved by the use of a **firewall,** a computer dedicated to screening access to the network from outside the network.
- Data sent over communications lines can be protected by **encryption,** the process of scrambling messages. The National Standards Institute has endorsed a process called **Data Encryption Standard (DES).**
- Personal computer security includes such measures as locking hardware in place; providing an appropriate physical environment; and using a **surge protector,** a device that prevents electrical problems from affecting data files.
- Files are subject to various types of losses and should be backed up on disk or tape.
- The security issue extends to the use of information about individuals that is stored in the computer files of credit bureaus and government agencies. The **Fair Credit Reporting Act** allows individuals to check the accuracy of credit information about them. The **Freedom of Information Act** allows people access to data that federal agencies have gathered about them. The **Federal Privacy Act** allows individuals access to information about them that is held not only by government agencies but also by private contractors working for the government. Individuals are also entitled to know how that information is being used. The **Video Privacy Protection Act** and the **Computer Matching and Privacy Protection Act** have extended federal protections.
- Some software, called **freeware,** is considered to be in the **public domain** because it is free. **Shareware** software is given away on the honor system, with the author asking for voluntary compensation.
- **Copyrighted software** costs money and must not be copied without permission from the manufacturer. Making illegal copies of copyrighted software is called **software piracy.**
- Software makers have tried in the past to protect their software with **copy protection,** a software or hardware block making it difficult or impossible to make copies. Today, many software publishers offer a **site license,** which permits a customer to make a limited number of copies of a given piece of software. **Concurrent licensing** allows a customer to use only a limited number of copies of a software product simultaneously.
- In 1992 Congress amended the **Copyright Act** to raise software piracy from a misdemeanor to a felony, with possible penalties of five years' jail time and $250,000 in fines.
- The industry organization called the **Software Publishers Association (SPA)** pursues software pirates.

Discussion Questions

1. Before accepting a particular patient, a doctor might like access to a computer file listing patients who have been involved in malpractice suits. Before accepting a tenant, the owner of an apartment building might want to check a file that lists people who have previously sued landlords. Should computer files be available for such purposes?
2. Discuss the following statement: "Some software is just too expensive for the average personal computer owner to buy. Besides, I only copy a friend's disk for personal use."
3. Why do some people consider computer viruses important? Discuss your answer from the point of view of the professional programmer, the MIS manager, and the hacker.

Student Study Guide

Multiple Choice

1. A computer crime in which money is embezzled in small amounts over time is
   a. the salami technique   c. the Trojan Horse
   b. blue-collar crime   d. data diddling
2. A computer dedicated to screening access to a network from outside the network:
   a. hot site   c. cold site
   b. vaccine   d. firewall
3. One safeguard against theft or alteration of data is the use of
   a. DES   c. identical passwords
   b. the Trojan Horse   d. data diddling
4. The legislation that prohibits government agencies and contractors from keeping secret personal files on individuals:
   a. Federal Privacy Act   c. Fair Credit Reporting Act
   b. Computer Abuse Act   d. Freedom of Information Act
5. A student who managed to modify her grades before they were entered into the computer file could be said to be
   a. piggybacking   c. using the salami technique
   b. data diddling   d. altering stored data
6. Computer crimes are usually
   a. easy to detect   c. prosecuted
   b. blue-collar crimes   d. discovered accidentally
7. The "what you are" criterion for computer system access involves
   a. a badge   c. biometrics
   b. a password   d. a magnetized card
8. The key factor in a computer installation that has met with disaster is the
   a. equipment replacement   c. loss of hardware
   b. insurance coverage   d. loss of processing ability
9. In anticipation of physical destruction, every computer organization should have a
   a. biometric scheme   c. disaster recovery plan
   b. DES   d. set of active badges
10. Software piracy includes
    a. badge theft   c. copyrighting
    b. program duplication d. data alteration
11. Authorization to make multiple software copies is called
    a. piggybacking   c. site licensing
    b. scavenging   d. copy protection
12. Secured waste, auditor checks, and applicant screening all aid
    a. data security   c. built-in software protection
    b. license protection   d. piracy detection
13. The weakest link in any computer system is the
    a. people in it   c. hardware
    b. password   d. software
14. A device that prevents electrical problems from affecting data files:
    a. site license   c. Trojan Horse
    b. hot site   d. surge protector
15. One form of built-in software protection for data is
    a. secured waste   c. applicant screening
    b. user profiles   d. auditor checks
16. A clip-on identification card with an embedded chip to signal its wearer's location:
    a. antivirus   c. active badge
    b. site license   d. consortium
17. An empty shell in which a company may install its own computer is called a
    a. restoration site   c. cold site
    b. hot site   d. hardware site
18. A program written when the programmer is employed by an organization is owned by
    a. the programmer   c. no one
    b. the state   d. the organization
19. Security protection for personal computers includes
    a. internal components   c. locks and cables
    b. software   d. all of these
20. The secret words or numbers to be typed in on a keyboard before any activity can take place are called
    a. biometric data   c. data encryptions
    b. passwords   d. private words
21. Another name for an antivirus:
    a. vaccine   c. worm
    b. Trojan Horse   d. DES
22. Another name for free software:
    a. encrypted software   c. copy protected software
    b. public domain software   d. shareware
23. A virus that replicates itself is called a
    a. bug   c. worm
    b. vaccine   d. bomb
24. A program whose sabotage depends on certain conditions is called a
    a. bug   c. worm
    b. vaccine   d. bomb
25. A person who gains illegal access to a computer system:
    a. hacker   c. worm
    b. software pirate d. zapper

True/False

T F 1. Most computer organizations cannot afford consortiums.
T F 2. A profile of a computer criminal suggests that he or she is likely to be regarded as an ideal employee.
T F 3. The Trojan Horse is an embezzling technique.
T F 4. If a computer crime is detected, prosecution is assured.
T F 5. Computer security is achieved by physically restricting access to the computer.
T F 6. Fingerprints are an example of biometrics.
T F 7. The actual loss of hardware is the major security problem due to its expense.

T F 8. A disaster recovery plan is a scheme devised in anticipation of major software piracy.
T F 9. It is legitimate to make a copy of software for backup purposes.
T F 10. Passwords, auditor checks, and separation of employee functions are data protection techniques.
T F 11. The science of studying individual body characteristics is called biometrics.
T F 12. Passwords are best changed annually.
T F 13. The spread of a vaccine is usually stopped by an antivirus.
T F 14. Data diddling is a criminal method whereby data is modified before it goes into a computer file.
T F 15. Unauthorized manipulation of programs falls into one of the three categories of computer crime.
T F 16. A mutual aid pact with another computer facility is one possibility for a disaster recovery plan.
T F 17. A key element of a disaster recovery plan is to establish priorities, that is, which programs must be up and running first.
T F 18. The Data Encryption Standard is a standardized list of passwords for software security.
T F 19. By allowing a programmer also to be a computer operator, a computer organization can improve security.
T F 20. Software can be patented.
T F 21. An organization using concurrent licensing can use unlimited copies of a software package concurrently.
T F 22. Most states have passed some form of computer crime law.
T F 23. A cold site is an environmentally suitable empty shell in which a company can install its own computer system.
T F 24. A firewall is copy protection added to software as a roadblock to prevent illegal copying.
T F 25. Shareware is software in the public domain.

Fill-In

1. What is an environmentally suitable empty shell into which a computer organization can put its computer system called? _____
2. What is a system of safeguards to protect a computer system and data from damage or unauthorized access called? _____
3. What is bypassing security systems with an illicitly acquired software package called? _____
4. What field is concerned with the measurement of individual body characteristics? _____
5. What is the name for a fully equipped computer center to be used in the event of a disaster? _____
6. What is the assurance to individuals that personal property is used properly called? _____
7. What is a person who gains access to a computer system illegally called? _____
8. Name a standardized public key by which senders and receivers can scramble and unscramble their messages. _____
9. What are the four categories of authorized access to a computer system?
    a. _____
    b. _____
    c. _____
    d. _____
10. What is a term for the unauthorized copying of software? _____
11. What law, passed in 1970, allows ordinary citizens to have access to data gathered by federal agencies? _____
12. Which law prohibits government agencies and private contractors hired by the government from keeping secret personal files? _____
13. The Fair Credit Reporting Act of 1970 gives people access to what information? _____
14. Software that is freely given away but with the hope of a contribution is called _____
15. Does a program belong to the employed programmer or the employing organization? _____
16. Which is potentially the least damaging type of security violation: hardware, software, or data? _____

Part Three ▼ Workplace Tools

17. What type of method is the salami technique? _____

18. What is placing an illicit program within a completed program, allowing unauthorized entry called? _____

19. What is a permit to make multiple copies of software called? _____

20. List the three general categories of computer crime.
    a. _____
    b. _____
    c. _____

21. Another name for a vaccine: _____

22. What kind of program transfers itself from computer to computer over a network? _____

23. What is the computer that monitors access to a network from outside the network? _____

24. What kind of license permits a customer to use only a limited number of copies of a software product simultaneously _____

25. What is the name for the identification card that can track the wearer's location? _____

Answers

Multiple Choice
1. a   6. d   11. c   16. c   21. a
2. d   7. c   12. a   17. c   22. b
3. a   8. d   13. a   18. d   23. c
4. a   9. c   14. d   19. d   24. d
5. b   10. b 15. b   20. b   25. a

True/False
1. T   6. T   11. T   16. T   21. F
2. T   7. F   12. F   17. T   22. T
3. F   8. F   13. F   18. F   23. T
4. F   9. T   14. T   19. F   24. F
5. F   10. T   15. T   20. T   25. F

Fill-In
1. cold site
2. security
3. zapping
4. biometrics
5. hot site
6. privacy
7. hacker
8. Data Encryption Standard (DES)
9. a. what you have
   b. what you know
   c. what you do
   d. what you are
10. software piracy
11. Freedom of Information
12. Federal Privacy Act
13. credit reports
14. shareware
15. the organization
16. hardware
17. embezzlement
18. trap door
19. site license
20. a. theft of computer time
    b. theft or damage to programs/data
    c. alteration of computer
21. antivirus
22. worm
23. firewall
24. concurrent license
25. active badge

PLANET INTERNET

Internet on the Internet

Magazine articles about the Internet usually supply a list of attractive Web sites, but little else of substance. Internet books usually have not caught up with the latest developments, which may have happened yesterday. But Net users have a ready source of current information: the Internet itself.

What kind of help is available? All kinds, from free online help navigating the Internet to the services of private paid consultants who advertise on the Net. You can even get the latest advice on choosing browser software; a good listing is at the WWW Client Software site.

Can I get help setting up my own Web site? Yes, but you need to make some arrangements yourself. To set up your own site, you need (1) to have access to a server and (2) to make your own home page. A *server* is a computer whose owner agrees to offer some hard disk space and also access to your link. Computer resources in educational, government, and business settings may be available to appropriate personnel. If you are an individual or a small business, however, you will probably use a private server and pay a monthly fee for the services provided.

A home page is prepared by you, possibly with some assistance, and then transferred to the server. You are then, theoretically, in business. But how will anyone know your site exists? You need to pass the word so that other sites will link to yours. Friends and colleagues who have sites may link to each other, but that is small potatoes. You need to request being added to some of the major directories. The Yahoo directory, for example, has an Add link that supplies an application form online. In general, once the existence of your link has been verified, your page is added. We assume that you also have browser software, with which to view your own home page and, of course, the rest of the Web.

Where can I get assistance to make a home page? There are several possibilities. Many colleges include home page creation as part of an Internet course. Private firms advertise courses to teach you the basics in a few hours. Resources to help you are right on the Net; our B/C site lists several references.

You can make a simple home page with word processing software that offers page preparation as an option. Since multimedia is a key aspect of the Web, you probably want to use attractive fonts and graphics, and possibly even sound or film clips. The Multimedia File Formats site, whose home page logo is shown here, was prepared specifically to assist beginners.

Serious business users often engage the services of consultants who can create a sophisticated home page. Such consultants usually offer a wide range of services related to the Internet. For example, the Open Business Markets consultants, whose Internet home page logo is shown here, combine technical expertise with business savvy to facilitate electronic commerce on the Internet. Many consulting firms also offer seminars.

Internet Exercises

1. **Structured exercise.** Begin with the B/C URL http://www.aw.com/bc/planet/ and link to the WWW Client Software site. Compare the features mentioned for the browsers you find there with browsers you have used.
2. **Freeform exercise.** Investigate the private consultants listed at the B/C site. You can probably leave e-mail messages with your questions. Compare services and prices for help with a home page or with server access or the total package.

Joyce Lindsay worked as a credit consultant for Nordstrom, a chain of stores selling high-quality clothing. Joyce had developed significant expertise over a period of years. Consider this example. A customer comes to the store, selects a $300 coat, and hands her charge card to the sales clerk. However, the customer's credit limit is just $1000, and she has existing unpaid credit of $875. Should the customer be allowed to charge the coat anyway, or should the clerk adhere strictly to the credit limit? This ticklish question is turned over to Joyce who, after quickly reviewing the customer's records, is able to grant the extra charge.

Although this seems like a system that works pretty well, Nordstrom recently converted the whole process to an expert system, a computer system in which the computer plays the role of expert. Why go to all that trouble and expense? Why not just stick with human experts? Well, there are problems with human experts. They are typically expensive, subject to biases and emotions, and they may even be inconsistent. However, the biggest problem is that the expertise of one individual is not readily available to multiple users at the same time. Finally, there have been occasions when experts have resigned or retired, leaving the company in a state of crisis. The computer, however, is ever present and just as available as the telephone.

When the new expert system was being developed at Nordstrom, computer specialists approached Joyce to ask her how she made her decisions. Some experts cling to the notion that their decisions are based on instinct, some kind of gut reaction. Study always reveals, however, that their "instincts" are based on a set of rules, possibly so embedded in their brains that the experts themselves are not even aware of it. Joyce was able to articulate most of her procedures to the computer specialist. In the example above the purchaser's records showed that she consistently paid her bill on time, that her average monthly balance was usually low, and that she had a good job. These insights, along with many other rules, became part of the new expert system.

What about Joyce? Is she now out of a job? No. With the installation of the computerized system, her role as expert has changed. She is now a consultant to the expert system, which needs constant updating and monitoring. Joyce has also been assigned some management responsibilities and, generally, has a more interesting job than she had before.

Chapter 12

Modern Trends in the Workplace
Expert Systems, Robotics, and Virtual Reality

LEARNING OBJECTIVES

- Understand the underlying concepts and terms of artificial intelligence
- Become acquainted with the fundamentals of expert systems, robotics, and virtual reality
- Understand how these fields of study have developed
- Understand the impact of these fields on business and everyday life

FROM THE LAB TO THE WORKPLACE

THE ARTIFICIAL INTELLIGENCE FIELD
- Early Mishaps
- How Computers Learn
- The Artificial Intelligence Debate
- Brainpower: Neural Networks
- The Famous Turing Test

THE NATURAL LANGUAGE FACTOR

EXPERT SYSTEMS
- Expert Systems in Business
- Building an Expert System
- The Outlook for Expert Systems

ROBOTICS
- Robots in the Factory
- Robot Vision
- Field Robots

VIRTUAL REALITY
- Travel Anywhere, But Stay Where You Are
- Getting Practical

Just a Bit Garbled

Recent advertisements for computer language translation software herald "completely automatic document translation," a process promised to be "fast and easy." But, as these examples demonstrate, automated translators have a way to go.

The city of Boston put out a weather warning to elderly citizens in English and Spanish: "Hypothermia means low body temperature. It is caused by exposure to cold." When the Spanish version was pumped through a computer translator, the English version came out "Hypothermia means to say temperature gets off the body and is caused by the exposition to the cold." A different translator was even worse: "Hypothermia wants to tell temperature lowered of the body and is caused by the (exposicion) to the fry."

From the Lab to the Workplace

Everyday items in the workplace, from ball point pens to masking tape to Post-it notes, started their existence in a research lab. In a loose sense computer scientists who develop cutting edge software are operating in a laboratory too. Their technologies, at least at first, often do not have obvious commercial applications. Furthermore, if a computer technology is little understood and owes its existence largely to government support, its proponents may be ridiculed. That is what happened to computer scientists whose specialty was *artificial intelligence*, a field of study relating computers to the human mind.

Today, subsets of artificial intelligence have practical applications. Before we examine these workplace applications, let us review the artificial intelligence background from which they sprang.

The Artificial Intelligence Field

Artificial intelligence (AI) is a field of study that explores how computers can be used for tasks that require the human characteristics of intelligence, imagination, and intuition. Computer scientists sometimes prefer a looser definition, calling AI the study of how to make computers do things that—at the present time—people can do better. The phrase "at the present time" is significant because artificial intelligence is an evolving science: As soon as a problem is solved, it is moved off the artificial intelligence agenda. A good example is the game of chess, once considered a mighty AI challenge. But now that most computer chess programs can beat most human competitors, chess is no longer an object of study by scientists and thus no longer on the artificial intelligence agenda.

Today the term *artificial intelligence* encompasses several subsets of interests (Figure 12-1):

- **Problem solving.** This area of AI includes a spectrum of activities, from playing games to planning military strategy.
- **Natural languages.** This facet involves the study of the person/computer interface in unconstrained English language.
- **Expert systems.** These AI systems present the computer as an expert on some particular topic.
- **Robotics.** This field involves endowing computer-controlled machines with electronic capabilities for vision, speech, and touch.

Although considerable progress has been made in these sophisticated fields of study, early successes did not come easily. Before we examine current advances in these areas, let us pause to consider some early moments in the development of artificial intelligence.

Early Mishaps

In the early days of artificial intelligence, scientists thought that the computer would experience something like an electronic childhood, in which it would gobble up the world's libraries and then begin generating new wisdom. Few people talk like this today because the problem of simulating intelligence is far more complex then just stuffing facts into the computer. Facts are useless without the ability to interpret and learn from them.

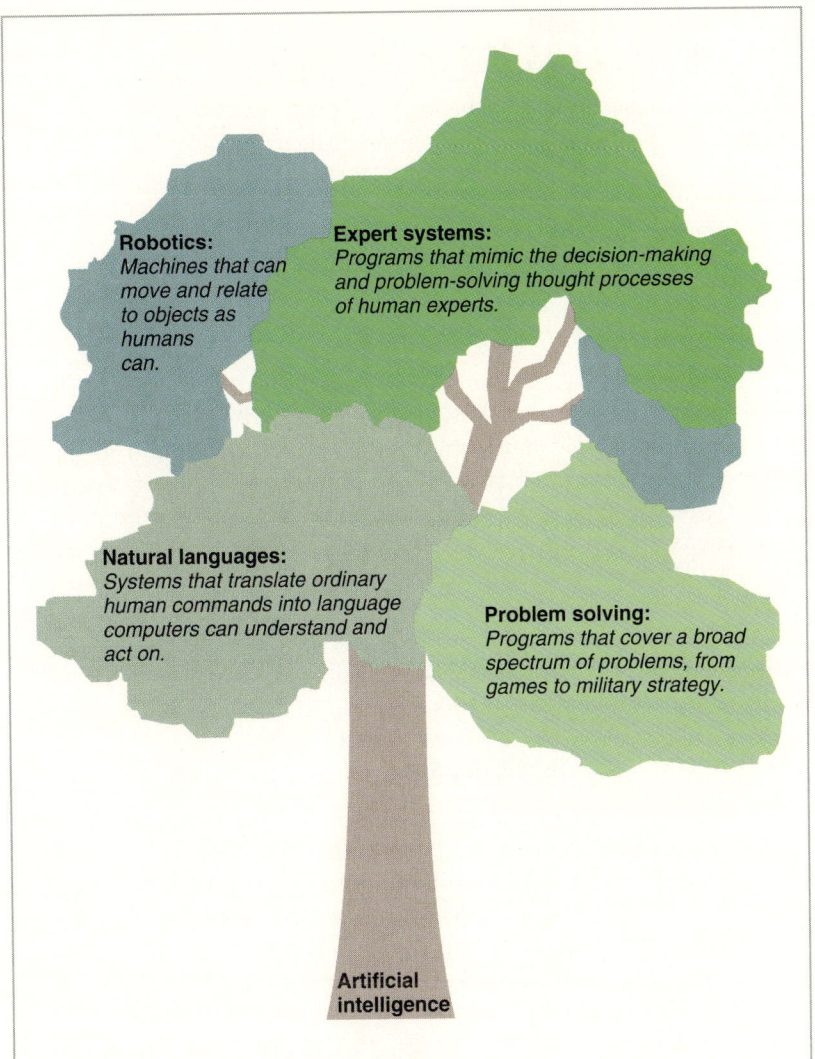

Figure 12-1 The artificial intelligence family tree.

An artificial intelligence failure on a grand scale was the attempt to translate human languages via the computer. Although scientists were able to pour vocabulary and rules of grammar into the computer, the literal word-for-word translations often produced ludicrous output. In one infamous example the computer was supposed to demonstrate its prowess by translating a phrase from English to Russian and then back to English. Despite the computer's best efforts, "The spirit is willing, but the flesh is weak" came back "The vodka is good, but the meat is spoiled." Recent language translation experiments, incidentally, have not been much more successful.

An unfortunate result of this widely published experiment was the ridicule of artificial intelligence scientists, considered dreamers who could not accept the limitations of a machine. Funding for AI research disappeared, plunging the artificial intelligence community into a slump from which it did not recover until expert systems emerged in the 1980s. Nevertheless, a hardy band of scientists continued to explore artificial intelligence, focusing on how computers learn.

How Computers Learn

The study of artificial intelligence is predicated on the computer's ability to learn—and to improve performance based on past errors. The two key elements of this process are called the knowledge base and the inference engine. A **knowledge base** is a set of facts and rules about those facts. An **inference engine** accesses, selects, and interprets a set of rules. The inference engine applies the rules to the facts to make up new facts—thus the computer has learned something new. Consider this simple example:

FACT: Amy is Ken's wife.
RULE: If X is Y's wife then Y is X's husband.

The computer—the inference engine—can apply the rule to the fact and come up with a new fact: Ken is Amy's husband. Although the result of this simplistic example may seem of little value, it is indeed true that the computer now knows two facts instead of just one. Rules, of course, can be much more complex and facts more plentiful, yielding more sophisticated results. In fact, artificial intelligence software is capable of searching through long chains of related facts to reach a conclusion—a new fact.

Further explanation of the precise way computers learn is beyond the scope of this book. However, we can use the learning discussion as a springboard to the question that most people ask about artificial intelligence: Can a computer really think?

The Artificial Intelligence Debate

To imitate the functioning of the human mind, a machine with artificial intelligence would have to be able to examine a variety of facts, address multiple subjects, and devise a solution to a problem by comparing new facts to its existing storehouse of data from many fields. So far, artificial intelligence systems cannot match a person's ability to solve problems through original thought instead of using familiar patterns as guides.

There are many arguments for and against crediting computers with the ability to think. Some say, for example, that computers cannot be considered intelligent because they do not compose like Beethoven or write like Shakespeare; the rejoinder is that neither do most ordinary human musicians or writers—you do not have to be a genius to be considered intelligent.

Look at it another way. Suppose you rack your brain over a problem, and then—Aha!—the solution comes to you all at once. Now, how did you do that? You do not know, and nobody else knows either. A big part of human problem solving seems to be that jolt of recognition, that ability to see things suddenly as a whole. Experiments have shown that people rarely solve problems by using step-by-step logic, the very thing that computers do best. Most modern computers still plod through problems one step at a time. The human brain beats computers at "Aha!" problem solving because it has millions of neurons working simultaneously. Now some scientists are taking that same approach with computers.

Brainpower: Neural Networks

A microprocessor chip is sometimes referred to as the "brain" of a computer. But, in truth, computers have not yet come close to matching the human brain, which has trillions of connections between billions of neurons. What is more, the most sophisticated conventional computer does

MAKING THE RIGHT CONNECTIONS

Power and Wealth

In this somewhat futuristic chapter, let us turn our attention to future issues relating to connectivity. The overriding issues in the information superhighway are as old as humanity itself: power and wealth. Control of information is power and access to information is wealth.

The basic question is this: Who will be in charge? In terms of access, will government or industry regulators insist that the superhighway be accessible to certain groups? Will access to the superhighway someday be defined by the courts as a right for everyone? Possibly, as some prefer, access will be determined by the invisible hand of the marketplace.

And who—or what—will control the information on the superhighway? At present, the question is wide open, as phone and cable and utility companies struggle for the upper hand. It is not hard to imagine other players in the fray. In particular, once its power is completely obvious to everyone, it seems inevitable that the government will somehow step in. Stay tuned.

---

not "learn" the same way the human brain learns. But let us consider an unconventional computer, one whose chips are actually designed to mimic the human brain. These computers are called **neural networks,** or, simply, neural nets.

If a computer is to function more like the human brain and less like an overgrown calculator, it must be able to experiment and to learn from its mistakes. Researchers are developing computers with a few thousand brain-like connections. Instead of the computer circuits following the usual step-by-step series of instructions, the circuits in a neural net computer form a grid, much like a nerve cell in the brain. The grid is used to recognize patterns. For instance, a neural network with optical sensors, shown in Figure 12-2, could be "trained" to recognize the letter *A*.

At best, today's neural networks consist of only a few thousand connections—still a far cry from the billions found in the human brain.

The Famous Turing Test

So, can a computer think or not? Listen to Alan Turing. Several decades ago, this English mathematician proposed a test of thinking machines. In the **Turing test,** a human subject is seated before two terminals that are connected to hidden devices. One terminal is connected to a terminal run by an unseen person, and the second terminal is connected to a computer. The subject is asked to guess, by carrying out conversations through the

Figure 12-2 Neural nets. This figure shows how a neural network with optical sensors is trained to recognize the letter *A*. (1) When the system makes an incorrect choice, the incorrect circuits are weakened. (2) When a correct choice is made, those circuits are strengthened. (3) After several attempts with different forms of the letter *A*, in which the correct circuits are repeatedly strengthened, the neural net is able to identify the letter accurately.

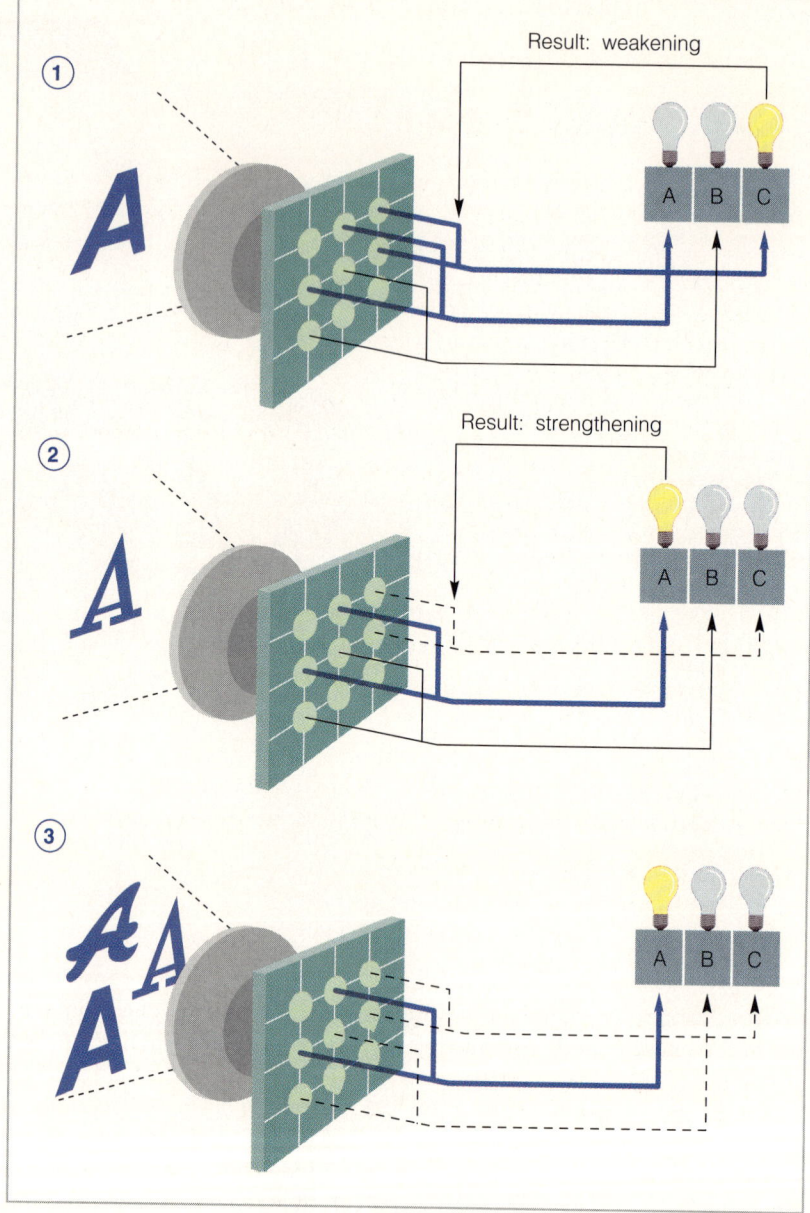

terminals, which is the person and which is the computer. If the human judge cannot tell the difference, the computer is said to have passed and is considered, for all practical purposes, a thinking machine. Every year, the Boston Computer Museum hosts a contest offering a $100,000 prize to anyone who writes a program for a computer that passes the Turing test. No one has ever won the full prize.

But perhaps we are asking the wrong question: Will a computer ever *really* think? One possible answer: Who cares? If a machine can perform a task really well, does it matter if it really thinks? Still another answer is: Yes, machines will really think, but not as humans do. They lack the sensitivity, appreciation, and passion that are intrinsic to human thought.

Meanwhile, scientists are getting rather good at developing related areas of artificial intelligence. We will focus on some of the more visible results of recent research in natural languages, expert systems, and robotics.

The Natural Language Factor

The language people use on a daily basis to write and speak is called a **natural language**. Natural languages are associated with artificial intelligence because humans can make the best use of artificial intelligence if they can communicate with computers in natural language. Furthermore, understanding natural language is a skill thought to require intelligence.

Some natural language words are easy to understand because they represent a definable item: *horse, chair, mountain*. Other words, however, are much too abstract to lend themselves to straightforward definitions: *justice, virtue, beauty*. But this kind of abstractness is just the beginning of the difficulty. Consider the word *hand* in these statements:

- Morgan had a hand in the robbery.
- Morgan had a hand in the cookie jar.
- Morgan is an old hand at chess.
- Morgan gave Sean a hand with his luggage.
- Morgan asked Marcia for her hand in marriage.
- All hands on deck!
- Look, Ma! No hands!

As you can see, natural language abounds with ambiguities; the word *hand* has a different meaning in each statement. In contrast, sometimes statements that appear to be different really mean the same thing:

- Alan sold Miguel a book for five dollars.
- Miguel bought a book for five dollars from Alan.
- Miguel gave Alan five dollars in exchange for a book.
- The book that Miguel bought from Alan cost five dollars.

It takes very sophisticated software (not to mention enormous computer memory) to unravel all these statements and see them as equivalent. A key function of the AI study of natural languages is to develop a computer system that can resolve ambiguities.

Feeding computers the vocabulary and grammatical rules they need to know is a step in the right direction. However, as you saw earlier in the account of the language translation fiasco, true understanding requires more: Words must be taken in context. Humans start acquiring a context for words from the day they are born. Consider this statement: Jack cried when Alice said she loved Bill. From our own context we can draw several possible conclusions: Jack is sad, Jack probably loves Alice, Jack probably thinks Alice doesn't love him, and so on. These conclusions may not be correct, but they are reasonable interpretations based on the context we supply. On the other hand, it would *not* be reasonable to conclude from the statement that Jack is a carpenter or that Alice has a new refrigerator.

One of the most frustrating tasks for AI scientists is providing the computer with the sense of context that humans have. Scientists have attempted to do this in regard to specific subjects and found the task daunting. For example, a scientist who wrote software so the computer could have a dialogue about restaurants had to feed the computer hundreds of facts that any small child would know, such as the fact that restaurants have food and that you are expected to pay for it.

Eliza

In the 1960s a computer scientist named Joseph Weizenbaum wrote a little program as an experiment in natural language. He named the program after Eliza Doolittle, the character in *My Fair Lady* who wanted to learn to speak proper English. The software allows the computer to act as a benign therapist who does not talk much but, instead, encourages the patient—the computer user—to talk.

The Eliza software has a storehouse of key phrases to be dragged out when triggered by the patient. For example, if a patient types "My mother never liked me," the software—cued by the word *mother*—can respond "Tell me more about your family." If there are no key words from the patient, the computer responds neutrally, with a phrase such as "I see" or "That's very interesting" or "Why do you think that?" If a patient gives yes or no answers, the computer may respond "I prefer complete sentences." With party tricks like these, the program is able to move along quite nimbly from line to line.

Weizenbaum was astonished to discover that people were taking his little program seriously, pouring out their hearts to the computer. In fact, what he viewed as misuse of the computer radicalized Weizenbaum, who spent the next several years giving speeches and writing articles against artificial intelligence.

Expert Systems

An **expert system** is a software package used with an extensive set of organized data that presents the computer as an expert on a particular topic. For example, a computer could be an expert on where to drill oil wells, or what stock purchase looks promising, or how to cook soufflés. The user is the knowledge seeker, usually asking questions in a natural—that is, English-like—language format. An expert system can respond to an inquiry about a problem with both an answer and an explanation of the answer. For example, an expert system about stock purchases could be asked if stocks of the Milton Corporation are currently a good buy. A possible answer is no, with backup reasons such as a very high price/earnings ratio or a recent change in top management. The expert system works by figuring out what the question means and then matching it against the facts and rules that it "knows" (Figure 12-3). These facts and rules, which reside on disk, originally come from human experts.

Expert Systems in Business

For years expert systems were no more than bold experiments, the exclusive property of the medical and scientific communities. Special programs could offer medical diagnoses or search for mineral deposits or examine chemical compounds. But in the early 1980s expert systems began to make their way into commercial environments. Consider these examples:

- The Campbell Soup Company has an expert system nicknamed Aldo, for Aldo Cimino, the human expert who knows how to fix cooking machines. Aldo was getting on in years and being run ragged, flying from plant to plant whenever a cooker went on the blink. Besides, how would the company manage when he retired? Now Aldo's knowledge has been distilled into an expert system that can be used by workers in any location.

Figure 12-3 An expert system on the job. This expert system helps Ford mechanics track down and fix engine problems.

- Factory workers at The Boeing Company use an expert system to assemble electrical connectors for airplanes. In the old days workers had to hunt through 20,000 pages of cross-referenced specifications to find the right parts, tools, and techniques for the job—approximately 42 minutes per search. The expert system lets them do the same thing in about 5 minutes.
- Employees at Coopers and Lybrand, an accounting firm, use an expert system called ExperTax to help clients with tax planning. The knowledge of tax planning experts is available to inexperienced accountants, and it is as close as their computers.
- The United Airlines terminal at O'Hare Airport in Chicago handles 400 flights per day, which must be distributed among 50 gates. Complications include the limitations of jumbo jets, which do not maneuver easily into some gates. Furthermore, both the weather and heavy runway traffic can affect how quickly planes can get in and out. Airline employees, who used to track planes on a gigantic magnetic board, now keep track of gate positions with an expert system that takes all factors into account (Figure 12-4).
- Johnson Wax, the makers of Raid® and other insecticides, designed an expert system to help employees wade through the product regulation maze. The system contains knowledge about all federal and local insecticide regulations and helps outline product requirements in order to meet the regulations.

The cost of an expert system can usually be justified in situations where there are few experts but great demand for knowledge. It is also worthwhile to have a system that is not subject to human failings, such as fatigue.

Building an Expert System

Some organizations choose to build their own expert systems to perform well-focused tasks that can easily be crystallized into rules. A simple example is a set of rules for a banker to use when making decisions about

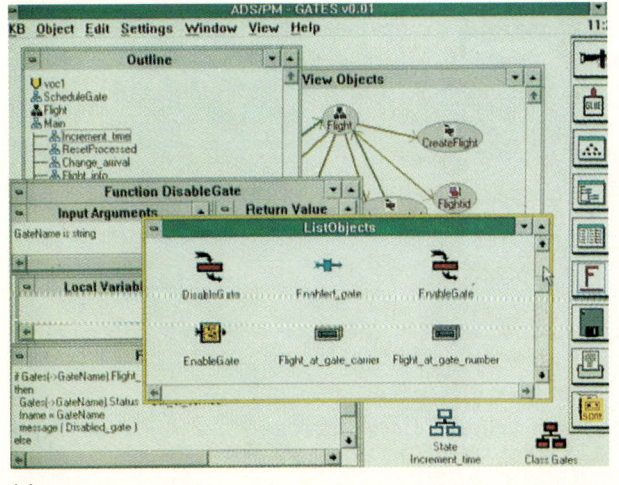
(a)

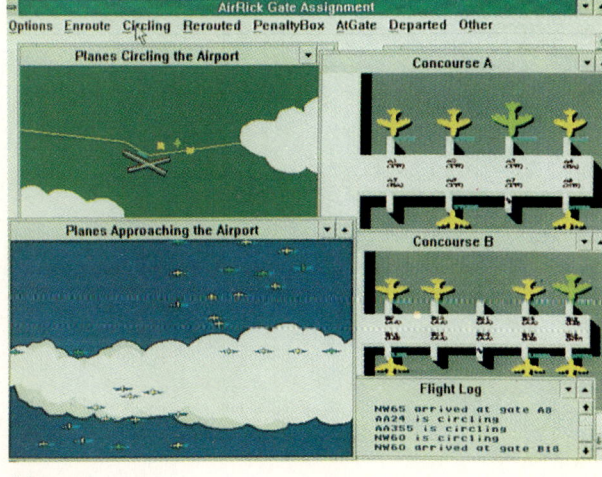
(b)

Figure 12-4 Airline scheduling program produced with the aid of an expert system. This system offers a graphical user interface to help solve a complex airport scheduling problem. (a) This screen illustrates the system's ability to display multiple views of objects and the relationships between them. (b) Various screen windows show planes circling the airport, the number of planes approaching the airport, gate information, and two concourses with planes at their gates.

whether to grant a loan. But very few organizations are capable of building an expert system from scratch. The sensible alternative is to buy an **expert shell,** a software package that consists of the basic structure used to find answers to questions. It is up to the buyer to fill in the actual knowledge on the chosen subject. You could think of the expert shell as an empty cup that becomes a new entity once it is filled—a cup of coffee, for instance, or a cup of sugar.

The most challenging task of building an expert system often is deciding who the appropriate expert is and then trying to pin down his or her knowledge. Experts often believe that much of their expertise is instinctive and thus find it difficult to articulate just why they do what they do. However, the expert is usually following a set of rules, even if the rules are only in his or her head. The person ferreting out the information, sometimes called a **knowledge engineer,** must have a keen eye and the skills of a diplomat. Sometimes cameras are used to observe the expert in action.

Once the rules are uncovered, they are formed into a set of IF-THEN rules: IF the customer has exceeded a credit limit by no more than 20 percent and has paid the monthly bill for six months, THEN extend further credit. After the system is translated into a computerized version, it is reviewed, changed, tested, and changed some more. This repetitive process can take months or even years. Finally, it is put into the same situations as the human expert would face, where it should give equal or better service but much more quickly.

The Outlook for Expert Systems

Some industry analysts feel that expert systems are beginning to mimic the analytic processes of humans and that, as a result, these programs border on true artificial intelligence. Putting together the facets we have discussed so far, a computer having artificial intelligence should understand the facts it knows, come up with new facts, and be able to engage in a wide-ranging conversation about them in a natural language. By these standards expert systems today are still rather dimwitted. Furthermore, each system has intelligence only in one specific area.

Expert systems will infiltrate companies department by department, much as personal computers did before them. Some expert systems are now available on personal computers. The main limitation of using an expert system on a personal computer is that the expert system requires a substantial amount of memory. Even so, it seems likely that more expert systems for personal computers will appear in the near future.

 # Robotics

Many people smile at the thought of robots, perhaps remembering the endearing C-3PO of *Star Wars* fame and its "personal" relationship with humans. But vendors have not made even a small dent in the personal robot market—the much-heralded domestic robots who wash windows have not yet become household staples. So, where are the robots today? Mainly in factories.

Robots in the Factory

Most robots are in factories, spray-painting and welding—and taking away jobs. The Census Bureau, after two centuries of counting people, has

ONE JUMP AHEAD

Robots Coming to Our Lives

Robots are journeying far from their original factory homes. Like computers before them, robots will soon be everywhere. Here are some examples.

Homer Hoover. Robots as vacuum cleaners have always been the ultimate robot joke, but now we can stop laughing. A robot is being developed that can "see" its way around the house, vacuuming as it goes, and avoid sucking up the cat.

Robo-Mow. Count on your neighbors to stare stupefied as a bright-green handle-less lawnmower cruises your yard unsupervised. You need only bury a wire two inches deep around the perimeter of your yard. Set Robo-mow within the border and it will continuously and silently cut grass.

My doctor the robot. If you have orthopedic surgery, you may find that a key player alongside the surgeon is a robot. For example, to make room for a hip implant, a robotic arm drills a long hole in a thigh bone. Robotic precision improves the implant, reduces pain after surgery, and speeds healing.

Lending a hand. Robots may soon be of significant use to the disabled. Researchers have already developed a robot for quadriplegics. The machine can respond to dozens of voice commands by answering the door, getting the mail, and even serving soup.

Road maintenance. In California, road signs may soon say "Robots at

Chip the robot.

Work." Robots use lasers to spot cracks in the pavement and dispense the right amount of patch material. Soon, robots will also be painting the road stripes.

Going bump in the night. Chip, the chunky errand boy on the night shift at Baltimore's Franklin Square Hospital, fetches medicine, late meals (shown here), medical records, and supplies. A robot who works for $6 an hour, Chip finds his way using sensitive whiskers and touch pads to "feel." Nurses love him because he saves them from having to run all over the hospital.

---

now branched out and today is counting robots as well. About 15,000 robots existed in 1985, a number that jumped to 50,000 just ten year later. What do robots do that merits all this attention?

A loose definition of *robot* is a type of automation that replaces human presence. But a **robot** is more completely defined as a computer-controlled device that can physically manipulate its surroundings. There are a wide variety of sizes and shapes of robots, each designed with a particular use in mind. Often these uses are functions that would be tedious or even dangerous for a human to perform. The most common industrial robots sold

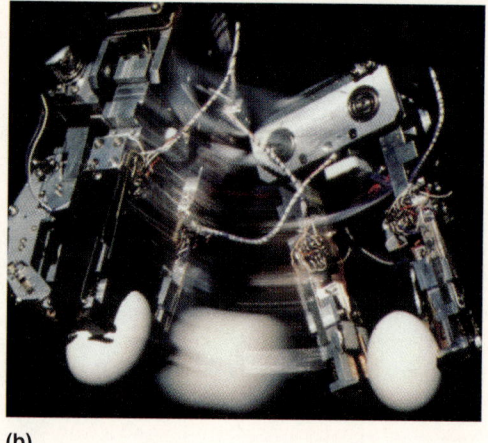

(a)   (b)

Figure 12-5 Industrial robots. (a) These standard robots are used in the auto industry to weld new cars. (b) This robot is not making breakfast. Hitachi uses the delicate egg, however, to demonstrate that its visual-tactile robot can handle fragile objects. The robot's sensors detect size, shape, and required pressure, attaining sensitivity almost equal to that of a human hand.

today are mechanical devices with five or six axes of motion, so they can rotate into proper position to perform their tasks (Figure 12-5).

We mentioned spray-painting and welding, but a more intelligent robot can adapt to changing circumstances. For example, with the help of a TV-camera eye, a robot can see components it is meant to assemble. It is able to pick them up, rearrange them in the right order, or place them in the right position before assembling them.

Robot Vision

Recently, **vision robots** have been taught to see in living color—that is, to recognize multicolored objects solely from their colors. This is a departure from the traditional approach, whereby robots recognize objects by their shapes (Figure 12-6), and from vision machines that "see" only a dominant color. For example, a robot in an experiment at the University of Rochester was able to pick out a box of Kellogg's Sugar Frosted Flakes from 70 other boxes. Among the anticipated benefits of such visual recog-

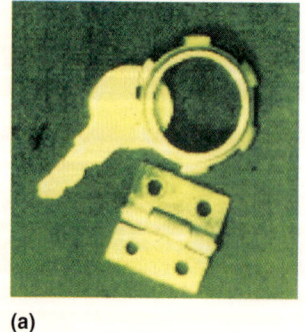
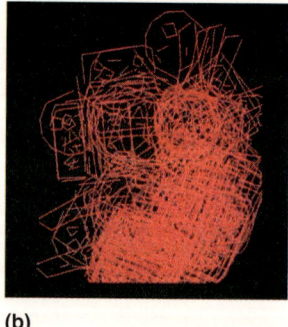
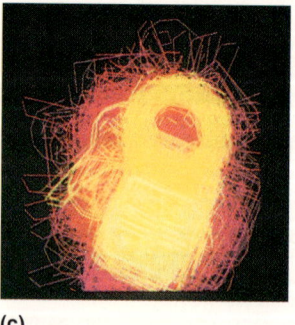
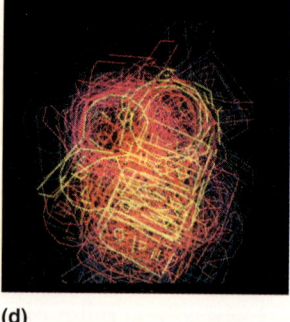

(a)   (b)   (c)   (d)

Figure 12-6 The seeing robot. Robots "see" by casting light beams on objects and identifying them by matching their shapes to those of already "known" objects. In this machine-vision sequence, (a) the object is seen by the robot, (b) the object is matched to known shapes, (c) inappropriate shapes are eliminated, and (d) the object is recognized.

Figure 12-7 Field robots. (a) Nicknamed Robotuna, scientists hope this undersea robot can map the ocean floor, track schools of real fish, or detect pollution—and then swim home with the data. (b) Can a robot really fly? Yes. Flying robots have both military and civilian uses. This Sentinel robot can soar up to 10,000 feet to spy on an enemy or to inspect high-voltage wires or to spot forest fires. (c) The robot called Spider checks gas tanks for cracks and sends computer images back to the ground, saving engineers from making a dangerous climb.

nition skills is supermarket checkout. You cannot easily bar code a squash, but a robot might be trained to recognize it by its size, shape, and color.

Field Robots

Just think of some of the places you would rather not be: inside a nuclear power plant, next to a suspected bomb, at the bottom of the sea, on the floor of a volcano, or in the middle of a chemical spill. But robots readily go to all those places. Furthermore, they go there to do some dangerous and dirty jobs. These days, **field robots**—robots "in the field"—inspect and repair nuclear power plants, dispose of bombs, inspect oil rigs for undersea exploration, explore steaming volcanos, clean up chemical accidents, and much more. As an example, an undersea robot ventured into the icy waters off Finland and scanned the sunken ferry *Estonia*, sending back pictures of its weakened bow, thought to be a cause of the disaster. Newer undersea robots are being designed to swim like fish (Figure 12-7a). Going in another direction, space researchers look forward to the day when "astrobots" can be stationed in orbit, ready to repair faulty satellites.

Only a few years ago there were just a handful of field robots commercially employed. Now there are hundreds, and soon there will be thousands. Field robots may be equipped with wheels, tracks, legs, fins, or even wings (Figure 12-7b). Field robots have been largely overshadowed by factory robots, mainly because they lack the independent glamour of their manufacturing counterparts; field robots must be remotely controlled by human operators. Now, however, the poor relative status of field robots is changing because enough computer power can be packed into a field robot to enable it to make most decisions independently. Field robots need all the power they can get. Unlike factory robots, which are bolted to the ground and blindly do the same tasks over and over again, field robots must often contend with a highly unstructured environment, such as changing terrain and changing weather.

Virtual Reality

The concept of **virtual reality** is to immerse a user in a computer-created environment so that the user physically interacts with that environment. This is made possible by sophisticated computers and optics that deliver to a user's eyes a three-dimensional scene in living color. The essence of the scene is a database used by a powerful computer to display graphic images. The virtual reality system can sense a user's head and body movements through cables linked to the headset and glove worn by the user. That is, sensors on the user's body send signals to the computer, which then adjusts the scene viewed by the user. Thus, the user's body movements can cause interaction with the virtual (artificial) world the user sees, and the computer-generated world responds to those actions (Figure 12-8).

Travel Anywhere, But Stay Where You Are

At the University of North Carolina, computer scientists have developed a virtual reality program that lets a user walk through an art gallery. A user puts on a head-mounted display, which focuses the eyes on a screen and shuts out the rest of the world. If the user swivels his or her head right, pictures on the right wall come into view; similarly, the user can view any part of the gallery by just making head movements. This action/reaction presents realistic continuing changes to the user. Although actually standing in one place, the user feels as if he or she is moving and wants to stop short as a pedestal appears in the path ahead. It is as if the user is actually walking around inside the gallery.

In another example scientists have taken data about Mars, sent back by space probes, and converted it to a virtual reality program. Information about hills, rocks, and ridges of the planet are used to create a Mars landscape, whose images are projected on the user's head screen.

Getting Practical

An embryonic technology such as virtual reality is filled with hype and promises. We must look to the practical commercial applications for real-world users to see where this technology might lead. Here are some applications under development:

- Wearing head mounts, consumers can browse for products in a "virtual showroom." From a remote location a consumer will be able to maneuver and view products along rows in a warehouse.
- Similarly, from a convenient office perch a security guard can patrol corridors and offices in remote locations.

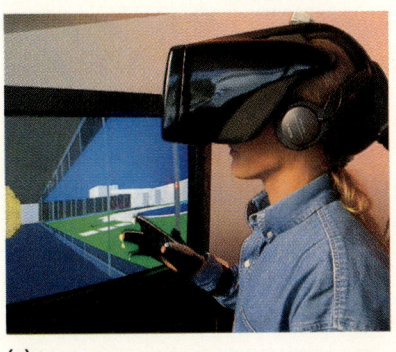

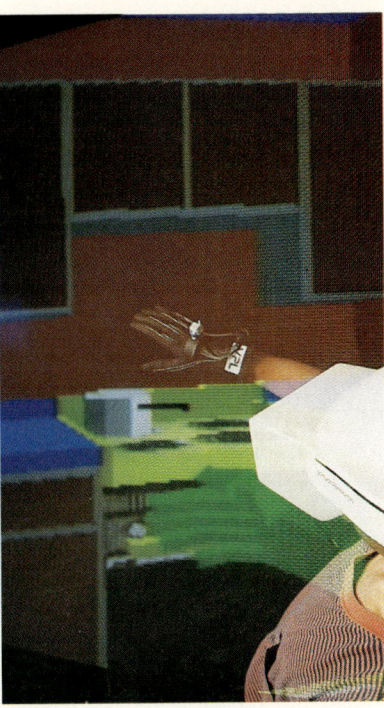

Figure 12-8 Virtual reality. (a) Users can "tour" a building by physically reacting—a turn of the head shows a different scene. (b) The data glove in the foreground has fiber optic sensors to interact with a computer-generated world. (c) Virtual reality technology can be used to let people who are in wheelchairs design their own apartments.

- Air traffic controllers may someday work like this: Microlaser scanner glasses project computer-generated images directly into the controller's eyes, immersing the controller in a three-dimensional scene showing all the aircraft in the area. To establish voice contact with the pilot of the plane, the controller merely touches the plane's image with a sensor-equipped glove.
- Using virtual reality headsets and gloves, doctors and medical students will be able to experiment with new procedures on simulated patients rather than real ones.

Any new technology has its drawbacks. In addition to rather clumsy physical equipment, today's virtual reality pioneers are faced with daunting costs. Many hurdles remain in the areas of software, hardware, and even human behavior before virtual reality can reach its full potential.

The immediate prospects for expert systems, robots, and virtual reality systems are growth and more growth. We can anticipate both increased sophistication and more diverse applications. The progress in the more esoteric applications of artificial intelligence will continue to be relatively slow. No one need worry that any computer can capture the wide-ranging sophistication of a human just yet. People are in charge of computers, not the other way around.

CHAPTER REVIEW

Summary and Key Terms

- **Artificial intelligence** (**AI**) is a field of study that explores how computers can be used for tasks that require the human characteristics of intelligence, imagination, and intuition. AI has also been described as the study of how to make computers do things that—at the present time—people can do better.
- Artificial intelligence is considered an umbrella term to encompass several subsets of interests, including problem solving, natural languages, expert systems, and robotics.
- In the early days of AI, scientists thought that it would be useful just to stuff facts into the computer; however, facts are useless without the ability to interpret and learn from them.
- An early attempt to translate human languages by providing a computer with vocabulary and rules of grammar was a failure because the computer could not distinguish the context of statements. This failure impeded the progress of artificial intelligence.
- The study of artificial intelligence is predicated on the computer's ability to learn—and to improve performance based on past errors.
- A **knowledge base** is a set of facts and rules about those facts. An **inference engine** accesses, selects, and interprets a set of rules. The inference engine applies rules to the facts to make up new facts.
- People rarely solve problems using the step-by-step logic most computers use. The brain beats the computer at solving problems, because it has millions of neurons working simultaneously.
- Computers whose chips are designed to mimic the human brain are called **neural networks**.
- In the **Turing test**, if a human judge cannot distinguish a human response from a computer response, the computer is said to be a thinking machine.
- **Natural languages** are associated with artificial intelligence because humans can make the best use of artificial intelligence if they can communicate with the computer in human language. Furthermore, understanding natural language is a skill thought to require intelligence. A key function of the AI study of natural languages is to develop a computer system that can resolve ambiguities.
- An **expert system** is a software package that is used with an extensive set of organized data and presents the computer as an expert on a specific topic. The expert system works by figuring out what the question means and then matching it against the facts and rules that it "knows."
- For years, expert systems were the exclusive property of the medical and scientific communities, but in the early 1980s they began to make their way into commercial environments.
- Some organizations choose to build their own expert systems to perform well-focused tasks that can easily be crystallized into rules, but few organizations are capable of building an expert system from scratch.
- Some users buy an **expert shell**, a software package that consists of the basic structure used to find answers to questions. It is up to the buyer to fill in the actual knowledge on the chosen subject.
- The person working to extract information from the human expert is sometimes called a **knowledge engineer.**
- A **robot** is a computer-controlled device that can physically manipulate its surroundings. Most robots are in factories.
- **Vision robots** traditionally recognize objects by their shapes or else "see" a dominant color. But some robots can recognize multicolored objects solely from their colors.
- **Field robots** inspect and repair nuclear power plants, dispose of bombs, inspect oil rigs for undersea exploration, put out oil well fires, clean up chemical accidents, and much more.

- **Virtual reality** immerses a user in a computer-created environment, so that the user physically interacts with the computer-produced three-dimensional scene.

Discussion Questions

1. Describe the differences in the way humans and machines learn.
2. Is it possible to create an expert system for doing term papers? Why or why not?
3. What kind of jobs are threatened by robots? Consider some jobs you may wish to have—now or in the future. Are the workers who perform them likely to be replaced by robots?

Student Study Guide

True/False

T F 1. Artificial intelligence has enjoyed wide respect from scientists and the government from its inception.

T F 2. *Artificial intelligence* is an umbrella term that covers many subjects.

T F 3. A knowledge base is a set of facts and rules about those facts.

T F 4. Early artificial intelligence scientists were called knowledge engineers.

T F 5. Early attempts to translate human language failed.

T F 6. An expert system is a kind of software.

T F 7. An expert shell is used by the end users of the expert system.

T F 8. Vision robots have sight capability that is significantly less than human capability.

T F 9. Field robots can do many tasks that are undesirable for humans.

T F 10. A knowledge engineer is a customer who uses an expert system.

T F 11. Using virtual reality, a user physically walks along hallways and from room to room.

T F 12. Most organizations who want expert systems do not attempt to build them from scratch.

T F 13. Using computers to mimic natural language is a relatively easy task for computer experts.

T F 14. Neural networks are chips designed to mimic the human brain.

T F 15. Computers generally solve problems in a step-by-step fashion.

T F 16. People generally solve problems in a step-by-step fashion.

T F 17. An underlying assumption of artificial intelligence is that the computer can learn.

T F 18. The medical and scientific communities were early users of expert systems.

T F 19. An inference engine is the same as a knowledge base.

T F 20. Understanding natural language is a skill thought to require intelligence.

T F 21. At one time AI scientists were the subject of ridicule.

T F 22. In general, a given word in a natural language has a single meaning.

T F 23. The Turing test proved conclusively that computers could match human brainpower.

T F 24. Most computer chess games can still be beaten by average human chess players.

T F 25. Most robots today are used in scientific experiments.

Fill-In

1. The generic term for using computers for jobs requiring human characteristics: _____

2. What is the term for software that can be used to build an expert system? _____

3. What is the software that interprets facts and rules? _____

4. What kind of robot can perform inspections and other dangerous tasks? _____

5. The person who extracts information from the human expert is called _____

6. Computers whose chips mimic the human brain are called _____

7. A device that can physically manipulate its surroundings is called _____

8. A set of facts and a set of rules about those facts is called _____

9. The language used by humans is called _____

10. What kind of robot can distinguish objects by their shape or color? _____

Answers

True/False

1. F	6. T	11. F	16. F	21. T
2. T	7. F	12. T	17. T	22. F
3. T	8. T	13. F	18. T	23. F
4. F	9. T	14. T	19. F	24. F
5. T	10. F	15. T	20. T	25. F

Fill-In

1. artificial intelligence
2. expert shell
3. inference engine
4. field robot
5. knowledge engineer
6. neural network
7. robot
8. knowledge base
9. natural language
10. vision robot

PLANET INTERNET

Eye on Business

The Internet was started by the military and long remained the province of the government and educational institutions. Businesses wondered if commercial enterprises were even permitted on the Net. The answer, a resounding yes, has lead to an explosion of activity. As the throbbing new center of the computing universe, the Net is attracting the brightest technical talent, the sharpest marketers, and the most ambitious entrepreneurs.

Who's out there? Good question, to which enterprising business people have responded with business directories. For example, the Commercial Site Index, whose logo of a world with stars representing sites is shown here, lets you peruse alphabetic listings of companies who have sites on the Net. If you prefer, you may request a search by company name. Formal directories, however, are not likely to include the very small businesses, the folks who go into business for themselves on the Internet.

A game anyone can play. And we do mean anyone, from agriculture to real estate to finance. In particular, even the smallest entrepreneur can get in on the act. Individuals can gain access to people and markets—even global markets—not readily available or even affordable elsewhere. For a minimum investment, far less than that needed for a physical store or office, you can have a server link and a smashing home page that exactly expresses the nature of your business. You can even alter the page as your business grows and changes.

But this isn't television. Despite the fantastic opportunities for sales of products and services, some commercial folks, especially retailers, do not take advantage of the Web's multimedia nature. If hawkers simply transfer their hype from traditional media such as television and magazines, the result may be hardly more than a bunch of print ads or boring catalogs. Furthermore, business people may continue to direct their pitch to a mass audience, failing to recognize Internet users as unique individuals who chose to look at their site. Savvy business users approach the Net with the idea of adding an interactive entertainment component to their pitch.

Taking care of business. Business people on the Net have special opportunities for customer service. FedEx, shown here, for instance, offers a home page where users can personally track a package they sent. But the big change for everyone is same-day response. When inquiries or orders are received over the Internet, the company should respond by e-mail, even just to acknowledge the order, within hours.

Investing. Money matters are taking a whole new turn as investors bravely go online. Since the Net is available any time from any place, the world's financial markets are never further away than the personal computer. Leaving behind what the major players can do, the individual investor can investigate markets, compare mutual funds, get up-to-date stock quotes, and more.

Internet Exercises

1. **Structured exercise.** Begin with the B/C URL http://www.aw.com/bc/planet/. Choose a business directory to see what kinds of sites are on the Net.
2. **Freeform exercise.** Assume that you want to start your own business using the Internet as your forum. Using a search engine, choose words related to that business to find sites that you can use as a source of information and inspiration.

In Chapter 6 we described the five steps of the programming process in a general way. We noted that the first step, defining the problem, is related to the larger arena of systems analysis and design, a subject we examined in Chapter 8. The second step involves planning the solution, and the last three steps—coding, testing, and documenting the program—are done in the context of a particular programming language, such as BASIC.

This appendix will look more closely at the planning phase, detailing the steps to help you understand how to develop program logic. First you will be introduced to three different approaches to program planning—*flowcharting, structure charts*, and *pseudocode*—and examples of each. Normally, a programmer would use only one or two to reach a solution. We present all three here, side by side, so you can compare them.

Flowcharts, which present a map of a solution, were the primary planning device for many years. They were favored over other methods because logic is easier to follow with pictures than with words. But flowcharts have some drawbacks: They are not easy to change, and they tend to be too detailed. However, there is a new wrinkle that makes flowcharts more palatable: They can be drawn and revised using flowcharting software.

A structure chart illustrates the structure of a program by depicting its parts as independent hierarchical modules. The resulting picture identifies a program's major functions at a high level, making it fairly easy to gain an overview quickly.

Pseudocode is easy to maintain. Since pseudocode is just words, it can be kept on a computer file and changed easily, using word processing or text editing. Although pseudocode is not a visual tool, it is nevertheless an effective vehicle for stating and following program logic. For these reasons, flowcharts have fallen out of favor and pseudocode has become popular. But flowcharts are often used as teaching devices, so we include them here.

In this appendix we will also examine another important topic: structured programming, an approach to programming that minimizes logic complexity. Let us begin with the pictures—flowcharts.

The Programming Process
Planning the Solution

LEARNING OBJECTIVES

- Understand fundamental flowcharting techniques and pseudocode
- Know the control structures of structured programming
- Understand the rationale for structured programming and its basic concepts

FLOWCHARTS
STRUCTURE CHARTS
PSEUDOCODE
STRUCTURED PROGRAMMING
  Sequence
  Selection
  Iteration
USING FLOWCHARTS, STRUCTURE CHARTS, AND PSEUDOCODE
  Example: Counting Salaries
  Example: Checking Customer Credit Balances
  Example: Determining Shift Bonuses
  Example: Computing Student Grades
PROGRAMMERS EMERGE
  Early Programming
  Expanding the Structured Programming Concept
  Modularity
  Is There a Future for Structured Programming?

Flowcharts

A **flowchart** presents a visual map of a program. The flowchart uses arrows to represent the direction of the program flow and boxes and other shapes to display actions. Note that in this discussion we are talking about a **logic flowchart,** a flowchart that represents the flow of logic in a program. A logic flowchart is different from a **systems flowchart,** which shows the flow of data through an entire computer system. We examined systems flowcharts in Chapter 8.

We will use the ANSI flowchart symbols introduced in Chapter 6 (see Figure 6-1). Templates of ANSI symbols (Figure A-1) are available in many office-supply stores and college bookstores and are helpful in drawing flowcharts by hand. The most common symbols you will use represent process, decision, connector, start/stop, input/output, and direction of flow. Now let us use flowcharting to show just what programming is all about.

Figure A-2 shows how you might flowchart a program to find the sum of all numbers between 1 and 100. There are a number of things to observe about this flowchart.

First, the program uses two places in the computer's memory as storage locations, or places to keep intermediate results. In one location is a counter, which might be like a car odometer: Every time a mile passes, the counter counts it as a 1. In the other location the computer stores a sum—that is, a running total of the numbers counted. The sum location will eventually contain the sum of all numbers from 1 through 100: $1 + 2 + 3 + 4 + 5 +\ldots + 100$.

Second, as we start the program, we must **initialize** the counter and the sum. When you initialize you set the starting values of certain storage locations, usually as the program execution begins. We will initialize the sum to 0 and the counter to 1.

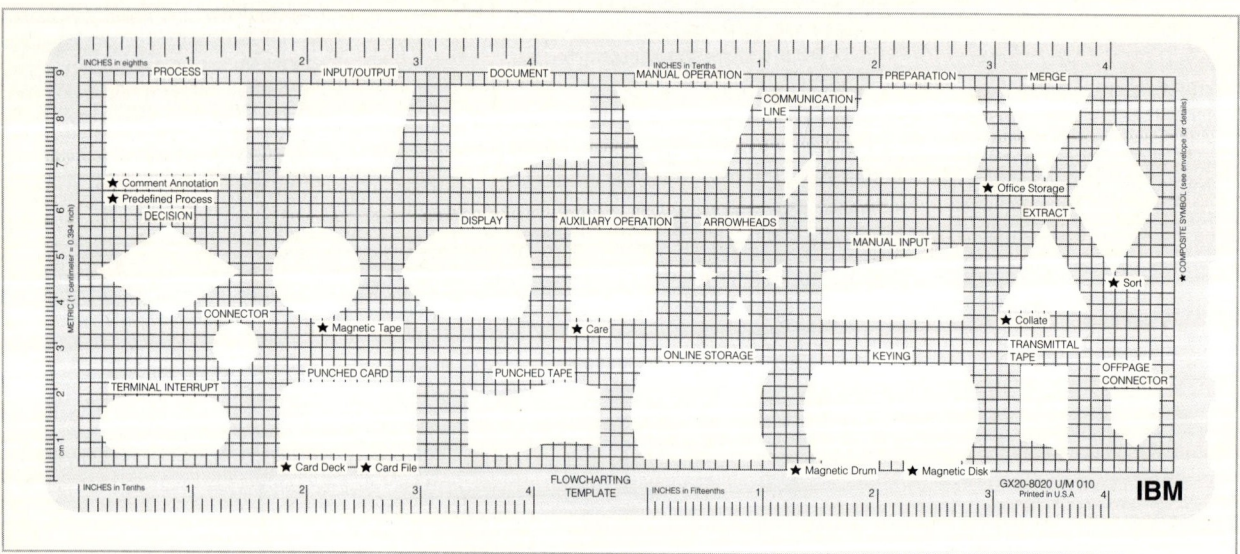

Figure A-1 A template containing standard ANSI flowcharting symbols. Templates like this one are used as drawing aids.

Third, note the looping. You add the counter to the sum and a 1 to the counter, and then you come to the decision diamond, which asks if the counter is greater than 100. If the answer is no, the computer loops back around and repeats the process. The decision box contains a **compare operation**; the computer compares two numbers and performs alternative operations based on the comparison. If the result of the comparison is yes, the computer produces the sum as output, as indicated by the print instruction.

A **loop**—also called an **iteration**—is the heart of computer programming. The beauty of the loop, which may be defined as the repetition of instructions under certain conditions, is that you, the programmer, have to describe certain instructions only once rather than describing them repeatedly. Once you have established the loop pattern and the conditions for concluding (exiting from) the loop, the computer continues looping and exits as it has been instructed to do. The loop is considered a powerful programming tool because the code is reusable; once written, it can be called upon many times. Notice also that the flowchart can be modified easily to sum the numbers from 1 through 1000 or 500 through 700 or some other variation. Now let us look at how structure charts are formed.

 ## Structure Charts

A **structure chart** graphically illustrates the structure of a program by showing independent hierarchical steps. This high-level picture identifies major functions that are the initial component parts of the structure chart. Each major component is then broken down into subcomponents, which are, in turn, broken down still further until sufficiently detailed components are shown. Since the components are pictured in hierarchical form, a drawing of this kind is also known as a **hierarchy chart**. A structure chart is easy to draw and easy to change, and it is often used to supplement or even to replace a logic flowchart.

An example of a structure chart is shown in Figure A-3. As the illustration shows, the top level of the structure chart gives the name of the program, Payroll process. The next level breaks the program down into its major functions: Read inputs, Compute pay, and Write outputs. One set of program statements performs each function. Each of these major functions is then subdivided further into smaller pieces. (We could break them down even further, but space does not permit it.)

The major functions are repeatedly subdivided into smaller pieces of manageable size. Each of these components is also, according to plan, as independent of the others as possible. For example, step 4.1, Write master, can be executed independently of any activity in step 4.3, Write paychecks.

We will use fairly small, concise structure charts in the examples in this appendix.

 ## Pseudocode

As you have learned, **pseudocode** is an English-like way of representing the solution to a problem. It is considered a "first draft" because the pseudocode eventually has to be translated into a programming language. Although pseudocode is like English and has some precision to it, it does not have the very definite precision of a programming language.

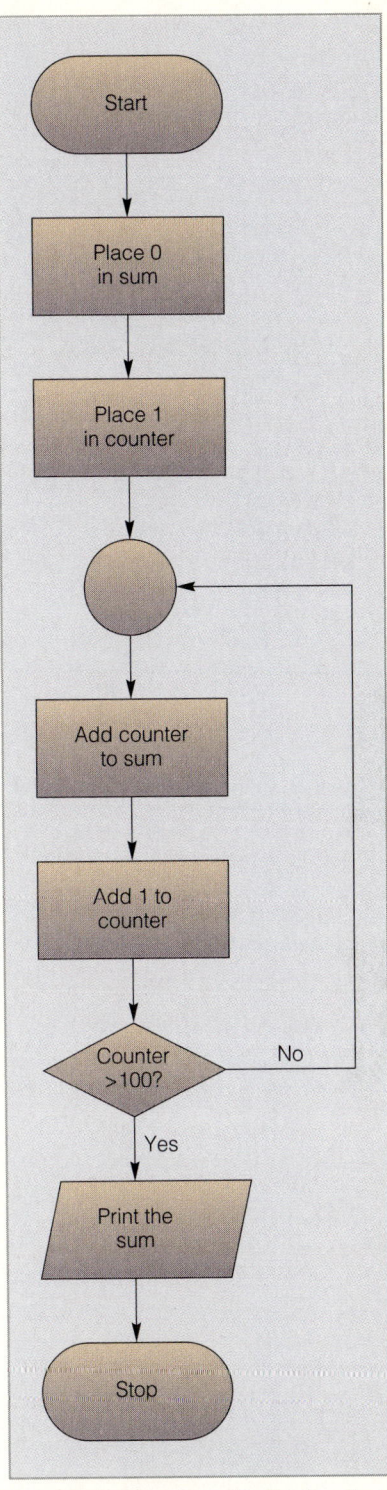

Figure A-2 A loop example. This flowchart uses a loop to find the sum of numbers from 1 through 100.

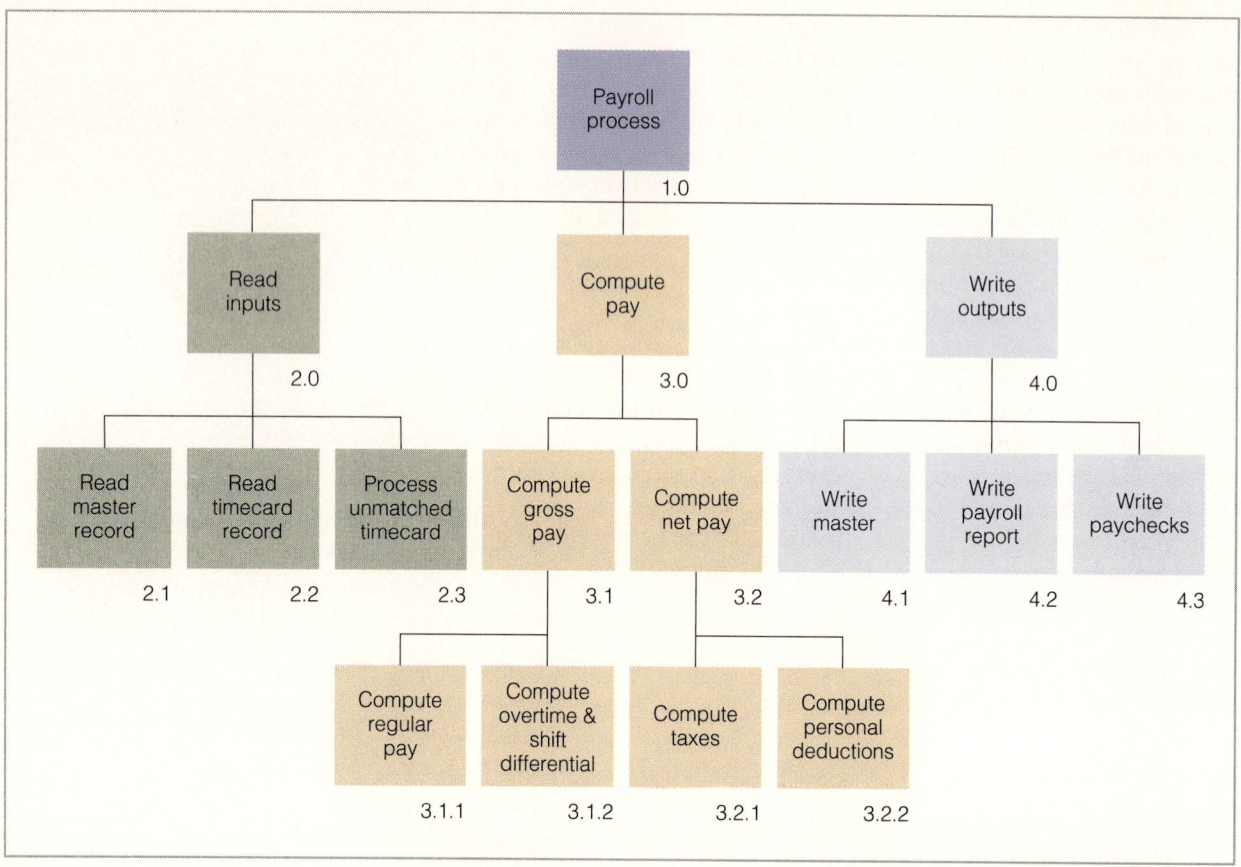

Figure A-3 A structure chart. The numbers outside the boxes refer to more detailed diagrams of these functions.

Pseudocode cannot be executed by a computer. When using pseudocode to plan a program, you can concentrate on the logic and not worry about the rules of a specific language. It is also easy to change pseudocode if you discover a flaw in your logic; once it is coded in a programming language, most people find that it is more difficult to change logic. Pseudocode can be translated into a variety of programming languages, such as Pascal or COBOL. It is helpful to introduce pseudocode in relation to flowcharting. Before doing so, however, let us consider structured programming.

 Structured Programming

Structured programming is a technique that emphasizes breaking a program into logical sections by using certain programming standards. Structured programming makes programs easier to write, check, read, and maintain. We will examine the rationale and concepts of structured programming more thoroughly later in this appendix. For now we will introduce some basic concepts of structure in this discussion of flowcharts, structure charts, and pseudocode. Note, however, that a programmer would use flowcharting, structure charts, or pseudocode to plan a solution. We present them together here so you can see how each method can be used to solve the same problem.

In a program, **control structures** control how the program executes. Structured programming uses a limited number of control structures to minimize the complexity of programs and thus to cut down on errors. There are three basic control structures in structured programming:

- Sequence
- Selection
- Iteration

These three are considered the basic building blocks of all program construction. You will see that we have used some of these structures already in Figure A-2.

Before we discuss each control structure in detail, it is important to note that each structure has only one **entry point** (the point where control is transferred to the structure) and one **exit point** (the point where control is transferred from the structure). This property makes structured programs easier to read and debug than unstructured programs.

Sequence

The **sequence control structure** is the most straightforward: One statement simply follows another in sequence. The left side of Figure A-4 shows the general format of a sequence control structure as it is used in flowcharting

Some Pseudocode Rules

Although pseudocode is not as formal as a programming language, many programmers follow rules like these:

Capitalize control words, such as IF and THEN.
- Write sequence statements in order, one under the other.
- Use IF-THEN-ELSE for decisions. Begin the decision with IF and end with ENDIF. THEN goes at the end of the IF line. If an ELSE is needed, align it with the IF and ENDIF. Indent the statements that go under THEN or ELSE.
- Use DOWHILE or DOUNTIL for iteration (looping). Indent the statements after the DO statement. End each DO with ENDDO, in the same margin as the DO.

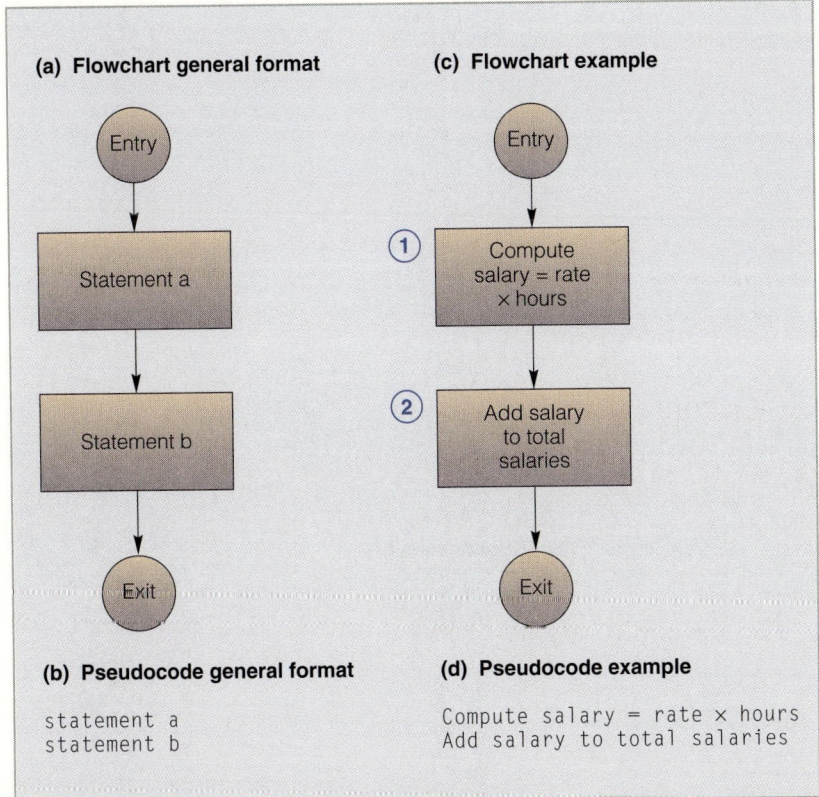

Figure A-4 Sequence. (a and b) The general format of the sequence control structure. (c and d) An example of a sequence control structure. To compute the total of the movie extra's wages, ① determine the extra's salary for that week by multiplying the hourly rate times the number of hours worked that week. ② Add that extra's salary to the total of other extras to find the total.

and in pseudocode. The right side of Figure A-4 shows an example of a sequence control structure: The two steps follow in sequence.

Selection

The **selection control structure** is used to make logical decisions. This control structure has two forms: IF-THEN-ELSE and IF-THEN. The IF-THEN-ELSE control structure works as follows: IF (a condition is true), THEN (do something), ELSE (do something different). For instance, IF the alarm clock goes off and it is a weekend morning, THEN just turn it off and go back to sleep, ELSE get up and go to work. Or, to use a more specific example, IF a student is a resident, THEN the fee equals number of credits times $450, ELSE fee is number of credits times $655. Figure A-5

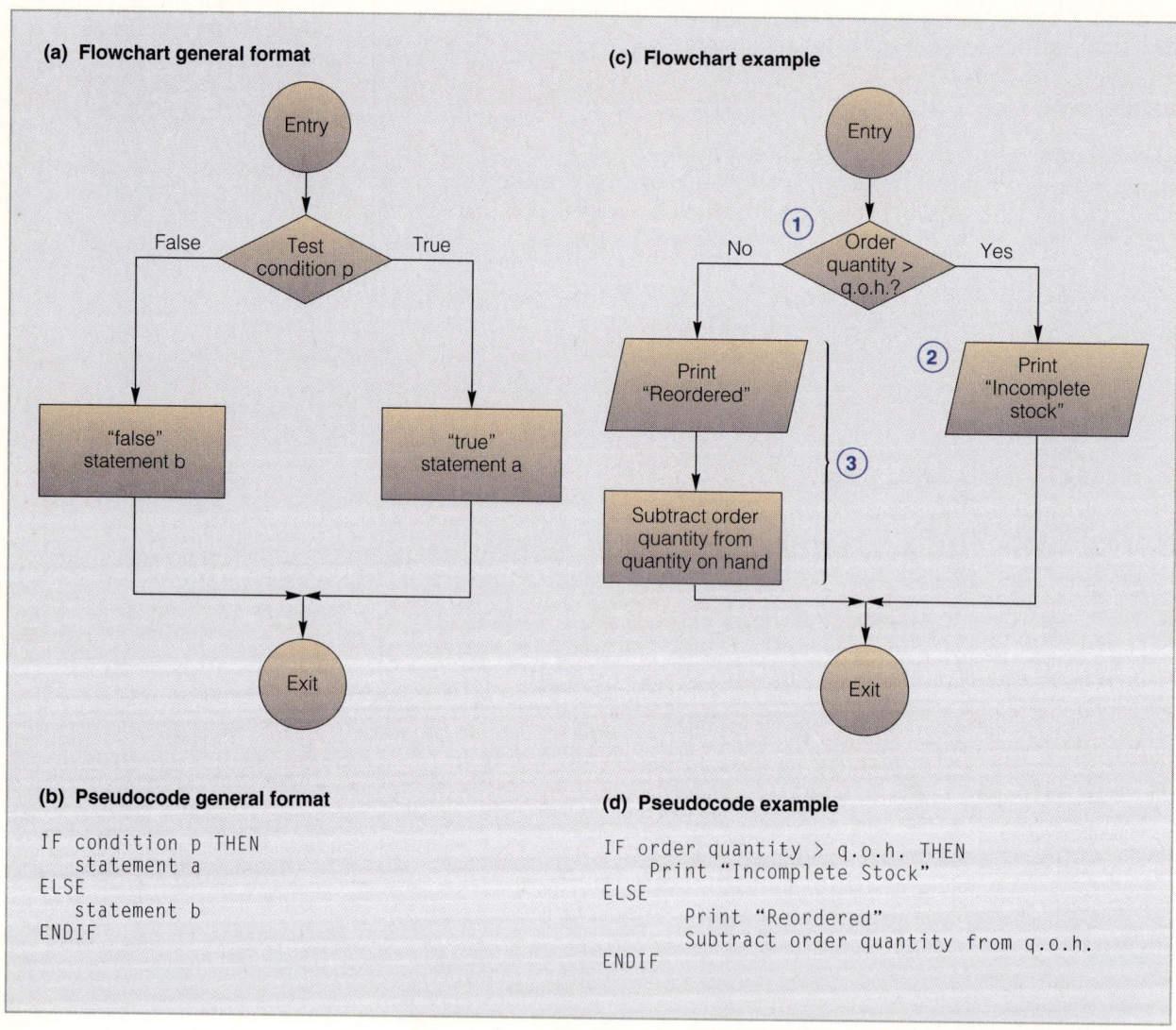

Figure A-5 IF-THEN-ELSE. (a and b) The general format of the IF-THEN-ELSE control structure. There can be one or more statements for each of the two paths, True and False. (c and d) An example of an IF-THEN-ELSE control structure. A trucker orders tires at a truck-tire warehouse. IF ① the quantity of the tires ordered is greater than the quantity of the tires on hand (q.o.h.) THEN ② the computer prints "Incomplete stock," ELSE ③ it prints "Reordered" and subtracts the quantity ordered from the quantity on hand.

shows the general format and an example of IF-THEN-ELSE in both a flowchart and in pseudocode.

IF-THEN is a special case of IF-THEN-ELSE. The IF-THEN selection is less complicated: IF the condition is true, THEN do something—but if it is not true, then do nothing. For example, IF the shift worked is shift 3, THEN add bonus of $50. Note that there will always be some action that results from using IF-THEN-ELSE; in contrast, IF-THEN may or may not produce action, depending on the condition. The IF-THEN variation is shown in Figure A-6.

Iteration

The **iteration control structure** is a looping mechanism. The only necessary iteration structure is the DOWHILE structure, which is shown in Figure A-7. An additional form of iteration is called DOUNTIL; DOUNTIL

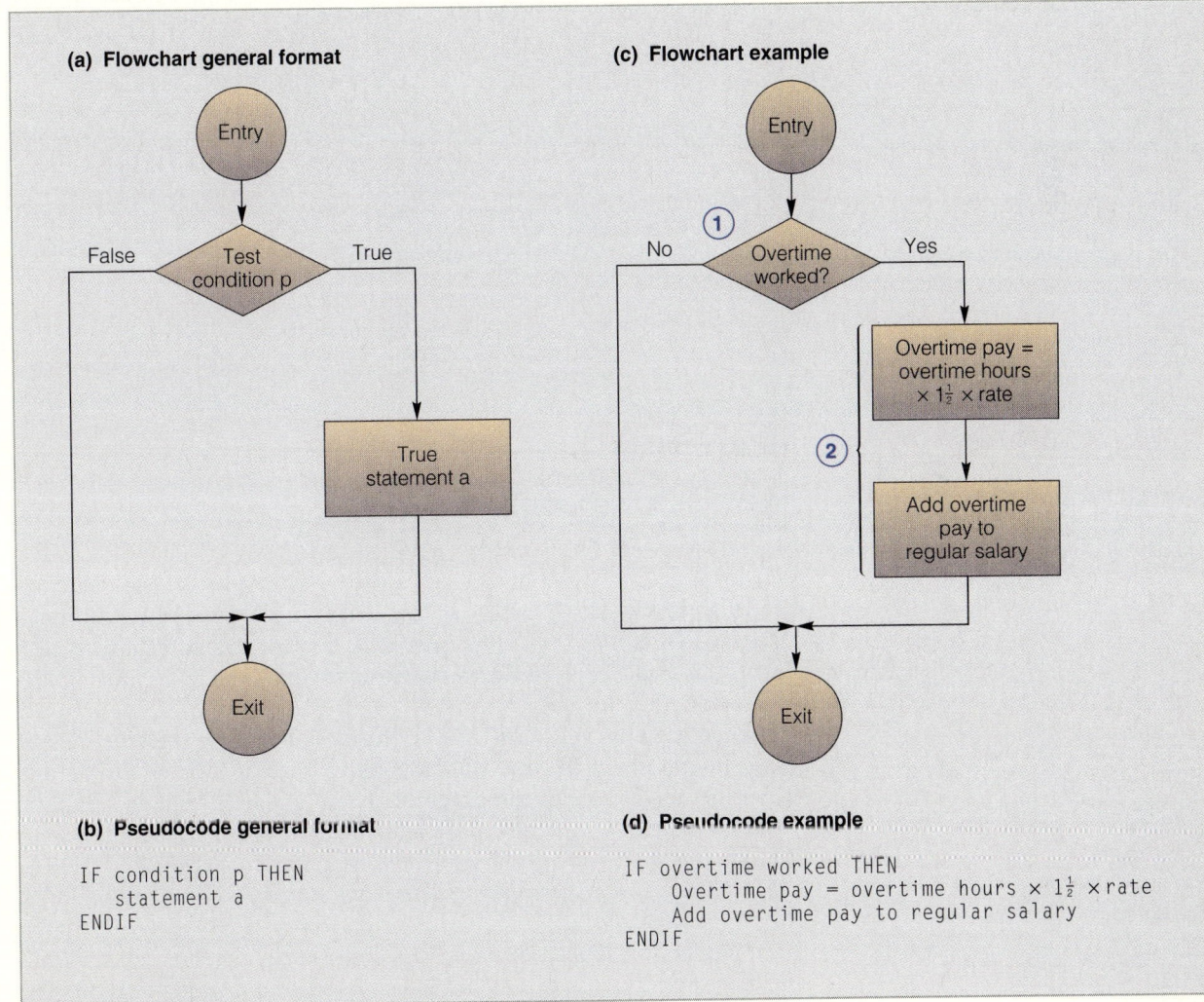

Figure A-6 IF-THEN. (a and b) The general format of the IF-THEN control structure, which is merely a special case of the IF-THEN-ELSE control structure. (c and d) An example of the IF-THEN control structure. IF ① a department store employee worked overtime THEN ② the program computes overtime pay by multiplying the overtime hours by 1½ times the hourly rate; the total is added to the employee's regular salary.

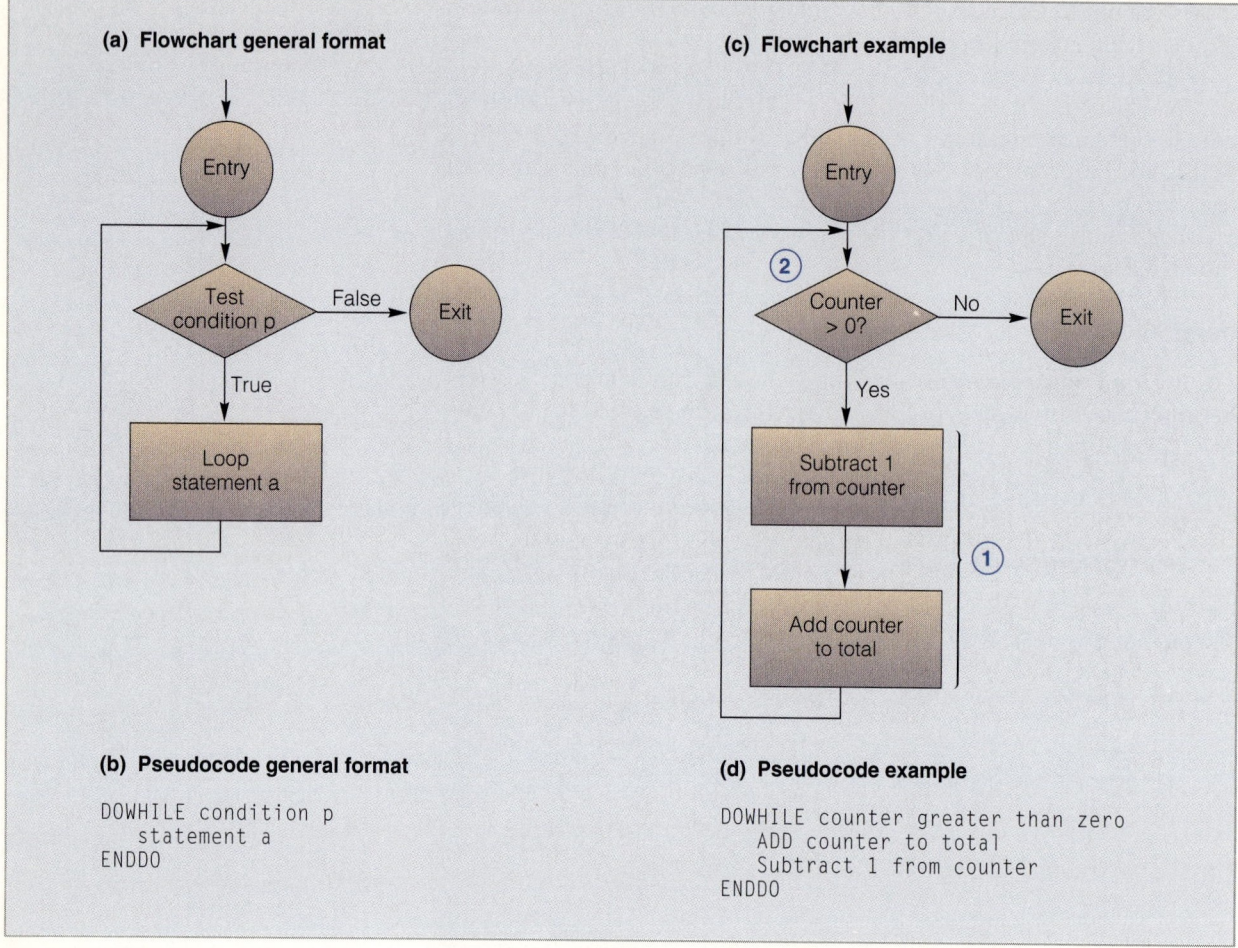

Figure A-7 DOWHILE. (a and b) The general format of the DOWHILE control structure. (c and d) An example of the DOWHILE control structure. DO ① add counter to total and subtract 1 from counter WHILE ② counter is greater than 0.

is really just a combination of sequence and DOWHILE. Although DOUNTIL is not one of the three basic control structures, it is convenient to introduce the DOUNTIL structure now, and it is shown in Figure A-8.

When looping, you must give an instruction to stop the repetition at some point; otherwise, you could theoretically go on looping forever and never get to the end of the program. There is a basic rule of iteration, which is related to structured programming: *If you have several statements that need to be repeated, a decision about when to stop repeating has to be placed either at the beginning of all the loop statements or at the end of all the loop statements.*

If you put the loop-ending decision at the beginning, it is called a **leading decision;** if you put it at the end, it is called a **trailing decision.** The position of the decision constitutes the basic difference between DOWHILE and DOUNTIL. As Figure A-7 shows, DOWHILE tests at the beginning of the loop; the diamond-shaped decision box is the first action of the loop process. The DOUNTIL loop tests at the end, as you can see in

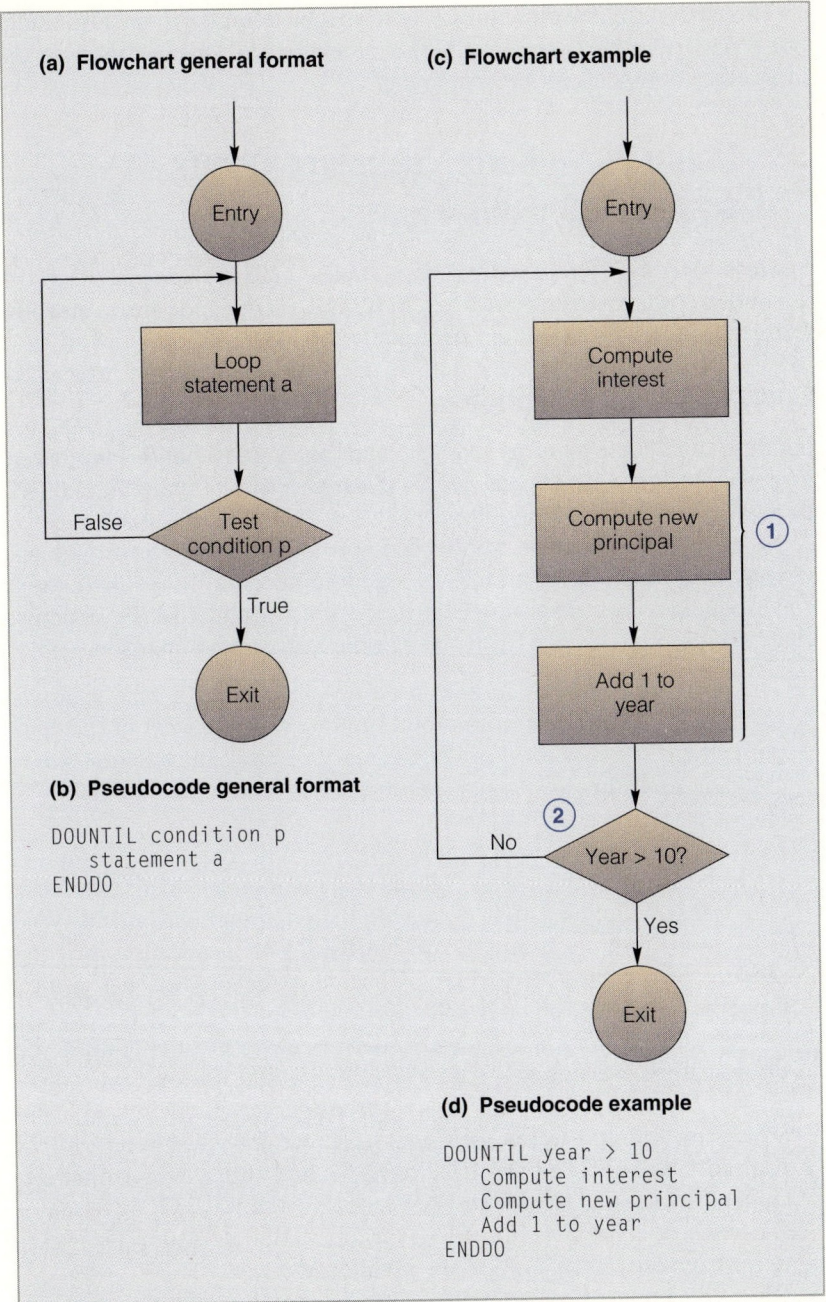

Figure A-8 DOUNTIL. (a and b) The general format of the DOUNTIL control structure. (c and d) Example of the DOUNTIL control structure. DO ① compute interest, compute new principal, and add the number 1 to year UNTIL ② the number of years is greater than 10.

Figure A-8. The DOUNTIL loop, by the way, guarantees that the loop statements are executed at least once, because the loop statements are executed before you make any test about whether to get out. This guarantee is not necessarily desirable, depending on the program logic. Also note that the test condition of DOUNTIL must be false to continue the loop; this is an important difference from the DOWHILE loop.

These basic control structures may seem a bit complex in the beginning, but it is worth taking your time to learn them. In the long run they are the most efficient models for programming.

Using Flowcharts, Structure Charts, and Pseudocode

Let us now consider four extended examples. In each example, solutions are shown in flowchart, structure chart, and pseudocode form. Keep in mind that normally you would select only one approach.

Example: Counting Salaries

Suppose you are the manager of a personnel agency that has 50 employees. You want to know how many people make over $30,000 a year, $20,000 to $30,000, and under $20,000.

Figure A-9 shows a solution to your problem. Let us go through the flowchart in Figure A-9a first. The circled numbers in the following text correspond to the circled numbers in the illustration. Use the structure chart and pseudocode in parts b and c of the figure for comparison.

① The program begins by initializing four counters to 0. The employee counter will keep track of the total number of employees in the company; the other counters—the high-salary counter, the medium-salary counter, and the low-salary counter—will count the numbers of employees in the salary categories.

② In the parallelogram-shaped input box, we indicate that the computer reads the salary at this point. A **Read** statement may be defined as code that brings something that is outside the computer into memory; to *read*, in other words, means to get. The Read statement causes the computer to get one employee's yearly salary; since the instruction is inside a loop, the computer will eventually get all salaries.

③ The first of the diamond-shaped decision boxes is a test condition that can go either of two ways—Yes or No. Note that if the answer to the question "Salary > $30,000?" is yes, then the computer will process this answer by adding 1 to the high-salary counter. If the answer is no, the computer will ask, "Salary < $20,000?"—and so on. Notice that one control structure, the IF-THEN-ELSE for the 20,000 decision, is entirely inside the IF-THEN-ELSE for the 30,000 decision; this is easy to see from the indentations in the pseudocode.

④ For every decision box, no matter what decision is made, program control flows to a connector. And, as the flowchart shows, each decision box has its own connector. Note that, in this case, each connector is directly below the decision box to which it relates.

⑤ Whatever the kind of salary, the machine adds 1 (for the employee) to the employee counter, and a decision box then asks, "Employee counter = 50?" (the total number of employees in the company).

⑥ If the answer is no (employee counter does not equal 50), the computer makes a loop back to the first connector and goes through the process again. Note that this is a DOUNTIL loop because the decision box is at the end rather than at the beginning of the computing process (DO keep processing UNTIL employee counter equals 50).

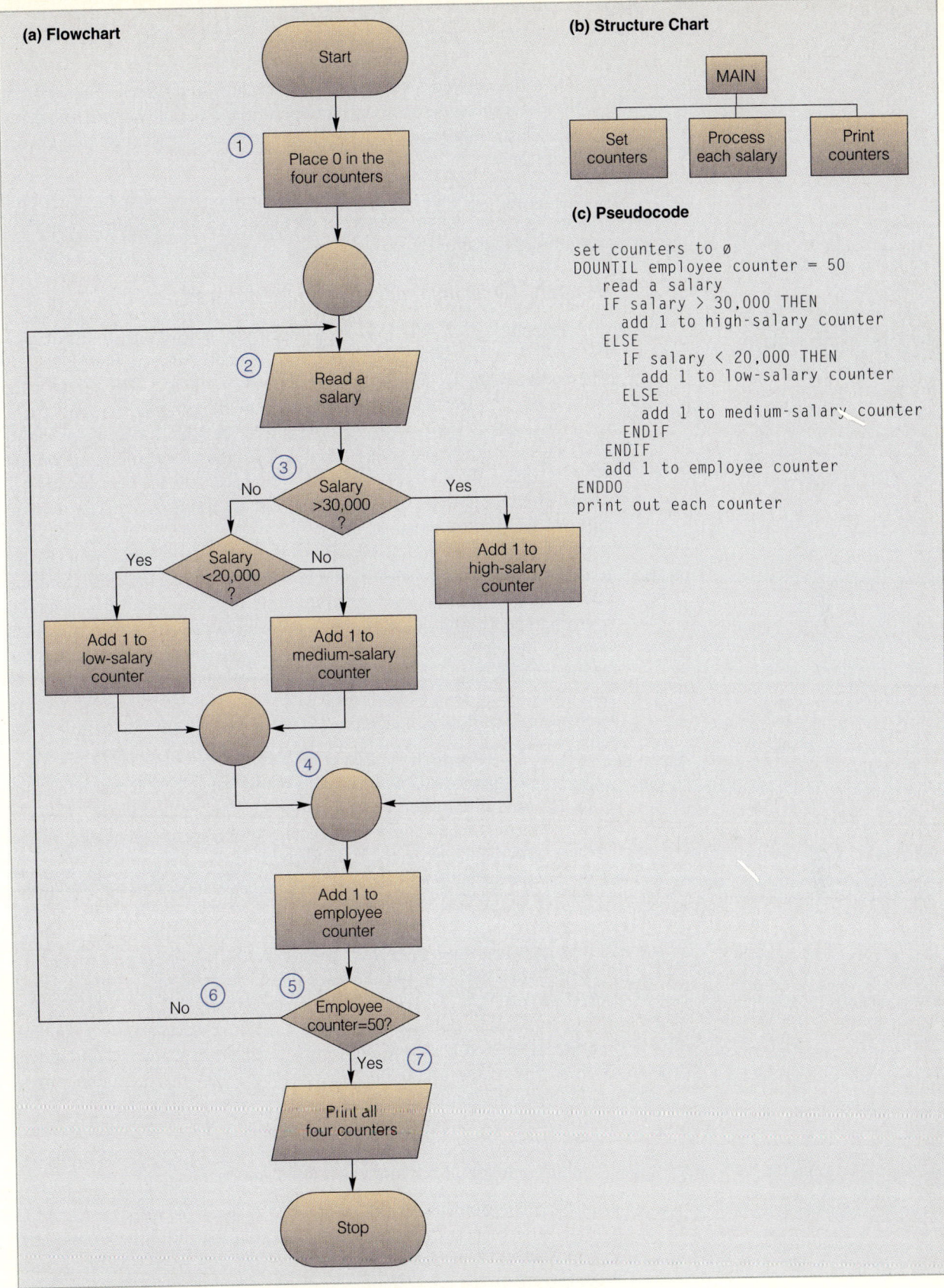

Figure A-9 Counting salaries. (a) A flowchart is used to detail the flow of the program logic in counting salaries. (b) A structure chart presents the same information in a hierarchical structure. (c) Pseudocode gets closer to the actual logic the program must follow.

⑦ When the answer is finally yes (employee counter does equal 50), the computer then goes to an output operation (a parallelogram) and prints the salary count for each of the three categories. The computing process then stops.

Review the flowchart and pseudocode and observe that every action is one of the three control structures we have been talking about: sequence, selection, or iteration.

Example: Checking Customer Credit Balances

In the example in this section, we will consider a flowchart that describes the process of checking a retail customer's credit balance (Figure A-10a). The structure chart (Figure A-10b) gives an overview. The pseudocode (Figure A-10c) provides another way to view the process. The file of customer records is kept on some computer-accessible medium, probably disk. This is a more true-to-life example than the previous salary example because, rather than a file with exactly 50 records, the file here contains an unknown number of records. The program has to work correctly no matter how many customers there are.

As store manager, you need to check the customer file and print out the record of any customer whose current balance exceeds the credit limit, so salesclerks will not ring up charge purchases for customers who have gone over their credit limits. (Recall that a record is a collection of related data items; a customer record would likely contain customer name, address, account number, and, as indicated, current balance and credit limit.) The interesting thing about this flowchart is that it contains the same input operation, "Read customer record," twice (see the parallelograms). We will see why this is necessary. Let us proceed through the flowchart:

① After reading the first customer record and proceeding through the connector, you have a decision box that asks, "Record received?" This is a test to see if you have run out of all customer records (which you probably would not have the first time through).

② If the answer is no (no record received), you have reached an **end of file**—meaning that no more records are in the file—and the process stops.

③ If the answer is yes (a record was received), the program proceeds to another decision box, which asks a question about the customer whose record you have just received: "Balance > limit?" This is an IF-THEN type of decision. If the answer is yes, then the customer is over the limit and, as planned, the computer prints the customer's record and moves on to the connector. If the answer is no, then the computer moves directly to the connector.

④ Now we come to the second Read statement, "Read customer record." Why are two such statements needed? Couldn't we just forget the second one and loop back to the first Read statement again? No.

The explanation lies in the rules of structure. As we stated, a loop requires a decision either at the beginning or at the end. If we omitted the second Read statement and looped back to the first Read statement, then the decision box to get us out of the loop ("Record received?") would be in the middle, not the beginning or the end of the loop. Why not put "Record received?" at the end? That strategy will not work either, because then the

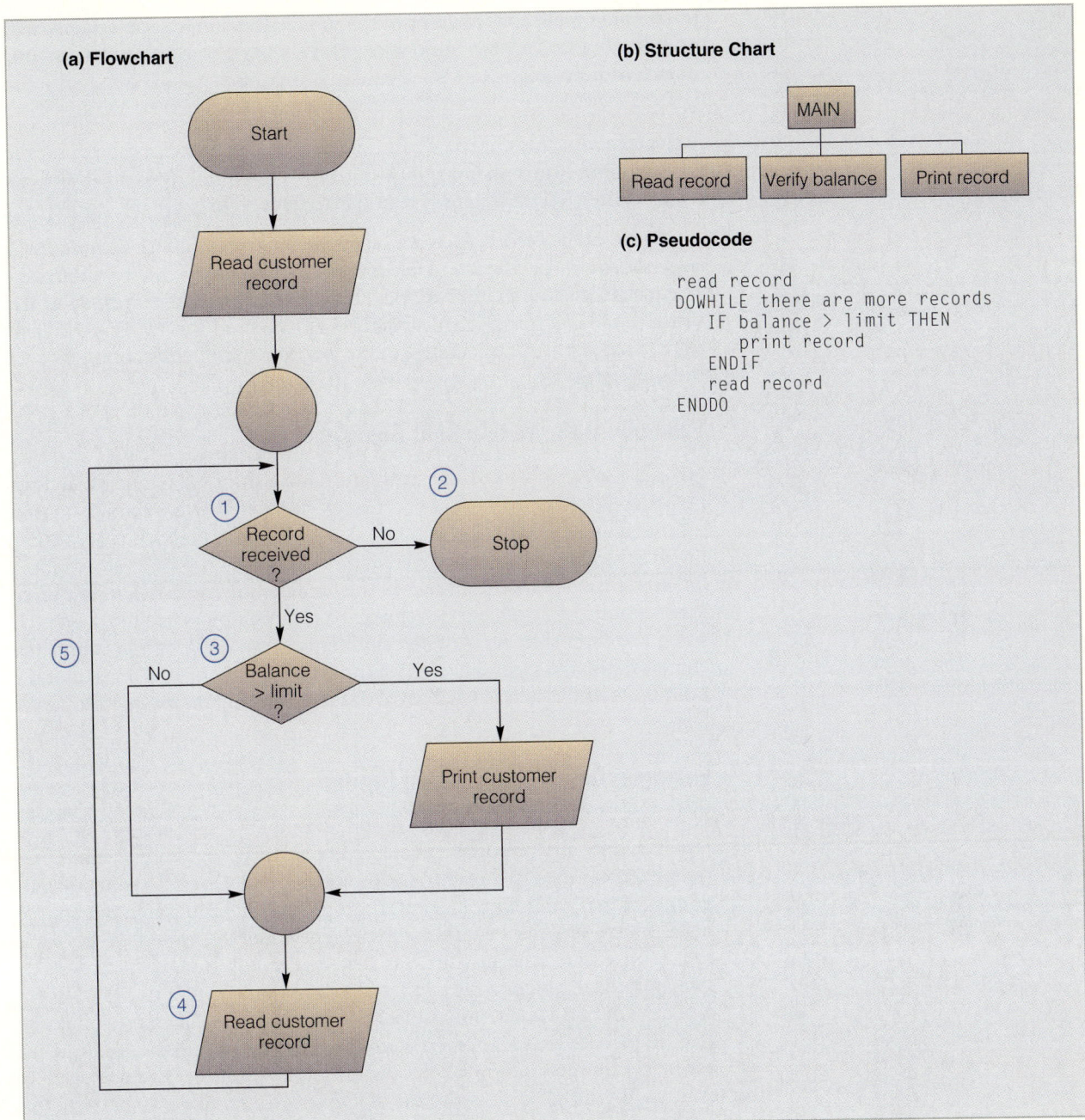

Figure A-10 Checking a credit balance. (a) A flowchart details the program logic in checking a customer's credit balance. (b) The structure chart and (c) the pseudocode each show a different method of detailing the same procedures.

instructions would tell the computer to do the processing before the program had ascertained if there were a record to process.

In summary: The decision box cannot go at the end, and the rules say it cannot be in the middle; therefore, the decision must go at the beginning of the processing. Thus, the only way to read a second customer record after the computer has read the first one is to have the second Read statement where you see it. The first Read statement is sometimes called the **priming read.** This concept of the double Read may seem complicated at first, but it is very important. Reviewing the description of this flowchart may help.

⑤ Next, the program loops back to the connector and repeats the process. Incidentally, this is a DOWHILE loop because the decision box is at the beginning rather than at the end of the computing process (DO keep processing WHILE records continue to be received).

Note that, as before, each action in the program is either a sequence, a selection, or an iteration. In fact, since you have now seen two totally different examples—counting salaries and checking credit balances—you can begin to see how the control structures can be used for different applications. That is, the subject matter of the program may change, but the structured programming principles remain the same.

Example: Determining Shift Bonuses

Here is a description of the problem whose solution is represented in Figure A-11. The problem concerns awarding employees bonuses based on the shift worked. The example is a little more elaborate because it involves moving data—employee number, name, and bonus—to a report line to set it up before printing. As the figures show, a first-shift employee gets a bonus of 5 percent of regular pay, but employees who work the second or third shift get a 10 percent bonus. Also, the program counts the employees on the second or third shifts—that is, it performs one count for both shifts. If the shift is not 1, 2, or 3, then the program produces an error message.

Example: Computing Student Grades

Now let us translate a flowchart—and accompanying structure chart and pseudocode—into a program. You could type this program on a computer terminal connected to a mainframe or minicomputer, or key it directly into your personal computer. The keyed program is the programmer's source code, which we described in Chapter 6. The computer would deliver back to you on a screen the answers you seek. Figure A-12 shows the flowchart, structure chart, pseudocode, program, and output.

The program is written in QBasic, a programming language similar to English in many ways. So, even with no knowledge of QBasic, you can get a general understanding of the program. The numbers to the far left column of the QBasic program in Figure A-12d are not part of the program and are included only for reference in this discussion. If you were entering this program to the computer you would not include these numbers.

We will first talk generally about QBasic commands and then discuss the logic and program related to student grades in detail. In a program, lines that begin with an apostrophe (') are called remarks or comments. These lines contain notes that are useful to the programmer and possibly other programmers. Comment lines are intended to be a programmer convenience only; they are not executed by the computer (and hence are not included in a flowchart, structure chart, or pseudocode). Beginning program comments usually describe what the program is supposed to do and list the variable names. Variable names are symbolic names of locations in

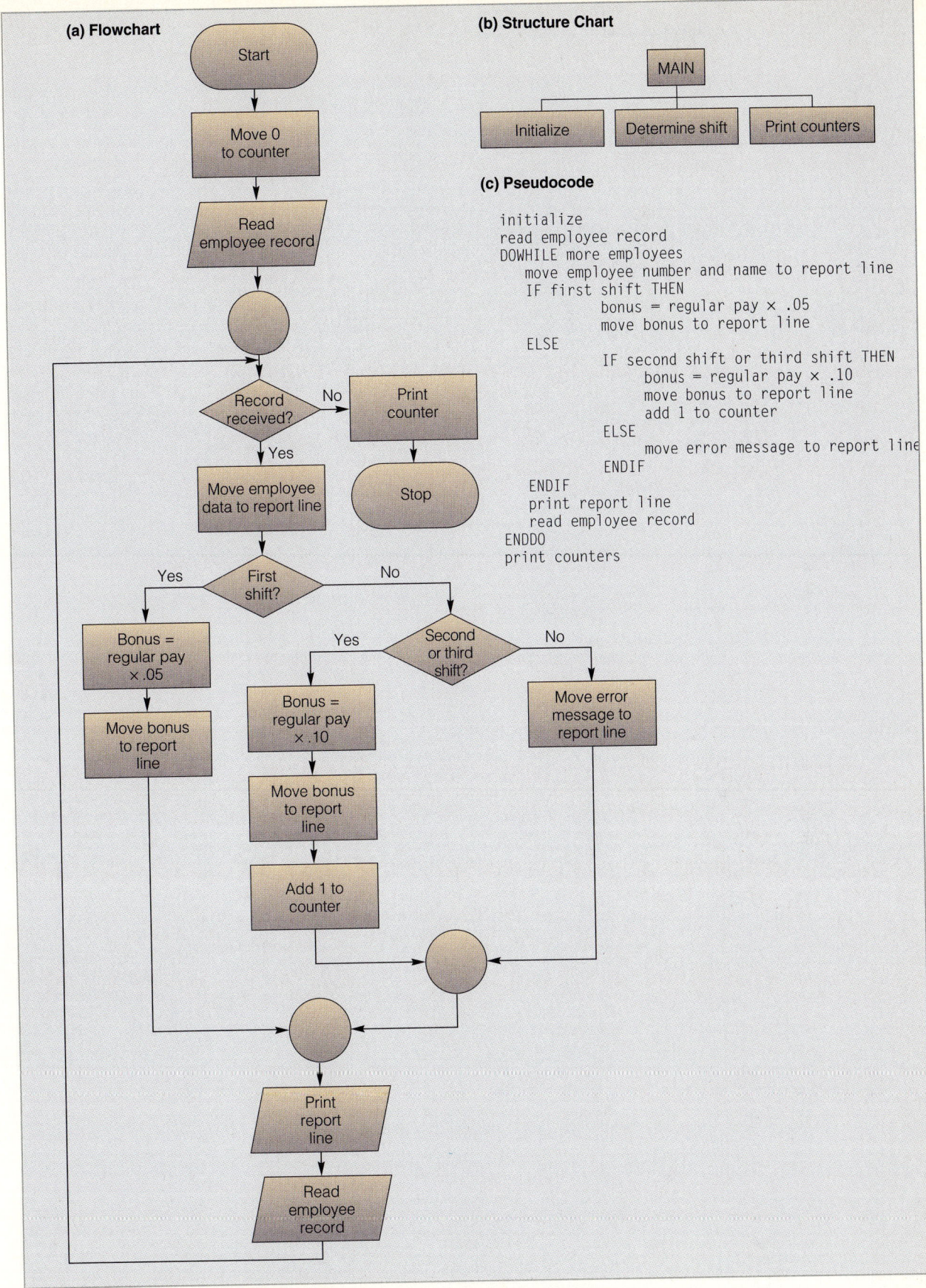

Figure A-11 Shift bonus. The logic for determining employee bonuses is shown as (a) a flowchart, (b) a structure chart, and (c) pseudocode.

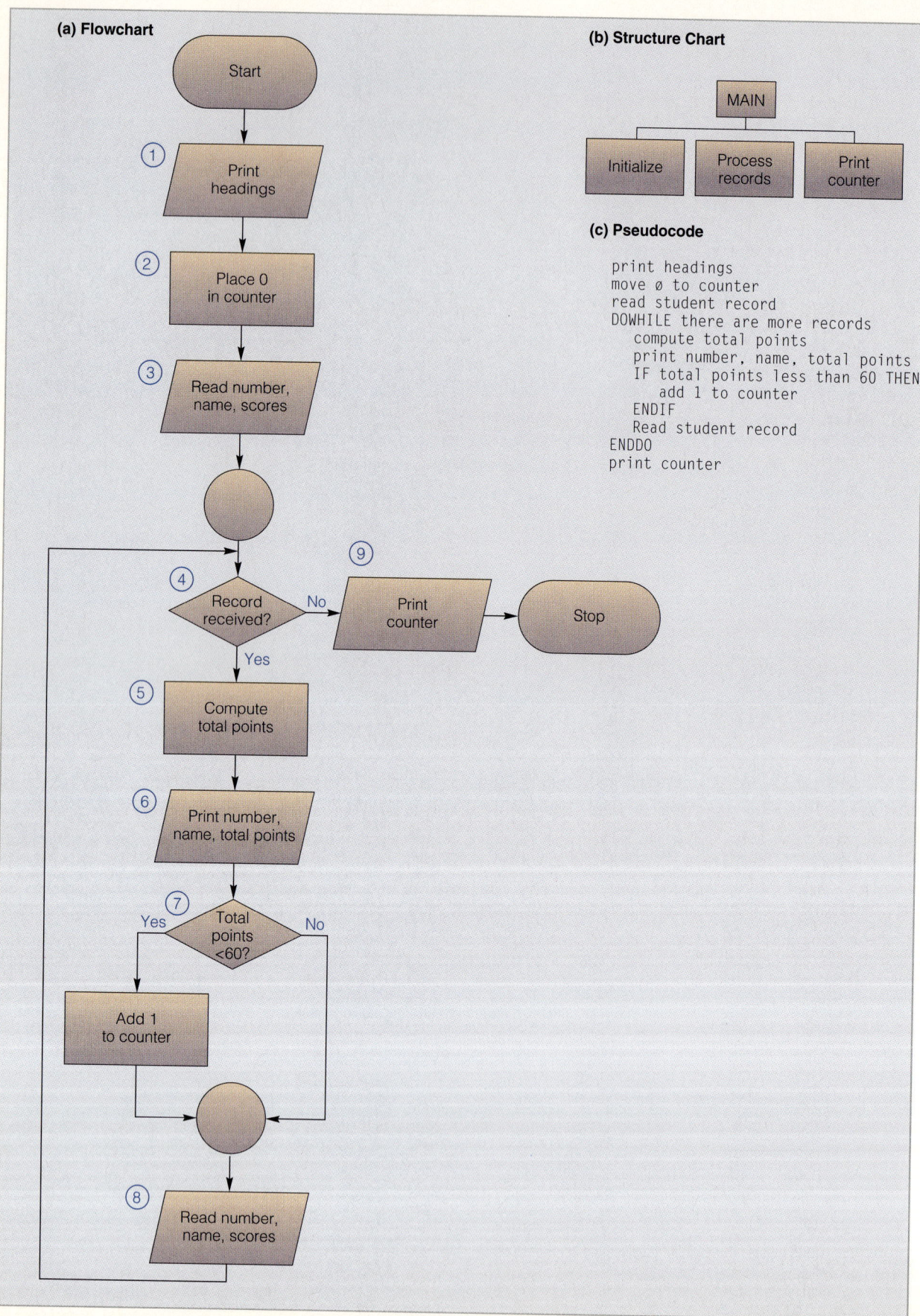

memory. As an alternative to an apostrophe, a comment line can begin with the word REM, which stands for remark.

A READ statement reads (gets) the data to be processed, as found in the DATA statements. A CLS statement clears the computer screen in preparation for expected program output. A PRINT statement tells the computer what message or data to output, in this case to the screen. The DO WHILE and LOOP statements together form a DOWHILE (leading decision) loop.

(d) QBasic program

```
1   '* ********************************************************************
2   '* Program to Compute Student Points
3   '*
4   '* This program reads, for each student, student number, student name, and 4 test scores.
5   '* These scores are to be weighted as follows:
6   '* Test 1:              20 percent
7   '* Test 2:              20 percent
8   '* Midterm:             25 percent
9   '* Final:               35 percent
10  '* The program computes and prints the total points for each student and also counts the
11  '* number of students whose total points fall below 60.
12  '*
13  '* Variable Names:      Description:
14  '* Number               Student number
15  '* Name$                Student name
16  '* Score1               Score on test 1
17  '* Score2               Score on test 2
18  '* Score3               Score on midterm
19  '* Score4               Score on final
20  '* Total                Total points for a student
21  '* Count                Count of students with total points less than 60
22  '*
23  '* ********************************************************************
24  '* Data to be processed
25  DATA 2164,Allen Schaab,60,64,73,78
26  DATA 2644,Martin Chan,80,78,85,90
27  DATA 3171,Christy Burner,91,95,90,88
28  DATA 5725,Craig Barnes,61,41,70,53
29  DATA 6994,Raoul Garcia,95,96,90,92
30  DATA 7001,Kay Mitchell,55,60,58,55
31  DATA 0,QUIT,0,0,0,0
32  '*
33  '* Program Initialization
34  CLS
35  PRINT
36  PRINT TAB(10); "Student Grade Report"
37  PRINT
38  PRINT "Student","Student","Total"
39  PRINT "Number","Name","Points"
40  PRINT
41  PRINT
42  Count=0
43  '* Report Processing
44  READ Number, Name$, Score1, Score2, Score3, Score4
45  DO WHILE Name$<>"QUIT"
46       Total = .20*Score1 + .20*Score2 + .25*Score3 + .35*Score4
47       PRINT Number, Name$, Total
48       IF Total < 60 THEN Count = Count + 1
49       READ Number, Name$, Score1, Score2, Score3, Score4
50  LOOP
51  PRINT
52  PRINT TAB(5); "Number of students with total points < 60:";Count
53  END
```

(e) Output

Student Grade Report

Student Number	Student Name	Total Points
2164	Allen Schaab	70.4
2644	Martin Chan	84.4
3171	Christy Burner	90.5
5725	Craig Barnes	56.5
6994	Raoul Garcia	92.9
7001	Kay Mitchell	56.8

Number of students with total points < 60: 2

Figure A-12 Student grades. The logic for determining student grades is shown in (a) a flowchart, (b) a structure chart, and (c) pseudocode. (d) The resulting QBasic program is presented, along with (e) the output produced by the program.

By the way, QBasic will automatically change QBasic keywords such as READ or PRINT to uppercase, regardless of how you type them.

Our problem is, first, to compute the student grades (ranging from 0 through 100) for a group of students, and, second, to count the number of students whose computed points are less than 60. The grade points are based on student performance on two tests, a midterm exam, and a final exam, the scores of which have been weighted in a certain way. Note the comment lines in the program (Figure A-12d). Lines 1 through 23 describe the program, including the weights to assign to the test scores, and also list the program variable names. Line 24 is also a comment, noting that the following lines contain data to be processed. As we will soon see, lines 25 through 31 will be needed once we get to the READ statement in the program logic.

Now we will consider the logic and program for computing student grades in detail. The circled numbers in the text correspond to the circled numbers in the flowchart, but you may follow the pseudocode if you prefer. From this point forward, corresponding statement numbers from the program follow in parentheses.

① **Print headings** (program lines 34 through 41). This statement refers to the headings on the output report (skip ahead momentarily to Figure A-12e to see what they will look like). Line 34 issues a command to clear the screen. Line 35 produces a blank line (so the report will not hug the top of the screen), followed by line 35, which tabs over 10 spaces and then prints the overall heading, "Student Grade Report." The fact that the words to be printed are in quotes means they will be printed as they appear within the quotes. After printing another blank line (line 37) for attractive spacing, lines 38 and 39 print the three column headings. The commas between the words cause a move to a new column; thus, observing line 38, the three words Student, Student, and Total will appear in three different columns. Similarly, from line 39, the words Number, Name, and Points will appear in three columns, directly under the words from line 38.

② **Place 0 in counter** (line 42). This line is not a form of input; this line causes the initialization process that is required here, at the outset. The counter will count the number of students who score fewer than 60 points, as you will see later.

③ **Read number, name, scores** (line 44). This is the priming read. The input data to be used for the read statements begins at line 25. The READ in line 44 is the first read statement, so it will read the record with the first set of data: 2164 will be placed in Number, Allen Schaab in Name$, and the four scores (60, 64, 73, and 78) placed in Score1 through Score4, respectively. Incidentally, the variable for name requires a $ at the end because it will contain alphanumeric data.

④ **Record received?** (line 45). This is the beginning of the DOWHILE loop: DO process the loop WHILE a valid record is received. In QBasic this loop is initiated by DO WHILE and ended by LOOP. The decision box asks if a valid record was read. How will the computer know this? Because, in the program, an extra phony record was included at the end; the word QUIT indicates "end of file." (In line 45, <> means not equal.) If a valid record *is* received (any name other than QUIT) then the statements within the loop are executed for that record. The statements within the loop are lines 46 through 49.

⑤ **Compute total points** (line 46). The scores are weighted 20 percent for the first test, 20 percent for the second test, 25 percent for the midterm, and 35 percent for the final exam. The total of these weighted scores gives the course grade. In the program these percentages are documented in remark statements (lines 4 through 9). The formula that totals the scores and incorporates the weightings is stated in line 46. Here the expression .20*Score1 means 20 percent times the first test score (Score1). In QBasic the asterisk symbol (*) is used as the multiplication symbol.

⑥ **Print number, name, total points** (line 47). Student number, name, and total points are output in three columns.

⑦ **Total points < 60?** (line 48). This decision box is given as an IF-THEN statement. If a student has fewer than 60 points, 1 is added to Count.

⑧ **Read number, name, scores** (line 49). As is required by the rules of structure, we have here an instance of a repeated READ statement. The first time this statement is encountered, the data for Martin Chan will be read, the next time through the loop the data for Christy Burner, and so on. Eventually, of course, the data for QUIT will be encountered. This is the last statement within the loop, so the LOOP statement (line 50) causes transfer to the top of the loop, line 45, where the data just received is examined to see if it is a valid record. For each of the valid records, the loop statements will be executed and a line of output will be produced.

⑨ **Print counter** (lines 51, 52). When the program reaches the end of the file (because the name read is QUIT), the program drops out of the loop to the lines *below* the word LOOP. The program now prints a blank spacing line and then (line 52) prints the total number of students who have fewer than 60 points.

All this is probably a bit confusing if you are a beginner. Practice helps.

Programmers Emerge

In the 1950s the programmer was hardly noticed, according to Edsger Dijkstra (pronounced "DIKE-stra"), one of the first to spotlight the importance of the programmer in computing. For one thing, the computers themselves were so large and so cantankerous to maintain that they attracted most of the attention. For another, Dijkstra said, "The programmer's somewhat invisible work was without any glamour: you could show the machine to visitors and that was several orders of magnitude more spectacular than some sheets of coding." Programmers flourished, nevertheless, as the demand for software grew.

In the 1960s hardware overreached software. The development of hardware and storage capabilities proceeded apace, but software development could not keep up. Hardware costs had decreased, but software projects ran over budget and did not meet their deadlines. And, when projects were finally completed, they often did not meet the users' needs.

In the 1970s software costs continued to rise. It was apparent that, if money was to be saved, software had to improve. No longer would programmers be allowed to produce programs that were casually tested or that only the programmer could read. The use of obscure coding in an attempt to shave a microsecond of computer time began to be discouraged. Programmers had to develop new, more organized approaches to the complex problems that, in the '70s, now seemed routine. The programmer's job was no longer "invisible work."

Early Programming

How did people go about programming in the early '60s? One computer scientist wrote: "Computer programming was so badly understood that hardly anyone even thought about proving programs correct; we just fiddled with a program until we knew it worked." Dijkstra, in fact, has nagged and cajoled programmers since the early years to think in advance instead of using a rearguard action for finding errors *after* a program is written.

Finding program errors after the fact was—and often still is—an accepted way of programming. That is, a programmer wrote a program that seemed to solve the problem, then the program was put to the test. As soon as an error turned up, that one was fixed. This would continue until, eventually, the programmer got the program working well enough to use. To Dijkstra, this seemed a shoddy way of doing things. "Program testing," he said, "is a very convincing way of demonstrating program errors but never their absence."

In 1966 C. Bohm and G. Jacopini introduced structured programming in a paper in *Communications of the ACM* (the journal of the Association for Computing Machinery). In their paper, which had been published previously in Italy, they proved mathematically that any problem solution could be constructed using only three basic control structures—the three structures that we have been calling *sequence*, *selection* (IF-THEN-ELSE), and *iteration*. It is interesting to note that the concept of structured programming has remained unchanged since it was proposed more than two decades ago.

These three control structures—sequence, selection, and iteration—were, of course, used before 1966. But other control structures were also used. The most notable was the transfer structure, also known as the GOTO. After Bohm and Jacopini proved the need for only the three basic control structures, however, the time had come to cut down on the number of GOTO statements.

The idea of structured programming was given a boost in March 1968, when Dijkstra published a now famous letter in *Communications of the ACM*. Under the heading "Go To Statements Considered Harmful," Dijkstra contended that using the GOTO statement was an invitation to making a mess of one's program and that reducing the number of GOTOs reduced the number of programming errors. GOTOs, he said, could be compared to pasta. If a person took a program and drew a line from each GOTO statement to the statement to which it transferred, the result would be a picture that looked like a bowl of spaghetti. Since then, people have referred to excessive GOTOs in a program as "spaghetti code."

Expanding the Structured Programming Concept

When the concept of program structure was first introduced, some people thought their programs would be structured if they simply got rid of GOTOs. There is more to it than that. Structured programming is a method of designing computer system components and their relationships to minimize complexity. So, in addition to limited control structures (again: sequence, selection, and iteration), two important aspects of structured programming are (1) top-down programming design and (2) module independence through coupling and cohesion. Before we describe these concepts, let us pause for an expanded formal definition: Structured programming is a set of programming techniques that uses a limited number of control structures, top-down design, and module independence.

One of the first steps in **top-down design** is to identify basic program functions. These functions are further divided into smaller and smaller subfunctions of more manageable size. These subfunctions are called modules. The structure charts we discussed earlier are good examples of top-down design, and using structure charts to plan programs is a means to achieve it.

Now let us look more closely at the way modules are planned.

Modularity

Computer professionals recognize that the way to efficient development and maintenance is to work on a program as a series of manageable pieces. The way that a program is divided has a significant effect on how it works. We have already noted, in the discussion of structure charts, that structured design involves organizing the pieces of a program in a hierarchical way. Lower-level components are called modules. Once converted to programmed form, a **module** is a set of logically related statements that performs a specific function.

One relationship between modules is called **coupling,** which is the measure of the strength of the relationship between two modules. Ideally, that relationship should be weak so that the modules are independent; then, if a change is made in one module, it will not affect other modules. Another relationship is called **cohesion,** the measure of the inner strength of an individual module. The best relationship here is a strong one; a module should have a single function, although that is not always possible. An example of a single function is the computation of withholding tax. This function would not be included with other functions, such as computing insurance deductions. Strong cohesion encourages module independence, which in turn makes future changes easier.

In addition, a module should have a **single entry** and a **single exit.** In a single-entry module, execution must always begin at the same place, usually at the beginning; the module can be entered at only a single point. Similarly, the module may be exited from only one place, as shown in Figure A-13. Keeping track of what is going on in a program is easier if there is only one way to get in and one way to get out of each module in it.

A module should be of manageable size. A single page of coded program instructions is often considered an ideal size.

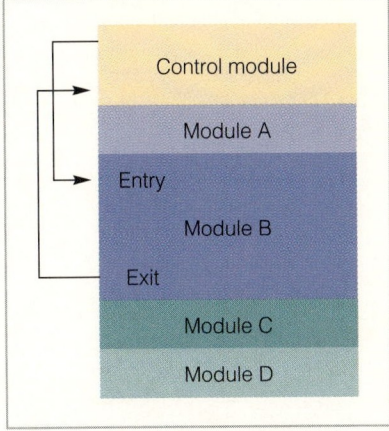

Figure A-13 Single entry, single exit. The program control module executes module B by transferring to its entry point. After the instructions in module B are executed, the module is exited via its exit point. Program control returns to the point of departure, in this case, the control module.

Is There a Future for Structured Programming?

We have focused on structure and its acceptance in the computer community. We have industry-wide agreement on this important issue and can consider it settled, right? No, not quite. Structured programming makes sense if you are using what is called a **procedural language**—that is, a language that presents a step-by-step process for solving a problem, as COBOL does. It is the steps—program statements—of the procedure that need structuring. But nothing stays the same for long in the computer industry. When the focus shifts from how to accomplish a task—the step-by-step way—to what you want to accomplish, **nonprocedural languages** are preferred. Object-oriented languages (discussed in Chapter 6) are the most popular nonprocedural languages. New, easy-to-use languages will continue to be introduced as the industry evolves.

▼   ▲   ▼

This chapter is useful to two types of audiences: students who expect to be users only and students who hope to be programmers. Students who will be users but not programmers can gain an appreciation of what it takes to plan a program. Even the simplest task can seem complex when expressed in step-by-step instructions. Those who will be programmers can understand that modern software is highly complex; they are unlikely to be bored.

CHAPTER REVIEW

Summary and Key Terms

- The planning phase of programming involves the steps necessary to help you understand how to develop program logic. Programmers and others can use flowcharting, structure charts, and pseudocode to plan a solution for a problem.
- **Flowcharts** are symbolic pictorial representations of step-by-step solutions to problems. They consist of arrows representing the direction a program takes and boxes and other symbols representing actions. A **logic flowchart** represents the flow of logic in a program; a **systems flowchart** shows the flow of data through an entire computer system.
- To **initialize** means to set the starting values of certain storage locations, usually as the program begins to execute.
- A **compare operation** occurs when the computer compares two numbers and performs alternative operations based on the comparison.
- A **loop**—also called an **iteration**—is defined as the repetition of instructions under certain conditions; once established, the loop is considered a powerful programming tool because the code is reusable.
- A **structure chart** graphically illustrates the structure of a program by showing independent hierarchical modules. It is also known as a **hierarchy chart**.
- **Pseudocode** is a way of representing the solution to a problem by using English to express the logic. It is considered a "first draft" for solving the problem because eventually it must be translated into a programming language.
- **Structured programming** is a technique that emphasizes breaking a program into logical sections by using certain universal programming standards. In more formal terms, structured programming is a set of programming techniques that includes a limited number of control structures, top-down design, and module independence.
- Structured programming uses **control structures** to handle execution. The **sequence control structure** involves one statement following another in sequence. A **selection control structure** can take one of two forms: IF-THEN-ELSE and IF-THEN. An **iteration control structure** is a looping mechanism that uses DOWHILE for **leading decisions** (at the beginning of the loop) and DOUNTIL for **trailing decisions** (at the end of the loop). Each structure has only one **entry point** (the point where control is transferred to the structure) and one **exit point** (the point where control is transferred from the structure). These characteristics make structured programs easier to read and debug than unstructured programs.
- A **Read** statement brings something that is outside the computer into memory; reading, in other words, means getting. The first Read statement is sometimes called the **priming read**. When you read to the **end of file**, there are no more records in the file, so the reading process stops.
- **Top-down design** identifies basic program functions before dividing them into subfunctions called modules. Structure charts illustrate top-down design.
- When converted to program form, a **module** is a set of logically related statements that performs a specific function.
- **Coupling** is the measure of the strength of the relationship between two modules. Weak coupling is ideal because a change in one module does not affect other modules. **Cohesion** is the measure of the inner strength of a module. Strong cohesion makes modules more independent, a characteristic that facilitates future changes. A module should also have a **single entry** and a **single exit** so that it is easier to keep track of the flow of logic in the program.
- A **procedural language,** such as COBOL, is a language that presents a step-by-step process of solving a problem. **Nonprocedural languages,** such as object-oriented languages, simply state what task is to be accomplished.

Although the story of computers has deep roots, the most fascinating part—the history of personal computers—is quite recent. The beginning of this history turns on the personality of Ed Roberts the way a watch turns on a jewel. It began when his foundering company took a surprising turn.

Like other entrepreneurs before him, Ed Roberts had taken a big risk. He had already been burned once, and now he feared being burned again. The first time, in the early 1970s, he had borrowed heavily to produce microprocessor-based calculators, only to have the chip producers decide to build their own product—and sell it for half the price of Ed's calculator.

Ed's new product was based on a microprocessor, too—the Intel 8080 —but it was a *computer*. A little computer. The "big boys" at the established computer firms considered computers to be industrial products; who would want a small computer? Ed was not sure, but he found the idea so compelling that he decided to make the computer anyway. Besides, he was so far in debt from the calculator fiasco that it did not seem to matter which project propelled him into bankruptcy. Ed's small computer and his company, MITS, were given a sharp boost by Les Solomon, who promised to feature the new machine on the cover of *Popular Electronics*. In Albuquerque, New Mexico, Ed worked frantically to meet the publication deadline, and he even tried to make the machine pretty, so it would look attractive on the cover (Figure B-1).

Making a good-looking small computer was not easy. This machine, named the Altair (after a heavenly "Star Trek" destination), looked like a flat box. In fact, it met the definition of a computer in only a minimal way: It had a central processing unit (on the chip), 256 characters (a paragraph!) of memory, and switches and lights on a front panel for input/output. No screen, no keyboard, no storage.

But the Altair was done on time for the January 1975 issue of *Popular Electronics,* and Roberts made plans to fly to New York to demonstrate the machine for Solomon. He sent the computer on ahead by railroad express. Ed got to New York, but the computer did not—the very first personal computer was lost! There was no time to build a new computer before the publishing deadline, so Roberts cooked up a phony version for the cover picture: an empty box with switches and lights on the front panel. He also placed an inch-high ad in the back of the magazine: Get your own Altair kit for $397.

Ed was hoping for perhaps 200 orders. But the machine—that is, the box—fired imaginations across the country. Two thousand customers sent checks for $397 to an unknown Albuquerque, New Mexico, company. Overnight, the MITS Altair personal computer kit was a runaway success.

Ed Roberts was an important player in the history of personal computers. Unfortunately, he never made it in the big time; most observers agree that his business insight did not match his technical skills. But other entrepreneurs did make it. In this appendix we will glance briefly at the early years of computers and then examine more recent history.

Appendix B

History and Industry
The Continuing Story of the Computer Age

LEARNING OBJECTIVES

- Understand the story of how computer technology unfolded, with particular emphasis on the "generations"
- Understand how people and events affected the development of computers
- Become familiar with the story of personal computer development

BABBAGE AND THE COUNTESS

HERMAN HOLLERITH:
    The Census Has Never Been the Same

WATSON OF IBM

THE START OF THE MODERN ERA

THE COMPUTER AGE BEGINS
    The First Generation, 1951–1958: The Vacuum Tube

    The Second Generation, 1959–1964: The Transistor

    The Third Generation, 1965–1970: The Integrated Circuit

    The Fourth Generation, 1971–Present: The Microprocessor

    The Fifth Generation: Onward

THE STORY OF PERSONAL COMPUTERS
    I Built It in My Garage

    The IBM PC Standard

    The Rise of Microsoft

Figure B-1 The Altair. The term *personal computer* had not been coined yet, so Ed Robert's small computer was called a "minicomputer" when it was featured on the cover of *Popular Electronics*.

 ## Babbage and the Countess

Born in England in 1791, Charles Babbage was an inventor and mathematician. When solving certain equations, he found the hand-done mathematical tables he used filled with errors. He decided a machine could be built that would solve the equations better by calculating the differences between them. He set about making a demonstration model of what he called a **difference engine** (Figure B-2). The model was so well received that in about 1830 he enthusiastically began to build a full-scale working version, using a grant from the British government.

However, Babbage found that the smallest imperfections were enough to throw the machine out of whack. Babbage was viewed by his own colleagues as a man who was trying to manufacture a machine that was utterly ridiculous. Finally, after spending its money to no avail, the government withdrew financial support.

Despite this setback, Babbage was not discouraged. He conceived of another machine, christened the **analytical engine,** which he hoped would perform many kinds of calculations. Although it was never built in his time, a model was eventually put together by his son. It was not until 1991 that a working version of the analytical engine was built and put on public display in London. It embodied five key features of modern computers:

- An input device
- A storage place to hold the number waiting to be processed
- A processor, or number calculator

Figure B-2 Charles Babbage's difference engine. Babbage's second difference engine was not completed in his lifetime. The one shown here was built in 1991 by the London Science Museum, according to Babbage's original design, in honor of Babbage's 200th birthday.

- A control unit to direct the task to be performed and the sequence of calculations
- An output device

If Babbage was the father of the computer, then Ada, the Countess of Lovelace, was the first computer programmer (Figure B-3). The daughter of English poet Lord Byron and of a mother who was a gifted mathematician, Ada helped develop the instructions for doing computations on the analytical engine. Lady Lovelace's contributions cannot be overvalued. She was able to see that Babbage's theoretical approach was workable, and her interest gave him encouragement. In addition, she published a series of notes that eventually led others to accomplish what Babbage himself had been unable to do. In her honor a modern programming language was named Ada.

Figure B-3 Ada, the Countess of Lovelace. Augusta Ada Byron, as she was known before she became a countess, was Charles Babbage's colleague in his work on the analytical engine and has been called the world's first computer programmer.

Herman Hollerith: The Census Has Never Been the Same

The hand-done tabulation of the 1880 United States census took seven and a half years. A competition was held to find some way to speed the counting process of the 1890 United States census. Herman Hollerith's tabulating machine won the contest. As a result of his system's adoption, an unofficial count of the 1890 population (62,622,250) was announced only six weeks after the census was taken.

The principal difference between Hollerith's and Babbage's machines was that Hollerith's machine used electrical rather than mechanical power (Figure B-4). Hollerith realized that his machine had considerable commercial potential. In 1896 he founded the successful Tabulating Machine Company, which, in 1924, merged with two other companies to form the International Business Machines Corporation—IBM.

Watson of IBM

For over 30 years, from 1924 to 1956, Thomas J. Watson, Sr., ruled IBM with an iron grip. Cantankerous and autocratic, supersalesman Watson made IBM a dominant force in the business machines market, first as a supplier of calculators, then as a developer of computers.

IBM's entry into computers was sparked by a young Harvard professor of mathematics, Howard Aiken. In 1936, after reading Lady Lovelace's notes, Aiken began to think that a modern equivalent of the analytical engine could be constructed. Because IBM was already such a power in the business machines market, with ample money and resources, Aiken worked out a careful proposal and approached Thomas Watson. In one of those make-or-break decisions for which he was famous, Watson gave him $1 million. As a result, the Harvard Mark I was born.

The Start of the Modern Era

Nothing like the **Mark I** had ever been built before. It was 8 feet high and 55 feet long, made of streamlined steel and glass, and it emitted a sound

Watson Smart? You Bet!

Just as computers were getting off the ground, Thomas Watson, Sr., saw the best and brightest called to arms in World War II. But he did not just bid his employees a sad adieu. He paid them. Each and every one received one quarter of his or her annual salary, in twelve monthly installments. The checks continued to arrive throughout the duration of the war. Every month those former employees thought about IBM and the generosity of its founder.

The result? A very high percentage of those employees returned to IBM after the war. Watson got his brain trust back, virtually intact. The rest is history.

Figure B-4 Herman Hollerith's tabulating machine. This electrical tabulator and sorter was used to tabulate the 1890 census.

during processing that one person said was "like listening to a roomful of old ladies knitting away with steel needles." Unveiled in 1944, the Mark I was never very efficient. But the enormous publicity it generated strengthened IBM's commitment to computer development. Meanwhile, technology had been proceeding elsewhere on separate tracks.

American military officials approached Dr. John Mauchly at the University of Pennsylvania and asked him to build a machine that would rapidly calculate trajectories for artillery and missiles. Mauchly and his student J. Presper Eckert relied on the work of Dr. John V. Atanasoff, a professor of physics at Iowa State University. During the late 1930s Atanasoff had spent time trying to build an electronic calculating device to help his students solve mathematical problems. He and an assistant, Clifford Berry, succeeded in building the first digital computer that worked electronically; they called it the **ABC**, for **Atanasoff-Berry computer** (Figure B-5).

After Mauchly met with Atanasoff and Berry in 1941, he used the ABC as the basis for the next step in computer development. From this association ultimately came a lawsuit, based on attempts to get patents for a commercial version of the machine Mauchly built. The suit was finally decided in 1974, when a federal court determined that Atanasoff had been the true originator of the ideas required to make an electronic digital computer actually work. (Some computer historians dispute this court decision.) Mauchly and Eckert were able to use the principles of the ABC to create the **ENIAC**, for **Electronic Numerical Integrator and Calculator**. The main

Figure B-5 The ABC. John Atanasoff and his assistant, Clifford Berry, developed the first digital electronic computer.

Figure B-6 Vacuum tubes. Vacuum tubes were used in the first generation of computers.

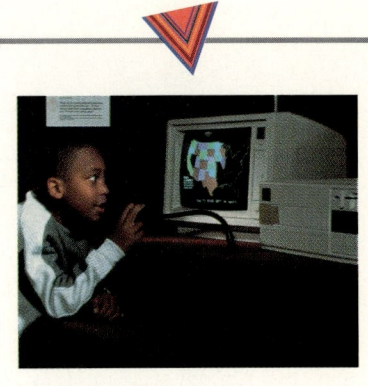

The Computer Museum

The Computer Museum in downtown Boston, Massachusetts, is the world's first museum devoted solely to computers and computing. The museum shows how computers have affected all aspects of life: science, business, education, art, and entertainment. Over half an acre of hands-on and historical exhibits chronicle the enormous changes in the size, capability, applications, and cost of computers over the past 40 years. Two mini-theaters show computer classics as well as award-winning computer-animated films.

The Computer Museum Store offers a large selection of such unique items as state-of-the-art silicon chip jewelry and chocolate "chips," as well as books, posters, and cassettes.

significance of the ENIAC is that, as the first general-purpose computer, it was the forerunner of the UNIVAC I, the first computer sold on a commercial basis.

The Computer Age Begins

The remarkable thing about the computer age is that so much has happened in so short a time. We have leapfrogged through four generations of technology in about 40 years—a span of time whose events are within the memories of many people today. The first three computer "generations" are pinned to three technological developments: the vacuum tube, the transistor, and the integrated circuit. Each has drastically changed the nature of computers. We define the timing of each generation according to the beginning of commercial delivery of the hardware technology. Defining subsequent generations has become more complicated because the entire industry has become more complicated.

The First Generation, 1951–1958: The Vacuum Tube

The beginning of the commercial computer age may be dated June 14, 1951. This was the date the first **UNIVAC—*Univ*ersal Automatic Computer**—was delivered to a client, the U.S. Bureau of the Census, for use in tabulating the previous year's census. The date also marked the first time that a computer had been built for business applications rather than for military, scientific, or engineering use. The UNIVAC was really the ENIAC in disguise and was, in fact, built by Mauchly and Eckert, who in 1947 had formed their own corporation.

In the first generation, **vacuum tubes**—electronic tubes about the size of light bulbs—were used as the internal computer components (Figure B-6). However, because thousands of such tubes were required, they generated a great deal of heat, causing many problems in temperature regulation and climate control. In addition, although all the tubes had to be working simultaneously, they were subject to frequent burnout—and the people operating the computer often did not know whether the problem was in the programming or in the machine.

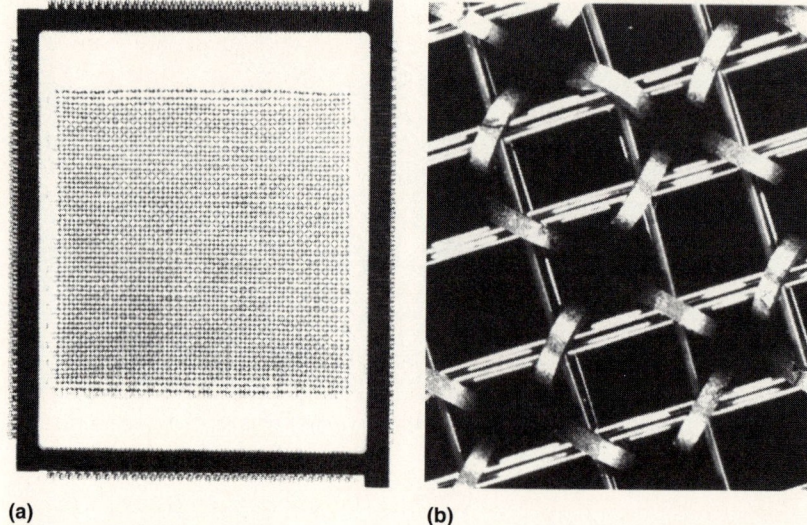

Figure B-7 Magnetic cores. (a) A 6- by 11-inch magnetic core memory. (b) Close-up of magnetic core memory. A few hundredths of an inch in diameter, each magnetic core was mounted on wires. When electricity passed through a wire on which a core was strung, the core could be magnetized as either on or off.

Another drawback was that the language used in programming was machine language, which uses numbers. (Present-day higher-level languages are more like English.) Using numbers alone made programming the computer difficult and time-consuming. The UNIVAC used **magnetic cores** to provide memory. These consisted of small, doughnut-shaped rings about the size of pinheads, which were strung like beads on intersecting thin wires (Figure B-7). To supplement primary storage, first-generation computers stored data on punched cards. In 1957 magnetic tape was introduced as a faster, more compact method of storing data.

The Second Generation, 1959–1964: The Transistor

Three Bell Lab scientists—J. Bardeen, H. W. Brattain, and W. Shockley—developed the **transistor,** a small device that transfers electric signals across a resistor. (The name *transistor* began as a trademark concocted from *trans*fer plus re*sistor*.) The scientists later received the Nobel prize for their invention. The transistor revolutionized electronics in general and computers in particular. Transistors were much smaller than vacuum tubes, and they had numerous other advantages: They needed no warm-up time, consumed less energy, and were faster and more reliable.

During this generation another important development was the move from machine language to **assembly languages**—also called **symbolic languages.** Assembly languages use abbreviations for instructions (for example, L for LOAD) rather than numbers. This made programming less cumbersome.

After the development of symbolic languages came **high-level languages,** such as **FORTRAN** (1954) and **COBOL** (1959). Both languages, still widely used today (in updated forms), are more English-like than assembly languages. High-level languages allowed programmers to give more attention to solving problems. Also, in 1962 the first removable disk pack was marketed. Disk storage supplemented magnetic tape systems and enabled users to have fast access to desired data.

All these new developments made the second generation of computers less costly to operate, and thus began a surge of growth in computer systems. Throughout this period computers were being used principally by business, university, and government organizations. They had not filtered down to the general public. The real part of the revolution was about to begin.

The Third Generation, 1965–1970: The Integrated Circuit

One of the most abundant elements in the earth's crust is silicon, a nonmetallic substance found in common beach sand as well as in practically all rocks and clay. The importance of this element to Santa Clara County, which is about 30 miles south of San Francisco, is responsible for the county's nickname: Silicon Valley. In 1965 Silicon Valley became the principal site for the manufacture of the so-called silicon chip: the integrated circuit.

An **integrated circuit** (IC) is a complete electronic circuit on a small chip of silicon. The chip may be less than $\frac{1}{8}$ inch square and contain thousands or millions of electronic components. In 1965 integrated circuits began to replace transistors in computers. The resulting machines were called third-generation computers. An integrated circuit was able to replace an entire circuit board of transistors with one chip of silicon much smaller than one transistor.

Integrated circuits are made of silicon because it is a **semiconductor**. That is, silicon is a crystalline substance that will conduct electric current when it has been "doped" with chemical impurities implanted in its lattice-like structure. A cylinder of silicon is sliced into wafers, each about 6 inches in diameter, and the wafer is etched repeatedly with a pattern of electrical circuitry. Several layers may be etched on a single wafer. The wafer is then divided into several hundred small chips, each with a complete circuit so tiny it is half the size of a human fingernail, yet under a microscope it looks as complex as a railroad yard.

The chips were hailed as a generational breakthrough because they had desirable characteristics: reliability, compactness, and low cost. Mass-production techniques have made possible the manufacture of inexpensive integrated circuits.

The beginning of the third generation was trumpeted by the IBM 360 series (named for 360 degrees—a full circle of service) in 1964. The System/360 family of computers, designed for both business and scientific use, came in several models and sizes. The "family of computers" concept made it possible for users to move to a more powerful machine without redoing the software that already worked on the current computer. The equipment housing was blue, leading to IBM's nickname, Big Blue.

The 360 series was launched with an all-out, massive marketing effort to make computers business tools—to get them into medium-size and smaller business and government operations where they had not been used before. Perhaps the most far reaching contribution of the 360 series was the decision to **unbundle** the software, that is, to sell the software sepa-

Better Late than Never

For years, Gilbert Hyatt made a claim that most of his colleagues derided: He said he had invented the microprocessor, the circuitry at the heart of all computers and consumer electronics. To everyone's astonishment, however, the U.S. Patent Office in 1990 granted Hyatt a patent that made him the official inventor of the microprocessor.

Intel Corporation had long claimed credit for the invention, pointing to Ted Hoff, a former Intel engineer, as the inventor. However, Hyatt had filed for the patent on what he called the "microcomputer" back in 1970—beating Intel's first microprocessor, the 4004, which came out in 1971. Early on, Hyatt had solicited venture capitalists to support his new company to manufacture his invention. He claims that one of them leaked information to Intel.

Since receiving his patent, Hyatt has negotiated licensing agreements with a number of companies, making him an overnight multimillionaire.

rately from the hardware. This approach lead to the creation of today's software industry.

Software became more sophisticated during this third generation. Several programs could run in the same time frame, sharing computer resources. This approach improved the efficiency of computer systems. Software systems were developed to support interactive processing, which used a terminal to put the user in direct contact with the computer. This kind of access caused the customer service industry to flourish, especially in areas such as reservations and credit checks.

Large third-generation computers began to be supplemented by minicomputers, which are functionally equivalent to full-size systems but are somewhat slower, smaller, and less expensive. These computers have become a huge success with medium-size and smaller businesses.

The Fourth Generation, 1971–Present: The Microprocessor

Through the 1970s computers gained dramatically in speed, reliability, and storage capacity, but entry into the fourth generation was evolutionary rather than revolutionary. The fourth generation was, in fact, an extension of third-generation technology. That is, in the early part of the third generation, specialized chips were developed for computer memory and logic. Thus, all the ingredients were in place for the next technological development—the general-purpose processor-on-a-chip, otherwise known as the **microprocessor,** which became commercially available in 1971.

Nowhere is the pervasiveness of computer power more apparent than in the explosive use of the microprocessor. In addition to the common applications of digital watches, pocket calculators, and personal computers, you can expect to find microprocessors in virtually every machine in the home or business—cars, copy machines, television sets, bread-making machines, and so on. Computers today are 100 times smaller than those of the first generation, and a single chip is far more powerful than ENIAC.

The Fifth Generation: Onward

The term *fifth generation* was coined by the Japanese to describe the powerful, "intelligent" computers they wanted to build by the mid-1990s. Later the term evolved to encompass several research fields related to computer intelligence: artificial intelligence, expert systems, and natural language.

But the true focus of this ongoing fifth generation is connectivity, the massive industry effort to permit users to connect their computers to other computers. The concept of the information superhighway has captured the imaginations of both computer professionals and everyday computer users.

 ## The Story of Personal Computers

Personal computers are the machines you can "get closest to," whether you are an amateur or a professional. There is nothing quite like having your very own personal computer. Its history is very personal too, full of stories of success and failure and of individuals with whom we can readily identify.

The Software Entrepreneurs

Ever thought you'd like to run your own show? Make your own product? Be in business for yourself? Entrepreneurs are a special breed. They are achievement oriented, like to take responsibility for decisions, and dislike routine work. They also have high levels of energy and a great deal of imagination. But perhaps the key is that they are willing to take risks.

Entrepreneurs often have still another quality—a more elusive quality—that is something close to charisma. This charisma is based on enthusiasm, and it allows them to lead people, form an organization, and give it momentum. Study these real-life entrepreneurs, noting their paths to glory and—sometimes—their falls.

Steve Jobs

Of the two Steves who formed Apple Computer, Steve Jobs was the true entrepreneur. Although they both were interested in electronics, Steve Wozniak was the technical genius, and he would have been happy to have been left alone to tinker. But Steve Jobs would not let him alone for a minute; he was always pushing and crusading. In fact, Wozniak had hooked up with an evangelist, and they made quite a pair.

When Apple was getting off the ground, Jobs wanted Wozniak to quit his job so he could work full-time on the new venture. Wozniak refused. His partner begged and cried. Wozniak gave in. While Wozniak built Apple computers, Jobs was out hustling, finding the best marketing man, the best venture capitalist, and the best company president. This entrepreneurial spirit paid off in a spectacular way as Apple rose to the top of the list of personal computer companies.

Bill Gates

When Bill Gates was a teenager, he swore off computers for a year and in his words, "tried to act normal." His parents, who wanted him to be a lawyer, must have been relieved when Bill gave up the computer foolishness and went off to Harvard in 1974. But Bill started spending weekends with his friend Paul Allen, dreaming about personal computers, which did not exist yet. When the MITS Altair, the first personal computer for sale, splashed on the market in January 1975, both Bill and Paul moved to Albuquerque to be near the action at MITS. But they showed a desire even then to chart their own course. Although they wrote software for MITS, they kept the rights to their work and formed their own company. Their company was called Microsoft.

When MITS failed, Gates and Allen moved their software company to their native Bellevue, Washington. They employed 32 people in 1980 when IBM came to call. Gates recognized the big league when he saw it and put on a suit for the occasion. Gates was offered a plum: the chance to develop the operating system (a crucial set of software) for IBM's soon-to-be personal computer. Although he knew he was betting the whole company, Gates never hesitated to take the risk. Gates purchased an existing operating system, which he and his crew reworked to produce MS-DOS—Microsoft Disk Operating System. It was this product that sent Microsoft on its meteoric rise.

Mitch Kapor

Kapor did not start out on a direct path to computer fame and riches. In fact, he wandered extensively, from being a disk jockey to piano teacher to counselor. He had done some programming, too, but did not like it much. But, around 1978, he found he did like fooling around with personal computers. In fact, he had found his niche.

In 1983 Kapor introduced a software package called Lotus 1-2-3, and there had never been anything like it before. Lotus added the term *integrated package* (now called an office suite) to the vocabulary; the phrase described the software's identity as a combination spreadsheet, graphics, and database program. Kapor's product catapulted his company to the top of the list of independent software makers in just two years.

Champions of Change

Entrepreneurs thrive on change. Jobs, Wozniak, and Kapor all left their original companies to start new companies. When Steve Jobs lost control of Apple Computer in 1986, he went out and started NeXT Computer, Inc. Stay tuned for future breakthroughs from these and other personal computer entrepreneurs.

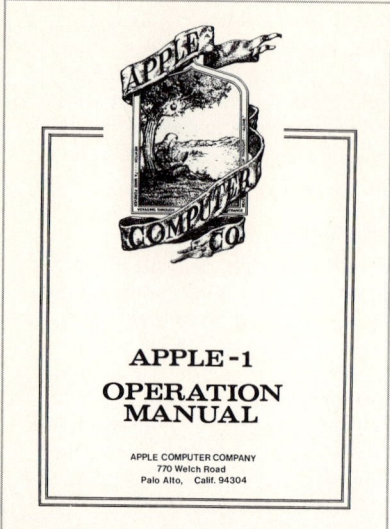

Figure B-8 Apple manual. Shown here is a collector's item: the very first manual for operation of an Apple computer. Unfortunately, the early manuals were a hodgepodge of circuit diagrams, software listings, and handwritten notes. They were hard to read and understand, enough to frighten away all but the most hardy souls.

Figure B-9 The IBM PC. Launched in 1981, this early IBM PC took just 18 months to rise to the top of the best-seller list.

I Built It in My Garage

As we noted in the beginning of the chapter, the very first personal computer was the MITS Altair, produced in 1975. But it was a gee-whiz machine, loaded with switches and dials—and no keyboard or screen. It took two teenagers, Steve Jobs and Steve Wozniak, to capture the imagination of the public with the first Apple computer. They built it in that time-honored place of inventors, a garage, using the $1300 proceeds from the sale of an old Volkswagen. Designed for home use, the Apple was the first to offer an easy-to-use keyboard and screen. Founded in 1977, Apple Computer was immediately and wildly successful. (Figure B-8 shows early documentation for the first commercial Apple computer.)

The first Apple computer, the Apple I, was not a commercial success. It was the Apple II that anchored the early years of the company. In fact, it was the combination of the Apple II and the spreadsheet software called VisiCalc, which we described in the opening story in Chapter 14, that caught the attention of the business community.

When Apple's stock was offered to the public in December 1980, it started a stampede among investors eager to buy in. Apple has introduced an increasingly powerful line of computers, including the currently popular Macintosh.

The other major player in those early years was Tandy Incorporated, whose worldwide chain of Radio Shack stores provided a handy sales outlet for the TRS-80 personal computer. Other manufacturers who enjoyed more than moderate success in the late 1970s were Atari and Commodore. Their numbers were to grow.

The IBM PC Standard

IBM announced its first personal computer in the summer of 1981. IBM captured the top market share in just 18 months, and even more important, its machine became the industry standard (Figure B-9). The IBM machine included innovations such as an 80-character screen line, a full upper- and lower-case keyboard, and the possibility of adding memory. IBM also provided internal expansion slots, so that peripheral equipment manufacturers could build accessories for the IBM PC. In addition, IBM provided hardware schematics and software listings to companies who wanted to build products in conjunction with the new PC. Many of the new products accelerated demand for the IBM machine. Even more important, many new companies sprang up just to support the IBM PC.

IBM made its computer from nonproprietary parts, opening the door for other manufacturers to do the same. Thus, other personal computer manufacturers emulated the IBM standard, producing IBM **clones,** copycat computers that can run software designed for IBM computers. Almost all the major personal computer manufacturers today—Compaq, Dell, Gateway, AST, and so many more—are IBM clones. In fact, clone computers now dominate the personal computer market, leaving IBM with a market share that is small when compared to its original success.

The Rise of Microsoft

In the history of the computer industry, the spotlight has been on the fast-changing hardware. However, personal computer users now focus more on the tremendous variety of software. The dominant force in personal computer software is the Microsoft Corporation.

Microsoft supplied the operating system—the underlying software—for the original IBM personal computer. This software, called MS-DOS, was used by IBM and by the IBM clones, permitting tiny Microsoft to grow quickly. Microsoft eventually presented more sophisticated operating systems, notably Windows 95. Although originally seen as merely the supplier of operating systems, a powerful but still small role, Microsoft went on to develop a variety of successful applications software. Microsoft has applications offerings for the standard applications: word processing, spreadsheets, database management, graphics, and communications. Furthermore, the company has made serious inroads into multimedia, online banking, and even movies. Microsoft is now the leading software company in the world.

▼ ▲ ▼

History is still being made in the computer industry, of course, and it is being made incredibly rapidly. A book cannot possibly pretend to describe all the very latest developments. Nevertheless, as we indicated earlier, the four areas of input—processing, output, and storage—describe the basic components of a computer system, whatever its date.

CHAPTER REVIEW

Summary and Key Terms

- Charles Babbage, a nineteenth-century mathematician, is called the father of the computer because of his invention of two computation machines. His **difference engine,** which could solve equations, led to another calculating machine, the **analytical engine,** which embodied the key parts of a computer system—an input device, a processor, a control unit, a storage place, and an output device. Countess Ada Lovelace, who helped develop instructions for carrying out computations on Babbage's device, is often called the first programmer.
- The first computer to use electrical power instead of mechanical power was Herman Hollerith's tabulating machine, which was used in the 1890 census in the United States. Hollerith founded a company that became the forerunner of International Business Machines Corporation (IBM).
- Thomas J. Watson, Sr., built IBM into a dominant force in the business machines market. He also gave Harvard professor Howard Aiken research funds with which to build an electromechanical computer, the **Mark I,** unveiled in 1944.
- John V. Atanasoff, with assistant Clifford Berry, devised the first digital computer to work by electronic means, the **Atanasoff-Berry Computer (ABC).**
- The **ENIAC (Electronic Numerical Integrator and Calculator),** developed by John Mauchly and J. Presper Eckert at the University of Pennsylvania in 1946, was the world's first general-purpose electronic computer.
- The first computer generation began June 14, 1951, with the delivery of the **UNIVAC (***Universal Automatic Computer***)** to the U.S. Bureau of the Census. First-generation computers required thousands of **vacuum tubes,** electronic tubes about the size of light bulbs. First-generation computers had slow input/output, were programmed only in machine language, and were unreliable. The main form of memory was **magnetic core.** Magnetic tape was introduced in 1957 to store data compactly.
- Second-generation computers used **transistors,** developed at Bell Laboratories. Compared to vacuum tubes, transistors were small, needed no warm-up, consumed less energy, and were faster and more reliable. During the second generation, **assembly languages,** or **symbolic languages,** were developed. They used abbreviations for instructions, rather than numbers. Later, **high-level languages,** such as **FORTRAN** and **COBOL,** were also developed. In 1962 the first removable disk pack was marketed.
- The third generation emerged with the introduction of the **integrated circuit (IC)**—a complete electronic circuit on a small chip of silicon. Silicon is a **semiconductor,** a substance that will conduct electric current when it has been "doped" with chemical impurities.
- With the third generation, IBM announced the System/360 family of computers, which made it possible for users to move up to a more powerful machine without redoing the software that already worked on the current computer. IBM **unbundled** the software, that is, sold it separately from the hardware. During this period more sophisticated software was introduced that allowed several programs to run in the same time frame and supported interactive processing, in which the user, by using a terminal, has direct contact with the computer.
- The fourth-generation **microprocessor**—a general-purpose processor-on-a-chip—grew out of the specialized memory and logic chips of the third generation.
- The term *fifth generation,* coined by the Japanese, evolved to encompass artificial intelligence, expert systems, and natural language. But the true focus of the fifth generation is connectivity, permitting users to connect their computers to other computers.
- The first personal computer, the MITS Altair, was produced in 1975. However, the first successful computer to include an easy-to-use keyboard and screen was offered by Apple Computer, founded by Steve Jobs and Steve Wozniak in 1977.

- IBM entered the personal computer market in 1981 and captured the top market share in just 18 months. Other manufacturers produced IBM **clones**, copycat computers that can run software designed for IBM computers. Clones now dominate the personal computer market.
- The worldwide leading software company is the Microsoft Corporation, which supplied the operating system for the original IBM personal computer and then went on to develop a variety of successful applications software.

Student Study Guide

Multiple Choice

1. Charles Babbage invented the
   a. Mark I
   b. abacus
   c. difference engine
   d. adding machine
2. The analytical engine used
   a. a punched card system
   b. electrical power
   c. vacuum tubes
   d. magnetic tape
3. The first computer to use electrical power was developed by
   a. Herman Hollerith
   b. John V. Atanasoff
   c. Thomas J. Watson
   d. Howard Aiken
4. The Mark I was built by
   a. Thomas Watson
   b. John V. Atanasoff
   c. Dr. John Mauchly
   d. Howard Aiken
5. The first digital computer to work electronically was the
   a. UNIVAC
   b. Mark I
   c. Atanasoff-Berry Computer
   d. analytical machine
6. The ENIAC, using ABC principles, was designed by
   a. Charles Babbage
   b. John V. Atanasoff
   c. Bell Laboratories
   d. Mauchly and Eckert
7. Hollerith's Tabulating Machine Company eventually became
   a. IBM
   b. Apple
   c. AT&T
   d. General Electric
8. First-generation computers were characterized by the use of the
   a. microprocessor
   b. transistor
   c. vacuum tube
   d. integrated circuit
9. Primary storage for the UNIVAC was via
   a. punched cards
   b. magnetic cores
   c. transistors
   d. disk packs
10. Transistors characterized a major improvement over
    a. semiconductors
    b. vacuum tubes
    c. disk packs
    d. the integrated circuit
11. Assembly languages, or symbolic languages, use
    a. binary code
    b. numbers
    c. abbreviations for instructions
    d. English words
12. COBOL and FORTRAN are examples of
    a. machine languages
    b. assembly languages
    c. high-level languages
    d. obsolete languages
13. Silicon chips of integrated circuits (ICs) marked the advent of the
    a. second generation
    b. third generation
    c. fourth generation
    d. fifth generation
14. IBM's System/360 family of computers was introduced during the
    a. 1950s
    b. second generation
    c. third generation
    d. 1990s
15. The fourth generation is identified by the
    a. minicomputer
    b. IBM System/360
    c. introduction of ICs
    d. microprocessor
16. The first Apple computer was built in
    a. a garage
    b. an apartment
    c. a warehouse
    d. a factory
17. Personal computers that can use software designed for the IBM PC are called
    a. cores
    b. clones
    c. emulators
    d. unbundled
18. The general-purpose processor-on-a-chip is otherwise known as the
    a. ENIAC
    b. pocket calculator
    c. microcomputer
    d. microprocessor
19. The leading software manufacturer is
    a. Apple
    b. MITS
    c. IBM
    d. Microsoft
20. The focus of the fifth generation is
    a. connectivity
    b. silicon
    c. symbolic languages
    d. memory chips

True/False

T F 1. The analytical engine embodied the five key concepts of a computer system.
T F 2. Countess Ada Lovelace could be considered the first computer programmer.
T F 3. Hollerith's tabulating machine was driven by mechanical power.
T F 4. Thomas G. Watson, Sr., developed the Mark I, an electromechanical computer.
T F 5. The ABC was built by Howard Aiken and Clifford Berry.
T F 6. The ENIAC had a large storage capacity and was a very flexible system.
T F 7. The ENIAC, made operational in 1946, was the world's first electronic digital computer.

T F   8. Thomas Watson was a dominating force at IBM.
T F   9. Charles Babbage and Ada Lovelace worked together to produce the ENIAC.
T F 10. The first generation, characterized by the vacuum tube, encompassed all computer development prior to 1960.
T F 11. The UNIVAC used thousands of vacuum tubes.
T F 12. A drawback of vacuum tubes was temperature control.
T F 13. Magnetic core was the dominant form of primary storage technology for two decades.
T F 14. The transistor of the second generation was faster and more reliable than the vacuum tube but used more energy.
T F 15. Both assembly languages and high-level languages were developed during the second generation.
T F 16. The first high-level language to receive widespread acceptance was COBOL.
T F 17. In 1962 the first removable disk pack was marketed, enabling users to have fast access to desired data.
T F 18. Integrated circuits, ICs, marked the advent of the third generation.
T F 19. Microprocessors led to the development of the personal computer.
T F 20. The Altair was developed by John Atanasoff.

Fill-In

1. What were the five key concepts of modern computers embodied in Charles Babbage's analytical engine?
   a. _____
   b. _____
   c. _____
   d. _____
   e. _____

2. The technology of the second generation: _____

3. Who is considered the world's first computer programmer? a. _____
   Whom did this person assist?
   b. _____

4. Who built IBM into a dominant force in the business machine market? _____

5. What machine was used in the U.S. Census of 1890? _____

6. Who were the two men to devise the first electronic digital computer? What did they call their machine?
   a. _____
   b. _____
   c. _____

7. To whom did Watson give research funds to build an electromechanical computer? What machine resulted from this funding? _____

8. What was the world's first general-purpose electronic digital computer? When was it made operational?
   a. _____
   b. _____

9. The first commercial computer, delivered in 1951: _____

10. What are the technological developments that mark the first, second, third, and fourth generations of the computer age?
    a. _____
    b. _____
    c. _____
    d. _____

11. What was the principal form of technology used for primary storage for the first two decades of the computer age? _____

12. What was introduced in 1957 as a faster, more compact method than punched cards and as a supplement to primary storage? _____

13. What was the first high-level language to receive widespread acceptance? _____

14. What was the first high-level language used for business programming? _____

15. What is the nonmetallic substance used in an integrated circuit? _____

16. What computer development generated the Nobel prize for its inventors? _____

17. Who founded Apple Computer?

    a. _____

    b. _____

18. What is the general-purpose processor-on-a-chip of the fourth generation? _____

19. What was the first personal computer called?
    _____

20. Who invented the first personal computer?
    _____

Test Answers

Multiple Choice

1. c	6. d	11. c	16. a
2. a	7. a	12. c	17. b
3. a	8. c	13. b	18. d
4. d	9. b	14. c	19. d
5. c	10. b	15. d	20. a

True/False

1. T	6. F	11. T	16. F
2. T	7. T	12. T	17. T
3. F	8. T	13. T	18. T
4. F	9. F	14. F	19. T
5. F	10. F	15. T	20. F

Fill-In

1. a. input
   b. processor
   c. control unit
   d. storage
   e. output
2. transistor
3. a. The Countess of Lovelace
   b. Charles Babbage
4. Thomas J. Watson, Sr.
5. Hollerith's tabulating machine
6. a. John Atanasoff
   b. Clifford Berry
   c. the ABC
7. a. Howard Aiken
   b. the Harvard Mark I
8. a. ENIAC
   b. 1946
9. UNIVAC I
10. a. vacuum tube
    b. transistor
    c. integrated circuit
    d. microprocessor
11. magnetic core
12. magnetic tape
13. FORTRAN
14. COBOL
15. silicon
16. transistor
17. a. Steve Jobs
    b. Steve Wozniak
18. the microprocessor
19. MITS Altair
20. Ed Roberts

Data can be represented in the computer in one of two basic ways: as **numeric data** or as **alphanumeric data.** The internal representation of alphanumeric data—letters, digits, special characters—was discussed in Chapter 2. Recall that alphanumeric data may be represented using various codes; ASCII is a common code. Alphanumeric data, even if all digits, cannot be used for arithmetic operations. Data used for arithmetic calculations must be stored numerically.

Data stored numerically can be represented as the binary equivalent of the decimal value with which we are familiar. That is, values such as 1050, 43218, and 3 that we input to the computer will be converted to the binary number system. In this appendix we shall study the binary number system (base 2) and two related systems, octal (base 8) and hexadecimal (base 16).

Appendix C

Number Systems

LEARNING OBJECTIVES

- Understand number bases, particularly bases 2, 8, and 16
- Know how to convert whole numbers among bases 10, 2, 8, and 16

NUMBER BASES
- Base 2: The Binary Number System
- Base 8: The Octal Number System
- Base 16: The Hexadecimal Number System

CONVERSIONS BETWEEN NUMBER BASES
- To Base 10 from Bases 2, 8, and 16
- From Base 10 to Bases 2, 8, and 16
- To Base 2 from Bases 8 and 16
- From Base 2 to Bases 8 and 16

Number Bases

A number base is a specific collection of symbols on which a number system can be built. The number base familiar to us is base 10, upon which the **decimal** number system is built. There are ten symbols—0 through 9—used in the decimal system.

Since society uses base 10, that is the number base most of us understand and can use easily. It would theoretically be possible, however, for all of us to learn to use a different number system. This number system could contain a different number of symbols and perhaps even symbols that are unfamiliar.

Base 2: The Binary Number System

Base 2 has exactly two symbols: 0 and 1. All numbers in the binary system must be formed using these two symbols. As you can see in column 2 of Table C-1, this means that numbers in the binary system become long quickly; the number 1000 in base 2 is equivalent to 8 in base 10. (When different number bases are being discussed, it is common practice to use the number base as a subscript. In this case we could say $1000_2 = 8_{10}$.) If you were to continue counting in base 2, you would soon see that the binary numbers were very long and unwieldy. The number 5000_{10} is equal to 10011100010000_2.

Table C-1	Number bases 10, 2, 8, 16: first values		
Base 10	Base 2	Base 8	Base 16
(Decimal)	(Binary)	(Octal)	(Hexadecimal)
0	0000	0	0
1	0001	1	1
2	0010	2	2
3	0011	3	3
4	0100	4	4
5	0101	5	5
6	0110	6	6
7	0111	7	7
8	1000	10	8
9	1001	11	9
10	1010	12	A
11	1011	13	B
12	1100	14	C
13	1101	15	D
14	1110	16	E
15	1111	17	F
16	10000	20	10

The size and sameness—all those zeros and ones—of binary numbers make them subject to frequent error when they are being manipulated by humans. To improve both convenience and accuracy, it is common to express the values represented by binary numbers in the more concise octal and hexadecimal number bases.

Base 8: The Octal Number System

The **octal** number system uses exactly eight symbols: 0, 1, 2, 3, 4, 5, 6, and 7. Base 8 is a convenient shorthand for base 2 numbers because 8 is a power of 2: $2^3 = 8$. As you will see when we discuss conversions, one octal digit is the equivalent of exactly three binary digits. The use of octal (or hexadecimal) as a shorthand for binary is common in printed output of main storage and, in some cases, in programming.

Look at the column of octal numbers in Table C-1. Notice that, since 7 is the last symbol in base 8, the following number is 10. In fact, we can count right through the next seven numbers in the usual manner, as long as we end with 17. Note, however, that 17_8 is pronounced "one-seven," not "seventeen." The octal number 17 is followed by 20 through 27, and so on. The last double-digit number is 77, which is followed by 100. Although it takes a little practice, you can see that it would be easy to learn to count in base 8. However, hexadecimal, or base 16, is not quite as easy.

Base 16: The Hexadecimal Number System

The **hexadecimal** number system uses exactly 16 symbols. As we have just seen, base 10 uses the familiar digits 0 through 9, and bases 2 and 8 use a subset of those symbols. Base 16, however, needs those ten symbols (0 through 9) and six more. The six additional symbols used in the hexadecimal number system are the letters A through F. So the base 16 symbols are: 0, 1, 2, 3, 4, 5, 6, 7, 8, 9, A, B, C, D, E, and F. It takes some adjusting to think of A or D as a digit instead of a letter. It also takes a little time to become accustomed to numbers such as 6A2F or even ACE. Both of these examples are legitimate numbers in hexadecimal.

As you become familiar with hexadecimal, consider the matter of counting. Counting sounds simple enough, but it can be confusing in an unfamiliar number base with new symbols. The process is the same as counting in base 10, but most of us learned to count when we were too young to think about the process itself. Quickly—what number follows 24CD? The answer is 24CE. We increased the rightmost digit by one—D to E—just as you would have in the more obvious case of 6142 to 6143. What is the number just before 1000_{16}? The answer is FFF_{16}; the last symbol (F) is a triple-digit number. Compare this with 999_{10}, which precedes 1000_{10}; 9 is the last symbol in base 10. As a familiarization exercise, try counting from 1 to 100 in base 16. Remember to use A through F as the second symbol in the teens, twenties, and so forth (... 27, 28, 29, 2A, 2B, 2C, 2D, 2E, 2F, 30, and so on).

Conversions Between Number Bases

It is sometimes convenient to use a number in a base different from the base currently being used—that is, to change the number from one base to another. Many programmers can nimbly convert a number from one base to another, among bases 10, 2, 8, and 16. We shall consider these conversion techniques now. Table C-2 summarizes the methods.

Table C-2 Summary conversion chart

From Base	To Base 2	To Base 8	To Base 16	To Base 10
2	—	Group binary digits by 3, convert	Group binary digits by 4, convert	Expand number and convert base 2 digits to base 10
8	Convert each octal digit to 3 binary digits	—	Convert to base 2, then to base 16	Expand number and convert base 8 digits to base 10
16	Convert each hexadecimal digit to 4 binary digits	Convert to base 2, then to base 8	—	Expand number and convert base 16 digits to base 10
10	Divide number repeatedly by 2; use remainders as answer	Divide number repeatedly by 8; use remainders as answer	Divide number repeatedly by 16; use remainders as answer	—

To Base 10 from Bases 2, 8, and 16

We present these conversions together because the technique is the same for all three.

Let us begin with the concept of positional notation. **Positional notation** means that the value of a digit in a number depends not only on its own intrinsic value but also on its location in the number. Given the number 2363, we know that the appearance of the digit 3 represents two different values, 300 and 3. Table C-3 shows the names of the relative positions.

Using these positional values, the number 2363 is understood to mean

```
2000
 300
  60
   3
2363
```

This number can also be expressed as:

$$(2 \times 1000) + (3 \times 100) + (6 \times 10) + 3$$

Appendix C ▼ Number Systems

Table C-3	Digit positions			
Digit	4th	3rd	2nd	1st (rightmost)
Position	Thousand	Hundred	Ten	Unit

We can express this expanded version of the number another way, using powers of 10. Note that $10^0=1$.

$$2363 = (2 \times 10^3) + (3 \times 10^2) + (6 \times 10^1) + (3 \times 10^0)$$

Once you understand the expanded notation, the rest is easy: You expand the number as we just did in base 10, but use the appropriate base of the number. For example, follow the steps to convert 61732_8 to base 10:

1. Expand the number, using 8 as the base:

$$61732 = (6 \times 8^4) + (1 \times 8^3) + (7 \times 8^2) + (3 \times 8^1) + (2 \times 8^0)$$

2. Complete the arithmetic:

$$61732 = (6 \times 4096) + (1 \times 512) + (7 \times 64) + (3 \times 8) + (2 \times 1)$$
$$= 24576 + 512 + 448 + 24 + 2$$

3. Answer: $61732_8 = 25562_{10}$

The same expand-and-convert technique can be used to convert from base 2 or base 16 to base 10. As you consider the following two examples, use Table C-1 to make the conversions. (For example, A in base 16 converts to 10 in base 10.)

Convert $C14A_{16}$ to base 10:

$$C14A_{16} = (12 \times 16^3) + (1 \times 16^2) + (4 \times 16^1) + (10 \times 16^0)$$
$$= (12 \times 4096) + (1 \times 256) + (4 \times 16) + (10 \times 1)$$
$$= 49482$$

So $C14A_{16} = 49482_{10}$.

Convert 100111_2 to base 10:

$$100111_2 = (1 \times 2^5) + (1 \times 2^2) + (1 \times 2^1) + (1 \times 2^0)$$
$$= 39$$

So $100111_2 = 39_{10}$.

From Base 10 to Bases 2, 8, and 16

These conversions use a simpler process but more complicated arithmetic. The process, often called the *remainder method*, is basically a series of repeated divisions by the number of the base to which you are converting. You begin by using the number to be converted as the dividend; succeeding dividends are the quotients of the previous division. The converted number is the combined remainders accumulated from the divisions. There are two points to remember:

1. Keep dividing until you reach a zero quotient.
2. Use the remainders in reverse order.

Consider converting 6954_{10} to base 8:

```
8|6954
8|869    2
8|108    5
8|13     4
8|1      5
0        1
```

Placing the remainders backward, $6954_{10} = 15452_8$.

Now use the same technique to convert 4823_{10} to base 16:

```
16|4823
16|3017
16|18    13 (= D)
16|1     2
0        1
```

The remainder 13 is equivalent to D in base 16. So $4823_{10} = 12D7_{16}$.

Convert 49_{10} to base 2:

```
2|49
2|24   1
2|12   0
2|6    0
2|3    0
2|1    1
0      1
```

Again placing the remainders in reverse order, $49_{10} = 110001_2$.

To Base 2 from Bases 8 and 16

To convert a number to base 2 from base 8 or base 16, convert each digit separately to three or four binary digits, respectively. Use Table C-1 to make the conversion. Leading zeros—zeros added to the front of the number—may be needed in each grouping of digits to fill out each to three or four digits.

Convert 4732_8 to base 2, converting each octal digit to a set of three binary digits:

4	7	3	2
100	111	011	010

So $4732_8 = 100111011010_2$. Notice that leading zeros were sometimes needed to make three binary digits from an octal digit: for octal digit 3, 11 became 011 and, for octal digit 2, 10 became 010.

Now convert $A046B_{16}$ to base 2, this time converting each hexadecimal digit to four binary digits:

A	0	4	6	B
1010	0000	0100	0110	1011

Thus $A046B_{16} = 10100000010001101011_2$.

From Base 2 to Bases 8 and 16

To convert a number from base 2 to base 8 or base 16, group the binary digits from the right in groups of three or four, respectively. Again use Table C-1 to help you make the conversion to the new base.

Convert 111101001011_2 to base 8 and base 16:

In the base 8 conversion, group the digits three at a time, starting on the right:

111   101   001   011
7   5   1   3

So $111101001011_2 = 7513_8$.

For the conversion to base 16, group the digits four at a time, starting on the right:

1111   0100   1011
F   4   B
$111101001011_2 = F4B_{16}$.

Sometimes the number of digits in a binary number is not exactly divisible by 3 or 4. You may, for example, start grouping the digits three at a time and finish with one or two "extra" digits on the left side of the number. In this case just add as many zeros as you need to the front of the binary number.

Consider converting 1010_2 to base 8. By adding two zeros to the front of the number to make it 001010_2, we now have six digits, which can be conveniently grouped three at a time:

001   010
1   2

So $1010_2 = 12_8$.

Glossary

A

Access arm A mechanical device that can access all the tracks of one cylinder in a disk storage unit.

Access time The time needed to access data directly on disk, consisting of seek time, head switching, and rotational delay.

Accumulator A register that collects the results of computations.

Acoustic coupler A modem that connects to a telephone receiver rather than directly to a telephone line.

Acquisition by purchase Buying an entire system for use by the organization, as opposed to designing a new system.

Active badge A badge that, imbedded with a computer chip, signals the wearer's location by sending out infrared signals, which are read by computers distributed throughout a building.

Active cell The cell currently available for use in a spreadsheet. Also called the current cell.

Ada A structured programming language, named for Countess Ada Lovelace, that encourages modular program design.

Address A number used to designate a location in memory.

Address register A register used to help locate where instructions and data are stored in memory.

ALGOL (ALGOrithmic Language) A language, developed primarily for scientific programming, that has limited file-processing capabilities.

Alphanumeric data Letters, digits, and special characters such as punctuation marks.

ALU See *Arithmetic logic unit*.

America Online (AOL) A major information utility that offers a variety of services.

Amplitude The height of the carrier wave in analog transmission. Amplitude indicates the strength of the signal.

Amplitude modulation A change of the amplitude of the carrier wave in analog data transmission to represent either the 0 bit or the 1 bit.

Analog transmission The transmission of data as a continuous electric signal in the form of a wave.

Analytical engine A historically significant machine designed by Charles Babbage that embodied the key characteristics of modern computers.

Analytical graphics Traditional line graphs, bar charts, and pie charts; used to illustrate and analyze data.

ANSI-COBOL A version of COBOL standardized in 1974 by the American National Standards Institute (ANSI).

ANSI American National Standards Institute.

Antivirus A computer program that stops the spread of a virus. Also called a vaccine.

AOL See *America Online*.

APL (A Programming Language) A language especially used for table handling.

Applications software Programs designed to perform specific tasks and functions.

Arithmetic/logic unit (ALU) Part of the central processing unit, the electronic circuitry of the ALU executes all arithmetic and logical operations.

Arithmetic operations Mathematical calculations the ALU performs on data.

Artificial intelligence The field of study that explores computer involvement in tasks requiring intelligence, imagination, and intuition.

ASCII (American Standard Code for Information Interchange) A coding scheme using 7-bit characters to represent data characters. A variation of the code, called ASCII-8, uses 8 bits per character.

Assembler program A translator program used to convert assembly language programs to machine language.

Assembly language A second-generation language that uses abbreviations for instructions, as opposed to only numbers. Also called symbolic language.

Asynchronous transmission Data transmission in which data is sent in groups of bits, with each group of bits preceded by a start signal and ended with a stop signal. Also called start/stop transmission.

Atanasoff-Berry Computer (ABC) The first electronic digital computer, designed by John V. Atanasoff and Clifford Berry, in the late 1930s.

ATM See *Automated teller machine*.

Attribute Regarding object-oriented programming, a fact related to an object.

Audio-response unit See *Voice synthesizer*.

Audit trail A method of tracing data from the output back to the source documents.

Automated teller machine (ATM) Input/output device connected to a computer used by bank customers for financial transactions.

Automatic reformatting In word processing, automatic adjustment of text to accommodate changes such as margin width.

Auxiliary storage Another name for secondary storage, which is storage for data and programs. Auxiliary storage is most often on disk.

Axis A reference line of a graph. The horizontal axis is the x-axis. The vertical axis is the y-axis.

B

Background In large computers, the memory area for running programs with low priorities. Contrast with foreground.

Backspace key The key used to delete text characters to the left of the cursor.

Backup system A method of storing data in more than one place to protect it from damage or loss.

Bandwidth The number of frequencies that can fit on one communications line or link at the same time, or the capacity of the link.

Bar code A standardized pattern of vertical marks that represents the Universal Product Code (UPC) that identifies a product.

Bar Code Reader A stationary photoelectric scanner that inputs bar codes by means of reflected light.

Bar graph A graph made up of filled-in columns or rows that represent the change of data over time.

BASIC (Beginner's All-purpose Symbolic Instruction Code) A high-level programming language that is easy to learn and use. Originally for beginners but now also used in business and by personal computer users any place.

Batch processing A data processing technique in which transactions are collected into groups, or batches, for processing.

BBS See *Bulletin board system*.

Binary Regarding number systems, the binary number system uses exactly two symbols, the digits 0 and 1.

Binary system A system in which data is represented by combinations of 0s and 1s, which correspond to the two states off and on.

Biometrics The science of measuring individual body characteristics; used in some security systems.

Bit A binary digit.

Block copy command In word processing, the command used to copy a block of text into a new location, leaving the text in its original location as well.

Block delete command In word processing, the command used to erase a block of text.

Block move command In word processing, the command used to remove a block of text from one location in a document and place it elsewhere.

Boldface Printed characters in darker type than the surrounding characters.

Bomb An application that sabotages a computer by triggering damage—usually at a later date. Also called a logic bomb.

Booting Loading the operating system into memory.

bpi See *Bytes per inch*.

Branch In a flowchart, the connection leading from the decision to one of two possible responses. Also called a path.

Bridge A device that recognizes and transmits messages to be sent to other similar networks.

Browser A program used to access the Internet.

Bulletin board system (BBS) Telephone-linked personal computers that provide public-access message systems.

Bus or bus line Electronic pathway for data travel among the parts of a computer. Also called data bus.

Business-quality graphics See *Presentation graphics*.

Bus network A network that has a single line to which each device is attached.

Byte Strings of bits (usually 8) used to represent one data character—a letter, digit, or special character.

Bytes per inch (bpi) An expression of the amount (density) of data stored on magnetic tape.

C

C A sophisticated programming language invented by Bell Labs in 1974.

C++ An object-oriented programming language; a version of C.

Cache A relatively small amount of very fast memory that stores data and instructions that are used frequently, resulting in improved processing speeds.

CAD/CAM See *Computer-aided design/computer-aided manufacturing*.

Candidates In systems analysis and design, alternative plans offered in the preliminary design phase of a project.

Carrier sense multiple access with collision detection (CSMA/CD) The line control method used by Ethernet. Each node has access to the communications line and can transmit if it hears no communication on the line. If two stations transmit simultaneously, they will wait and retry their transmissions.

Carrier wave An analog signal used in the transmission of electric signals.

CASE See *Computer-aided software engineering*.

Cathode ray tube (CRT) The most common type of computer screen.

CD-ROM See *Compact disk read-only memory*.

Cell The intersection of a row and a column in a spreadsheet. Entries in a spreadsheet are stored in individual cells.

Cell address In a spreadsheet, the column and row coordinates of a cell. Also called the cell reference.

Cell contents The label, value, formula, or function contained in a spreadsheet cell.

Cell reference In a spreadsheet, the column and row coordinates of a cell. Also called the cell address.

Centering Placing a line of text midway between the left and right margins.

Centralized data processing Keeping hardware, software, storage, and computer access in one location.

Central processing unit (CPU) Electronic circuitry that executes stored program instructions. It consists of two parts: the control unit and the arithmetic/logic unit. The CPU processes raw data into meaningful, useful information.

CGA Stands for color graphics adaptor. An early color screen standard with 320x200 pixels.

Change agent The role of the systems analyst in overcoming resistance to change within an organization.

Character A letter, number, or special character such as $.

Characters per inch (cpi) An expression of the amount (density) of data stored on magnetic tape.

Chief Information Officer (CIO) Manager of an MIS department.

Circuit One or more conductors through which electricity flows.

CISC See *Complex instruction set computer.*

Class Regarding object-oriented programming, object class contains the characteristics that are unique to that class.

Client 1. An individual or organization contracting for systems analysis. 2. In a client/server network, a program on the personal computer that allows that node to communicate with the server.

Client/server A network setup that involves a server computer, which controls the network, and clients, other computers that access the network and its services. In particular, the server does some processing, sending the client only the portion of the file it needs or possibly just the processed results. Contrast with *File server.*

Clip art Illustrations already produced by professional artists for public use. Computerized clip art is stored on disk and can be used to enhance a graph or document.

Clock A component of the CPU that produces pulses at a fixed rate to synchronize all computer operations.

Clone A personal computer that can run software designed for either IBM or Macintosh personal computers.

Clustered-bar graph Bar graph comparing several different but related sets of data.

CMOS See *Complementary metal oxide semiconductor.*

Coaxial cable Bundles of insulated wires within a shielded enclosure. Coaxial cable can be laid underground or undersea.

COBOL (COmmon Business-Oriented Language) An English-like programming language used primarily for business applications.

CODASYL (COnference of DAta SYstem Languages) The organization of government and industrial representatives that introduced the programming language COBOL.

Cohesion Regarding structured programming, a measure of the inner "strength" of a program module.

Cold site An environmentally suitable empty shell in which a company can install its own computer system.

Collaborative software See *Groupware.*

Collision The problem that occurs when two records have the same disk address.

COM See *Computer output microfilm.*

Command A name that invokes the correct program or program segment.

Compact disk read-only memory (CD-ROM) Optical data storage technology using disk formats identical to audio compact disks.

Compare operation An operation in which the computer compares two data items and performs alternative operations based on the comparison.

Compiler A translator that converts the symbolic statements of a high-level language into computer-executable machine language.

Complementary metal oxide semiconductor (CMOS) A semiconductor device that does not require a large amount of power to operate. The CMOS is often found in devices that require low power consumption, such as portable computers.

Complex instruction set computer (CISC) A CPU design that contains a large number of instructions of varying kinds, some of which are rarely used. Contrast with *RISC.*

CompuServe A major information utility that offers a variety of services.

Computer A machine that accepts data (input) and processes it into useful information (output). A computer system requires four main aspects of data handling—input, processing, output, and storage.

Computer-aided design/computer-aided manufacturing (CAD/CAM) The use of computers to create two- and three-dimensional pictures of products to be manufactured.

Computer-aided software engineering (CASE) Software that provides an automated means of designing systems.

Computer conferencing A method of sending, receiving, and storing typed messages within a network of users.

Computer Fraud and Abuse Act A law passed by Congress in 1984 to fight computer crime.

Computer literacy The awareness and knowledge of, and the capacity to interact with, computers.

Computer Matching and Privacy Protection Act Legislation that prevents the government from comparing certain records in an attempt to find a match.

Computer operator A person who monitors and runs the computer equipment in a large system.

Computer output microfilm (COM) Computer output produced as very small images on sheets or rolls of film.

Computer programmer A person who designs, writes, tests, and implements programs.

Computer system A system that has a computer as one of its components.

Computing Services A department that manages computer resources for an organization. Also called Information Services or Management Information Systems.

Concurrent licensing A software licensing agreement in which a customer is permitted to use only a limited number of copies of a software product simultaneously.

Concurrently With reference to the execution of computer instructions, in the same time frame but not simultaneously. See also *Multiprogramming.*

Conditional replace A word processing function that asks the user whether to replace text each time the program finds a particular item.

Connector A symbol used in flowcharting to connect paths.

Consortium A joint venture to support a complete computer facility to be used in an emergency.

Continuous word system A speech recognition system that can understand sustained speech so users can speak normally.

Control structure A pattern for controlling the flow of logic in a program. The three basic control structures are sequence, selection, and iteration.

Control unit The circuitry that directs and coordinates the entire computer system in executing stored program instructions. Part of the central processing unit.

Coordinating In systems analysis, orchestrating the process of analyzing and planning a new system by pulling together the various individuals, schedules, and tasks that contribute to the analysis.

Copyright Act Legislation that makes software piracy a felony, with possible penalties of five years of jail time and $250,000 in fines.

Copy protection Software or hardware that makes it difficult or impossible to make unauthorized copies of software.

Copyrighted software Software that costs money and must not be copied without permission from the manufacturer.

Coupling Regarding structured programming, a measure of the strength of the relationship between program modules.

CPU See *Central processing unit*.

CRT See *Cathode ray tube*.

CSMA/CD See *Carrier sense multiple access with collision detection*.

Current cell The cell currently available for use in a spreadsheet. Also called the active cell.

Current drive The disk drive currently being used by the computer system. Also called the default drive.

Cursor An indicator on the screen; it shows where the next user-computer interaction will be. Also called a pointer.

Cursor movement keys Keys on the computer keyboard that allow the user to move the cursor on the screen.

Custom software Software that is specifically tailored to user needs.

Cylinder A set of tracks on a magnetic disk, one from each platter, vertically aligned. These tracks can be accessed by one positioning of the access arm.

Cylinder method A method of organizing data on a magnetic disk. This method organizes data vertically, which minimizes seek time.

D

DASD See *Direct-access storage device*.

DAT See *Digital audio tape*.

Data Raw input to be processed by a computer.

Database An organized collection of interrelated files stored together with minimum redundancy. Specific data items can be retrieved for various applications.

Database management system (DBMS) A set of programs that creates, manages, protects, and provides access to the database.

Data collection device A device that allows direct data entry in such places as factories and warehouses.

Data communications The process of exchanging data over communications facilities.

Data communications systems Computer systems that transmit data over communications lines, such as public telephone lines or private network cables.

Data Encryption Standard (DES) The standardized public key by which senders and receivers can scramble and unscramble messages sent over data communications equipment.

Data entry operator A person who enters data for computer processing.

Data flow diagram (DFD) A diagram that shows the flow of data through an organization.

Data item Data in a relational database table.

Data point Each dot or symbol on a line graph. Each data point represents a value.

Data transfer The transfer of data between memory and the place on the disk track—from memory to the track if writing, from the track to memory if reading

Data transfer rate The speed with which data can be transferred to or from a disk and a computer.

Date field In database file structure, a field that is used for dates and is automatically limited to eight characters, including slashes that separate the month, day, and year.

DBMS See *Database management system*.

DDP See *Distributed data processing*.

Debugging The process of detecting, locating, and correcting mistakes in a program.

Decision box The standard diamond-shaped box used in flowcharting; it indicates a decision.

Decision logic table See *Decision table*.

Decision support system (DSS) A computer system that supports managers in nonroutine decision-making tasks. A DSS involves a model, a mathematical representation of a real-life situation.

Decision table A standard table of the logical decisions that must be made regarding potential conditions in a given system. Also called a decision logic table.

Default drive The disk drive currently being used. Also called the current drive. This is the disk drive to which commands refer in the absence of any specified drive.

Default settings The settings automatically used by a program unless the user specifies otherwise, thus overriding them.

Delete key The key used to delete the text character at the cursor location or a text block that has been selected or marked.

Demodulation The reconstruction of the original digital message after analog transmission.

Density The amount of data stored on magnetic tape; expressed in number of characters per inch (cpi) or bytes per inch (bpi).

Dependent variable Output of a computerized model, so called because it depends on the inputs.

DES See *Data Encryption Standard*.

Desk-checking A programming phase in which a programmer mentally checks the logic of a program to ensure that it is error-free and workable.

Desktop publishing Using a personal computer, special software, and a laser printer to produce very high-quality documents that combine text and graphics. Also called electronic publishing.

Desktop publishing program A software package for designing and producing professional-looking documents.

Also called a page composition program or page makeup program.

Detail design A systems design subphase in which the system is planned in detail, including the details of output, input, files and databases, processing, and controls and backup.

DFD See *Data flow diagram*.

Diagnostics Error messages provided by the compiler as it translates a program. Diagnostics inform the user of programming language syntax errors.

Difference engine A historically significant machine designed by Charles Babbage to solve polynomial equations by calculating the successive differences between them. See also his other machine, *Analytical engine*.

Digital audio tape (DAT) A high-capacity tape that records data using a method called helical scan recording, which places the data in diagonal bands that run across the tape rather than down its length.

Digital transmission The transmission of data as distinct on or off pulses.

Digitizing tablet A graphics input device that allows the user to create images. It has a special stylus that can be used to draw or trace images, which are then converted to digital data that can be processed by the computer.

Direct access Immediate access to a record in secondary storage, usually on disk.

Direct-access storage device (DASD) A storage device, usually disk, in which a record can be accessed directly.

Direct-connect modem A modem connected directly to the telephone line.

Direct conversion A system conversion in which the user simply stops using the old system and starts using the new one.

Direct file organization An arrangement of records so each is individually accessible.

Direct file processing Processing that allows the user to access a record directly by using a record key.

Disaster recovery plan Guidelines for restoring computer processing operations if they are halted by major damage or destruction.

Discrete word system A speech recognition system limited to understanding isolated words.

Disk drive A machine that allows data to be read from a disk or written on a disk.

Diskette A single disk, made of flexible mylar, on which data is recorded as magnetic spots. Older diskettes are 5 ¼ inches in diameter, but newer computers use the 3 ½-inch diskette, which has a hard plastic jacket, a convenient size, and greater storage capacity.

Disk pack A stack of magnetic disks assembled together.

Displayed value The calculated result of a formula or function in a spreadsheet cell.

Distributed data processing (DDP) A data processing system in which processing is decentralized, with the computers and storage devices in dispersed locations.

Documentation 1. Related to a program, a detailed written description of the programming cycle and specific facts about the program. 2. The instruction manual for packaged soft-

ware. 3. In systems analysis and design, the written records of all phases of the systems development life cycle.

Dot-matrix printer A printer that constructs a character by activating a matrix of pins to produce the shape of a character on paper.

Download In a networking environment, to receive data files from another computer, probably a larger computer or a host computer. Contrast with *Upload*.

DRAM See *Dynamic random-access memory*.

DSS See *Decision support system*.

Dynamic random-access memory (DRAM) Memory chips that are periodically regenerated, allowing the chips to retain the stored data.

E

EDI See *Electronic data interchange*.

EFT See *Electronic fund transfer*.

EGA Stands for enhanced graphics adaptor. A color screen standard with 640 × 350 pixels.

Electronic data interchange (EDI) A set of standards by which companies can electronically exchange common business forms such as invoices and purchase orders.

Electronic fund transfer (EFT) Paying for goods and services by transferring funds electronically.

Electronic mail (e-mail) Sending messages from one terminal or computer to another.

Electronic publishing Using a personal computer, special software, and a laser printer to produce very high-quality documents that combine text and graphics. Also called desktop publishing.

Electronic spreadsheet A computerized worksheet used to organize data into rows and columns for analysis.

E-mail See *Electronic mail*.

Encapsulation Regarding object-oriented programming, the containment of both data and its related instructions in the object.

Encryption The process of encoding data to be transmitted via communications links, so that the contents of the data is protected from unauthorized people.

End of file The point in a program or module where all records on a file have been read.

End-user The person who buys and uses computer software or who has contact with computers.

ENIAC (Electronic Numerical Integrator and Computer) The first general-purpose electronic computer, which was built by Dr. John Mauchly and J. Presper Eckert, Jr., and was first operational in 1946.

Entry point Regarding structured programming, the point in a module where control is transferred. Each module has only one entry point.

Equal-to condition (=) A logical operation in which the computer compares two numbers to determine equality.

Erase head The head in a magnetic tape unit that erases any data previously recorded on the tape before recording new data.

Ergonomics The study of human factors related to computers.

ESS See *Executive support system*.

Ethernet A popular type of local area network that uses a bus topology.

E-time The execution portion of the machine cycle; E-time includes the execute and store operations.

Event-driven Refers to multiprogramming; programs share resources based on events that take place in the programs. One program is allowed to use a particular resource (such as the central processing unit) to complete a certain activity (event) before relinquishing the resource to another program.

Executive support system (ESS) A decision support system for senior-level executives who make decisions that affect an entire company.

Exit point Regarding structured programming, the point in a module from which control is transferred. Each module has only one exit point.

Expansion slots The slots inside a computer that allow a user to insert additional circuit boards.

Expert shell Software having the basic structure to find answers to questions; the questions themselves can be added by the user.

Expert system Software that presents the computer as an expert on some topic.

Exploded pie chart A pie chart with a "slice" that is separated from the rest of the chart.

F

Facsimile technology (fax) The use of computer technology to send digitized graphics, charts, and text from one facsimile machine to another.

Fair Credit Reporting Act Legislation that allows individuals access to credit records and gives them the right to challenge them.

Fax See *Facsimile technology*.

Fax modem A modem that allows the user to transmit and receive faxes without interrupting other applications programs, as well as the usual modem functions.

Feasibility study The first phase of systems analysis, in which planners determine if and how a project should proceed. Also called a system survey or a preliminary investigation.

Federal Privacy Act Legislation stipulating that government agencies cannot keep secret personnel files and that individuals can have access to all government files, as well as those of private firms contracting with the government, that contain information about them.

Fiber optics Technology that uses glass fibers that can transmit light as a communications link to send data.

Field A set of related characters.

Field name In a database, the unique name describing the data in a field.

Field robot A robot that is used on location to inspect nuclear plants, dispose of bombs, clean up chemical spills, and so forth.

Field type In a database, a category describing a field and determined by the kind of data the field will accept. Common field types are character, numeric, date, and logical.

Field width In a database or spreadsheet, the maximum number of characters that can be contained in a field.

Fifth generation A term that has evolved to encompass several research fields related to computer intelligence: artificial intelligence, expert systems, and natural language.

File 1. A repository of data. 2. A collection of related records. 3. In word processing, a document created on a computer.

File server A network relationship in which an entire file is sent to a node, which then does its own processing. Also, the network computer exclusively dedicated to making files available on a network. Contrast with *Client/server*.

File transfer software In a network, software used to transfer files from one computer to another. See also *Download* and *Upload*.

Find command In word processing, the ability to locate certain text in the document. Also called the Search command.

Find-and-replace function A word processing function that finds and changes each instance of a repeated item.

Firewall As part of computer security, a computer dedicated to screening access to a network from outside the network.

Flash memory Nonvolatile memory chips.

Floppy disk A flexible magnetic diskette on which data is recorded as magnetic spots.

Flowchart The pictorial representation of an orderly step-by-step solution to a problem.

Font A complete set of characters in a particular size, typeface, weight, and style.

Font library A variety of type fonts stored on disk. Also called soft fonts.

Footer In word processing, the ability to place the same line, with possible variations such as page number, on the bottom of each page.

Footnote In word processing, the ability to make a reference in a text document to a note at the bottom of the page.

Foreground In large computers, an area in memory for programs that have a high priority. Contrast with *Background*.

Format 1. The process of preparing a disk to accept data. 2. The specifications that determine the way a document or worksheet appears on the screen or printer.

Formula In a spreadsheet, an instruction placed in a cell to calculate a value.

FORTH A programming language designed for real-time control tasks.

FORTRAN (FORmula TRANslator) The first high-level programming language, introduced in 1954 by IBM; it is scientifically oriented.

Fourth-generation language A nonprocedural language. Also called a 4GL, or a very high-level language.

Freedom of Information Act Legislation that allows citizens access to personal data gathered by federal agencies.

Freeware Software that is free; said to be in the public domain.

Frequency The number of times an analog signal repeats during a specific time interval.

Frequency modulation The alteration of the carrier wave frequency to represent 0s and 1s.

Front-end processor A communications control unit designed to relieve the central computer of some communications tasks.

Full-duplex transmission Data transmission in both directions at once.

Full justification In word processing, making both the left and right margins even.

Function A built-in spreadsheet formula.

Function keys Special keys programmed to execute commonly used commands; the commands vary according to the software being used.

G

Gantt chart A bar chart commonly used to depict schedule deadlines and milestones, especially in systems analysis and design.

Gateway A collection of hardware and software resources to connect two dissimilar networks, allowing computers in one network to communicate with those in the other.

GB See *Gigabyte*.

General-purpose register A register used for several functions, such as arithmetic and addressing purposes.

Generic operating system An operating system that works with more than one manufacturer's computer system. Also called a portable operating system.

Gigabyte (GB) One billion bytes.

Grammar/style program A word processing program that identifies common grammatical and writing errors.

Graphical user interface (GUI) An image-based computer interface in which the user sends directions to the operating system by selecting icons from a menu or manipulating icons on the screen by using a pointing device such as a mouse.

Graphics Pictures or graphs.

Graphics adapter board A circuit board that enables a computer to display pictures or graphs as well as text. Also called a graphics card.

Graphics card See *Graphics adapter board*.

Greater-than condition (>) A comparison operation that determines if one value is greater than another.

Groupware Software that lets a group of people develop or track a project together, usually including electronic mail, networking, and database technology. Also called collaborative software.

GUI See *Graphical user interface*.

H

Hacker 1. An enthusiastic, largely self-taught computer user. 2. Currently, a person who gains access to computer systems illegally, usually from a personal computer.

Half-duplex transmission Data transmission in either direction, but only one way at a time.

Halftone As used in desktop publishing, a reproduction of a black-and-white photograph; it is made up of tiny dots.

Hard copy Printed paper output.

Hard disk A metal platter coated with magnetic oxide that can be magnetized to represent data. Hard disks are usually in a pack, often in a sealed module.

Hardware The computer and its associated equipment.

Hashing Applying a formula to a record key to yield a number that represents a disk address. Also called randomizing.

Hayes compatible Modems that use the standard originated for Hayes brand modems.

Head crash The result of a read/write head touching a disk surface and causing all data to be destroyed.

Header In word processing, the ability to place the same line, with possible variations such as page number, on the top of each page.

Head switching In reading or writing a disk, activation of a particular read/write head over a particular track.

Helical recording Placing data on tape by placing it in tracks that run diagonally across the tape.

Hexadecimal Regarding number systems, the hexadecimal number system uses exactly 16 symbols, digits 0 through 9, and the letters *A* through *F*.

Hierarchy chart See *Structure chart*.

High-level languages English-like programming languages that are easier to use than older symbolic languages.

Host computer The central computer in a network, to which other computers, and perhaps terminals, are attached.

Hot site For use in an emergency, a fully equipped computer center with hardware, communications facilities, environmental controls, and security.

I

Icon A small picture on a computer screen; it represents a computer activity.

Imaging Using a scanner to convert a drawing, photo, or document to an electronic version that can be stored and reproduced when needed. Once scanned, text documents may be processed by optical recognition software so that the text can be manipulated.

Impact printer A printer that forms characters by physically striking the paper.

Implementation The phase of a systems analysis and design project that includes training, equipment conversion, file conversion, system conversion, auditing, evaluation, and maintenance.

Indent In word processing, widening the margin for certain text.

Independent variable Input to a computerized model, called independent because it can change.

Indexed file organization The combination of sequential and direct file organization.

Indexed file processing A method of file organization that represents a compromise between sequential and direct methods. Indexed processing stores records in the file in sequential order, but the file also contains an index of keys; the address associated with the key is then used to locate the record on the disk.

Indexed processing See *Indexed file processing*.

Inference engine Related to the field of artificial intelligence, particularly how computers learn; a process that accesses, selects, and interprets a set of rules. The inference engine applies rules to the facts to make up new facts.

Information Input data that has been processed by the computer; data that is organized, meaningful, and useful.

Information Center A company unit that offers employees computer and software training, help in getting data from other computer systems, and technical assistance.

Information Services A department that manages computer resources for an organization. Also called Computing Services or Management Information Systems.

Information utility A commercial consumer-oriented communications system, such as America Online, CompuServe, or Prodigy, that offers a variety of services.

Inheritance Regarding object-oriented programming, the property meaning that an object in a subclass automatically possesses all the characteristics of the class to which it belongs.

Initialize Set the starting values of certain storage locations, often at the beginning of program execution.

Ink-jet printer A nonimpact printer that forms output text or images by spraying ink from jet nozzles onto the paper.

Input Raw data that is put into the computer system for processing.

Input device A device that puts data in computer-understandable form and sends it to the processing unit.

Input requirements In systems design, the plan for input medium and content, and forms design.

Inquiry A user request for information from a computer.

Instance Regarding object-oriented programming, a specific occurrence of an object.

Instruction set The commands that a CPU understands and is capable of executing. Each type of CPU has a fixed group of these instructions, and each set usually differs from that understood by other CPUs.

Integrated circuit A complete electronic circuit on a small chip of silicon.

Internal font A font built into the read-only memory (a ROM chip) of a printer.

Internal modem A modem on a circuit board. An internal modem can be installed in a computer by the user.

Internal storage The electronic circuitry that temporarily holds data and program instructions needed by the CPU. Also called memory, main memory, primary memory, primary storage, and main storage.

Internet A public communications network once used primarily by businesses, governments, and academic institutions but now also used by individuals via various private access methods.

Interrupt In multiprogramming, a condition that temporarily suspends the execution of an individual program.

Interview In systems analysis, talking to anyone connected to an existing system for the purpose of data gathering.

Iteration The repetition of program instructions under certain conditions. Also called a loop.

Iteration control structure A looping mechanism.

I-time The instruction portion of the machine cycle; I-time includes the fetch and decode operations.

J

Joystick A graphics input device that allows fingertip control of figures on a CRT screen.

Justification In word processing, aligning text along left or right margins, or both.

K

K or KB See *Kilobyte*.

Kerning In word processing or desktop publishing, adjusting the space between characters to create a more attractive or readable appearance.

Key A unique identifier for a record.

Keyboard A common computer input device similar to the keyboard of a typewriter.

Kilobyte (K or KB) 1024 bytes.

Knowledge base Related to the field of artificial intelligence, particularly how computers learn; a set of facts and rules about those facts.

Knowledge-based system In an artificial intelligence environment, a system that uses a natural language to access a knowledge base.

Knowledge engineer Related to building an expert system, the person working to extract information from the human expert.

L

Label In a spreadsheet, data consisting of a string of text characters.

LAN See *Local area network*.

Laptop computer A small portable computer. Also called a notebook computer.

Laser printer A nonimpact printer that uses a light beam to transfer images to paper.

LCD See *Liquid crystal display*.

Leading In word processing or desktop publishing, the vertical spacing between lines of type.

Leading decision The loop-ending decision that occurs at the beginning of a DO-WHILE loop.

Legend In regard to a graph, text that explains the colors, shading, or symbols used to label the data points.

Less-than condition (<) A logical operation in which the computer compares values to determine if one is less than another.

Librarian A person who catalogs processed computer disks and tapes and keeps them secure.

Light pen A graphics input device that allows the user to interact directly with the computer screen.

Line graph A graph made by connecting data points with a line, particularly useful for showing trends over time.

Line spacing In word processing, the amount of space between lines of text in a document; single spacing and double spacing are common.

Link A physical data communications medium.

Link/load phase A phase that takes the machine language object module and adds necessary prewritten programs to produce output called the load module; the load module is executable.

Liquid crystal display (LCD) The flat display screen found on some laptop computers.

LISP (LISt Processing) A programming language designed to process nonnumeric data; popular for writing artificial intelligence programs.

Load module An executable version of a program.

Local area network (LAN) A network designed to share data and resources among several computers, usually personal computers in a limited geographical area, such as an office or a building.

Logical field A field used to keep track of true and false conditions.

Logical operations Comparing operations. The ALU is able to compare numbers, letters, or special characters and take alternative courses of action depending on the result of the comparison.

Logic chip A central processing unit on a chip, generally known as a microprocessor but called a logic chip when used for some special purpose, such as controlling some under-the-hood action in a car.

Logic error A flaw in the logic of a program.

Logic flowchart A flowchart that represents the flow of logic in a program.

LOGO A programming language designed for use in schools to teach problem-solving skills.

Loop The repetition of program instructions under certain conditions. Also called iteration.

M

Machine cycle Combination of I-time and E-time, the steps used by the central processing unit to execute instructions.

Machine language The lowest level of language; it represents data and instructions as 1s and 0s.

Magnetic core Flat doughnut-shaped metal used as an early memory device.

Magnetic disk An oxide-coated disk on which data is recorded as magnetic spots.

Magnetic-ink character recognition (MICR) A method of machine-reading characters made of magnetized particles. A common application is checks.

Magnetic tape A magnetic medium with an iron-oxide coating that can be magnetized. Data is stored on the tape as extremely small magnetized spots.

Magnetic tape unit A data storage unit used to record data on and retrieve data from magnetic tape.

Magneto-optical (MO) A hybrid disk that has the high-volume capacity of an optical disk but can be written over as a magnetic disk. It uses both a laser beam and a magnet to properly align magnetically sensitive metallic crystals.

Mainframe A large computer that has access to billions of characters of data and is capable of processing large amounts of data very quickly. Notably, mainframes are used by such data-heavy customers as banks, airlines, and large manufacturers.

Main memory The electronic circuitry that temporarily holds data and program instructions needed by the CPU. Also called memory, primary memory, primary storage, main storage, and internal storage.

Main storage The electronic circuitry that temporarily holds data and program instructions needed by the CPU. Also called memory, main memory, primary memory, primary storage, and internal storage.

Management Information System (MIS) A set of formal business systems designed to provide information for an organization.

Management Information Systems A department that manages computer resources for an organization. Also called Computing Services or Information Services.

Margin In word processing, the unused space on the right and left sides of a document, and on the top and bottom of a document.

Mark To define a block of text before performing block commands.

Mark I An early computer; built in 1944 by Harvard professor Howard Aiken.

Master file A semipermanent set of records.

MB See *Megabyte*.

Megabyte (MB) One million bytes. The unit often used to measure memory or storage capacity.

Megaflops One million floating-point operations per second. One measure of a computer's speed.

Megahertz (MHz) One million cycles per second. Used to express microprocessor speeds.

Memory The electronic circuitry that temporarily holds data and program instructions needed by the CPU. Also called main memory, primary memory, primary storage, main storage, and internal storage.

Memory management The process of allocating memory to programs and keeping the programs in memory separate from one another.

Memory protection In a multiprogramming system, the process of keeping a program from straying into other programs in memory.

Menu An on-screen list of choices.

Message Regarding object-oriented programming, a command telling what—not how—something is to be done, which activates the object.

Method Regarding object-oriented programming, instructions that tell the data what to do. Also called an operation.

MHz See *Megahertz*.

MICR See *Magnetic-ink character recognition*.

MICR inscriber A device that adds magnetic characters to a document, in particular, the amount of a check.

Microcomputer A relatively small and inexpensive type of computer, usually used by an individual in a home or office setting. Also called a personal computer.

Microfiche Sheets of film (4 by 6 inches) that can be used to store computer output.

Microprocessor A general-purpose processor on a chip.

Microsecond One-millionth of a second.

Micro-to-mainframe link Connection between microcomputers and mainframe computers.

Microwave transmission Line-of-sight transmission of data signals through the atmosphere from relay station to relay station.

Millisecond One-thousandth of a second.

Minicomputer A computer with storage capacity and power less than a mainframe's but greater than a personal computer's.

MIPS Millions of instructions per second. A measure of how fast a central processing unit can process information.

MIS See *Management Information System*.

MIS manager A person, familiar with both computer technology and the organization's business, who runs the MIS department.

MITS Altair Generally considered the first personal computer, offered as a kit to computer hobbyists in 1975.

Model 1. A type of database, each type representing a particular way of organizing data. The three database models are hierarchical, network, and relational. 2. In a DSS, an image of something that actually exists or a mathematical representation of a real-life system.

Modem Short for modulate/demodulate. A device that converts a digital signal to an analog signal or vice versa. Used to transfer data between computers over analog communications lines.

Modula-3 A Pascal-like language designed for writing systems software.

Modulation Using a modem, the process of converting a signal from digital to analog.

Module Regarding structured programming, a set of logically related statements that perform a specific function.

Monochrome A computer screen that displays information in only one color, usually green, on a black background.

Monolithic Refers to the inseparable nature of memory chip circuitry.

Mouse A handheld computer input device whose rolling movement on a flat surface causes corresponding movement of the cursor on the screen. Also, a mouse button can be clicked to make selections from choices on the screen.

Motherboard Inside the personal computer housing, a board that holds the main circuits of the computer hardware, including the central processing unit.

Multimedia Software that typically presents information with text, illustrations, photos, narration, music, animation, and film clips—possible because the high-volume capacity of optical disks can accommodate photographs, film clips, and music. To use multimedia software, you must have the proper hardware: a CD-ROM drive, a sound card, and speakers.

Multiple-range graph A graph that plots the values of more than one variable.

Multiprocessing Using more than one central processing unit, a computer can run multiple programs simultaneously, each using its own processor.

Multiprogramming A large computer operating system feature under which different programs from different users compete for the use of the central processing unit; these programs are said to run concurrently.

Multitasking A feature of an operating system, in which several programs can be running in the same time frame and thus compete concurrently for the use of the central processing unit.

N

Nanosecond One-billionth of a second.

Natural language A programming language that resembles human language.

Network A computer system that uses communications equipment to connect two or more computers and their resources.

Network interface card (NIC) A circuit board that can be inserted into a slot inside a personal computer to allow it to send and receive messages on a local area network (LAN).

Network manager A person designated to manage and run a computer network.

Network operating system (NOS) An operating system designed to let computers on a network share resources such as hard disks and printers. A NOS supports resource sharing, data security, troubleshooting, and administrative control.

Neural network A computer whose chips are designed to mimic the human brain.

NIC See *Network interface card*.

Node A device, usually a personal computer, that is connected to a network.

Noise Electrical interference that causes distortion when a signal is being transmitted.

Nonimpact printer A printer that prints without striking the paper.

Nonprocedural language A language that states what task is to be accomplished but does not state the steps needed to accomplish it.

NOS See *Network operating system*.

Notebook computer A small portable computer. More often called a laptop computer.

O

Object Regarding object-oriented programming, a self-contained unit that contains both data and related facts and functions—the instructions to act on that data.

Object module A machine-language version of a program; it is produced by a compiler or assembler.

Object-oriented programming (OOP) A programming approach that uses objects, self-contained units that contain both data and related facts and functions—the instructions to act on that data.

OCR See *Optical character recognition*.

OCR-A The standard typeface for characters to be input by optical character recognition.

Octal Regarding number systems, the octal number system uses exactly eight symbols, digits 0 through 7.

Office automation The use of technology to help achieve goals in an office. Often associated with data communications.

OMR See *Optical mark recognition*.

Online In a data communications environment, a direct connection from a terminal to a computer or from one computer to another.

OOP See Object-oriented programming.

Open Systems Interconnection (OSI) A set of communications protocols defined by the International Standards Organization (ISO) that has been endorsed as a standard by the United Nations.

Operating environment An operating system designed as a shell so the user does not have to memorize or look up commands.

Operating system A set of programs that lies between applications software and the computer hardware, through which a computer manages its own resources.

Operation Regarding object-oriented programming, instructions that tell the data what to do. Also called a method.

Optical character recognition (OCR) A computer input method that uses a light source to read special characters and convert them to electrical signals to be sent to the computer.

Optical disk Storage technology that uses a laser beam to store large amounts of data at relatively low cost.

Optical mark recognition (OMR) A computer input method that uses a light source to recognize marks on paper and convert them to electrical signals to be sent to the computer.

Optical read-only memory (OROM) An optical medium that can be read, but not written to, by the user.

Optical recognition system A category of computer input method that uses a light source to read optical marks, optical characters, handwritten characters, and bar codes and convert them to electrical signals to be sent to the computer.

Organization chart A hierarchical diagram depicting lines of authority within an organization, usually mentioning people by name and title.

OROM See *Optical read-only memory*.

OSI See *Open Systems Interconnection*.

Output Raw data that has been processed by the computer into usable information.

Output device A device, such as a printer, that makes processed information available for use.

Output requirements In systems design, the plan for output medium and content, types of reports needed, and forms design.

Outsourcing Assigning the design and management of a new or revised system to an outside firm, as opposed to developing such a system in house.

P

Packaged software Software that is packaged and sold in stores. Also called commercial software.

Page composition Adding type to a layout. In desktop publishing, the software may be called a page composition program.

Page composition program A software package for designing and producing professional-looking documents that combine text and graphics. Also called a page makeup program or desktop publishing program.

Page frame The space in main memory in which to place a page.

Page layout In publishing, the process of arranging text and graphics on a page.

Page makeup program A software package for designing and producing professional-looking documents that combine text and graphics. Also called a page composition program or desktop publishing program.

Pages Equal-size blocks into which a program is divided to be placed into corresponding noncontiguous memory spaces called page frames. See also *Page frame*.

Page table The index-like table with which the operating system keeps track of page locations.

Pagination In word processing, options for placing the page number in various locations on the document page.

Paging The process of dividing a program into equal-size pages, keeping program pages on disk, and calling them into memory as needed.

Pan To move the cursor across a spreadsheet or a database to force into view fields that do not fit on the initial screen.

Parallel conversion A method of systems conversion in which the old and new systems are operated simultaneously until the users are satisfied that the new system performs to their standards; then the old system is dropped.

Parallel processing Using many processors, each with its own memory unit, that work at the same time to process data much more quickly than with the traditional single processor. Contrast with serial processing.

Participant observation A form of observation in which the systems analyst temporarily joins the activities of the group.

Partition A separate memory area that can hold a program, as part of a memory management technique that simply divides memory into separate areas. Also called a region.

Pascal A structured, high-level programming language named for Blaise Pascal, a seventeenth-century French mathematician.

Path In a flowchart, the connection leading from the decision box to one of two possible responses. Also called a branch.

Peer-to-peer network A network setup in which there is no controlling server computer; all computers on the network share programs and resources.

Pen-based computer A small portable computer that accepts handwritten input on a screen with a penlike stylus. Also called personal digital assistant.

Peripheral equipment All the input, output, and secondary storage devices attached to a computer.

Personal computer A relatively small and inexpensive type of computer, usually used by an individual in a home or office setting. Also called a microcomputer.

Personal computer manager The manager in charge of personal computer use.

Personal digital assistant (PDA) A small portable computer that accepts handwritten input on a screen. Also called a pen-based computer.

PgDn key The key used to advance the document one full screen.

PgUp key The key used to back up to the previous screen.

Glossary

Phase 1. In data transmission, the relative position in time of one complete cycle of a carrier wave. 2. In systems analysis and design, a portion of the systems development life cycle (SDLC).

Phased conversion A systems conversion method in which the new system is phased in gradually.

Picosecond One-trillionth of a second.

Pie chart A pie-shaped graph used to compare values that represent parts of a whole.

PILOT A programming language used most often to write computer-aided instruction in various subjects.

Pilot conversion A systems conversion method in which a designated group of users try the system first.

Pixel A picture element on a computer display screen; a pixel is merely one dot in the display.

PL/I (Programming Language One) A free-form and flexible programming language designed as a compromise between scientific and business programs.

Plot area The area in which a graph is drawn, that is, the area above the x-axis and to the right of the y-axis.

Plotter A graphics output device that can draw hard-copy graphics output in the form of maps, bar charts, engineering drawings, and even two- or three-dimensional illustrations.

Point A typographic measurement equaling approximately 1/72 inch.

Pointer An indicator on a screen; it shows where the next user-computer interaction will be. Also called a cursor.

Point-of-sale (POS) terminal A terminal used as a cash register in a retail setting. It may be programmable or connected to a central computer.

Polymorphism In object-oriented programming, when an individual object receives a message it knows how, using its own methods, to process the message in the appropriate way for that particular object.

Pop-up menu A submenu that originates from a menu selection on the bottom of the screen.

Portable operating system An operating system that works with more than one manufacturer's computer system. Also called a generic operating system.

Positional notation In dealing with number systems, the value of a digit in a number depends not only on its own intrinsic value but also on its location in the number. For example, the digit two has more value in the number 234 than in the number 762.

POS terminal See *Point-of-sale (POS terminal)*.

Preliminary design The subphase of systems design in which the new system concept is developed.

Preliminary investigation The first phase of the systems analysis and design life cycle, in which planners determine if and how a project should proceed. Also called a feasibility study or a system survey.

Presentation graphics Sophisticated business graphics. Presentation graphics programs include a library of symbols and drawings called clip art. Also called a business-quality graphics program.

Primary memory The electronic circuitry that temporarily holds data and program instructions needed by the CPU. Also called memory, primary storage, main storage, internal storage, and main memory.

Primary storage The electronic circuitry that temporarily holds data and program instructions needed by the CPU. Also called memory, primary memory, main storage, internal storage, and main memory.

Priming read The first read statement in a program.

Printer A device for generating computer-produced output on paper.

Printer spacing chart A chart used to determine and show a report format.

Print preview In word processing and some other applications, the ability to review a page or page of a document on the screen as it will be printed.

Procedural language A language used to present a step-by-step process for solving a problem.

Process 1. The computer action required to convert input to output. 2. An element in a data flow diagram that represents actions taken on data: comparing, checking, stamping, authorizing, filing, and so forth.

Process box In flowcharting, a rectangular box that indicates an action to be taken.

Processor The central processing unit (CPU) of a computer.

Prodigy A major information utility that offers a variety of services.

Program A set of step-by-step instructions that directs a computer to perform specific tasks and produce certain results. More generically called software.

Programmable read-only memory (PROM) Chips that can be programmed with specialized tools called ROM burners.

Programmer/analyst A person who performs systems analysis functions in addition to programming.

Programming language A set of rules that can be used to tell a computer what operations to do. There are many different programming languages.

Project management software Software that allocates people and resources, monitors schedules, and produces status reports.

PROLOG (PROgramming LOGic) A language popular for natural language programming.

PROM See *Programmable read-only memory*.

Prompt A signal that the computer or operating system is waiting for data or a command from the user.

Proprietary operating system An operating system used exclusively with the computer hardware for which it was written.

Protocol A set of rules for the exchange of data between a terminal and a computer or between two computers.

Prototype A limited working system or subset of a system that is developed to test design concepts.

Pseudocode An English-like way of representing the solution to a problem, especially structured programming control structures.

Public domain software Software that is free and not copyrighted.

Pull-down menu A menu of choices that appears, as a window shade is pulled down, when an initial menu choice is made.

Q

Questionnaire In the data-gathering phase of systems analysis, a source of facts to be input as data.

Queues Areas on disk in which programs waiting to be run are kept.

Query languages A variation on fourth generation languages, which can be used to retrieve data from databases.

R

Ragged right margin In word processing, the unjustified right margin of a document, or nonalignment of text at the right edge of a document.

RAID See *Redundant array of inexpensive disks*.

RAM See *Random-access memory*.

Random-access memory (RAM) Memory that provides temporary storage for data and program instructions.

Randomizing Applying a formula to a key to yield a number that represents a disk address. Also called hashing.

Range A group of one or more cells, arranged in a rectangle, that a spreadsheet program treats as a unit.

Raster-scan technology A video display technology. The back of the screen display has a phosphorous coating, which will glow whenever it is hit by a beam of electrons.

Read To bring data outside the computer into memory.

Read-only media Media recorded on by the manufacturer that can be read from but not written to by the user

Read-only memory (ROM) Memory containing data and programs that can be read but not altered. Data remains after the power is turned off.

Read/write head An electromagnet that reads the magnetized areas on magnetic media and converts them into the electrical pulses that are sent to the processor.

Real storage That part of memory that temporarily holds part of a program pulled from virtual storage.

Real-time processing Processing in which the results are available in time to affect the activity at hand.

Record 1. A set of related fields. 2. In a database relation, one row.

Reduced instruction set computer (RISC) A computer that offers only frequently used instructions. Since fewer instructions are offered, this is a factor in improving the computer's speed. Contrast with *CISC*.

Redundant array of inexpensive disks (RAID) Secondary storage that uses several connected hard disks that act as a unit. Using multiple disks allows manufacturers to improve data security, access time, and data transfer rates.

Refresh To maintain the image on a CRT screen by reforming the screen image at frequent intervals to avoid flicker. The frequency is called the scan rate; 60 times per second is usually adequate to retain a clear image.

Region A separate memory area that can hold a program, as part of a memory management technique that simply divides memory into separate areas. Also called a partition.

Register A temporary storage area for instructions or data.

Relation A table in a relational database model.

Relational database A database in which the data is organized in a table format consisting of columns and rows.

Relational model A database model that organizes data logically in tables.

Relational operator An operator (such as <, >, or =) that allows a user to make comparisons and selections.

Removable hard disk cartridge A supplemental hard disk, that, once filled, can be replaced with a fresh one.

Replication The ability of groupware to copy data from user to user throughout the network, so that all users have the most recent version of the database.

Report Program Generator (RPG) A problem-oriented language designed to produce business reports.

Resolution The clarity of a video display screen or printer output.

Resource allocation The process of assigning resources to certain programs.

Response time The time between a typed computer request and the response of the computer.

Reverse video The feature that highlights on-screen text by switching the usual text and background colors.

Ring network A "circle" of point-to-point connections between computers at local sites. A ring network does not contain a central host computer.

RISC See *Reduced instruction set computer*.

Robot A computer-controlled device that can physically manipulate its surroundings.

ROM See *Read-only memory*.

ROM burner A specialized device used to program progammable read-only memory (PROM) chips.

Rotational delay For disk units, the time it takes for a record on a track to revolve under the read/write head.

Round-robin scheduling A system of having users take turns using the processor.

RPG See *Report Program Generator*.

S

Sampling In systems analysis, collecting a subset of data relevant to the system under study.

Sans serif A typeface that is clean, with no serif marks.

Satellite transmission Data transmission from earth station to earth station via communications satellites.

Scanner A device that uses a light source to read text and images directly into the computer. Scanners can be of two varieties, handheld and desktop.

Scan rate The number of times a CRT screen is refreshed in a given time period. A scan rate of 60 times per second is usually adequate to retain a clear screen image.

Screen A television-like output device that can display information.

Scrolling A feature that allows the user to move to and view any part of a document on the screen. Used especially in word processing, spreadsheets, and databases.

SDLC See *Systems development life cycle*.

Sealed module A disk drive containing the disks, access arms, and read/write heads sealed together. Also called a Winchester disk.

Search command In word processing, the ability to locate certain text in the document. Also called the Find command.

Secondary storage Additional storage, often on disk, for data and programs. Secondary storage is separate from the CPU and memory. Also called auxiliary storage.

Sector method A method of organizing data on a disk in which each track is divided into sectors that hold a specific number of characters. Data on the track is accessed by referring to the surface number, track number, and sector number where the data is stored.

Security A system of safeguards designed to protect a computer system and data from deliberate or accidental damage or access by unauthorized persons.

Seek time The time required for an access arm to position over a particular track on a disk.

Selection control structure A control structure used to make program logic decisions.

Semiconductor A crystalline substance that conducts electricity when it is "doped" with chemical impurities.

Semiconductor storage Data storage on a silicon chip.

Sequence control structure A control structure in which one statement follows another in sequence.

Sequential file organization The arrangement of records in ascending or descending order by a certain field called the key.

Sequential file processing Processing in which records are usually in order according to a key field.

Serial processing Processing in which a single processor can handle just one instruction at a time. Contrast with parallel processing.

Serif Small marks added to letters of a typeface; the marks are intended to increase readability of the type face. Contrast with *Sans serif*.

Server 1. In a client/server network arrangement, the computer that controls and manages the network and its services; the server usually has hard disks that hold files needed by users on the network. 2. A computer used to access the Internet; it has special software that uses the Internet protocol.

Service program A prewritten program that performs routine file-handling tasks, such as file conversion and sort-merges. Also called a utility program.

Shareware Software that is given away free, although the maker hopes that satisfied users will voluntarily pay for it.

Shell An operating environment layer that separates the operating system from the user.

SIMM See *Single in-line memory module*.

Simplex transmission Transmission of data in one direction only.

Simulation The use of computer modeling to reach decisions about real-life situations.

Single entry Regarding structured programming, the unique point where execution of a program module begins.

Single exit Regarding structured programming, the unique point where termination of a program module occurs.

Single in-line memory module (SIMM) A board containing memory chips that can be plugged into a computer expansion slot.

Single-range bar graph A bar graph that plots the values of only one variable.

Single-range graph A graph that plots the values of only one variable.

Sink In a data flow diagram, a destination for data going outside the system.

Site license A license permitting a customer to make multiple copies of a piece of software.

Smalltalk An object-oriented language that supports a particularly visual system.

Soft copy Computer-produced output displayed on a screen.

Soft font A font that can be downloaded from the font library on disk with a personal computer to a printer.

Software Instructions that tell a computer what to do. Also called programs.

Software piracy Unauthorized copying of computer software.

Software Publishers Association (SPA) An organization representing software makers. A key function of the organization is reducing software piracy.

Source In a data flow diagram, an origin outside the system.

Source data automation The use of special equipment to collect input data and send it directly to the computer.

Source document An instrument, usually paper, containing data to be prepared as input to a computer.

Source module A program as originally coded, before being translated into machine language.

Source program listing The printed version of a program as the programmer wrote it, usually produced as a byproduct of compilation.

SPA See *Software Publishers Association*.

Speech recognition Converting input data given as the spoken word to a form the computer can understand.

Speech recognition device A device that accepts the spoken word through a microphone and converts it into digital code that can be understood by a computer.

Speech synthesis The process of enabling machines to talk to people.

Spelling checker program A word processing program that checks the spelling of words in a document.

Spooling A process in which files to be printed are placed temporarily on disk.

Spreadsheet A worksheet divided into rows and columns that can be used to analyze and present business data.

SRAM See *Static random access memory*.

Stacked-bar graph A bar graph in which all data common to a given row or column appears stacked in one bar.

Star network A network consisting of one or more computers connected to a central host computer.

Start/stop symbol An oval symbol used to indicate the beginning and end of a flowchart.

Start/stop transmission Asynchronous data transmission.

Static random-access memory (SRAM) A type of RAM that requires a continuous current to hold data. SRAM is usually faster, but larger and more expensive than dynamic RAM. Contrast with *Dynamic random-access memory* (DRAM).

Storage register A register that temporarily holds data taken from or about to be sent to memory.

Structure chart A chart that illustrates the top-down design of a program and is often used to either supplement or replace a logic flowchart.

Structured interview An interview in which only planned questions are used.

Structured programming A set of programming techniques that includes a limited number of control structures, top-down design, and module independence.

Style In word processing, the way a typeface is printed, for example, in italic.

Submenu An additional set of options related to a prior menu selection.

Supercomputer The largest and most powerful category of computers.

Supermicro A computer that combines the compactness of a desktop computer with power that almost equals that of a mainframe.

Supermini A minicomputer at the top of its class in terms of capacity and price.

Supervisor program An operating system program that controls the entire operating system and calls in other operating system programs from disk storage as needed.

Supply reel A reel that has tape with data on it or on which data will be recorded.

Surge protector A device that prevents electrical problems from affecting data files.

SVGA Stands for super VGA. A superior screen standard with 800x600 pixels or 1024x768 pixels.

Symbolic address The meaningful name for a memory location. Instead of just a number, for example, a symbolic address should be something meaningful, such as NAME or SALARY.

Symbolic language A second-generation language that uses abbreviations for instructions. Also called assembly language.

Synchronous transmission Data transmission in which characters are transmitted together in a continuous stream.

Synonym The name for a record's disk address, produced by a hashing scheme, that is the same as a pre-existing address for a different record.

Syntax The rules of a programming language.

Syntax errors Errors in the use of a programming language.

Synthesis by analysis Speech synthesis in which a device analyzes the input of an actual human voice, stores and processes the spoken sounds, and reproduces them as needed.

Synthesis by rule Speech synthesis in which a device applies linguistic rules to create an artificial spoken language.

System An organized set of related components established to perform a certain task.

System journal A file whose records represent real-time transactions.

System requirements A detailed list of the things a particular system must be able to do, based on the results of the systems analysis.

Systems analysis A phase of the systems development life cycle, to study an existing system to determine how it works and how it does or does not meet user needs, with an eye to improving the system.

Systems analyst A person who plans and designs computer systems.

Systems design A phase of the systems development life cycle, to develop a plan for a new or revised system, based on the results of the systems analysis phase.

Systems development A phase of the systems development life cycle, whose activities include programming and testing.

Systems development life cycle (SDLC) The multiphase process required for creating or revising a computer system.

Systems flowchart A drawing that depicts the flow of data through some part of a computer system.

System software All programs related to coordinating computer operations, including the operating system, programming language translators, and service programs.

System survey The first phase of systems analysis, in which planners determine if and how a project should proceed. Also called a feasibility study or a preliminary investigation.

System testing A testing process in which the development team uses test data to determine whether programs work together satisfactorily.

T

Take-up reel A reel that always stays with the magnetic tape unit.

Tape drive The drive on which reels of magnetic tape are mounted when their data is ready to be read to or written on by the computer system.

TCP/IP See Transmission Control Protocol/Internet Protocol.

Telecommuting Using telecommunications and computers at home as a substitute for working at an office outside the home.

Teleconferencing A system of holding conferences by linking geographically disbursed people together through computer terminals or personal computers.

Teleprocessing A system in which terminals are connected to the central computer via communications lines.

Template 1. In desktop publishing, a predetermined page design that lets a user fill in text and art. 2. In a spreadsheet program, a worksheet that has already been designed for the solution of a specific type of problem, so that a user need only fill in the data.

Terminal A device that consists of an input device (usually a keyboard), an output device (usually a screen), and a communications link to the computer.

Terminal emulation software Data communications software that makes a personal computer act like a terminal, so it can communicate with a larger computer.

Text block A continuous section of text in a document that has been marked or selected.

Text editor Software that is somewhat like a word processing program, used by programmers to create a program file.

Thesaurus program With a word processing program, this program provides a list of synonyms and antonyms for a selected word in a document.

Time-driven Time sharing is said to be time-driven because each user is given a time slice in which the computer works on that user's tasks before moving on to another user's tasks.

Time-sharing A special case of multiprogramming in which several people use one computer at the same time.

Time slice In time sharing, a period of time, a fraction of a second, during which the computer works on a user's tasks.

Title The caption on a graph that summarizes the information in the graph.

Token passing The protocol for controlling access to a Token Ring network. A special signal, or token, circulates from node to node, allowing the node that "captures" the token to transmit data.

Token Ring network A network protocol that uses token passing to send data over the shared network cable. A computer that wants to send a message must capture the token before sending.

Top-down design Regarding structured programming, a design technique that identifies basic program functions before dividing them into subfunctions called modules.

Topology The physical layout of a local area network.

Touch screen A computer screen that accepts input data by letting the user point at the screen to select a choice. The finger touching the screen interrupts the light beams on the monitor edge, pinpointing the selected screen location.

Track On magnetic disk, one of many data-holding concentric circles.

Trackball A ball used as an input device; it can be hand manipulated to cause a corresponding movement of the cursor on the screen. Trackballs are often built in on portable computers.

Trailing decision The loop-ending decision that occurs at the end of a DO-UNTIL loop.

Transaction file A file that contains all changes to be made to the master file: additions, deletions, and revisions.

Transaction processing The technique of processing transactions one at a time, in the order in which they occur.

Transistor A small device that transfers electrical signals across a resistor.

Translator Software, typically a compiler, that converts a program into the machine language the computer can understand.

Transmission Control Protocol/Internet Protocol (TCP/IP) A standardized protocol permitting different computers to communicate via the Internet.

Transponder A device in a communications satellite that receives a transmission from earth, amplifies the signal, changes the frequency, and retransmits the data to a receiving earth station. The transponder makes sure that the stronger outgoing signals do not interfere with the weaker incoming signals.

Trojan horse An application that covertly places destructive instructions in the middle of a legitimate program but appears to do something useful.

Turing test A test proposed to determine whether a computer can be said to be a thinking machine: If a human judge, working with two different terminals, one connected to a human and one to a computer, cannot tell which is which.

Twisted pairs Wires twisted together in an insulated cable. Twisted pairs are frequently used to transmit data over short distances. Also called wire pairs.

Typeface A set of characters—letters, symbols, and numbers—of the same design.

Type size The size, in points, of a typeface.

U

Unbundle Sell software separately from the hardware on which it will run.

Underlining Underscoring text.

Unit testing Testing an individual program by using test data.

UNIVAC I (Universal Automatic Computer I) The first computer built for business purposes.

Universal Product Code (UPC) A code number unique to a product. The UPC code is the bar code on the product's label.

UNIX A generic multiuser, time-sharing operating system developed in 1971 at Bell Labs.

Unstructured interview An interview in which questions are planned in advance, but the questionnaire can deviate from the plan.

Update To keep files current by changing data as appropriate.

Updating in place The ability to read, change, and return a record to its same place on the disk.

Upload In a networking environment, to send a file from one computer to another, usually to a larger computer or a host computer. Contrast with Download.

User A person who uses computer software or has contact with computer systems.

User friendly A term to refer to software that is easy for a novice to use.

User involvement The involvement of users in the systems development life cycle.

User's guide An instruction manual that holds the printed software documentation.

Utility program A prewritten program that performs routine file-handling tasks, such as copying and sorting files. Also called a service program.

V

Vaccine A computer program that stops the spread of a virus. Also called an antivirus.

Vacuum tube An electronic tube used as a basic component in the first generation of computers.

Value In a spreadsheet, data entered into a cell.

Variable 1. On a graph, the items that the data points describe. 2. In a program, a name assigned to a memory location, whose contents can vary.

Vector An arrow—a line with directional notation—used in a data flow diagram.

Vertical centering In word processing, adjusting the top and bottom margins so that text is midway between the top and the bottom of the page.

Vertical market A market consisting of a group of similar customers.

Vertical market software Software for a group of similar customers, such as accountants or doctors.

Very high-level language A fourth-generation language.

VGA Stands for video graphics array. A common screen standard with 640x480 pixels.

Videoconferencing Computer conferencing combined with cameras and wall-size screens.

Video graphics Computer-produced animated pictures.

Video Privacy Protection Act Legislation that prohibits video vendors from revealing what videos their customers rent.

Virtual memory See *Virtual storage*.

Virtual reality A system in which a user is immersed in a computer-created environment, so that the user physically interacts with the computer-produced three-dimensional scene.

Virtual storage A technique of memory management in which part of the application program is stored on disk and is brought into memory only as needed. The secondary storage holding the rest of the program is considered virtual storage.

Virus A set of illicit instructions that passes itself on to other programs in which it comes in contact.

Vision robot A robot that can recognize an object by its shape or color.

Voice input Using the spoken word as a means of entering data to a computer.

Voice mail A system in which a spoken message is digitized and stored in the recipient's voice mailbox. Later the recipient can dial the mailbox, and the system delivers the message in audio form.

Voice output device See *Voice synthesizer*.

Voice synthesizer A device that converts data in main storage to vocalized sounds understandable to humans. Also called an audio-response unit and voice output device.

Volatile Subject to loss when electricity is interrupted or turned off. Data in semiconductor storage is volatile.

Volume testing The testing of a program or a system by using real data in large amounts.

W

Walkthrough A process in which a group of programmers —your peers—review your program and offer suggestions in a collegial way.

WAN See *Wide area network*.

Wand reader An input device that scans the special letters and numbers on price tags in retail stores and sends that input data to the computer. Often connected to a point-of-sale terminal in a retail store.

Weight In word processing or desktop publishing, the variation in the visual heaviness of a typeface; for example, words look much heavier when in boldface type.

"What-if" analysis The process of changing one or more spreadsheet values and observing the resulting calculated effect.

Wide area network (WAN) A network of geographically distant computers and terminals. Contrast with *Local area network*.

Winchester disk A disk drive in which the disks, access arms, and read/write heads are combined in a sealed module.

Wire pairs Wires twisted together in an insulated cable. Wire pairs are frequently used to transmit data over short distances. Also called twisted pairs.

Word The number of bits that constitute a common unit of data, as defined by the computer system.

Word processing Computer-based creating, editing, formatting, storing, retrieving, and printing of a text document.

Word wrap A word processing feature that automatically starts a word at the left margin of the next line if there is not enough room for it on the line.

World Wide Web (the Web) An Internet subset of sites with text, images, and sounds linked together to peruse related topics.

WORM See *Write-once, read-many media*.

Worm A program that spreads and replicates over a network.

Write-once, read-many media (WORM) Media that can be written on only once; then they become read-only media.

WYSIWYG An acronym for *what you see is what you get*—a phrase used in word processing to describe the ability of the program to show, on the screen, text styles such as italic and boldface. Non-wysiwyg programs cannot show the actual style on the screen but only on the paper printout.

X

x-axis The horizontal reference line of a graph, often representing units of time.

Y

y-axis The vertical reference line of a graph, usually representing values or amounts, such as dollars, staffing levels, or units sold.

Z

Zone recording Involves dividing a disk into zones to take advantage of the storage available on all tracks, by assigning more sectors to tracks in outer zones than to those in inner zones.

Credits

Frontispiece
©Mark Yankus

Table of Contents
viii, (left) ©Diane Fenster; (middle) ©Carini/The Image Works; (right) Courtesy of HSC Software; **ix,** (left) ©Bob Thomason/TSI; (middle) ©David Young/Tom Stack & Associates; (right) ©Nancy Brown/The Image Bank; **x,** Courtesy of International Business Machines; **xii,** Courtesy of International Business Machines; **xiii,** ©L.D. Gordon/The Image Bank; **xiv,** Courtesy of Intergraph; **xv,** ©Charles Thatcher/TSI; **xvi,** ©Larry Dale Gordon/The Image Bank; **xvii,** (left) ©Dan McCoy/Rainbow; (right) ©HMS Images/The Image Bank; **xviii,** Courtesy of Intergraph; **xix,** ©Dan McCoy/Rainbow; **xx,** ©Joseph Drivas/Image Bank; (right) Harry Bartlett/Lightsources.

Introduction
Text page 2: Fig.1, (top)©R. Ian Lloyd/Westlight, (btm. left)©C. Michael Krasowitz/FPG, (btm. right)©Jon Riley.TSI. **3:** Fig. 2, (btm)© C. Stephen Simpson/FPG, (left)©Image Bank, (mid. top)©David Hanover/TSI, (right)©Barros &Barros/The Image Bank. **4:** Fig. 3, ©Alan Levenson/TSI. **5:** Fig.4, ©Barros & Barros/Image Bank. **6:** Fig. 5, (top left) ©Bill Frymire, (top right) Courtesy of Pixar, (btm) ©Tom Cushwa, (mid. right) Courtesy of Pixar. **7:** Fig. 6,©The Image Bank; Fig. 7, ©Frank Siteman/Rainbow; Fig. 8, ©Ken Eward/Biografx. **8:** Fig. 9, ©Joe Sohn/Uniphoto; Fig. 10, Image produced by NCSA using Wavefront's Visualizer ™; Fig. 11©Blair Seits/Photo Researchers, Inc. **9:** Fig.12, Courtesy of Microsoft Corporation; Fig. 14, ©Hank Morgan/Photo Researchers, Inc.; **10:** Fig.15, (left)©Jeremy Mason/SPL/Photo Researchers, Inc.; (right) ©Mason Morfit /FPG; Fig. 16, Courtesy of Intergraph.

Part and Chapter Openers
©Joseph Maas

Chapter 1
18: 1-2, Courtesy of IBM; **19:** 1-3a, Peter Steiner, 1-3b, ©Shambroom/Photo Researchers, Inc.; **20:** 1-4a, ©The Stock Solution; 1-4b, Courtesy of Texas Instruments; Margin Note, Courtesy of Timex; **21:** 1-5a, ©David R. Frazier/Photo Researchers, Inc., 1-5b, Courtesy of International Business Machines; 1-5c, ©Damien Lovegrove / SPL/Photo Researchers, Inc.; 1-5d, Courtesy of 3M; **22:** ©Diane Fenster; **23:**1-6a, © Ed Kashi, 1-6b; Courtesy of International Business Machines; 1-6c, Courtesy of Digital Equipment Corporation; 1-6d, Courtesy of International Business Machines; **24:** 1-7, Courtesy of the Minnesota Supercomputer Center; **26:** 1-8 (btm. right) Courtesy of Compaq Computer Corporation; (left) ©Ron Rovtar/FPG; (top right) ©Arthur Tilley/FPG; 1-9, Courtesy pf GRID Systems Corporation; **28:** 1-11, ©Steven Underwood; **29:** ©James Porto/FPG; **32:** 1-13, (left) Courtesy of Claris Corporation; (right) Courtesy of Micrografx, Inc. **33:** ©Stephen Collins/ Photo Researchers, Inc.

Chapter 2
44: ©Gred Pease/TSI; **51:** 2-8, Courtesy of International Business Machines; 2-9, Courtesy of Intel Corporation; **53:** Courtesy of Medicard/Advantage Data, Inc.
Text credit page 44: Making the Right Connection: Based on an original interview done in 1995 with Randy Katz, Project Director with Advanced Research Projects Agency (ARPA) of the Department of Defense.

Chapter 3
66: 3-1, Courtesy of Unisys Corporation; **67:** 3-2a, Courtesy of Compaq Corporation; 3-2b, Courtesy of MacDonald's Corporation, 3-2c, Courtesy of International Business Machines;**70:** 3-4a, ©David Mallory Jones/Uniphoto; 3-4b, ©David Bishop/PhotoTake NYC; **71:** Courtesy of Microsoft Corporation; **73:** 3-8a, Courtesy of Spectra-Physics; 3-8b, ©Tony Freeman/Photo Edit; **75:** 3-10a, Courtesy of Logitech; 3-10b, Courtesy of Hewlett Packard; **76:** 3-12, Courtesy of NCR Corporation; Margin Note, Courtesy of Nordic Track; **77:** Courtesy of International Business Machines; **78:** ©Hank Morgan/Photo Researchers, Inc.; **79:** 3-15a, Courtesy of NEC; 3-15b&c, Courtesy of International Business Machines; **80:** Courtesy of Hewlett Packard; **83:** 3-11, Kodak, MRC box, ©Telegraph Color Library/FPG; **84:** 3-18a, Minnesota Supercomputer Center; 3-18b&c, Courtesy of Crystal Graphics; **85:** 3-19a&b, ©New Line Cinema/Motion Picture & TV Archive; 3-20a, ©Hank Morgan/Science Source/Photo Researchers, Inc.; 3-20b,©Berenguier/Jerrican/Photo Researchers, Inc.; **86:** 3-21,©P.Gontier/The Image Works; 3-22, ©Ed Kashi; **87:** ©Guy Aroch.

Chapter 4
98: 4-3a,©Frederick Bodin/Benjamin/Cummings; 4-3b, Courtesy of BASF Corporation; 4-3c, KAO Inforsystems Company; **99:** 4-4a, Courtesy of Ancodyne, Inc.; 4-4c, Courtesy of Seagate Technology Corporation; **100:** Courtesy ofQuantum Corporation; **102:** Courtesy of PLI; **103:** Courtesy of Pioneer Communications of America; **106:** 4-11a, Courtesy of International Business Machines; 4-11b, Courtesy of 3M Corporation; **107:** ©John Curtis/Benjamin/Cummings; **108:** Courtesy of PSION; **113:** NASA; **115:** (left) ©Lawrence Migdale/TSI.

Chapter 5
127: Margin Note, Courtesy of Esther Dyson; 5-1, Courtesy of Dell Corporation; **128:** Courtesy of International Business Machines; **131:** Courtesy of Hayes Modem; **134:** ©Rick Reinhard/FPG; **136:** 5-10, ©Telegraph Color Library/FPG; Margin Note, NASA; **143:** Courtesy of International Business Machines; **144:** 5-16, ©Davis Young-Wolff/Photo Edit; 5-17, ©HMS Images/The Image Bank; **145:** ©Bruce Jaffe/Gamma Liaison.; **146:** 5-19a, ©F. Pedrick/Image Works; 5-19b, ©Eugen Gebhardt/FPG; **147:** ©Joseph Maas; **149:** NASA.

Chapter 6
165: Courtesy of the VA Medical Center, Columbus, OH; **171:** ©David Frazier/Photo Researchers, Inc.; **180:** Department of the Navy; **192:** Courtesy of Microsoft Corporation; **193:** ©Steve Dunwall/The Image Bank.

Chapter 7
205: ©Matthew McVay/Stock Boston; **210:** Courtesy of Microsoft Corporation; **215:** ©Mark Yankus; **219:** ©Robert Visser/Greenpeace.

Chapter 8
232: Courtesy of William Jepson, UCLA Graduate School of Architecture and Urban Planning; **237:** ©Jim Pickerell/Image Works; **253:** Courtesy of Scitor; **257:** Courtesy of Crystal Graphics

Chapter 9
271: 9-1 (top right) ©Gabe Palmer/Mugshots/The Stock Market, (btm. right) Courtesy of Compaq Computer Corporation, (btm. left) ©Howard Grey/TSI, (top left) ©Arthur Tilley/FPG; **274:** Courtesy of the Lotus Development Corpration; **275:** Courtesy of ON Technology; **277:** Courtesy of the Karl Eller Graduate School of Management, University of Arizona; **279:** 9-5a, Courtesy of Stephen C. Cande, Stephen Lewis/Lamont-Doherty Geological Observatory, Joyce Miller and Scott Fergus-

CR-1

son/URI; 9-5b, Courtesy of Crystal Graphics; 9-5c, ©Andy Sacks/Tony Stone Images. 281: ©Herb Simms; 282: ©Frank Herholdt/TSI.

Chapter 10
295: ©Ron Chapple/FPG; 296: Courtesy of Dell Corporation; 298: Courtesy of Primavera; 299: Courtesy of Microsoft Corporation; 302: ©Peter Wiant

Chapter 11
316: Courtesy of Eyedentify; 323: Courtesy of Dell Corporation; 324: ©Mark Scott/FPG; 328: Courtesy of Corel; 329: Courtesy of Software Publishers Association.

Chapter 12
341: Pamela Hobbs; 344: ©Ed Kashi/PhotTake NYC 345: 12-4a&b, Courtesy of AION Development Corporation; 347: Courtesy of Transitions Research Corporation; 348: 12-5a, ©Andy Sacks/TSI; 12-5b, Courtesy of Japan Airlines; 12-6a-d, Courtesy of Thinking Machines Corporation; 349: 12-7a, ©Dave Barrett/MIT; 12-7b, Cortesy of Control Data Corporation; 12-7c,©Fujifotos/The Image Works; 351: 12-8a, ©P. Howell/Gamma -Liaison; 12-8b, Courtesy of VPL Research; 12-8c, ©David Sutton.

Appendix B
B-2: B-1, Reprinted from Popular Electronics, January, 1975. ©1975 Ziff-Davis Publishing Co.; B-2, Courtesy of Science Museum Library, London; B-3: D. Bohl/The Computer Museum, Boston; B-4: B-4, Courtesy of International Business Machines; B-5, Iowa State University of Science and Technology; Margin Note, Courtesy of International Business Machines; B-5: B-6, Courtesy of International Business Machines; Margin Note, The Computer Museum, Boston; B-6: Courtesy of International Business Machines; B-9: (left) ©Reportage Stock; (middle) ©Matthew McVay/Stock Boston; (right) Courtesy of EFF; photo by Seth Resnick; B-10: B-8, Margaret and Jerry Wozniak; B-9, Courtesy of International Business Machines.

Microchips Gallery
Opener: ©Joseph Drivas/Image Bank; 2: Courtesy of International Business Machines; 3: ©Precision Visuals International/Science Photo Library/Photo Researchers, Inc.; 4: ©Robert Holmgren; 5: Courtesy of Motorola, Inc.; 6: AT&T Archives; 7: Courtesy of International Business Machines; 8: Courtesy of Sperry Corporation; 9: Courtesy of TRW, Inc.; 10: Courtesy of Micron Technology Inc.; 11: ©Dan McCoy/Rainbow; 12: ©J. Burgess/SPL/Photo Researchers, Inc.; 13: AT&T Archives; 14-16: Courtesy of Hewlett Packard; 17: ©Astrid &Hanns-Frieder Michler/SPL/Photo Researchers, Inc.; 18: Courtesy of Advanced Micro Devices Inc.; 19: Courtesy of National Semiconductor; 20: Telegraph Colour Library/FPG; 21: Courtesy of Advanced Micro Devices, Inc.; 22: Courtesy of Quantum Corporation

Computers at Work Gallery
Opener: ©John Abbott; 1: ©Robert Frerck/TSI; 2: ©John Abbott; 3: ©Window Phone; 4: ©Curt Fischer/TSI; 5: ©Churchill & Klehr; 6: ©TMartha Everson/Lightsources Stock; 7: ©The Image Bank; 8: ©Harry Bartlett/Lightsources Stock;9: ©John Abbott; 10: ©W. Herbert Simms; 11: ©Seth Resnick/Lightsources Stock; 12, 13: ©John Abbott; 14: ©HMS Images/The Image Bank; 15: ©John Abbott; 16: ©Hank Morgan/Rainbow; 17: ©Andy Freeberg; 18: ©Hans Ulrich; 19: Courtesy of Pixar; 20: ©J. Deathe; 21, 22: Courtesy of Ray Dream, Inc.; 23: ©Tim Arroyo; 24: ©Mark Yankus; 25: ©Chris Purcell; 26: ©T. Tracy/FPG; 27: ©Bob Sacha; 28: ©Photo Researchers, Inc.; 29: ©Hank Morgan/Science Source/Photo Researchers, Inc.

Multimedia Gallery
Opener: The Ultimate Frank Lloyd Wright by Brian Preiss for Microsoft Home; 2: Courtesy of Microsoft Corporation; 3: (top, middle) Art Gallery, Microsoft Home, Courtesy of Microsoft Corporation; (bottom) Fred Bodin/Benjamin/Cummings; 4: 1, 2, Courtesy of Microsoft Corporation; 3-5, Courtesy of Corbis Publishing; 6-8, Courtesy of Microsoft Corporation; 5: Courtesy of Microsoft Corporation; 6: 19, 20, Courtesy of Brøderbund Software, Inc.; 21-24, Courtesy of Microsoft Corporation; 25, Courtesy of Turner Publishing Inc., and Swifte International, Ltd.; 7: 26-30, Courtesy of Microsoft Corporation; 31-33, Courtesy of Brøderbund Software, Inc.; 8: Courtesy of Microsoft Corporation.

Computer Graphics Gallery
1: Ron Richey; 2: Jose Simancas; 3: Cecil Rice; 4: John Stephens/Electronic Photo Imaging; 5: Jose Figueiredo; 6: Luiz Eduardode Oliveira; 7: Helen Golden; 8: Karin Schminke; 9: Dorothy Krause; 10: Pamela Hobbs; 11: Diane Fenster; 12: Bonnie Lhotka; 13: Wendy Grossman; 14: HSC Software/KAI's Power Tools; 15: Wendy Grossman; 16: Lydia M. Cabico; 17: Ray Trygstad; 18: Richard Swier; 19: Wil Dawson; 20: Jeff Bisch; 21: Diane Fenster; 22: Stephen Arscott and Chris Ticar; 23-26: Katrin Eisman; 27: XAOS.

Buyer's Guide
1: ©John Reis/PhotoLink; 2: Courtesy of Apple Corporation; 3: (top left) Courtesy of International Business Machines; (top right) Courtesy of International Business Machines; (bottom left) Courtesy of Dell Corporation; (bottom right) Courtesy of NEC; 4: (top left) Courtesy of Microcomputer Accessories; (bottom left) Courtesy of Gateway 2000; (right) Courtesy of International Business Machines; 5:(top left) Courtesy of International Business Machines; (middle left) Courtesy of Hewlett Packard; (middle bottom) Courtesy of Canon Computer Systems; (top right) Courtesy of QMS Corporation; (bottom right) Courtesy of QMS Corporation; 6: (bottom left) Courtesy of Sparc; (top right) Courtesy of Microsoft Corporation; 7: (top) Courtesy of International Business Machines; (bottom) Courtesy of Inmac; 8: ©John Curtis/Benjamin/Cummings; 9: ©Richard Tauber/Benjamin/Cummings.

Index

NOTE: Page numbers in italics indicate figures or tables separated from accompanying text.

< (less-than symbol), as relational operator, 44
= (equal sign), as relational operator, 44
> (greater-than symbol), as relational operator, 44
8-bit word length, 50
16-bit word length, 50
32-bit word length, 50
64-bit word length, 50
486DX/486SX chips, 53–54
8088 chip, evolution to Pentium from, 55

A drive, 207
A> prompt, 207
ABC (Atanasoff-Berry computer), B-4
Access arm, 98, 99, 100
    access time and, 113
    cylinder method and, 102–103
Access control, 314–315, 319
Access methods, data organization and, 110–113
Access problems, on Internet, 265
Access time, 113
Accounting
    data collection devices used for, 76
    software packages for, 283
    vertical market software for, 281, *282*
Accumulator, 45
ACM (Association for Computing Machinery), 170
Acoustic coupler modems, 130
Acquisition by purchase, systems design and, 243
Active badges, 314
Ada, 184, 186
    applications used for, *178*
    program and sample output, *187*
Ada, Countess of Lovelace, 186, B-3
Addressable graphics, 79
Addresses, 45, 48–49
    for records, 111, *112*
    symbolic, 48–49
Address register, 45
Administrative control, in network operating system, 212
Advertising
    computer graphics used in, 83, *84*
    Internet and, 151
    software for, 284
Aerospace companies, large computers used by, 25, 270
Agent software, 297
Agriculture, computer use in, 8
AI. *See* Artificial intelligence
Aiken, Howard, B-3
Airlines, computer use by, 7
    flash chips, 54
    mainframes, 25
    online reservations, 277
    speech recognition, 76
Air traffic control
    expert systems and, 345
    virtual reality and, 351
ALGOL, 189
Allen, Paul, B-9
Alphanumeric data, representation of, C-0
Altair computer, B-0, B-2, B-9, B-10
Alt key, 68, 69
ALU. *See* Arithmetic/logic unit
American Ballet Theater, computer use by, 281
American National Standards Institute (ANSI)
    ANSI-COBOL and, 180
    Data Encryption Standard (DES), 322
    flowchart symbols, 166, *167*, *250*, A-2
    OCR-A typeface, 72

American Standard Code for Information Exchange (ASCII), 50
America Online (AOL), 147–148, 150
    Internet access through, 149
    Macintosh tips, 150
    magazines available on, 129
Amplitude of carrier wave, 129, *130*
Amplitude modulation (AM), 130
Analog transmission, 129–130
    conversion of digital signals to, 130, *131*
Analytical engine, B-2 to B-3
Andreesen, Marc, 123
Animation, 83, 84, *85*. *See also* Graphics
ANSI. *See* American National Standards Institute
Antipiracy hotline, 329
Antivirus software, 319–320
APL programming language, *189*
Apple Computer, Inc., rise of, B-9, B-10
Apple I computer, B-10
Apple II computer, B-10
    VisiCalc and, B-10
Apple Macintosh. *See* Macintosh operating system; Macintosh tips
Applications software
    See also Software; Spreadsheet software
    custom vs. packaged, 28–29
    defined, 28
    desktop publishing, 31
    electronic agents, 297
    graphics, 32
    for high-level managers, 296–299
    illegal copying of, 326–329
    information center and, 302
    Macintosh computers and, 204
    memory management with, 216–217
    Microsoft and, B-11
    operating systems and, 203–205
    for project management, 253–254
    security and copyright issues, 317
    time-sharing, 216
    Windows 95 and, 210–212
    word processing, 31, 284
    workplace tools, 277–286
Applications software files, 208
Architecture, computers used in, 6
Architecture of microprocessors, 50–51
Arithmetic/logic unit (ALU), 42, 43–44
    in machine cycle, 46, 47
    memory and, 45
    in microprocessor, 50–51
Arithmetic operations, 43, C-0. *See also* Mathematical operations
Arpanet, 39
Arrow keys, 68, 69
Art
    See also Graphics
    on CD-ROM, *105*
    computer-generated, 6, *24*
Artificial intelligence (AI), 338–343
    computer learning and, 340, 341
    defined, 338
    early problems with, 338–339
    expert systems, 336, 338, *339*, 344–346
    human brain processes and, 340–341
    LISP used for, *189*
    natural languages, 338, *339*, 343
    robotics, 338, *339*, 346–350
ASCII code, 50
ASCII-8 code (extended ASCII), 50
ASM (Association for Systems Management), 170
Assembler programs, 171–172
Assembly languages, 171–172, *173*
    computer development and, B-6
Association for Computing Machinery (ACM), 170
Association for Systems Management (ASM), 170

Association of Women in Computing (AWC), 170
Asynchronous transmission, 131–132
AT&T Bell Laboratories. *See* Bell Laboratories
Atanasoff, Dr. John V., B-4
Atanasoff-Berry computer (ABC), B-4
Atari computers, B-10
ATMs (automatic teller machines), 145
Attributes of objects, 189
Audio-response units, 82
Auditing
    as data security measure, 318
    for illegal software copies, 329
    in systems implementation phase, 257–258
Audit trail, 257–258
Australian National University site, 93
Automated teller machines (ATMs), 145
Automobile design, computer use in, *4*, 24
Automobile junk yards, batch and transaction processing in, 116–117
Automobile manufacturing, robots used in, *348*
Automobile repair shops
    expert systems used in, *344*
    vertical market software for, 282
Automobiles, computers in, 7, 50
Auxiliary storage. *See* Secondary storage
AWC (Association of Women in Computing), 170

Babbage, Charles, B-2 to B-3
Baby Think It Over, 87
Background, in memory management, 216–217
Backspace key, 68, 69
Backup computer sites, 316
Backup media, magnetic tape as, 21–22, 105, 107, *108*
Backup systems, 107, 323
    copying software for, 328
    in detail design, 251
    for disaster recovery, 316
    liability and, 317
    resistance to, 308
Badges, as security device, 314
Balance the Budget site, 307
Balloon Help, 204
Bandwidth, 135
Banking applications, 7
    batch and transaction processing in, 116
    computer crime and, 311, 312
    delayed output in, 67
    direct file processing in, 111
    disaster loss and, 316
    electronic fund transfers (EFTs), 145, 219
    half-duplex transmission in, 132–133
    instant output in, 66
    large computers used in, 25, 270
    MICR in, 70–71
Bar code reader, 18, *19*, 72–73
Bar codes, 7, 72–73
    direct input from, *67*
    instant output from, 66
    ubiquity of, 20
Bardeen, J., B-6
Base 2. *See* Binary number system
Base 8. *See* Octal number system
Base 10. *See* Decimal number system
Base 16. *See* Hexadecimal number system
Baseball stadiums, handheld computer use in, 126
BASIC, 182
    for education and business applications, *178*
    as general-purpose language, 173
    object-oriented version of, 191
    program and sample output, *184*
    QBasic program example, A-14, A-17 to A-19
Batch processing, 113, *115*
    system controls for, 250–251
Batch programs

I-1

in event-driven environment, 215
   memory management with, 217
BBSs (bulletin board systems), 145
B/C URL, 63
Beauty salons, vertical market software for, 283–284
Behavior problems, on Internet, 265
Bell Laboratories
   transistors developed at, B-6
   UNIX and, 220
Benjamin/Cummings Publishing Company URL, 63
Berners-Lee, Tim, 93
Berry, Clifford, B-4
Binary (on/off) system, 49–50
Binary/decimal equivalents, 49
Binary number system, C-2 to C-3
   conversions to and from, C-4 to C-7
Biochips, 56
Biometrics, as security device, 314–315
Bits, 49
Blood banks, bar codes used in, 73
Bob, 210
Body chips, 52
The Boeing Company, expert systems used by, 345
Bohm, C., A-20
Bombs, 312
Booting the system, 205
Borland's Turbo Pascal, 182, 184, *185*
Boston Computer Museum, B-5
   Turing test and, 342
Boston Computer Society, 284
Bpi. *See* Bytes per inch
Brattain, H. W., B-6
Break-even software, 280
Bridge, in LAN, 140
Bristol, University of, Philosophy Department home page, 93
Browsers, 63, 149, 151
   FAQs about, 123
"Bugs," 169
Bulletin board systems (BBSs), 145
Business applications
   See also Networking; *specific types of software*; Workplace
   computer characteristics and, 3–4
   computer types and, 22–25, 26
   expert systems, 344–345
   future trends, 147
   graphics. *See* Business graphics
   on Internet, 151, 284–285, 355
   for laptops, 2, *26*, 295
   object technology in, 191–192
   for personal computers, 268
   programming languages for, 177–178, 180, 182, 186, *189*
   speech recognition in, 76
Business contacts, software tools, 284–285
Business graphics, 6, 83–84
   presentation (business-quality), 285
Bus lines
   defined, 53
   parallel processing and, 56
   processor speed and, 53–54
Bus network, *138*, 139
   Ethernet as, 142
Buying a computer, 40
Byron, Augusta Ada. *See* Ada, Countess of Lovelace
Bytes, 49
Bytes per inch (bpi), on magnetic tape, 104

C, 186
   applications used for, *178*
   as general-purpose language, 173, 186
C++, 186, *188*
   as object-oriented language, 190–191
   program and sample output, *188*
C> prompt, 205
Cables
   communications links, 134–135
   in LANs, 139
Cache, 54
CAD/CAM (computer-aided design/computer-aided manufacturing), 85–86
Calculations. *See* Arithmetic operations

Campbell Soup Company, expert system used by, 344
Candidates, in systems design, 243–244
Caps Lock key, 68
Career Mosaic, 199
Career planning
   Internet resources, 199
   MIS trends, 304
   network managers, 301
   personal computer managers, 300–301
   personal computer use and, 272
   for programmers, 162, 169–170, 192–193
Carrey, Jim, 85
Carrier sense multiple access with collision detection (CSMA/CD), 142
Carrier wave, 129, *130*
Cascade virus, 320, *321*
Case, Steve, 150
CASE (computer-aided software engineering) tools, 246–247, *248*
Cash Management System, natural language in, 177
Cathode ray tube (CRT), 78–79
CB radio, as half-duplex transmission, 132
CD-ROM disks, 103
   as secondary storage devices, 43
CD-ROM drives, 103–104
   in MS-DOS configuration, 206
Census tabulation
   Hollerith's machine for, B-3, B-4
   UNIVAC for, B-5
Centralized computers, 26, 126
Centralized data processing, 126
Central processing unit (CPU), 16, 17, *18*, 19–20, 42–56
   arithmetic/logic unit (ALU), 42, 43–44
   control unit, 42, 43
   data representation in, 49–50
   instruction sets and, 47–48
   in machine cycle, 46–47
   memory and, 42–43, 45, 52
   in microprocessor, 50–51
   multiprogramming and, 209–210, 214–215
   multitasking and, 210
   program execution by, 46–48
   registers, 44–45
   speed and power in, 53–56
   time-sharing and, 215–216
CERN site, 93
CGA (color graphics adapter), 79
Change agent, systems analyst as, 231
Characters
   ASCII coding scheme, 50
   as bytes, 49
   in data organization, 108, *109*
Characters per inch (cpi), on magnetic tape, 104
Charge-card transactions, delayed output in, 67
Charts. *See* Business graphics
Check bits, in synchronous transmission, 132
Checking accounts, MICR use with, 70–71. *See also* Banking applications
Chess games, artificial intelligence and, 338
Chief information officer (CIO), 33, 293
Chinese keyboard, 67
Chips
   See also Microprocessors
   defined, 50
   evolution of, 55
   future trends, 56
   semiconductor storage, 51
   size of, 51
CHKDSK command (MS-DOS), 206
Chrysler Corporation, CAD/CAM used by, 85–86
CIO (chief information officer), 33, 293
CIS (Computer Information Systems), 33
CISCs (complex instruction set computers), 55
Class, in object-oriented programming, 190, *191*
Classification of computers, 22–25
Client, of systems analyst, 230
Client/server networking, 128, 140, 141
   operating system for, 212
Clip art, copyright and, 328
Clock, 47, 51
Clock cycles, 47
Clock speed, 53
Clones, market impact of, B-10
CLS command (MS-DOS), 206

CLS statement, in QBasic program, A-17
CMOS design, 51
Coaxial cable, 134
   in LANs, 139
COBOL, 180, 182
   See also High-level languages
   for business applications, *178*, 180
   4GLs vs., 174
   as general-purpose language, 173
   Grace M. Hopper and, 180
   object-oriented versions of, 191
   program and sample output, *183*
COBOL 85, 180
CODASYL, COBOL introduced by, 180
Coding, in programming process, 166
Coding schemes, 50
Cohesion, between program modules, A-21
Cold site, for disaster recovery, 316
Colleges, computer requirements in, 8
Collision
   on Ethernet network, 142
   as hashing result, 111
Color graphics adapter (CGA), 79
Color printers, 81
COM (computer output microfilm), 82
Commands, MS-DOS, 205, 206
Comment lines, in programs, A-14
Commercial communications services. *See* Information utilities
Commercial Site Index, 355
Commercial software. *See* Packaged software
Commodore computers, B-10
Communications, 126–153
   See also Communications links; Data communications; Local Area Networks; Wide Area Networks
   defined, 32
   future trends, 147
   network configurations, 126–129
   networking applications, 142–148. *See also* Internet
   network topologies, 138–139
   "paperless office" and, 83
   protocols, 136–138
   software tools, 277–278
   Windows 95 and, 212
Communications links, 133–136, *137*
   coaxial cables, 134
   defined, 133
   fiber optics, 134–135
   microwave transmission, 135
   mixing and matching, 136, *137*
   satellite transmission, 135–136
   wire pairs (twisted pair), 134
Communications media, 128
Communications satellites, 135, 136
Community computing, 232
Compact disk read-only memory. *See* CD-ROM disks
Compaq 386, 80386 chip in, 55
Compare operation, in programming, A-3
Comparison operations. *See* Logical operations
Compiler, in programming process, 168
Complementary metal oxide semiconductor (CMOS) design, 51
Complex instruction set computers (CISCs), 55
Compression utility programs, for increasing disk space, 102
CompuServe Information Service, 147
Computer
   See also Computer systems
   classifications, 22–25
   defined, 16–17
   impact on workplace of, 273–276, 277
Computer-aided design, 4, 6
Computer-aided design/computer-aided manufacturing (CAD/CAM), 85–86
Computer-aided instruction, PILOT used in, *189*
Computer-aided software engineering (CASE) tools, 246–247, *248*
Computer art, 6, *24*
Computer chips. *See* Chips; Microprocessors
Computer crime, 310–314
   law enforcement and, 310–311, 312, 314
   protecting data from, 318–319
   types of, *313*
Computer Fraud and Abuse Act (1986), 314

Index

Computer games. *See* Games
Computer graphics. *See* Graphics
Computer Information Systems (CIS), 33
Computer literacy, defined, 3–4
Computer Matching and Privacy Protection Act (1988), 325–326
Computer operators, 33
Computer output microfilm (COM), 82
Computer professionals, 33, 170
   malpractice insurance for, 317
   organizations of, 170
Computer Professionals for Social Responsibility, 265
Computer programmers. *See* Programmers
Computer screen. *See* Screen
Computer systems, 15–35
   body chips, 52
   buying, 40
   central processing unit (CPU), 16, 17, *18*, 19–20, 42–56
   components of, 16, *16*. *See also specific components*
   defined, 230
   disaster recovery plans for, 316–317
   front-end processors, 127
   on information superhighway, 44
   input devices, 16, 17, *18*, *19*
   instruction sets and, 47–48
   output devices, 16, 17, *18*, 20
   peripheral equipment, 22
   secondary storage devices, 16, 17, *18*, 20–22, 26
   security issues. *See* Security
Computer time, theft of, 312
Computer user groups, 284
Computer users, 32–33
Computer viruses. *See* Viruses
Computing Services departments, 33
Concurrent licensing, 329
Concurrently run programs, 210, 214–215
Concurrent processing, 214
Conditions, logical operations and, 43
Conferences, computerized, 276, *277*
Congressional Record, online access to, 129
Connectivity, 27–28, 29
   future trends, 341, B-8
   Information Superhighway, 44
   "paperless office" and, 83
   unknown consequences of, 257
Consortiums, for disaster recovery, 316
Consumer Reports, online access to, 129
Contacts, software tools for, 284–285
Continuous word systems, 76
Control (Ctrl) key, 68
Control Data 7600 mainframe computer, 23
Controlling, as management function, 292, 293
Control structures, A-5
   examples using, A-10 to A-19
   introduction of, A-20
   iteration, A-7 to A-9
   selection, A-6 to A-7
   sequence, A-5 to A-6
Control unit, 42, 43
   in machine cycle, 46, *47*
   memory and, 45
   in microprocessors, 50
   registers and, 44
Coopers and Lybrand, expert system used by, 345
COPY command (MS-DOS), 206
Copying software, 326–329
   legal copying, 327, 328
   site license for, 328–329
Copy protection, 328
Copyright Act, software and, 329
Copyrighted software, 317, 328
   clip art, 328
Cost control
   data collection devices used for, 76
   as personal computer manager role, 300
   polling software for, 304
Cost reduction, computers used for, 5
Coupling, of program modules, A-21
Cpi. *See* Characters per inch
C prompt, 205
CPU. *See* Central processing unit
Cray supercomputer, 23
Crime detection software, 280

Criminal activities, 310–314
CRT (cathode ray tube), 78–79
CSMA/CD (carrier sense multiple access with collision detection), 142
Ctrl (Control) key, 68
Current drive. *See* Default drive
Cursor, methods for moving, 68
Customer service, software tools for, 284
Customizing packaged software, 243
Custom software, 28
Cyber Rights Home Page, 265
Cyberzone, for college students, 199
Cycle World, online access to, 129
Cylinder method of data organization, 102–103

DASD (direct-access storage device), 111
DAT (digital audio tape), 104
Data
   See also Input
   access and ethics, 326
   access help source, 302
   access methods, 110, 111–113
   access time, 113
   addresses, 48–49
   backing up. *See* Backup systems
   bits, bytes, and words, 49–50
   coding schemes for, 50
   conversion of to new system, 256
   criminal activities with, 312
   defined, 19
   graphical presentation of. *See* Business graphics
   organization of, 108–110
   processing. *See* Data processing
   remote access to, 295
   replication of, 275
   security issues, 318–319. *See also* Security
   in system journals, 251
Data analysis, by systems analyst, 239–241
Database management systems (DBMSs), 31
Databases
   defined, 108–109
   in detail systems design, 250, *251*
   groupware, 274–276
   natural languages used with, 177
   query languages used with, 175
   workplace uses for, 280–281
Data collection devices, 74, 75
Data communications, 26–28
Data communications systems, 126
Data Encryption Standard (DES), 322
Data entry operators, 33
Data files, 208
   See also Files
   conversion of to new system, 256
Data flow diagrams (DFDs), 239–240
Data gathering, in systems analysis, 237–239
Data processing, 113–117
   batch processing, 113, *115*
   batch and transaction combined, 116–117
   centralized, 126
   data entry operators and, 33
   distributed (DDP), 26–27, 126
   teleprocessing, 126
   transaction processing, 114, *116*
Data Processing Management Association (DPMA), 170
Data representation, 49–50
Data retrieval, workplace software tools, 280–281
Data security. *See* Security
DATA statements, A-17
Data storage
   as computer characteristic, 5
   CPU and, 42–43
   in registers, 44–45
   types of, 42–43
   workplace software tools, 280–281
Data transfer, 113
Data transfer rates, 113, 131, *132*
Data transmission, 129–133
   asynchronous vs. synchronous, 131–132
   digital vs. analog, 129–130
   modems and, 130–131
   simplex, half-duplex, and full-duplex, 132–133
Date, millennium problems, 164
DBMSs. *See* Database management systems
DDP. *See* Distributed data processing

Debugging, in programming process, 169
Decentralized computers, 26
Decimal number system, C-2
   conversions to and from, C-3 to C-6
   decimal/binary equivalents, *49*
Decision making, computers used for, 5
Decision-making software, workplace uses of, 279–280
Decision support systems (DSSs), 296–298
Decision tables (decision logic tables), 241
Decoding, in program execution, 46
Default (current) drive, 206–207
DEL (Delete) command (MS-DOS), 206
Delete (Del) key, 68, 69
Demodulation, 130, *131*
Density, of magnetic tape, 104
Dependent variables, in computer model, 297–298
DES (Data Encryption Standard), 322
Design
   business graphics, 278
   CAD/CAM, 85–86
   computers used for, *2*
   by systems analysts, *233*, 243–252
Desk-checking
   laser printers and, 81
   in programming process, 168
Desktop publishing (DTP)
   software packages, 31
   workplace uses for, 284
Detail design, by systems analyst, 243, 247–252
DFDs (data flow diagrams), 239–240
Diagnostics, in programming process, 168
Diagrams. *See* Business graphics
Dictation, speech recognition devices for, 76
Diddling, 312
Difference engine, B-2
Digital audio tape (DAT), 104
Digital Equipment Corporation (DEC), VAX minicomputer, 23
Digital transmission, 129
   conversion to analog, 130, *131*
Digitizing tablet, for graphics input, 86
Dijkstra, Edsger, A-19, A-20
DIR (Directory) command (MS-DOS), 206
Direct-access storage device (DASD), 111
Direct-connect modems, 131
Direct conversion, to new system, 256
Direct file processing (direct access), 110, 111, *112*
Directing, as management function, 292, 293
Disabled, computer use by, 11
   robot helpers, 347
   speech recognition, 76
Disaster recovery plan, 316–317
Disaster vulnerability, 315–316
Discrete word systems, 76
Disk drives
   backup systems for, 107
   default (current) drive, 206–207
   defined, 21, 98
   external hard disk drives, 102
   MS-DOS configurations, 205–207
   removable cartridge hard disk drives, 98, 100, 102
Diskettes, 97–98
   advantages of, 97–98
   as backup media, 107
   care of, 97
   data files on, 208
   data organization on, 100, *101*, 102
   formatting (MS-DOS), 206
   as magnetic media, 96–97
   as secondary storage devices, 17, *18*, 20, *21*, 43
Disk packs, *21*, 98
   cylinder method on, 102–103
   data organization on, 100, *101*, 102, 103
   introduction of, B-7
Display. *See* Screen
Distributed data processing (DDP), 26–27, 126
DNA modeling, 7
Documentation, 28, 29
   in programming process, 169
   in systems analysis phase, 237–239, 242
   in systems development phase, 254
   in systems implementation phase, 255
DOS. *See* MS-DOS operating system
Dot-matrix printers, 79, *80*, 81

DOUNTIL control structure, A-7 to A-9
  example, A-10, *A-11*
DOWHILE control structure, A-7, A-8
  examples, *A-13*, A-14, A-16, A-17, A-18
DO WHILE statement, in QBasic program, A-17, A-18
Downloading a file, 139
DPMA Data Processing Management Association), 170
DRAM (dynamic RAM), *51*, *52*
DSSs (decision support systems), 296–298
Duplex transmission methods, 132–133
DX vs. SX chips, 53–54
Dynamic RAM (DRAM), *51*, *52*
Dyson, Esther, 127

Earthquake prediction, graphics software used in, 278, *279*
Earth stations, 135
Eckert, J. Presper, B-4, B-5
Eckert-Mauchly Computer Corporation, Grace M. Hopper and, 180
Economics, information as element of, 2
EDI (electronic data interchange), 145
Education, computer use in, 8
  computer classes, 14
  minicomputers, 25
  OMR (mark sensing), 72
  PILOT use, *189*
  programming languages used in, 178
EFTs. *See* Electronic fund transfers
EGA (enhanced graphics adapter), 79
Egypt Interactive Home Page, 93
8-bit word length, 50
8088 chip, evolution to Pentium from, 55
Electronic agent software, 297
Electronic data interchange (EDI), 145
Electronic fund transfers (EFTs), 145, 219
  criminal activity with, 311
Electronic mail (e-mail), 28, 142–143
  addresses of "rich and famous," 142
  educational uses of, 193
  excessive, 265
  flaming by, 265
  future trends, 273
  Internet and, 149, 151
  privacy issues, 327
Electronic malls, 159
Electronic Numerical Integrator and Calculator (ENIAC), B-4 to B-5
Electronic spreadsheets, *30*, *31*
Eliza, 343
E-mail. *See* Electronic mail
Embezzlement, 311, 312, *313*
Employees, data security and, 318–319
Encapsulation of objects and instructions, 189
Encryption, 322
End key, 68, 69
End of file, A-12
End-users, defined, 16
Energy companies, computer use by, 7
Engineering applications, 2, 4
  FORTRAN used for, 178
  graphics used in, *85*
  mainframe computers used in, 25
  personal computers used in, 25
Enhanced graphics adapter (EGA), 79
ENIAC computer, B-4 to B-5
Enter key, 68, 69
Entrepreneurs, computer development and, B-9
Entry point, to control structure, A-5
Environmental applications
  delayed output in, 67
  simulations, *10*
Environmental awareness, online transactions and, 219
Environmental requirements, 322
  systems implementation and, 256
Equal sign (=), as relational operator, 44
Equal-to conditions, 43
Equipment conversion, in systems implementation phase, 255–256
ERASE command (MS-DOS), 206
Erase head, on magnetic tape unit, 104, *106*
Error checking
  in programming process, 168–169
  in synchronous transmission, 132
Error messages, syntax errors, 168–169

Esc (Escape) key, *68*
ESS (executive support system), 299
Ethernet, 142
Ethical issues, 326–329
  data access, 326–327
  software copying, 327–329
E-time (execution time), *46*, 47
Evaluation, in systems implementation phase, 258
Event-driven programs, 214–215
Execution time (E-time), *46*, 47
Executive support system (ESS), 299
Exit point, from control structure, A-5
ExperTax, 345
Expert shell software, 346
Expert systems, 344–346
  advantages of, 336
  as AI field, 338, *339*, 346
  building, 345–346
  defined, 344
External modems, 131
Eye, as identification source, 315, *316*

Facsimile technology, 144
Factories. *See* Industrial applications
Fair Credit Reporting Act (1970), 324–325
FAQs (frequently asked questions), 123
Farming, computer use in, 8
Fax modems, 144
Fax technology, 144
Feasibility study, by system analyst, 231, *233*, 234–237
Federal Express
  bar code use by, 73
  on Internet, 355
Federal Privacy Act (1974), 325
Fiber optic cable, 134–135
  in LANs, 139
"Fiche" (microfiche), 82
Field robots, 349–350
Fields
  in data organization, 108, *109*
  key, 109, 111, 112
Fifth-generation computers, B-8
Fifth-generation languages. *See* Natural languages
File conversion, in systems implementation phase, 255–256
File names, Windows 95 and, 210–211
Files
  access methods, 110, *112*, 113
  backing up, 323
  copying, 206
  in data flow diagram, 240
  data processing methods, 113–117
  defined, 108
  deleting from disk, 206
  in detail systems design, 250, *251*
  downloading and uploading, 139
  encryption software for, 322
  listing, 206
  MS-DOS commands for, 206
  in network organization, 128
  organization methods, 111–112
  renaming, 206
  types of, 208
File server, in LAN, 140, 141
File Transfer Protocol (FTP), 123
File transfer software, in wide area network, 139
Financial trading
  See also Stock analysis
  Internet and, 355
  personal computers used in, 25
Firewall, 321–322
First-generation computers, B-5 to B-6
Flags, on Internet, 227
Flamers, in newsgroups, 307
Flaming, 265
Flash memory, 54
Flight simulators
  games, 9
  used in training, 11
Floppy disks. *See* Diskettes
Flowcharts
  advantages and disadvantages of, A-0
  examples, *179*, A-11, A-13, A-15, A-16
  iteration control structure for, A-7 to A-9
  in programming process, 165–166, *167*, A-2 to A-3
  selection control structure for, A-6 to A-7

sequence control structure for, A-5 to A-6
  systems, 250, *251*
Focus, *176*
Foreground, in memory management, 216
Foreign languages
  on Internet, 93
  translation problems, 338, 339, 343
Forklift, instant output from, 66
FORMAT command (MS-DOS), 206
FORTH, *189*
FORTRAN, 178, 180
  See also High-level languages
  error messages, 169
  as general-purpose language, 173
  program and sample output, *181*
  for scientific applications, *178*
4GLs (very high-level languages). *See* Fourth-generation languages
486DX/486SX chips, 53–54
Fourth-generation computers, B-8
Fourth-generation languages (4GLs), *172*, 174–175, *176*
  prototyping with, 244
  query languages, 175
  training in, 192–193
Fractals, *24*, 227
Fraud. *See* Computer crime
Freedom of Information Act (1970), 325
Freeware, 327
Frequency of carrier wave, 130
Frequency modulation (FM), 130
Frequently asked questions (FAQs), 123
Friends of the Earth, 307
Frog Dissection Kit, 289
Front-end processor, 127
FTP (File Transfer Protocol), 123
Full-duplex transmission, 133
Function keys, *68*
Future trends
  communications and networking, 147, 153
  connectivity, 341
  microprocessors, 56
  MIS and, 304
  programming languages, 192–193, A-21 to A-22
  secondary storage devices, 117
  structured programming, A-21 to A-22
  telephone and e-mail messages, 273
  unknown consequences and, 257
  in workplace computing, 286

Games, 9
  as management problem, 299
  as training tool, 304
Gantt chart, 252–253
Gates, Bill, 126, 205, B-9
Gateway, in LAN, 140
GB. *See* Gigabytes
General-purpose register, 45
  getting around on, 63
Generic operating systems, 220
Gigabytes (GB), 49
Gopher, 123
GOTO statements, A-20
Government agencies
  computer use by, 8
  Internet and, 149, 307
  as mainframe users, 25
  minicomputers used by, 25
  as supercomputer users, 23
Graphical user interface (GUI)
  on Macintosh computers, 204
  in Windows operating system, 208–209
Graphics, 83–87
  See also Graphics applications
  advantages of, 32
  CAD/CAM, 85–86
  fractals, *24*
  input devices, 86–87
  on Internet, 227
  as output, *20*
  output devices, 87
  presentation, 285
  screen display of, 79
  workplace software applications, 278–279
Graphics adapter board, 79
Graphics applications, 6
  See also Graphics

in business. *See* Business graphics
in medicine, 7, 9
personal computers used for, 25
in weather forecasting, 8
Graphics card, 79
Graphics programs, 32
Greater-than conditions, 43
Greater-than-or-equal-to conditions, 43
Greater-than symbol (>), as relational operator, 44
"Green" computers, 16
Groupware (collaborative software), 274–276
impact on management of, 294–295
GUI (graphical user interface), 204, 208–209

Hackers
criminal activity by, 310–311
password protection and, 319
personal computer manager and, 301
Half-duplex transmission, 132–133
Handheld computers
See also Pen-based computers
flash chips in, 54
in food services, 126
Handwritten characters, optical recognition of, 73–74
Hard copy, 79
Hard disk drive, in MS-DOS configuration, 205, 206–207
Hard disks, 98–100
access time on, 113
backup systems for, 107
capacity of, 98, 102
data organization on, 100, *101*, 102–103
external, 102
as magnetic media, 96–97
RAID groups, 100, *101*
removable hard disk cartridges, 98, 100, 102
as secondary storage devices, 20–21, 43
space considerations, 94, 108
Hardware, 16–28
defined, 16
input devices, *16*, 17, *18*, *19*
output devices, *16*, 17, *18*, 20
peripheral equipment, 22
plug and play features, 212
secondary storage devices, *16*, *18*, 20–22
systems implementation and, 255–256
types of computers, 22–25, *26*
Harvard Mark I. *See* Mark I computer
Hashing (randomizing), 111
Hayes-compatible modems, 131
Head crash, 98
Head switching, 113
Health and medicine. *See* Medical applications
Helical scan recording, 104
Help sources
information centers, 302
on Internet, 123
Macintosh Balloon Help, 204
Hewlett-Packard laser printers, *80*
Hexadecimal number system, C-3, C-2
conversions to and from, C-4 to C-7
Hierarchy charts, A-3, *A-4*
High-level languages, 172–173
computer development and, B-6 to B-7
High-level managers
decision-making software for, 296–299
on management pyramid, 292
planning role of, 292–293
History of computers, B-0, B-2 to B-11
Hoff, Ted, B-8
Hollerith, Herman, tabulating machine by, B-3, *B-4*
Holographic storage, 117
Home computers. *See* Personal computers
Home key, *68*, 69
Home offices, computers and, 8
Home pages, 63
preparing your own, 335
résumés on, 199
Home robots, 347
Homes, computer use in, 3, 9
Hopper, Grace M., 169, 180
Hot computer, in wide area network, 139
Hot list, in WWW browsers, 123
Hot site, for disaster recovery, 316

html (HyperText Markup Language), 123
http (HyperText Transfer Protocol), 123
Human body chips, 52
Hyatt, Gilbert, B-8
HyperText Markup Language (html), 123
HyperText Transfer Protocol (http), 123

IBM
Ada used by, 186
APL language introduced by, 189
computer history and, B-3, B-4
FORTRAN introduced by, 178
Microsoft and, B-9
Token Ring networks and, 142
IBM 360 series computers, B-7
IBM AT, 80286 chip in, 55
IBM clones. *See* IBM-compatible computers
IBM-compatible computers
impact of, B-10
operating systems for, 203–204, 205–208
IBM computers
assembly language for, *173*
chips in, 55
machine language for, *172*
IBM PC
8088 chip in, 55
impact of, B-10
Icons, 69, *70*
in graphical user interface, 204, 208, 209
on Internet, 227
ICs. *See* Integrated circuits
Identification numbers, as security device, 314
IF-THEN control structure, A-6, A-7
examples, A-12, A-13, A-17, A-19
IF-THEN-ELSE control structure, A-6 to A-7
example, A-10, *A-11*
IF-THEN rules, in expert systems, 346
Images, on Internet, 227
Imaging, for data input, 74, 76
Impact Online, 307
Impact printers, 81
Implementation phase, in systems development life cycle, 233, 254–258
Independent variables, in computer model, 297–298
Indexed file processing (indexed processing), 110, 112
Industrial Age, 2
Industrial applications
CAD/CAM, 85–86
data collection devices in, 74, 76
expert systems, 344, 345
mainframes in, 25
networking example, 151–153
robots in, *10*, 346–348
Inference engine, 340
Information
See also Output
defined, 19
as element of economics, 2
management and. *See* Management Information Systems
as power, 294
Information Age, 1–11
Information centers, 302
Information resource managers, 293
Information Services departments, 33, 293, 302
Information Superhighway, 44
control issues on, 341
Internet and, 151
on-ramp preparations, 124
satellites and, 136
Information utilities, 146–148
future trends, 147
Internet access through, 149
making business contacts on, 285
Inheritance, in object-oriented programming, 190, *191*
Initializing, A-2
Ink-jet printers, 81
Input, 67–78
CPU and, 42
defined, *16*, 17
by keyboard, 67, 68–69
in machine cycle, 46
methods, 18–19. *See also specific methods*

by mouse, 68–69, *70*
source data automation methods, 70–78
by trackball, 69–70, *71*
user views of, 66–67
Input data files, 208
Input devices, *16*, 17, *18*, *19*
program execution and, 46
Input requirements, in detail design, 248–249
Insert (Ins) key, 68, 69
Instance, in object-oriented programming, 189
Instruction set, 47–48
CISC computers and, 55
RISC computers and, 54–55
Instruction time (I-time), 46
Insurance, computer malpractice, 317
Insurance companies
distributed data processing used by, 27
large computers used by, 25, 27, 270
Integrated circuits (ICs), computer development and, B-7 to B-8. *See also* Chips
Integrated software, Lotus 1-2-3 and, B-9
Intel Corporation
Ada used by, 186
chips described, 55
microprocessor and, B-8
Intel 8080 microprocessor, in first home computer, B-0
Intellect (query language), 175
Intel Pentium microprocessor, 50
Interactive Employment Network, 199
Internal modems, 131
Internal Revenue Service
Internet access to, 307
OCR used by, 74
privacy issues and, 324
Internal storage. *See* Memory
International Business Machines. *See* IBM
International Collegiate Programming Contest, 171
International Standards Organization (ISO), communications protocol standards, 137–138
Internet, 148–149, 151
access to, 63, 149
appeal of, 39
business uses of, 151, 284–285, 355
career resources on, 199
defined, 28, 39, 148
educational uses of, 193
e-mail addresses, 142
FAQs on, 123
free and not-free information on, 289
getting around on, 63, 123, 149, 151
as global village, 93
government resources on, 307
graphics on, 227
help sources on, 123, 335
Information Superhighway, 44
newsgroups on, 307
nonprofit organizations on, 307
problems on, 265
shopping on, 159
space photos on, 113
structured and freeform exercises for, 63
useful resources on, 307
Windows 95 and, 212
Interrupts, in multiprogramming, 214
Interviews, by systems analyst, 237–238
Inventory, devices for, 76
ISO. *See* International Standards Organization
Iteration (loop), in programming, A-3
Iteration control structure, A-5, A-7 to A-9
examples, A-17, A-18, A-19
I-time (instruction time), 46

Jacopini, G., A-20
Japan's Hottest Links (JWeb), 93
Jerusalem B virus, 320
Job Monster Board, 199
Jobs
applicant screening for, 318
computer knowledge as requirement for, 272
computer professionals, 33
Internet resources, 199
programming as career, 162, 169–170, 192–193
robots and, 346–347
Jobs, Steve, B-9, B-10
Jobweb, 199

Johnson Wax, expert system used by, 345
Joystick, as graphics input device, 87
"Junkies," 301
JWeb (Japan's Hottest Links), 93

K (kilobytes), 49
Kaplan Education Center, 199
Kapor, Mitch, B-9
KB (kilobytes), 49
Kemeny, John, 182
Keyboard
    as input device, 17, *18*, *19*, 67
    standard layout, 68–69
    unusual types of, 67
Key field, 109
    in direct file processing, 111
    in indexed file processing, 112
    in sequential file processing, 111
"Killer app," search for, 107
Kilobytes (KB or K), 49
Kiosks, touch-screen, 77, 78
Knowledge base, 340
Knowledge-based systems, natural languages in, 175–176
Knowledge engineers, 346
Kurtz, Thomas, 182

Lands' End, computer use by, 64, 66
Languages
    data representation codes, 50
    on Internet, 93
    translation software problems, 338, 339, 343
LAN managers, 301
LANs. *See* Local area networks
Laptop computers, 25, *26*
    See also Notebook computers
    business uses of, 2, *26*, 295
    CMOS design used for, 51
    screens on, 79
    security issues, 323
    trackballs on, 69–70
    virtual office and, 281
Laser printers, *80*, 81
Law enforcement, computer use in, 7
    See also Computer crime
    batch and transaction processing, 116
    crime detection software, 280
LCD (liquid crystal display) screens, 79
Leading decision, in iteration control structure, A-8, A-17
Legal services, software tools, 281
Less-than conditions, 43
Less-than-or-equal-to conditions, 43
Less-than symbol (<), as relational operator, 44
Letters, as bytes, 49
LEXIS, 281
Librarians, computer role of, 33
Library applications, 2
Library of Congress, on Internet, 307
Light pen, for graphics input, 86–87
Line printers, 81
Link/load phase, in programming process, 169
Links, at Web sites, 63
Liquid crystal display (LCD) screens, 79
LISP, *189*
Load module, in programming process, 169
Loan amortization software, 279–280
Local area networks (LANs), 27, 126, *127*, 139–142
    components of, 139–140
    defined, 139
    geographical areas for, 128
    protocols, 141–142
    types of, 140–141
Local governments, online access to, 232
Logical operations, 43
Logic chips, 50. *See also* Microprocessors
Logic errors, 169
Logic flowchart, A-2 to A-3
    systems flowchart vs., 250
LOGO, *189*
Loop
    in iteration control structure, A-7 to A-9
    in programming, A-3
LOOP statement, QBasic program examples, A-17, A-18, A-19
Lotus 1-2-3, B-9

Louvre, mirror site for, 93
Lovelace, Ada, Countess of, 186, B-3
Low-level managers, computers and, 292, 293
Lurkers, in newsgroups, 307

Machine cycle, 46–47
    expressed in MHz, 53
Machine language, 171, *172*
    assembler programs for, 171–172
    compilers for, 172
    in early computers, B-6
Macintosh operating system, 200, 203, 204
    as operating environment, 208
Macintosh tips
    for America Online, 150
    applications interface, 204
    increasing hard disk capacity, 102
Macintosh Toolbox, 204
Magazines, online, 129
Magnetic code, on everyday items, 20
Magnetic core memory, B-6
Magnetic disks, 96–103
    as secondary storage media, 20
Magnetic-ink character recognition (MICR), 70–71
Magnetic tape, 21–22, *21*, 104–105, *106*
    punched cards replaced by, B-6
Magnetic tape units, 104–105, *106*
    as backup system, 105, 107, *108*
Magneto-optical (MO) disks, 104
Mail delivery, bar code use in, 73
Mailing list software, 284. *See also* Database management systems
Mail-order houses
    direct access used by, 111
    input and output data, 64, 66
    as mainframe users, 25
    systems analysis in, 239
    vertical market software for, 283
Mainframe computers, 23, 24–25
    memory capacity of, 49
    in networks, 27
    operating systems for, 203, 213–220
Main memory. *See* Memory
Main storage. *See* Memory
Maintenance, in systems implementation phase, 258
Malpractice insurance, 317
Management
    classic functions in, 292–293
    computer systems and, 290
    impact of groupware on, 294–295
    system analyst's report to, 242
Management Information Systems (MIS), 292–304
    data security measures, 318–319
    decision support systems, 296–298
    defined, 293
    executive support system (ESS), 299
    future trends, 304
    information centers, 302
    network or LAN manager roles, 301
    new management model and, 294–296
    personal computer management, 299–304
Management Information Systems (MIS) departments, 33, 293
Management pyramid, 292, 294
Mandelbrot, Benoit, 227
Manufacturing applications. *See* Industrial applications
Marketing, Internet and, 151
Mark I computer, B-3
    Grace M. Hopper and, 180
    original "bug" in, 169
Mark sensing (optical mark recognition), 72
The Mask, 85
Master file, in batch processing, 114, *115*
Mathematical applications, FORTRAN used for, 178
Mathematical operations, processor speed and, 53. *See also* Arithmetic operations
Mathison, Nora, 281
Mauchly, Dr. John, B-4, B-5
MB (megabytes), 49
McCaw, Craig, 126
McCaw Cellular Communications, Teledesic Corporation and, 136
McDonald's, keyboard used by, 67
Medical applications, 9

body chips, 52
    graphics software in, 6, 279
    MediCards, 53
    robots in, 347
    speech recognition in, 76
    by students, 66–67
    vertical market software, 281
    virtual reality in, 351
MediCards, 53
Meeting Meter, 276
Meetings, computerized, 276, 277
Megabytes (MB), 49
Megaflop, processor speed expressed as, 53
Megahertz (MHz), processor speeds as, 53
Memory, 45
    addresses in, 48–49
    augmenting, 52
    bus lines and, 53–54
    cache, 54
    capacity of, 49
    components of, 51–53
    CPU and, 42–43, 45
    defined, 19–20, 45
    flash, 54
    random-access (RAM), 51–52
    read-only (ROM), 52–53
    registers and, 44–45
    secondary storage and, 17, *18*
    semiconductor storage, 51
    sharing, 215–218
    virtual (virtual storage), 217–218
Memory management, defined, 216
Memory protection, 218
Menus, in graphical user interface, 204, 208–209
Messages, in object-oriented programming, 190
Message systems. *See* Bulletin board systems; Electronic mail; Voice mail
Meteorology, computer use in, 8, 277–278
Methods, in object-oriented programming, 189
MHz. *See* Megahertz (MHz)
MICR (magnetic ink character recognition), 70–71
MICR inscriber, 71
Microcomputers. *See* Personal computers
Microfiche, 82
Microfilm, computer output (COM), 82
Microphone, as input device, 67, 77
Microprocessors
    computer development and, B-8
    in everyday items, 20, 43, B-8
    future trends, 56
    in personal computers, 50–51, B-8
    size of, 51
    speed and power of, 53–56
Microseconds, processor speed in, 53
Microsoft Corporation
    rise of, B-9, B-10 to B-11
    Teledesic Corporation and, 136
Microsoft MS-DOS. *See* MS-DOS operating system
Microsoft QuickBASIC, 182
Microsoft Visual BASIC, 191, *192*
Microsoft Windows. *See* Windows operating system
Microsoft Windows 95 operating system, 205, 210–212
Micro-to-mainframe links, 27
Microwave transmission, 135
Middle-level managers, computers and, 292, 294
Military
    Ada used in, 178, 186
    computer development and, B-4
    robots used by, *349*
Millennium programming problems, 164
Million floating-point operations per second (megaflop), 53
Million instructions per second (MIPS), 53
Milliseconds
    access time in, 113
    processor speed in, 53
Minicomputers, 23, 25
    development of, B-8
    distributed data processing and, 126
    in networks, 27
MIPS, processor speed expressed as, 53
Mirror sites, 93
MIS. *See* Management Information Systems
Misinformation, as Internet problem, 265
MIS managers, 293

MITS Altair computer, B-0, B-2, B-9, B-10
MO (magneto-optical) disks, 104
Models
 in decision support systems, 297–298
 in executive support systems, 299
Modems, 130–131
 defined, 130
 in network design, 127, *128*
 speed considerations, 128, 131, *132*, 151
 types of, 130–131
Modula-3, *189*
Modulation/demodulation, 130, *131*
Modules, in structured programming, A-21
Monitor. *See* Screen
Monochrome screen, 79
Monolithic chips, defined, 51
Monster Board, 199
Mosaic, 123
Motherboard, 50
Mouse, *70*
 as input device, *18*, *19*, 68–69
Movies, video graphics used in, 84, *85*
MS-DOS operating system, 200, 204, 205–208
 commands, 205, 206
 impact of, B-11
 origins of, B-9
 Windows 95 and, 210–211, B-11
Multimedia, 104
 applications, *105*
 fiber optics and, 135
 "killer app" search, 107
 sound output in, 82–83
 on World Wide Web, 123
Multimedia File Formats site, 335
Multiprocessing, multiprogramming vs., 214
Multiprogramming, 209–210
 memory management with, 216–217
 multiprocessing vs., 214
Multitasking, 210
Musicians, graphics software used by, 278
Music output, 82–83
 on CD-ROM, *105*

Nanoseconds, processor speed in, 53, 54
National Geographic, online access to, 129
Natural languages, *172*, 175–177
 as AI field, 338, *339*, 343
 PROLOG and, *189*
Nature Conservancy, 219
Netiquette, 265
Netizens, 63
Netscape, *149*
Network cable, in LANs, 139
Networking
 See also Networks
 example, 151–153
 future trends, 147, 153
 large computers used in, 270
 management and, 33, 294, 295, 301
 personal computers and, 29, 268, 271
 remote users and, 295
 unknown consequences of, 257
 uses for, 142–148. *See also* Internet
Network-interface card (NIC), 140
Network managers, 33, 294, 295, 301
Network operating system (NOS), 212
Networks, 27
 See also Networking
 communications links, 133–136, *137*
 complexity of, 151–153
 configurations, 126–129
 data transmission on, 129–133
 defined, 126
 geographical area and, 128
 resource allocation on, 213–220
 security issues, 321–322
 topologies, 138–139
 Windows 95 and, 212
Neural networks (neural nets), 340–341, *342*
News, on Internet, 307
Newsgroups, 307
NeXT Computer, Inc., B-9
NIC (network-interface card), 140
Nodes, in networks, 138–139
Noise, in wire-pair links, 134
Nonimpact printers, 81–82
Nonprocedural languages, 174, A-22

Nonprofit organizations, on Internet, 307
Nonresident programs, in operating system, 202, 203
Nordic Track, imaging used by, 76
Nordstrom, expert systems used by, 336
NOS (network operating system), 212
Notebook computers, flash chips in, 54. *See also* Laptop computers
Notes, as groupware, 274–276
Number bases, C-2 to C-3
 conversions between, C-3 to C-7
 defined, C-2
Number codes, as security device, 314
Numbers
 as output, 20
 representation of. *See* Number systems
Number systems, C-0 to C-7
 binary, C-2 to C-3
 hexadecimal, C-3
 octal, C-3
Numerical output, 20
Numeric data, representation of. *See* Number systems
Numeric keys, *68*, *69*
Num Lock key, *68*
Num Lock mode, *69*
Nygaard, Dr. Kristin, 189

Object, defined, 188–189
Object module, in programming process, 169
Object-oriented programming (OOP), 186–192
 business applications, 191–192
 classes in, 190
 classes and subclasses in, 190
 messages in, 190
 object examples, 189–190, *191*
 reuse of classes in, 190
Object-oriented programming languages, 190–191, A-22
Observation, in systems analysis, 239
Obsolescence of computers, 16
OCR (optical character recognition), 72
OCR-A typeface, 72
Octal number system, C-3, C-2
 conversions to and from, C-4 to C-7
Office automation, defined, 142
Office suites, Lotus 1-2-3 and, B-9
OMR (optical mark recognition), 72
Online Career Center, 199
Online processing. *See* Transaction processing
Online services. *See* Information utilities
On/off (binary) system, 49–50
OOP. *See* Object-oriented programming
Open Business Markets consultants, 335
Open Systems Interconnection (OSI) protocols, 137–138
Operating environments, 208–212
Operating systems, 28, 200–221
 defined, 202
 functions of, 202–203
 on large computers, 213–220
 for networks, 212
 proprietary vs. generic (portable), 220
 service programs and, 220
 shells added to, 208–212
 standards, 220
 supervisor programs in, 202
 in system files, 208
 virus danger to, 320, *321*
Operational managers, 292, 293
Operations, in object-oriented programming, 189
Optical character recognition (OCR), 72
Optical disks, 103–104
 CD-ROM, 103
 magneto-optical (MO) disks, 104
 multimedia software on, 104
 photos processed on, 74, *75*
 as secondary storage devices, *21*
Optical mark recognition (OMR), 72
Optical recognition, 71–74, *75*. *See also specific methods*
Organization chart, 234–235
Organizations, of computer professionals, 170, 265
Organizing, as management function, 292, 293
OSI. *See* Open Systems Interconnection protocols
Output, 20, 78–87
 as computer component, *16*, 17

CPU and, *42*
 defined, *16*, 20
 graphics, 83–87
 in machine cycle, 47
 user views of, 66–67
Output data files, 208
Output devices, *16*, 17, *18*, 20
 computer output microfilm (COM), 82
 printers, 79–82
 screen, 78–79
 voice, 82
Output requirements, in detail design, 248, *249*
Outsourcing, 243
Overview of computer systems, 15–35

P5 chip. *See* Pentium microprocessor
P6 chip, 55
Packaged (commercial) software, 28–29
Page Down/Page Up keys (PgDn/PgUp), *68*, *69*
Page frames, 217
Page table, 218
Paging, in memory management, 217
Paine Webber, model training approach of, 303–304
Paper handling, computers and, 7
"Paperless office," 83
Parallel conversion, to new system, 256
Parallel processing, 55–56
Parenting, computerized tryout of, 87
Partitions, in memory management, 216
Pascal, 182, 184
 applications used for, *178*
 object-oriented version of, 191
 program and sample output, *185*
Password, as security device, 314, 319
Pause key, *68*, *69*
Payroll applications
 data collection devices in, 74
 input and output data in, 67
PCs. *See* Personal computers
PDAs (personal digital assistants). *See* Pen-based computers
Peer-to-peer networks, 141
Pen-based computers, 25, *26*, 67. *See also* Hand-held computers
Pentium microprocessor, 50, *51*, 55
People, online access to, 129
Performance
 bus width and, 53–54
 caching and, 54
 flash memory and, 54
 parallel processing, 55–56
 processor speed and, 53
 RISC technology and, 54–55
Performance enhancement, computers used in, 11
Performing arts, software tools used in, 280–281
Peripheral equipment
 defined, 22
 plug and play features, 212
Permission, Internet issues, 227
Personal computer junkies, 301
Personal computer managers, 300–301
Personal computers (PCs), *23*, 25
 chips in, 50–53
 controlling acquisition of, 301
 data communications with, 28, 29, 268
 expert systems on, 346
 historical overview, B-0, B-8 to B-11
 impact of, 272
 as job requirement, 272
 multitasking on, 210
 network operating system for, 212
 in networks, 29, 139, 268, 271
 operating system overviews, 203–205, 208–212
 security for, 322–323
 workplace issues, 268, 299–304
Personal digital assistants (PDAs). *See* Pen-based computers
Personnel issues
 data security, 318, 319
 privacy, 325
PgDn/PgUp (Page Down/Page Up) keys, *68*, *69*
Phased conversion, to new system, 256
Phone directory, on CD-ROM, 104
Phones. *See* Telephones
Photonics, 56
Photos, optical recognition of, 74, *75*

Picoseconds, processor speed in, 53
Piggybacking, 312
PILOT, 189
Pilot conversion, to new system, 256
Pixels, 79
Planet Access, 63
Planning
  for disaster recovery, 316–317
  as management function, 292, 293
PL/I, 189
Plotter, for graphics output, 86, 87
Plug and play concept, Windows 95 and, 212
Pointer, in graphical user interface, 204. See also Cursor
Pointing, mouse used for, 18
Point-of-sale (POS) terminal, 72
  bar code use with, 72–73
  wand reader use with, 72
Polling software, 304
Polymorphism, in object-oriented programming, 190, 191
Popular Electronics, Altair featured in, B-0, B-2
Pop-up menus, in graphical user interface, 209
Portable computers. See Laptop computers; Pen-based computers
Portable operating systems, 220
Positional notation, number base conversions and, C-4
Postal Service, bar code use by, 73
POS terminal. See Point-of-sale terminal
Power, processing speed and, 53–56
Power outages, 315–316
  surge protection and, 323
Preliminary design, by systems analyst, 243–244
Preliminary investigation, by systems analyst, 231, 233, 234–237
Presentation graphics, 285
Presentations, in system analysis and design, 244, 245
Presentation software, 285
Primary memory. See Memory
Primary storage. See Memory
Priming read, A-13 to A-14
Printers, 79–82
  as output devices, 17, 18, 20, 79
  sharing, 218, 219–220
Printer spacing chart, in detail design, 248, 249
Print Screen key, 68, 69
Print spooling, 218, 219–220
PRINT statement, in QBasic program, A-17
Privacy issues, 323–326
  e-mail and, 327
  legislation and, 324–326
  in workplace, 326
Problem definition, by system analyst, 235–236
Problem solving, as AI field, 338, 339
Procedural languages
  4GLs as, 174
  structured programming and, A-21 to A-22
Processes, in data flow diagram, 240
Processing
  See also Data processing
  as computer component, 16, 17
  defined, 16
  parallel, 55–56
  serial, 55
  system controls for, 250–251
Processing speed, 4, 50, 53–56
  workplace impact of, 272–273
Processor. See Central processing unit
Prodigy, 147, 148
Productivity
  4GLs and, 174
  computer use and, 5, 270, 271
Productivity (task-oriented) software, 30–32
Professional services, future trends, 147. See also Computer professionals
Program execution, CPU and, 46–48
Programmable read-only memory (PROM) chips, 52–53
Programmer/analysts, 230
Programmers, 33, 162, 169–170
  Ada, Countess of Lovelace, B-3
  defined, 16
  education required for, 170, 192–193

Programming, 164–170
  addresses in, 48–49
  as career, 162, 169–170, 192–193
  contest, 171
  examples, A-10 to A-19
  flowcharts in, 165–166, 167, A-2 to A-3
  historical background, A-19 to A-21
  maintenance tasks, 258
  object-oriented, 186–192
  pseudocode in, 165, 166, A-5, A-3 to A-4
  reasons for learning, 164
  "spaghetti code" concept, A-20
  structure charts in, A-3, A-4
  structured, A-4 to A-10, A-20 to A-22
  in systems development phase, 254
  tasks involved in, 164–169
Programming applications, walking systems, 165
Programming languages, 170–186, 187–188, 189
  See also specific languages
  applications of, 178
  choosing among, 177–178
  in coding process, 166
  computer development and, B-6, B-7, B-8
  defined, 164
  future trends, 192–193, A-21 to A-22
  levels of, 171–177
  object-oriented, 190–191
  syntax, 166
  training in, 192–193
Programs, defined, 16, 164. See also Software
Project Gutenberg, 307
Project management software, 253–254, 285–286
PROLOG, 189
PROM (programmable read-only memory) chips, 52–53
Prompt, in MS-DOS, 205
Property management software, 280
Proprietary operating systems, 220
Protocols, 128, 136–138
  defined, 136
  LAN, 141–142
  standards, 137–138
  TCP/IP, 148–149
Prototype, in systems design, 244, 246
Prototyping tools, 244
Pseudocode
  advantages of, A-0
  examples, A-11, A-13, A-15, A-16
  iteration control structure for, A-7 to A-9
  for programming examples, 179
  in programming process, 165, 166, A-3 to A-4
  rules for using, A-5
  selection control structure for, A-6 to A-7
  sequence control structure for, A-5 to A-6
Public domain software, 327
Pull-down menus, in graphical user interface, 204, 208–209
Punched cards, B-6

QBasic, program example written in, A14, A-17 to A-19
Query-by-Example (QBE), 175
Query languages, 175
Questionnaires, in systems analysis, 238–239
Queues, in memory management, 217
QuickBASIC, 182

Radio Shack TRS-80 computer, B-10
RAID (redundant array of inexpensive disks), 100, 101
Railroads, computer use by, 7, 11
RAM (random-access memory), 51–52
  dynamic (DRAM), 51, 52
  static (SRAM), 52
Randomizing (hashing), 111
Raster-scan technology, 79
Read-only media, 103
Read-only memory (ROM), 52–53
Read statements, A-10
  priming read, A-13 to A-14
  in QBasic program, A-17, A-18, A-19
Read/write head, 98, 99
  access time and, 113
  on magnetic tape unit, 104, 106
Real storage, 217
Real-time processing, 114. See also Transaction processing

Receiving device, in network design, 127, 128
Records
  in data organization, 108, 109
  direct access to, 111
  in indexed file processing, 112
  sequential access to, 111
Recycling of computers, 16
Red Cross, bar code use by, 73
Reduced instruction set computers (RISCs), 54–55
Redundant array of inexpensive disks (RAID), 100, 101
Refreshing of screen, 79
Regions, in memory management, 216
Registers (temporary storage areas), 44–45
Relational operators, 44
Reliability of computers, 5
REM (remark), in program, A-17
Remainder method, in number base conversions, C-5 to C-6
Remote access, 295
  security issues, 321–322
Removable hard disk cartridges, 98, 100, 102
RENAME command (MS-DOS), 206
Replication of data, in groupware, 275
Researchers
  graphics software used by, 278, 279
  legal research software, 281
Resident programs, in operating system, 202, 203
Resolution of screen, 79
Resource allocation, 213–220
Response time, in time-sharing system, 216
Restaurants, touch screens used by, 77
Retailing applications, 7
  data processing methods in, 114, 116
  minicomputers used in, 25
  network configuration in, 127
  OCR-A and POS terminals, 72
Retinal pattern, as security device, 315, 316
Return key. See Enter key
Ring network, 138–139
  Token Ring protocol, 142
RISC technology (reduced instruction set computers), 54–55
Ritchie, Dennis, 186, 220
Roberts, Ed, B-0, B-2
Robotics, 10, 346–350
  as AI field, 338, 339
Robots, defined, 347
ROM (read-only memory), 52–53
ROM burners (PROM chips), 52–53
Rotational delay, 113
Round-robin scheduling, 216
RPG (Report Program Generator), 189
Rules
  in artificial intelligence, 340
  in expert systems, 346

Salami technique, 312
Sales personnel
  remote access to data for, 295
  software tools for, 285
Sampling, in systems analysis, 239
SAT (Scholastic Aptitude Test), 33
Satellite transmission, 135–136
Scanners, 74, 75
Scanning, 18–19, 74, 76
Scan rate, 79
Scavenging, 312
Scheduling
  Gantt chart for, 252–253
  groupware for, 275
  project management software for, 253–254
Scholastic Aptitude Test (SAT), 33
Scientific American, online access to, 129
Scientific applications, 10
  programming languages for, 178, 189
Screen
  color standards, 79
  CRT, 78–79
  liquid crystal display (LCD), 79
  monochrome, 79
  as output device, 17, 18, 20, 78–79
Screen capture software, 227
Screen reports, in detail design, 248, 249
Screen savers, 81
Scroll Lock key, 68, 69
SDLC. See Systems development life cycle

Sealed modules, disk packs in, 98
Search engines, 159
Secondary storage
　backup systems, 107
　benefits of, 96
　CPU and, 42
　defined, 20, 43
Secondary storage devices, 16, 18, 20–22
　access methods on, 111–112
　access time on, 113
　centralized vs. decentralized systems, 26
　as computer component, 16, 17
　computer development and, B-6, B-7
　data organization on, 108–110
　flash memory and, 54
　future trends, 117
　magnetic disks, 96–103
　magnetic tape storage, 104–105, 106
　optical disks, 103–104
　sharing of, 218–219
　space considerations, 94, 108
Second-generation computers, B-6 to B-7
Sector method of data organization, 100, 101
Sectors, zone recording on, 101, 102
Security, 314–323
　See also Backup systems
　access control, 314–315
　audit trail for, 257–258
　computer crime problems, 310–314
　criminal "tricks," 312
　for data, 318–319
　data encryption for, 322
　defined, 314
　disaster vulnerability, 315–316
　firewalls for, 321–322
　groupware and, 276
　for networks, 212, 321–322
　personal computer manager and, 300–301
　personnel issues, 318, 319
　RAID groups and, 100
　remote access and, 295
　removable hard disk cartridges for, 98–99
　software ownership and, 317
　surge protection for, 322–324
　WORM disks for, 103
Seek time, 113
Selection control structure, A-5, A-6 to A-7
　example, A-9 to A-10, A-12, A-13, A-17, A-19
Semiconductors, integrated circuits as, B-7
Semiconductor storage, 51
Sending device, in network design, 127, 128
Sequence control structure, A-5 to A-6
Sequential file processing, 111
Serial processing, 55
Server
　for Internet access, 149, 335
　in LANs, 140, 141
Service programs (utilities), 220
Shareware, 327
Shell software, for expert systems, 346
Shift key, 68
Shipping, computer use in, 7
　bar codes, 73
　computer crime and, 311
　data collection devices, 76
Shockley, W., B-6
Shoemaker-Levy comet, 113
Shopping
　agent software for, 297
　future trends, 147
　on Internet, 159
　virtual reality, 350
Shredders, 318
Sight recognition, 77–78
Signature, as security device, 314
Silicon chips. See Chips
Silicon Valley, B-7
SIMM board (single in-line memory module), 52
Simplex transmission, 132, 133
Simulations, 298
Single entry/single exit, in program modules, A-21
Single in-line memory module (SIMM board), 52
Sink, in data flow diagram, 240
Site licensing, 328–329
Sites, on World Wide Web, 63
16-bit word length, 50
64-bit word length, 50

Smalltalk, as object-oriented language, 191
Smart e-mail software, 143
Smithsonian Magazine, online access to, 129
Social issues, in Internet use, 265
Social Security Administration, EFT used by, 145
Soft copy, 79
Software, 28–32
　See also Applications software; Operating systems; Program execution; Programming
　access restriction in, 319
　acquisition by purchase, 243
　antipiracy hotline, 329
　categories of, 28–30
　collaborative (groupware), 274–276
　concurrent licensing of, 329
　copyright issues, 317, 328
　defined, 16, 28
　diskettes for delivery of, 98
　encryption, 322
　entrepreneurs, B-9, B-10 to B-11
　ethical issues, 326–327
　illegal copying of, 326–329
　integrated, B-9
　in preliminary systems design, 243
　for prototyping, 244
　public domain, 327
　security issues, 317, 328
　shareware, 327
　site licensing of, 328–329
　system, 28, 202
　task-oriented, 30–32
　unbundling of, B-7 to B-8
　vertical market, 281–284
　viruses in, 319–321
　workplace tools, 277–286
Software industry, IBM and origins of, B-7 to B-8
Software piracy, 328, 329
Software Publishers Association (SPA), piracy fought by, 329
Solomon, Les, B-0
Sound card, 83
　for multimedia, 104
Source, in data flow diagram, 240
Source data automation, 70
Source document, translation to input required for, 67
Source module, in programming process, 169
SPA. See Software Publishers Association
Space program
　access to photos from, 113
　computer use in, 10
　field robots and, 349
　virtual reality and, 350
"Spaghetti code," A-20
SPAudit program, 329
Speakers, for multimedia, 104
Special effects, supercomputers used for, 24
Speech recognition devices, 76, 77
　problems with, 78
Speech synthesis, 82
Speed of computers, 4, 50, 53–56
　workplace impact of, 272–273
Spooling, for printer sharing, 218, 219–220
Sports statistics software, 280
Spreadsheet software, 30, 31
　as "killer app," 107
　workplace uses of, 279–280
SQL (Structured Query Language), 175
SRAM (static RAM), 52
Staffing, as management function, 292, 293
Standard operating system, UNIX as, 220
Standards, protocols, 137–138
Star network, 138
Start/stop method (asynchronous transmission), 131–132
Static RAM (SRAM), 52
Statistics, sports statistics software, 280
Status lights, on keyboard, 68
Stock analysis
　See also Financial trading
　expert systems used in, 344
　supercomputers used in, 24
Stock management
　communications software for, 278
　computer crime and, 311
Stop bits, 132

Storage
　defined, 17
　primary. See Memory
　real, 217
　sharing, 218–219
　temporary storage areas (registers), 44–45
　virtual (virtual memory), 217–218
　workplace software tools, 280–281
Storage capability, as computer characteristic, 5
Storage devices. See Secondary storage devices
Storage register, 45
Strategic managers, 292
Structure charts, A-3, A-4
　advantages of, A-0
　examples, A-11, A-13, A-15, A-16
　top-down design in, A-21
Structured programming, A-4 to A-10
　future trends, A-21 to A-22
　introduction of, A-20
　modularity in, A-21
　top-down design in, A-21
Structured Query Language (SQL), 175
Subclasses, in object-oriented programming, 190, 191
Supercomputers, 23–24
　fractals created with, 24
SuperDoubler, 102
Supermarkets, bar codes used by, 72–73
Supermicros, 25
Superminis, 25
Super VGA (SVGA), 79
Supervisor program, 202
Supply reel, on magnetic tape unit, 105, 106
Supreme Court decisions, on Internet, 307
Surge protectors, 322–323
Surgery, computer use in, 9
SVGA (super VGA), 79
SX vs. DX chips, 53–54
Symantec Corporation, SuperDoubler by, 102
Symbolic addresses, 48–49
Symbolic languages. See Assembly languages
Symbols, as bytes, 49
Synchronous transmission, 128, 132
Synonyms, as hashing result, 111
Syntax, 166
　checking for errors, 168–169
Synthesis by analysis, for voice output, 82
Synthesis by rule, 82
System, defined, 230
System conversion, in systems implementation phase, 256–257
System files, 208
System journals, 251
System library, in programming process, 168
System requirements, determined by systems analyst, 242
System software, defined, 28, 202
Systems analysis, 232, 233, 237–242
　See also Systems development life cycle
　C used in, 178
　defined, 230
　impetus for, 230, 231
　in programming process, 165
Systems analysts, 230
　as change agents, 231
　preliminary investigation by, 231, 233, 234–237
　problems encountered by, 258–259
　on programming career ladder, 170
　role of summarized, 33, 228, 230, 231, 259
　skills required of, 231, 232
　system design by, 233, 243–252
　systems analysis by, 232, 233, 237–242
Systems control, in detail design, 250–251
Systems design, defined, 230
Systems development, 233, 252–254
Systems development life cycle (SDLC), 232–258
　CASE tools and, 246–247, 248
　implementation phase, 233, 254–258
　overview of, 232–234
　preliminary investigation, 231, 233, 234–237
　systems analysis phase, 232, 233, 237–242
　systems design phase, 233, 243–252
　systems development phase, 233, 252–254
Systems flowchart, 250, 251, A-2
Systems processing, in detail design, 250, 251
System survey, 231, 233, 234–237
System testing, 254

Tab key, 68
Tabulating Machine Company, B-3
Tactical managers, 292, 293
Take-up reel, on magnetic tape unit, 105, *106*
Tandy Incorporated, TRS-80 computer by, B-10
Tape backup systems, 21–22, 105, 107, *108*
Tape drives, 21–22
Tape storage. *See* Magnetic tape
Task bar, in Windows 95, 211
Task-oriented management, 295–296
Task-oriented software, 30–32
Tax planning, expert system for, 345
TCP/IP (Transmission Control Protocol/Internet Protocol), 148–149
Teamwork
 groupware and, 274–276, 294–295
 management changes with, 295–296
Technical support, as information center role, 302
Technical writers, user's manuals by, *255*
Technology overload, personal computer manager and, 300
Telecommuting, 146, 147
Teleconferencing, 144
Teledesic Corporation, 136
Telephone directory, CD-ROM, 104
Telephones
 as communications links, 133
 flash chips in, 54
 full-duplex transmission on, 133
 as input devices, 76, 77
 voice mail systems, 143–144, 273
 voice output on, 82
Teleprocessing, 126
Television
 computer graphics used in, 83, 84–85
 future trends, 147
 on information superhighway, 44
 as simplex transmission, 132
Temporary storage areas (registers), 44–45
Terminal emulation software, 139
Terminals, 19, 79
Terrorist threats, 315
Testing
 Internet help resources, 199
 OMR used in, 72
 in programming process, 166, 168–169, A-20
 in systems development, 254
Text editor, in programming process, 166
Theft problems, 312, *313*
 computer or component theft, 322
 data security measures, 318–319
 laptop computers and, 323
 software and, 326–329
Third-generation computers, B-7 to B-8
Third-generation languages, 172
 4GLs vs., 174
32-bit word length, 50
Thompson, Ken, 220
TIME command (MS-DOS), 206
Time-driven programs, 215
Time magazine, online version of, 129
Time-sharing, 215–216
Time slices, 215, 216
Token passing, 142
Token Ring networks, 142
Top-down design, in structured programming, A-21
Top managers. *See* High-level managers
Topologies, network, 128, 138–139
Touch screens, 77, *78*
Trackball, 69–70, *71*
Tracks, *96*, 97
 in cylinder data organization, 102–103
 sectors on, 100, *101*
 zone recording on, 100, *101*, 102
Trailing decision, in iteration control structure, A-8
Training
 importance of, 303
 as information center role, 302
 for personal computer users, 302–304
 remote users and, 295
 by systems analysts, 254–255
Training applications, 11
Transaction file, 114, *115*
Transaction log, as security measure, 318
Transaction processing, 110, 114, *116*
Transfer structure (GOTO statements), A-20
Transistors, computer development and, B-6 to B-7

Translating, in programming process, 168–169
Translation software problems, 338, 339, 343
Translators, 168
 assembler programs, 171–172
 compilers, 172
Transmission Control Protocol/Internet Protocol (TCP/IP), 148–149
Transponder, 135
Transportation, computer use in, 7, 11
Trapdoor, 312
Travel agencies, output used by, 78
Trojan Horse, 312
Troubleshooting, in network operating system, 212
TRS-80 computer, B-10
Turbo C++, program and sample output, *188*
Turbo Pascal, 182, 184, *185*
Turing, Alan, 341
Turing test, 341–342
Twisted pairs (wire pairs), 134
 in LANs, 139
Typefaces, OCR-A, 72
Typing, keyboard used for, 18

Unbundled software, impact of, B-7 to B-8
Uniform Resource Locators. *See* URLs
United Airlines, expert system used by, 345
Unit testing, in systems development, 254
UNIVAC I computer, B-5
 Grace M. Hopper and, 180
Universal Automatic Computer. *See* UNIVAC I computer
Universal Product Code (UPC), 72
Universities, computer requirements in, 8
University of Bristol Philosophy Department home page, 93
UNIX, 220
UPC (Universal Product Code), 72
Updating of master file, in batch processing, 114, *115*
Updating in place, 111
Uploading a file, 139
Urban planning applications, 232
URLs (Uniform Resource Locators), 63
 FAQs about, 123
U.S. Census
 Hollerith's tabulating machine for, B-3, *B-4*
 UNIVAC bought by, B-5
U.S. Congress
 Computer Fraud and Abuse Act by, 314
 Internet access to, 307
 privacy legislation, 324–326
U.S. Department of Defense
 Ada and, 184, 186
 COBOL and, 180
Useless Pages Web site, 265
Usenet, 307
User friendly, defined, 29
User groups, 284
User involvement
 in systems analysis, 231
 in training, 303–304
User's manual, in systems implementation phase, *255*
U.S. Postal Service, bar code use by, 73
Utilities. *See* Service programs
Utility companies, computer use by, 7

Vaccines, for computer viruses, 319–320
Vacuum tubes, computer development and, B-5 to B-6
Vandalism, 315
VAX minicomputer, *23*
Vectors, in data flow diagrams, 240
VER command (MS-DOS), 206
Vertical market software, 281–284
Very high-level languages (4GLs), 172, 174–175, *176*
VGA (video graphics array), 79
Videoconferencing, 144–145
Video graphics, 84–85
Video graphics array (VGA), 79
Video Privacy Protection Act (1988), 325
Videotape rental stores, vertical market software for, 282

Virtual office, 281
Virtual reality, 350–351
Virtual storage (virtual memory), 217–218
Viruses, 319–321
 computer crime examples, 312
 defined, 319
VisiCalc, B-10
Vision robots, 348–349
Visual BASIC, 191, *192*
Voice input, 67, 76, 77
 problems with, 78
Voice mail, 143–144, 273
Voice output, 82
Voice recognition, as security device, 315
Voice synthesizers, 82
Volatile memory, 51
Volume testing, in systems development, 254

Walkthrough, in programming process, 168
Wal-Mart
 EDI used by, 145
 kiosks used by, 77
Wand reader, 18, 72
WANs. *See* Wide area networks
Watson, Thomas J., Sr., B-3, *B-4*
Weather forecasting, computer use in, *8*, 277–278
Web. *See* World Wide Web (WWW)
WebMuseum site (Paris), 93
Web sites, 63
Weizenbaum, Joseph, 343
WESTLAW, 281
"What if...?" analysis, decision support systems for, 296–298
White House, 149
The Whole Internet Catalog, 63, 199
Wide area networks (WANs), 128, 139
Winchester disk, 98
Windows, in graphical user interface, 204
Windows operating system, 200, 204
 large computers and, 209–210
 as operating environment, 208
 overview of, 208–209
Windows 95 operating system, 205, 210–212
Wireless networks, 139–140
Wire pairs (twisted pairs), 134
 in LANs, 139
Wirth, Niklaus, 182
Word output, 20
Word processing software, 31, 284
Words, size of, 50
Workplace
 computer game playing at, 299
 future trends, 286
 impact of computers on, 273–276, 277
 impact of groupware on, 274–276, 294–295
 intangible issues in, 310
 management issues in. *See* Management Information Systems
 personal computers in, 268, 270–273
 security, privacy, and ethics issues in, 310. *See also specific issues*
 software tools used in, 277–286
 virtual office, 281
Worksheet, *See also* Spreadsheet software
World Communities site, 93
World Wide Web (WWW), 63
 CERN site and, 93
 FAQs, 123
 navigating. *See* Browsers
WORM (write-once, read-many) media, 103
Worms, damage by, 319
Wozniak, Steve, B-9, B-10
Write-once, read-many (WORM) media, 103
Written documents, in systems analysis, 237
WWW. *See* World Wide Web
WWW Client Software site, 335

Xerox Palo Alto Research Center, GUI created at, 204

Yahoo, 63
 adding sites to, 335
Yankee Doodle virus, 320

Zapping, 312
Zone recording, 100, *101*, 102